Reading Essentials
Life Science

The ***McGraw·Hill*** *Companies*

 **Education**

Send all inquiries to:
McGraw-Hill Education
8787 Orion Place
Columbus, OH 43240-4027

ISBN: 978-0-07-889385-8
MHID: 0-07-889385-2

Printed in the United States of America.

2 3 4 5 6 7 8 9 10 REL 15 14 13 12 11

Table of Contents

To the Student iv

Nature of Science Scientific Explanations .... NOS 1

Chapter 1 Classifying and Exploring Life 1

Chapter 2 Cell Structure and Function 17

Chapter 3 From a Cell to an Organism 39

Chapter 4 Reproduction of Organisms 53

Chapter 5 Genetics 67

Chapter 6 The Environment and Change Over Time 83

Chapter 7 Bacteria and Viruses 103

Chapter 8 Protists and Fungi 121

Chapter 9 Plant Diversity 137

Chapter 10 Plant Processes and Reproduction 153

Chapter 11 Animal Diversity 173

Chapter 12 Animal Structure and Function 189

Chapter 13 Animal Behavior and Reproduction 207

Chapter 14 Structure and Movement 225

Chapter 15 Digestion and Excretion 241

Chapter 16 Respiration and Circulation 259

Chapter 17 Immunity and Disease 283

Chapter 18 Control and Coordination 299

Chapter 19 Reproduction and Development 319

Chapter 20 Matter and Energy in the Environment 331

Chapter 21 Populations and Communities 349

Chapter 22 Biomes and Ecosystems 365

Chapter 23 Using Natural Resources 387

To the Student

In today's world, a knowledge of science is important for thinking critically, solving problems, and making decisions. But understanding science can sometimes be a challenge.

Reading Essentials takes the stress out of reading, learning, and understanding science. This book covers important concepts in science, offers ideas for how to learn the information, and helps you review what you have learned.

In each lesson you will find:

Before You Read

- **What do you think?** asks you to agree or disagree with statements about topics that are discussed in the lesson.

Read to Learn

This section describes important science concepts with words and graphics. In the margins you can find a variety of study tips and ideas for organizing and learning information.

- **Study Coach and Mark the Text** offer tips for finding the main ideas in the text.
- **Foldables® Study Organizers** help you divide the information into smaller, easier-to-remember concepts.
- **Reading Check** questions ask about concepts in the lesson.
- **Think It Over** elements help you consider the material in-depth, giving you an opportunity to use your critical thinking skills.
- **Visual Check** questions relate specifically to the art and graphics used in the text. You will find questions that get you actively involved in illustrating the information you have just learned.
- **Math Skills** reinforces the connection between math and science.
- **Academic Vocabulary** defines important words that help you build a strong vocabulary.
- **Word Origin** explains the English background of a word.
- **Key Concept Check** features ask the Key Concept questions from the beginning of the lesson.
- **Interpreting Tables** includes questions or activities that help you interact with the information presented.

After You Read

This final section reviews key terms and asks questions about what you have learned.

- The **Mini Glossary** assists you with science vocabulary.
- Review questions focus on the key concepts of the lesson.
- **What do you think now?** gives you an opportunity to revisit the *What do you think?* statements to see if you changed your mind after reading the lesson.

See for yourself—***Reading Essentials*** makes science enjoyable and easy to understand.

Scientific Explanations

Understanding Science

Read to Learn

What is science?

The last time that you watched squirrels play in a park or in your yard, did you realize that you were practicing science? Every time you observe the natural world, you are practicing science. **Science** *is the investigation and exploration of natural events and of the new information that results from those investigations.*

When you observe the natural world, you might form questions about what you see. While you are exploring those questions, you probably use reasoning, creativity, and skepticism to help you find answers to your questions. These behaviors are the same ones that scientists use in their work and that other people use in their daily lives to solve problems. ✓

Scientists use a reliable set of skills and methods in different ways to find answers to questions. After reading this chapter, you will have a better understanding of how science works, the limitations of science, and scientific ways of thinking. In addition, you will recognize that when you practice science at home or in the classroom, you probably use scientific methods to answer questions just as scientists do.

Branches of Science

No one person can study all the natural world. Therefore, people tend to focus their efforts on one of the three fields or branches of science—life science, Earth science, or physical science. Then people or scientists can seek answers to specific problems within one field of science.

Life Science Biology, or life science, is the study of all living things. For example, a forest ecologist is a life scientist who studies interactions in forest ecosystems. Biologists ask questions such as the following: How do plants produce their own food? Why do some animals give birth to live young and others lay eggs? How are reptiles and birds related?

Key Concepts 

- What is scientific inquiry?
- What are the results of scientific investigations?
- How can a scientist prevent bias in a scientific investigation?

Study Coach

Building Vocabulary Write each vocabulary term in this lesson on an index card. Shuffle the cards. After you have studied the lesson, take turns picking cards with a partner. Each of you should define the term using your own words.

✓ **Reading Check**

1. Describe the behaviors scientists use in their work.

Earth Science The study of Earth, including Earth's landforms, rocks, soil, and forces that shape Earth's surface, is Earth science. Earth scientists ask questions such as the following: How do rocks form? What causes earthquakes? What substances are in soil?

Physical Science The study of chemistry and physics is physical science. Physical scientists study the interactions of matter and energy. They ask questions such as these: How do substances react and form new substances? Why does a liquid change to a solid? How are force and motion related?

Scientific Inquiry

As scientists study the natural world, they ask questions about what they observe. To find the answers to these questions, they use certain skills, or methods. The figure below and at the bottom of the next page shows a sequence of the skills that a scientist might use in an investigation. Sometimes, not all of these skills are performed in an investigation or are performed in this order. Scientists practice scientific inquiry—a process that uses a variety of skills and tools to answer questions or to test ideas about the natural world. ✓

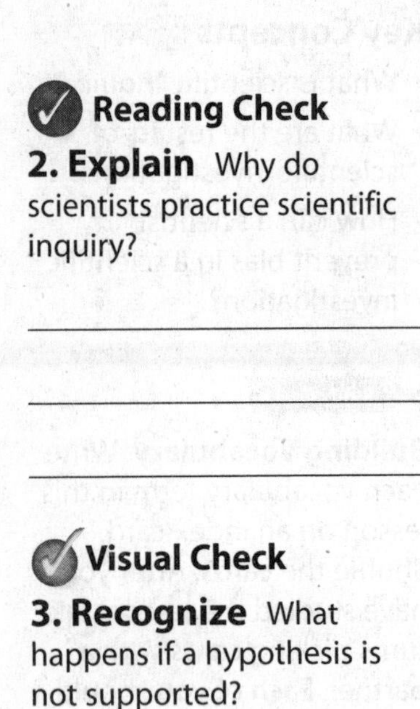

Reading Check
2. Explain Why do scientists practice scientific inquiry?

Visual Check
3. Recognize What happens if a hypothesis is not supported?

Ask Questions

Like a scientist, you use scientific inquiry in your life, too. Suppose you decide to plant a vegetable garden. As you plant, you water some seeds more than others. You weed part of the garden and mix fertilizer into some of the soil.

Steps in Scientific Investigation

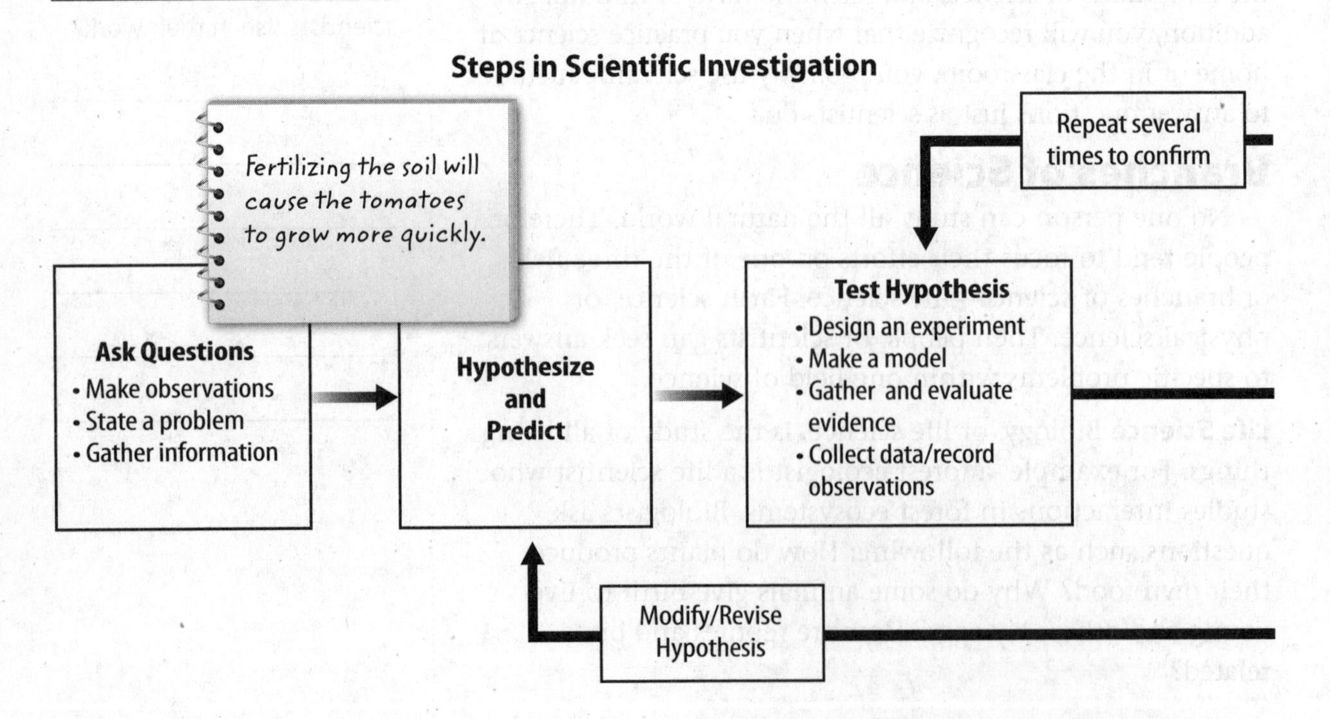

Observation After a few weeks, you observe that some plants are growing better than others. An **observation** *is using one or more of your senses to gather information and take note of what occurs.* Observations often are the beginning of the process of inquiry and can lead to questions such as "Why are some plants growing better than others?"

Inferring As you are making observations and asking questions, you recall that plants need water and light to grow. You infer that perhaps some vegetables are receiving more water or sunlight than others and, therefore, are growing better. *An* **inference** *is a logical explanation of an observation that is drawn from prior knowledge or experience.*

Hypothesize

After making observations and inferences, you are ready to develop a hypothesis and investigate why some vegetable plants are growing better than others. *A possible explanation about an observation that can be tested by scientific investigations is a* **hypothesis.** Your hypothesis might be: Some plants are growing larger and more quickly than others because they are receiving more water and sunlight. Or, your hypothesis might be: The plants that are growing quickly have received fertilizer because fertilizer helps plants grow.

Predict

After you state a hypothesis, you might make a prediction to help you test your hypothesis. *A* **prediction** *is a statement of what will happen next in a sequence of events.* For instance, based on your hypotheses, you might predict that if some plants receive more water, sunlight, or fertilizer, then they will grow larger.

Reading Check
4. Consider How does the process of inquiry usually begin?

Reading Check
5. Analyze What is the purpose of a hypothesis?

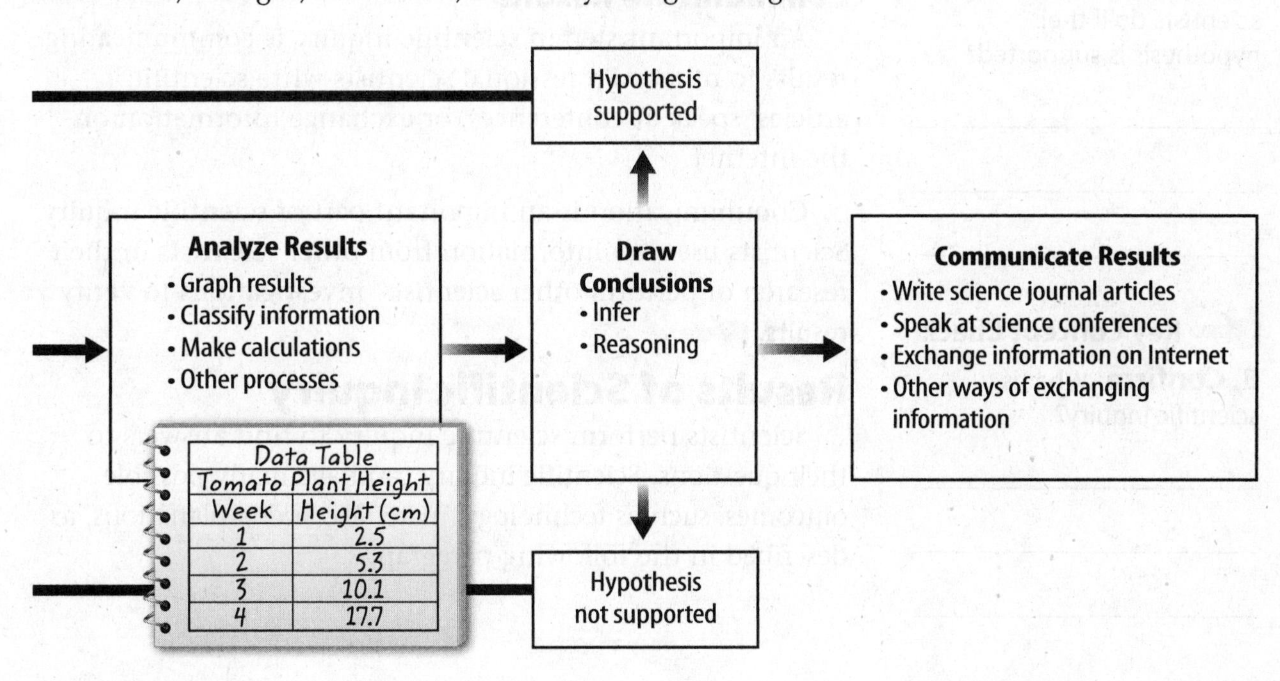

Test Your Hypothesis

When you test a hypothesis, you often are testing your predictions. For example, you might design an experiment to test you hypothesis on the fertilizer. You set up an experiment in which you plant seeds and add fertilizer to only some of them. Your prediction is that the plants that get the fertilizer will grow more quickly. If your prediction is confirmed, your hypothesis is supported. If your prediction is not confirmed, you hypothesis might need revision.

Analyze Results

As you are testing your hypothesis, you are probably collecting data about the plants' growth rates and how much fertilizer each plant receives. At first, it might be difficult to recognize patterns and relationships in data. Your next step might be to organize and analyze your data. You can create graphs, classify information, or make models and calculations. After the data are organized, you can more easily study the data and draw conclusions. ✓

Draw Conclusions

Now you must decide whether your data support your hypothesis and then draw conclusions. A conclusion is a summary of the information gained from testing a hypothesis. You might make more inferences when drawing conclusions. If your hypothesis is supported, you can repeat your experiment several times to confirm your results. If your hypothesis is not supported, you can modify it and repeat the scientific inquiry process. ✓

Communicate Results

An important step in scientific inquiry is communicating results to others. Professional scientists write scientific articles, speak at conferences, or exchange information on the Internet.

Communication is an important part of scientific inquiry. Scientists use new information from other scientists in their research or perform other scientists' investigations to verify results.

Results of Scientific Inquiry

Scientists perform scientific inquiry to find answers to their questions. Scientific inquiry can have many possible outcomes, such as technology, materials, and explanations, as described in the following paragraphs.

Reading Check

6. Relate What are three ways to organize data?

______

______

______

Reading Check

7. Explain What should scientists do if their hypothesis is supported?

______

______

______

Key Concept Check

8. Confirm What is scientific inquiry?

______

______

______

Technology New technology is one possible outcome of scientific inquiry. **Technology** *is the practical use of scientific knowledge, especially for industrial or commercial use*. Televisions, MP3 players, and computers are examples of technology.

New Materials The creation of new materials is another possible outcome of an investigation. For example, scientists have developed a bone bioceramic. A bioceramic is a natural calcium-phosphate mineral complex that is part of bones and teeth. This synthetic bone mimics natural bone's structure. Its porous structure allows a type of cell to grow and develop into new bone tissue.

The bioceramic can be shaped into implants that are treated with certain cells from the patient's bone marrow. It then can be implanted into the patient's body to replace missing bone.

Possible Explanations Many times, scientific investigations answer the questions who, what, when, where, and how. For example, who left fingerprints at a crime scene? When should fertilizer be applied to plants? What organisms live in rain forests?

Scientific Theory and Scientific Laws

Scientists often repeat scientific investigations many times to verify, or confirm, that the results for a hypothesis or a group of hypotheses are correct. This can lead to a scientific theory.

Scientific Theory The everyday meaning of the word *theory* is "an untested idea or an opinion." However, in science, *theory* has a different meaning. *A* **scientific theory** *is an explanation of observations or events based on knowledge gained from many observations and investigations*.

For example, about 300 years ago, scientists began looking at samples of trees, water, and blood through the first microscopes. They noticed that all these organisms were made of tinier units, or cells. As more scientists observed cells in other organisms, their observations became known as the cell theory.

The cell theory explains that all living things are made of cells. A scientific theory is assumed to be the best explanation of observations unless it is disproved. The cell theory will continue to explain the makeup of all organisms until an organism is discovered that is not made of cells.

FOLDABLES

Make a two-column chart book to organize your notes on scientific investigations.

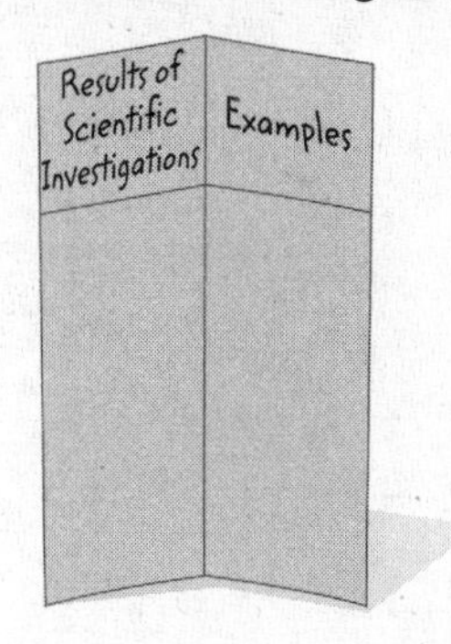

Key Concept Check

9. Define What are the results of scientific investigations?

Reading Check

10. Point Out What can change a scientific theory?

Scientific Laws Scientific laws are different from societal laws, which are an agreement on a set of behaviors. *A* **scientific law** *describes a pattern or an event in nature that is always true.* A scientific theory might explain how and why an event occurs. But a scientific law states only that an event in nature will occur under specific conditions.

Example of Scientific Law The law of conservation of mass states that the mass of materials will be the same before and after a chemical reaction. This scientific law does not explain why this occurs—only that it will occur. The table below compares a scientific theory and a scientific law.

Comparing Scientific Theory and Scientific Law

Scientific Theory	Scientific Law
A scientific theory is based on repeated observations and scientific investigations.	Scientific laws are observations of similar events that have been observed repeatedly.
If new information does not support a scientific theory, the theory will be modified or rejected.	If new observations do not follow the law, the law is rejected.
A scientific theory attempts to explain why something happens.	A scientific law states that something will happen.
A scientific theory usually is more complex than a scientific law and might contain many well-supported hypotheses.	A scientific law usually is based on one well-supported hypothesis that states that an event will occur.

Interpreting Tables

11. Consider How do scientific theories and laws compare?

Reading Check

12. Summarize Why is it important to question information in the media?

Key Concept Check

13. State How can a scientist minimize bias in a scientific investigation?

Skepticism in Media

When you see scientific issues in the media, such as newspapers, radio, television, and magazines, it is important to be skeptical. When you are skeptical, you question information that you read or hear, or events you observe. Is the information truthful? Is it accurate? It also is important that you question statements made by people outside their area of expertise and claims that are based on vague statements.

Evaluating Scientific Evidence

Critical thinking is an important skill in scientific inquiry. **Critical thinking** *is comparing what you already know with the information you are given in order to decide whether you agree with it.* Identifying and minimizing bias also is important when conducting scientific inquiry. To minimize bias in an investigation, sampling, repetition, and blind studies can be helpful, as shown in the figure on the next page.

Identifying and Preventing Bias

❶ Sampling
A method of data collection that involves studying small amounts of something in order to learn about the larger whole is sampling. A sample should be a random representation of the whole.

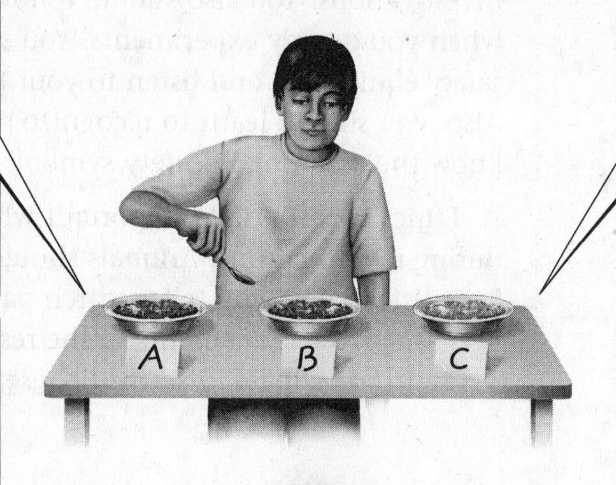

❸ Blind Study
A procedure that can reduce bias is a blind study. The investigator, subject, or both do not know which item they are testing. Personal bias cannot affect an investigation if participants do not know what they are testing.

❷ Bias
It is important to reduce bias during scientific investigations. Bias is intentional or unintentional prejudice toward a specific outcome. Sources of bias in an investigation can include equipment choices, hypothesis formation, and prior knowledge.

Suppose you were part of a taste test for a new cereal. If you knew the price of each cereal, you might think that the most expensive one tastes the best. This is a bias.

Most Popular Brand $2.00
Mediocre Brand $1.50
Least Popular Brand $0.50

❹ Repetition
If you get different results when you repeat an investigation, then the original investigation probably was flawed. Repetition of an experiment helps reduce bias.

Science cannot answer all questions.

You might think that any question can be answered through a scientific investigation. But some questions cannot be answered using science.

For example, science cannot answer a question such as, Which paint color is the prettiest? Questions about personal opinions, values, beliefs, and feelings cannot be answered scientifically. However, some people use scientific evidence to try to strengthen their claims about these topics.

Visual Check
14. Identify What are some sources of bias?

____

____

____

____

Safety in Science

Scientists follow safety procedures when they conduct investigations. You also should follow safety procedures when you do any experiments. You should wear appropriate safety equipment and listen to your teacher's instructions. Also, you should learn to recognize potential hazards and know the meaning of safety symbols.

ACADEMIC VOCABULARY
ethics
(noun) rules of conduct or moral principles

Ethics are especially important when using living things during investigations. Animals should be treated properly. Scientists also should tell research participants about the potential risks and benefits of the research. Anyone can refuse to participate in scientific research.

After You Read

Mini Glossary

critical thinking: comparing what you already know with the information you are given in order to decide whether you agree with it

hypothesis: a possible explanation about an observation that can be tested by scientific investigations

inference: a logical explanation of an observation that is drawn from prior knowledge or experience

observation: using one or more of your senses to gather information and take note of what occurs

prediction: a statement about what will happen next in a sequence of events

science: the investigation and exploration of natural events and of the new information that results from those investigations

scientific law: a pattern or an event in nature that is always true

scientific theory: an explanation of observations or events based on knowledge gained from many observations and investigations

technology: the practical use of scientific knowledge, especially for industrial or commercial use

1. Review the terms and their definitions in the Mini Glossary. Write a sentence that describes the importance of critical thinking.

2. Use the graphic organizer below to list the three branches of science.

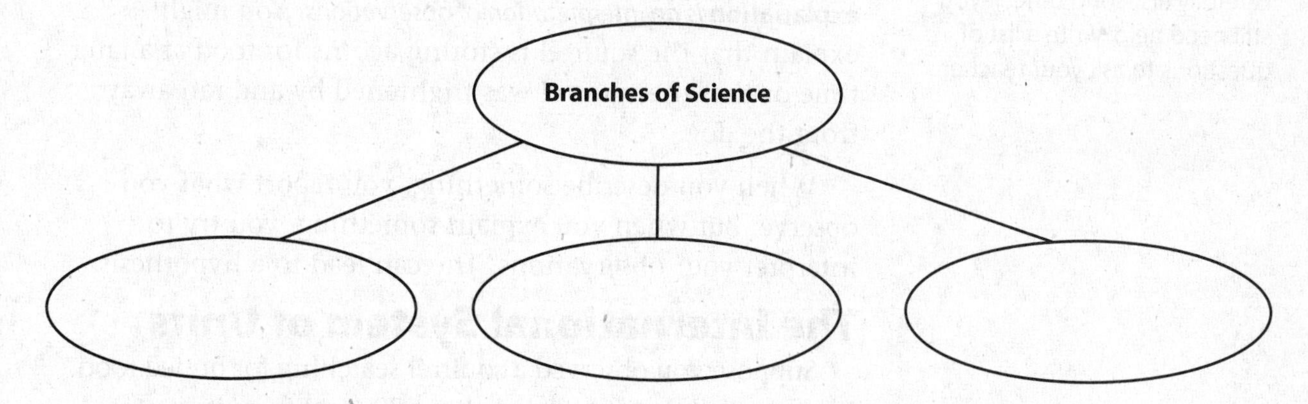

3. Discuss what types of questions cannot be answered scientifically and why.

Log on to ConnectED.mcgraw-hill.com and access your textbook to find this lesson's resources.

Scientific Explanations

Measurement and Scientific Tools

Key Concepts 

- What is the difference between accuracy and precision?
- Why should you use significant digits?
- What are some tools used by life scientists?

Mark the Text

Sticky Notes As you read, use sticky notes to mark information that you do not understand. Read the text carefully a second time. If you still need help, write a list of questions to ask your teacher.

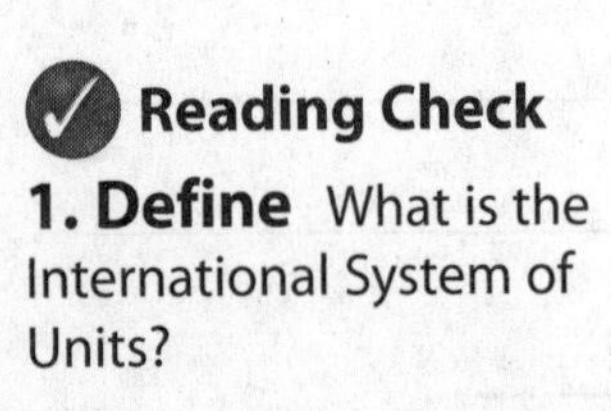

Reading Check

1. Define What is the International System of Units?

Read to Learn

Description and Explanation

How would you write a description of a squirrel's activity? *A* **description** *is a spoken or written summary of observations.*

Your description might include information such as the squirrel buried five acorns near a large tree or that the squirrel climbed the tree when a dog barked. A qualitative description uses your senses (sight, sound, smell, touch, taste) to describe an observation. *A large tree* is a qualitative description.

A quantitative description uses numbers to describe the observation. *Five acorns* is a quantitative description. You can use measuring tools, such as a ruler, a balance, or a thermometer to make quantitative descriptions.

How would you explain the squirrel's activity? *An* **explanation** *is an interpretation of observations*. You might explain that the squirrel is storing acorns for food at a later time or that the squirrel was frightened by and ran away from the dog.

When you describe something, you report what you observe. But when you explain something, you try to interpret your observations. The can lead to a hypothesis.

The International System of Units

Suppose you observed a squirrel searching for buried food. You recorded that it traveled about 200 feet from its nest.

Someone who measures distances in meters might not understand how far the squirrel traveled. The scientific community solved this problem in 1960. It adopted *an internationally accepted system for measurement called the* **International System of Units (SI).**

SI Base Units and Prefixes

Like scientists and many others around the world, you probably use the SI system in your classroom. All SI units are derived from the seven base units listed in the table on the left below. For example, the base unit for length, or the unit most commonly used to measure length, is the meter. You have probably made measurements in kilometers or millimeters before. Where do these units come from?

A prefix can be added to a base unit's name to indicate either a fraction or a multiple of that base unit. The prefixes are based on powers of ten, such as 0.01 and 100, as shown below on the right. One centimeter (cm) is one one-hundredth of a meter and a kilometer (km) is 1,000 meters. ✓

✓ Reading Check

2. Name What is added to the name of a base unit to indicate a fraction or a multiple of the base unit?

______

SI Base Units

Quantity Measured	Unit (Symbol)
Length	meter (m)
Mass	kilogram (kg)
Time	second (s)
Electric current	ampere (A)
Temperature	Kelvin (K)
Substance amount	mole (mol)
Light intensity	candela (cd)

Prefixes

Prefix	Meaning
Mega– (M)	1,000,000 (10^6)
Kilo– (k)	1,000 (10^3)
Hecto– (h)	100 (10^2)
Deka– (da)	10 (10^1)
Deci– (d)	0.1 (10^{-1})
Centi– (c)	0.01 (10^{-2})
Milli– (m)	0.001 (10^{-3})
Micro– (μ)	0.000 001 (10^{-6})

Conversion It is easy to convert from one SI unit to another. You multiply or divide by a power of ten. You also can use proportion calculations to make conversions. For example, a biologist measures an Emperor goose in the field. Her triple-beam balance shows that the goose has a mass of 2.8 kg. She could perform the calculation below to find the goose's mass in grams, x. ✓

$$\frac{x}{2.8\ \text{kg}} = \frac{1{,}000\ \text{g}}{1\ \text{kg}}$$

$$(1\ \text{kg})x = (1{,}000\ \text{g})(2.8\ \text{kg})$$

$$x = \frac{(1{,}000\ \text{g})(2.8\ \cancel{\text{kg}})}{1\ \cancel{\text{kg}}}$$

$$x = 2{,}800\ \text{g}$$

Notice that the answer has the correct units.

Interpreting Tables

3. Identify What unit measures mass?

______

✓ Reading Check

4. State How do you convert one SI unit to another?

______

______

______

FOLDABLES®

Make a horizontal two-tab book with a top tab to compare precision and accuracy.

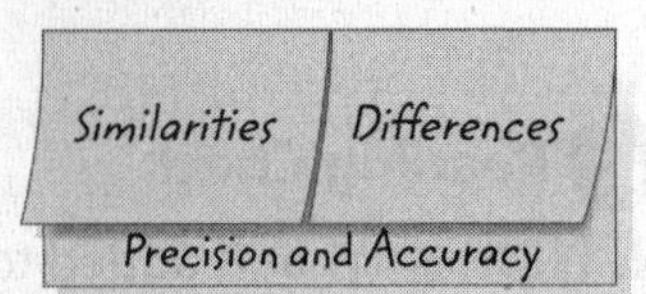

Precision and Accuracy

Suppose your friend Simon tells you that he will call you in one minute, but he calls you a minute and a half later. Sarah tells you that she will call you in one minute, and she calls exactly 60 seconds later. What is the difference? Sarah is accurate, and Simon is not. **Accuracy** *is a description of how close a measurement is to an accepted or true value*. However, if Simon always calls about 30 seconds later than he says he will, then Simon is precise. **Precision** *is a description of how similar or close measurements are to each other*, as shown in the figure below.

Accuracy and Precision

Accurate

An arrow in the center indicates high accuracy.

Precise but not accurate

Arrows far from the center indicate low accuracy. Arrows close together indicate high precision.

Accurate and precise

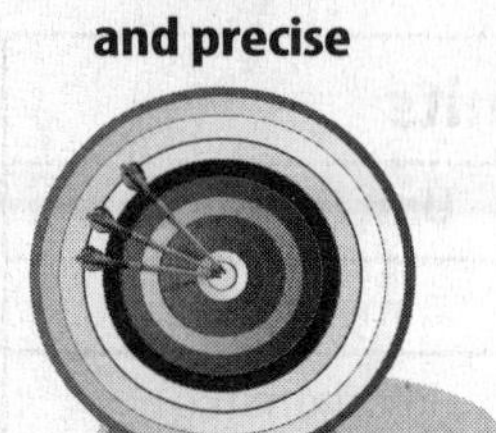

Arrows in the center indicate high accuracy. Arrows close together indicate high precision.

Not accurate or precise

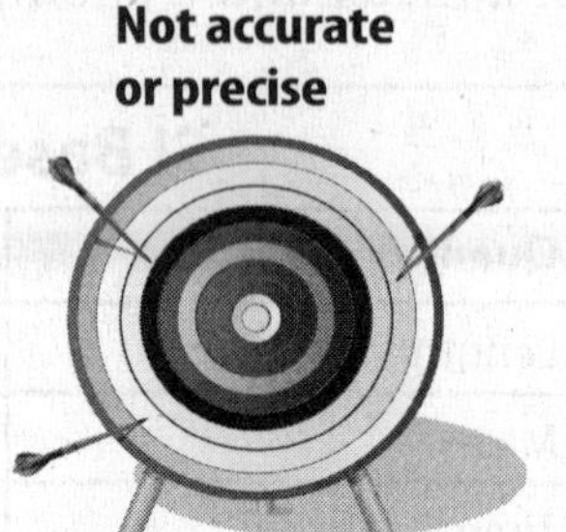

Arrows far from the center indicate low accuracy. Arrows far apart indicate low precision.

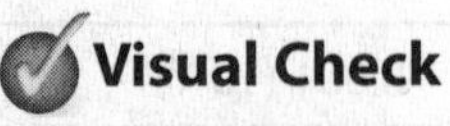

Visual Check

5. Interpret What do arrows close together in the target indicate?

____

____

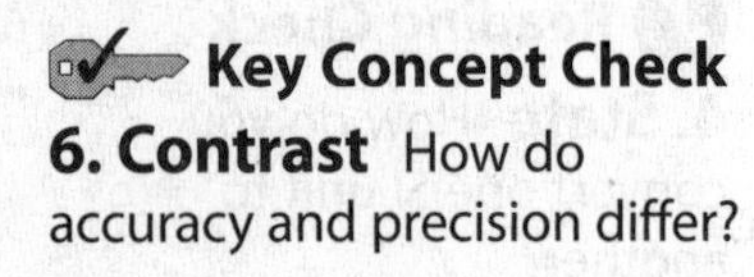

Key Concept Check

6. Contrast How do accuracy and precision differ?

____

____

____

The table at the top of the next page illustrates the difference between precise and accurate measurements.

Students were asked to find the melting point of sucrose, or table sugar. Each student took three temperature readings and calculated the mean, or average, of his or her data.

Refer to the table at the top of the next page. As the recorded data in the table shows, student A had the most accurate data. That student's melting point mean, 184.7°C, is closest to the scientifically accepted melting point, 185°C. Although not accurate, student C's measurements are the most precise because they are similar in value.

Student Melting Point Data			
	Student A	**Student B**	**Student C**
Trial 1	183.5°C	190.0°C	181.2°C
Trial 2	185.9°C	183.3°C	182.0°C
Trial 3	184.6°C	187.1°C	181.7°C
Average	184.7°C	186.8°C	181.6°C
Sucrose Melting Point (accepted value) 185°C			

Measurement and Accuracy

The tools used to take measurements can limit the accuracy of the measurements. Suppose you are measuring the temperature at which sugar melts, and the thermometer's measurements are divided into whole numbers. If your sugar sample melts between 183°C and 184°C, you can estimate the temperature between these two numbers. But, if the thermometer's measurements are divided into tenths, and your sample melts between 183.2°C and 183.3°C, your estimate between these numbers would be more accurate.

Significant Digits

In the second example above, you know that the temperature is between 183.2°C and 183.3°C. You could estimate that the temperature is 183.25°C. When you take any measurement, you know some digits for certain and you estimate some digits. **Significant digits** *are the number of digits in a measurement that are known with a certain degree of reliability.* The significant digits in a measurement include all digits you know for certain, plus one estimated digit. Therefore your measurement of 83.25°C would contain five significant digits. The figure below shows how to round to three significant digits. Since the ruler is divided into tenths, you know the rod is between 5.2 cm and 5.3 cm. You can estimate that the rod is 5.25 cm.

Measurement Tools: Known and Estimated Digits

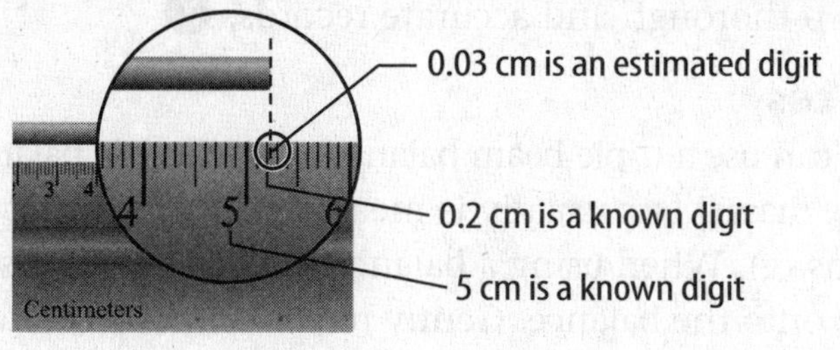

Visual Check

7. Analyze Why were student B's measurements imprecise compared to the measurements of student C?

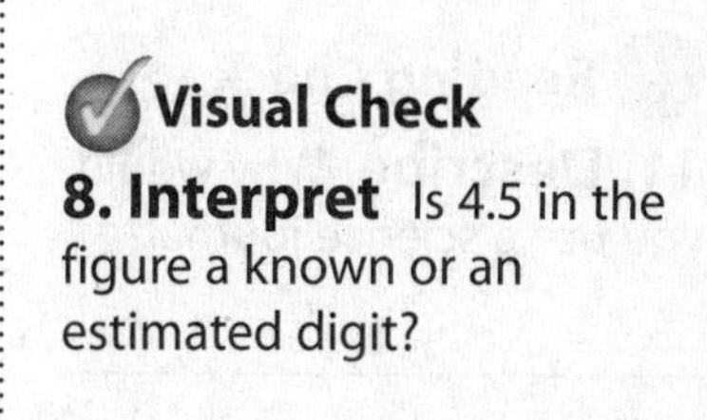

Visual Check

8. Interpret Is 4.5 in the figure a known or an estimated digit?

The number 5,281 has four significant digits. Rule 1, in the table to the right, states that all nonzero numbers are significant.

9. Significant Digits
Use the rules in the table to determine the number of significant digits in each of the following numbers: 2.02; 0.0057; 1,500; and 0.500.

____

____

____

Key Concept Check
10. Explain Why should you use significant digits?

____

____

____

____

Reading Check
11. Describe Why would you use a science journal?

____

____

____

____

Significant Digits

Rules

1. All nonzero numbers are significant.
2. Zeros between nonzero digits are significant.
3. Final zeros used after the decimal are significant.
4. Zeros used solely for spacing the decimal are not significant. The zeros indicate only the position of the decimal.

* The bold numbers in the examples are the significant digits.

Example Number	Significant Digits	Applied Rules
1.234	4	1
1.2	2	1
0.0**23**	2	1, 4
0.**200**	3	1, 3
1,002	4	1, 2
3.07	3	1, 2
0.00**1**	1	1, 4
0.0**12**	2	1, 4
50,600	3	1, 2, 4

Using significant digits lets others know how certain your measurements are. The rules for using significant digits are shown in the table above.

Scientific Tools

Scientific inquiry often requires the use of tools. Scientists, including life scientists, might use the tools listed below and on the next page. You might use one or more of them during a scientific inquiry, too.

Science Journal

In a science journal, you can record descriptions, explanations, plans, and steps used in a scientific inquiry. A science journal can be a spiral-bound notebook or a loose-leaf binder. It is important to keep your science journal organized so you can find information when you need it. Make sure you keep thorough and accurate records.

Balances

You can use a triple-beam balance or an electric balance to measure mass. Mass usually is measured in kilograms (kg) or grams (g). When using a balance, do not let objects drop heavily onto the balance. Gently remove an object after you record its mass.

Thermometer

A thermometer measures the temperature of substances. The Kelvin (K) is the SI unit for temperature. However, in the science classroom, you measure temperature in degrees Celsius (°C). Use care when you place a thermometer into a hot substance so you do not burn yourself. Handle glass thermometers gently so they do not break. If a thermometer does break, tell your teacher immediately. Do not touch the broken glass or the thermometer's liquid. Never use a thermometer to stir anything. ✓

Glassware

Laboratory glassware is used to hold, pour, heat, and measure liquids. Most labs have many types of glassware. For example, flasks, beakers, petri dishes, test tubes, and specimen jars are used as containers. To measure the volume of a liquid, you use a graduated cylinder. The unit of measure for liquid volume is the liter (L) or milliliter (mL). ✓

Compound Microscope

Microscopes enable you to observe small objects that you cannot observe with just your eyes. Usually, two types of microscopes are in science classrooms—dissecting microscopes and compound light microscopes. Microscopes have either a single eyepiece or two eyepieces to observe a magnified image of a small object or an organism.

Microscopes can be damaged easily. It is important to follow your teacher's instructions when carrying and using a microscope.

Computers—Hardware and Software

Computers process information. In science, you can use computers to compile, retrieve, and analyze data for reports. You also can use them to create reports and other documents, to send information to others, and to research information.

The physical components of computers, such as monitors and keyboards, are called hardware. The programs that you run on computers are called software. These programs include word processing, spreadsheet, and presentation programs. When scientists write reports, they use word processing programs. They use spreadsheet programs for organizing and analyzing data. Presentation programs can be used to explain information to others. ✓

Reading Check

12. State In a science classroom, what unit of measure do you use for temperature?

Reading Check

13. Identify Which unit of measure do you use for liquid volume?

Reading Check

14. Name What are the physical components of a computer called? What are computer programs called?

Tools Used by Life Scientists

Life scientists often use the tools described below.

Magnifying Lens

A magnifying lens is a hand-held lens that magnifies, or enlarges, the image of an object. It is not as powerful as a microscope and is useful when great magnification is not needed. Magnifying lenses also can be used outside the lab where microscopes might not be available.

Reading Check
15. Describe how you might use a magnifying lens.

Slide

To observe an item using a compound light microscope, you first must place it on a thin, rectangular piece of glass called a slide. You must handle slides gently to avoid breaking them.

Dissecting Tools

Scientists use dissecting tools, such as scalpels and scissors, to examine tissues, organs, or prepared organisms. Dissecting tools are sharp, so always use extreme caution when handling them.

Pipette

A pipette is similar to a eyedropper. It is a small glass or plastic tube used to draw up and transfer liquids.

Key Concept Check
16. Name What are some tools used by life scientists?

After You Read

Mini Glossary

accuracy: a description of how close a measurement is to an accepted or true value

description: a spoken or written summary of observations

explanation: an interpretation of observations

International System of Units (SI): the internationally accepted system for measurement

precision: a description of how similar or close measurements are to each other

significant digits: the number of digits in a measurement that are known with a certain degree of reliability

1. Review the terms and their definitions in the Mini Glossary. Write a sentence that explains the importance of accuracy in measurement.

___

___

2. Complete the following table to identify and describe SI base units.

Unit (symbol)	Quantity Measured
ampere (A)	
	light intensity
meter (m)	
	temperature
mole (mol)	

3. Explain the difference between a description and an explanation.

___

___

Log on to ConnectED.mcgraw-hill.com and access your textbook to find this lesson's resources.

Nature of Science

LESSON 3

Scientific Explanations

Case Study

Key Concepts
- How do independent and dependent variables differ?
- How is scientific inquiry used in a real-life scientific investigation?

Study Coach

Identify the Main Ideas As you read, organize your notes in two columns. In the left column, write the main idea of each paragraph. In the right column, write details that support each main idea. Review your notes to help you remember the details of the lesson.

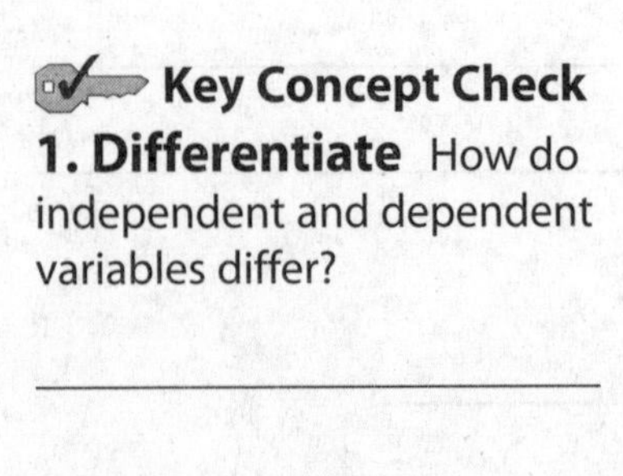

Key Concept Check
1. Differentiate How do independent and dependent variables differ?

Read to Learn

Biodiesel from Microalgae

For several centuries, fossil fuels have been the main sources of energy for industry and transportation. However, scientists have shown that burning fossil fuels negatively affects the environment. Also, some people are concerned about eventually using up the world's reserves of fossil fuels.

During the past few decades, scientists have explored using protists to produce biodiesel. Biodiesel is a fuel made primarily from living organisms. Protists are a group of tiny organisms that usually live in water or moist environments. Some of these protists are plantlike because they make their own food using a process called photosynthesis. Microalgae are plantlike protists.

Designing a Controlled Experiment

Scientists use scientific inquiry to investigate the use of protists to make biodiesel. They designed controlled experiments to test their hypotheses. The tables in this lesson are examples of how scientists practiced inquiry and the skills you read about in Lesson 1.

A controlled experiment is a scientific investigation that tests how one variable affects another variable. *A* **variable** *is any factor in an experiment that can have more than one value.* Controlled experiments have two types of variables. *The* **dependent variable** *is the factor measured or observed during an experiment. The* **independent variable** *is the factor that you want to test. It is changed by the investigator to observe how it affects a dependent variable.* **Constants** *are the factors in an experiment that remain the same.*

A controlled experiment has an experimental group and a control group. The experimental group is used to study how a change in the independent variable changes the dependent variable. The control group contains the same factors as the experimental group, but the independent variable is not changed. Without a control, it is difficult to know whether your experimental observations result from the variable you are testing or from another factor.

Biodiesel

The idea of running diesel engines on fuel made from plant or plantlike sources is not entirely new. Rudolph Diesel invented the diesel engine. He used peanut oil to demonstrate how the engine worked. However, once petroleum was introduced as a diesel fuel source, it was preferred over peanut oil because of it was cheaper. ✓

Oil-rich food crops such as soybeans can be used to produce biodiesel. However, some people are concerned that crops grown for fuel use will replace crops grown for food. If farmers grow more crops for fuel, then the amount of food available worldwide will be reduced. Because of food shortages in many parts of the world, replacing food crops with fuel crops is not a good solution. ✓

Aquatic Species Program

In the late 1970s, the U.S. Department of Energy began funding its Aquatic Species Program (ASP) to investigate ways to remove air pollutants. Coal-fueled power plants produce carbon dioxide (CO_2), a pollutant, as a by-product. In the beginning, the study examined all aquatic organisms that use CO_2 during photosynthesis—their food-making process. These included large plants, commonly known as seaweeds, plants that grow partially underwater, and microalgae.

Scientists hoped that these organisms might remove excess CO_2 from the atmosphere. During the studies, however, the project leaders noticed that some microalgae produced large amounts of oil. The program's focus soon shifted to using microalgae to produce oils that could be processed into biodiesel. The scientists' observations and prediction are summarized in the table below. When referring to the examples in tables in this lesson, recall that a hypothesis is a tentative explanation that can be tested by scientific investigation. A prediction is a statement of what someone expects to happen next in a sequence of events.

Observation and Prediction

Scientific investigations often begin when someone observes an event in nature and wonders why or how it occurs.

Observation	Prediction
While testing microalgae to discover if they would absorb carbon pollutants, ASP project leaders saw that some species of microalgae had high oil content.	If the right conditions are met, then plants and plantlike organisms can be used as a source of fuel.

✓ **Reading Check**

2. Name What did Rudolph Diesel use as fuel?

____

____

✓ **Reading Check**

3. Explain Why is there a concern that crops grown for fuel use will replace crops grown for food?

____

____

____

Interpreting Tables

4. Relate How do scientific investigations often begin?

____

____

Which Microalgae?

Microalgae are microscopic organisms that live in marine (salty) or freshwater environments. Like many plants and plantlike organisms, they use photosynthesis and make sugar. The process requires light energy. Microalgae make more sugar than they can use. They convert excess sugar to oil. Scientists focused on these microalgae because their oil then could be processed into biodiesel. ✓

Scientists started by collecting and identify promising microalgae species. The search focused on microalgae in shallow, inland, saltwater ponds. Scientists predicted that these microalgae were more resistant to changes in temperature and salt content in the water.

✓ **Reading Check**

5. Define What are microalgae?

Design an Experiment and Collect Data
One way to test a hypothesis is to design an experiment and collect data.
The ASP scientists developed a rapid screening test to discover which microalgae species produced the most oil.

Interpreting Tables

6. State What was the goal of the screening test?

By 1985, a test was in place for identifying microalgae with high oil content. Two years later, 3,000 microalgae species had been collected. Scientists checked these samples for tolerance to acidity, salt levels, and temperature. From the samples, 300 species were selected. Of these 300 species, green algae and diatoms showed the most promise. However, it was obvious that no one strain was going to be perfect for all climates and water types.

Hypotheses and Predictions
During a long investigation, scientists form many hypotheses and conduct many tests.
Hypothesis: Microalgae species in shallow, saltwater ponds are most resistant to variations in temperature and salt content.
Prediction: Microalgae species most resistant to variations in temperature and salt content will be the most useful species in producing biodiesel.
Hypothesis: Microalgae grown with inadequate amounts of nitrogen alter their growth processes and produce more oil.
Independent Variable: amount of nitrogen available
Dependent Variable: amount of oil produced
Constants: the growing conditions of algae (temperature, water quality, exposure to the Sun, etc.)

Interpreting Tables

7. Identify What is the dependent variable in this example?

Oil Production in Microalgae

Scientists also began researching how microalgae produce oil. Some studies suggested that starving microalgae of nutrients, such as nitrogen, could increase the amount of oil they produced. However, starving the microalgae also caused them to be smaller, resulting in no overall increase in oil production.

Outdoor Testing v. Bioreactors

By the 1980s, the ASP scientists were growing microalgae in outdoor ponds in New Mexico. However, outdoor conditions were very different from those in the laboratory. The cooler outdoor temperatures resulted in smaller microalgae. Native algae species also invaded the ponds, forcing out the high-oil-producing laboratory microalgae species.

The scientists continued to focus on growing microalgae in open ponds. Many scientists still believe that these open ponds are better for producing large quantities of biodiesel from microalgae. But some researchers are now growing microalgae in closed glass containers called bioreactors. Inside these bioreactors, organisms live and grow under controlled conditions. This method avoids many of the problems associated with open ponds. However, bioreactors are more expensive than open ponds.

A biofuel company in the western United States has been experimenting with a low-cost bioreactor. A scientist at the company explained that they examined the ASP program and hypothesized that they could use long plastic bags instead of closed glass containers. However, microalgae grown in plastic bags are more expensive to harvest.

Why So Many Hypotheses?

According to Dr. Richard Sayre, a biofuel researcher, all the ASP research was based on forming hypotheses. He says, "It was hypothesis-driven. You just don't go in and say 'Well, I have a feeling this is the right way to do it.' You propose a hypothesis. Then you test it." Dr. Sayre added, "Biologists have been trained over and over again to develop research strategies based on hypotheses. It's sort of ingrained into our culture. You don't get research support by saying, 'I'm going to put together a system, and it's going to be wonderful.' You have to come up with a question. You propose some strategies for answering the question. What are your objectives? What outcomes do you expect for each objective?"

Reading Check

8. Explain Why didn't starving microalgae of nutrients provide an overall benefit?

____

____

____

____

Reading Check

9. Relate What is a disadvantage of bioreactors?

____

____

____

Reading Check

10. Describe Why is it important for a scientific researcher to develop a good hypothesis?

____

____

____

____

Increasing Oil Yield

Scientists from a biofuel company in Washington State thought of another way to increase oil production. Researchers knew microalgae use light energy, water, and carbon dioxide and they make sugar. The microalgae eventually convert sugar into oil. The scientists wondered if they could increase microalgae oil production by distributing light to all microalgae, including those below the surface.

Interpreting Tables

11. State How did scientists think they could get microalgae to produce more oil?

Hypothesis and Prediction
Scientists hypothesize that they can increase microalgae oil production by distributing light to greater depths.
Hypothesis: If the top layer of microalgae blocks light from reaching microalgae beneath them, then they produce less oil because light is not distributed evenly to all the microalgae.
Prediction: If light is distributed more evenly, then more microalgae will grow, and more biodiesel will be produced.

Bringing Light to Microalgae

Normally microalgae grow near the surface of a pond. Any microalgae about 5 cm below the pond's surface will grow less. Why is this? First, water blocks light from reaching deep into the pond. Second, microalgae at the top of a pond block light from reaching microalgae below them. Only the top part of a pond is productive. ✓

✓ Reading Check

12. State Where do microalgae normally grow within a pond?

Experimental Group

Researchers decided to assemble a team of engineers to design a light distribution system. Light rods distribute artificial light to microalgae in a bioreactor. The bioreactor controls the environmental conditions that affect how the microalgae grow. These conditions include temperature, nutrient levels, carbon dioxide level, airflow, and light. ✓

✓ Reading Check

13. Identify In the experimental group, what variables are controlled in the bioreactor?

Data from their experiments showed scientists how their microalgae in well-lit environments grow compared to how microalgae grow in dimmer environments. Using solar data for various parts of the country, the scientists concluded that the light rod would significantly increase microalgae growth and oil production in outdoor ponds. These scientists next plan to use the light-rod growing method in outdoor ponds.

Field Testing

Scientists plan to take light to the microalgae instead of moving microalgae to light. Dr. Jay Burns is chief microalgae scientist at a biofuel company. He said, "What we are proposing to do is to take the light from the surface of a pond and distribute it throughout the depth of the pond. Instead of only the top 5 cm being productive, the whole pond becomes productive."

Note that research scientists and scientists in the field rely on scientific methods and scientific inquiry to solve real-life problems. When a scientific investigation lasts for several years and involves many scientists, such as this study, many hypotheses can be tested. Some hypotheses are supported, and other hypotheses are not.

Regardless of which hypotheses are supported, information is gathered and lessons are learned. Hypotheses are refined and tested many times. This process of scientific inquiry results in a better understanding of the problem and the possible solutions.

Analysis and Conclusion
Scientists tested their hypothesis, collected data, analyzed the data, and drew conclusions.
Analyze Results: The experimental results showed that microalgae would produce more oil using a light-rod system than by using just sunlight.
Draw a Conclusion: The researchers concluded that the light-rod system greatly increased microalgae oil production.

Another Way to Bring Light to Microalgae

Light rods are not the only way to bring light to microalgae. Paddlewheels can be used to keep changing the location of the microalgae. Paddlewheels continuously rotate microalgae to the surface so the organisms are exposed to more light.

Reading Check

14. Summarize What is the benefit of the light-distribution system?

Interpreting Tables

15. Express What did the researchers conclude would increase algae yield?

Key Concept Check

16. Explain Describe three ways in which scientific inquiry was used in this case study.

Why Grow Microalgae?

Although the focus of this case study is microalgae growth for biodiesel production, growing microalgae has other benefits.

Some of the benefits of growing algae are shown in the figure below. Power plants that burn fossil fuels release carbon dioxide into the atmosphere. Evidence indicates that this contributes to global warming. During photosynthesis, microalgae use carbon dioxide and water, release oxygen, and produce sugar, which they convert to oil. Not only do microalgae produce a valuable fuel, they also remove pollutants from and add oxygen to the atmosphere.

Visual Check

17. Analyze What is used as a feedstock for microalgae?

____

Cultivating Microalgae

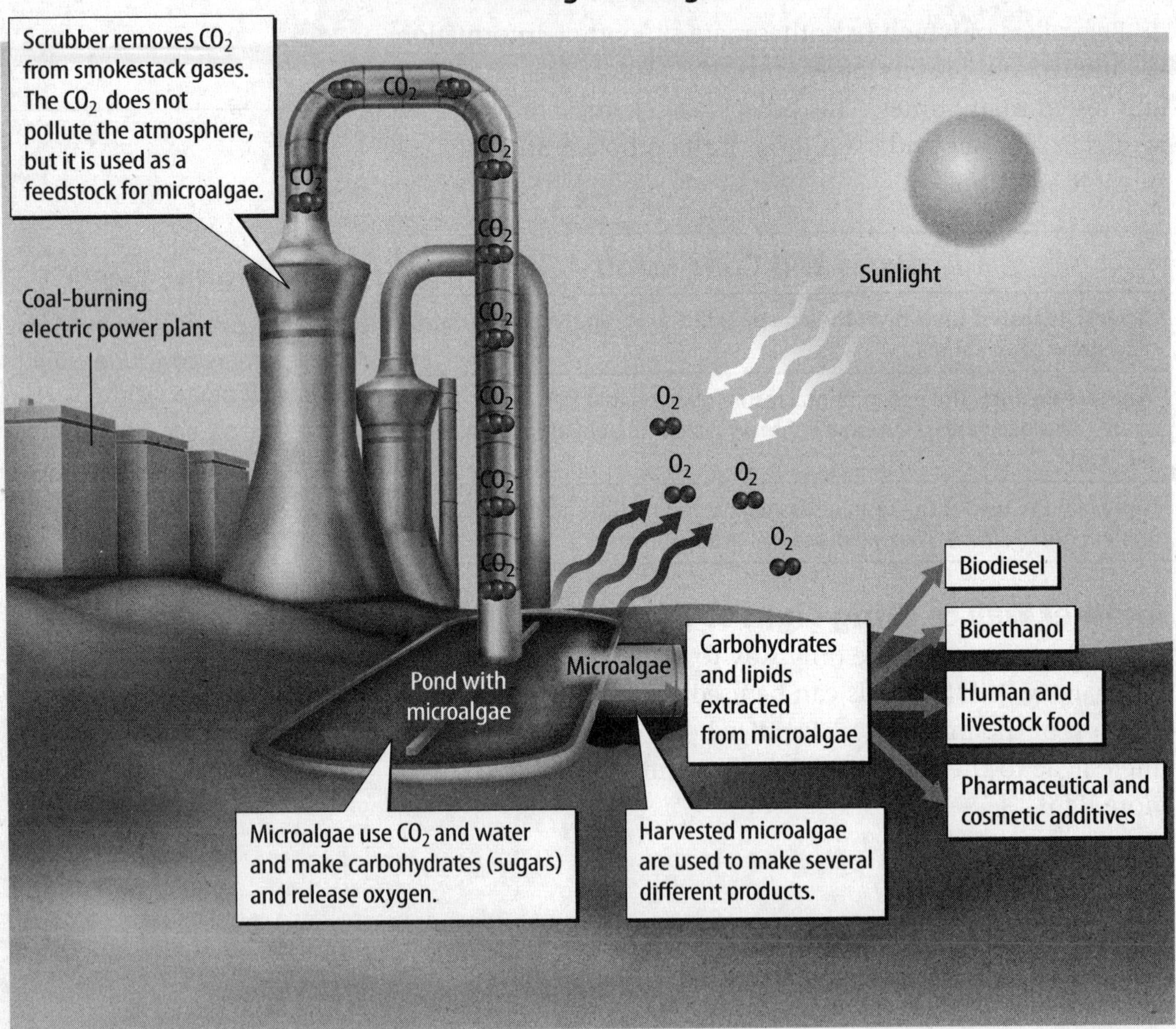

Are microalgae the future?

Scientists face many challenges in their quest to produce biodiesel from microalgae. For now, the costs of growing microalgae and extracting their oils are too high to compete with petroleum-based diesel. However, the combined efforts of government-funded programs and commercial biofuel companies might one day make microalgae-based biodiesel an affordable reality in the United States. ✓

New Plants One company in Israel has a successful test plant in operation. Plans are underway to build a large-scale industrial facility to convert carbon dioxide gases released from an Israeli coal-powered electrical plant into useful microalgae products. If this technology performs as expected, microalgae cultivation might occur near coal-fueled power plants in other parts of the world, too. ✓

Drawing Conclusions Currently, scientists have no final conclusions about using microalgae as a fuel source. As long as petroleum remains relatively inexpensive and available, it probably will remain the preferred source of diesel fuel. However, if petroleum prices increase or availability decreases, new sources of fuel will be needed. Biodiesel made from microalgae might be one of the alternative fuel sources used. ✓

✓ **Reading Check**

18. Assess What is preventing algae-based biodiesel from competing with petroleum-based diesel?

____

____

✓ **Reading Check**

19. Predict If the technology is successful, what might happen to algae cultivation?

____

____

____

✓ **Reading Check**

20. State What might cause a demand for biodiesel made from microalgae?

____

____

____

After You Read

Mini Glossary

constant: a factor in an experiment that remains the same

dependent variable: the factor measured or observed during an experiment

independent variable: a factor that you want to test and that is changed by the investigator to observe how it affects a dependent variable

variable: any factor in an experiment that can have more than one value

1. Review the terms and their definitions in the Mini Glossary. Write a sentence that describes the use of variables in controlled experiments.

____

____

2. Complete the following flowchart that shows the ASP's initial study of pollution control.

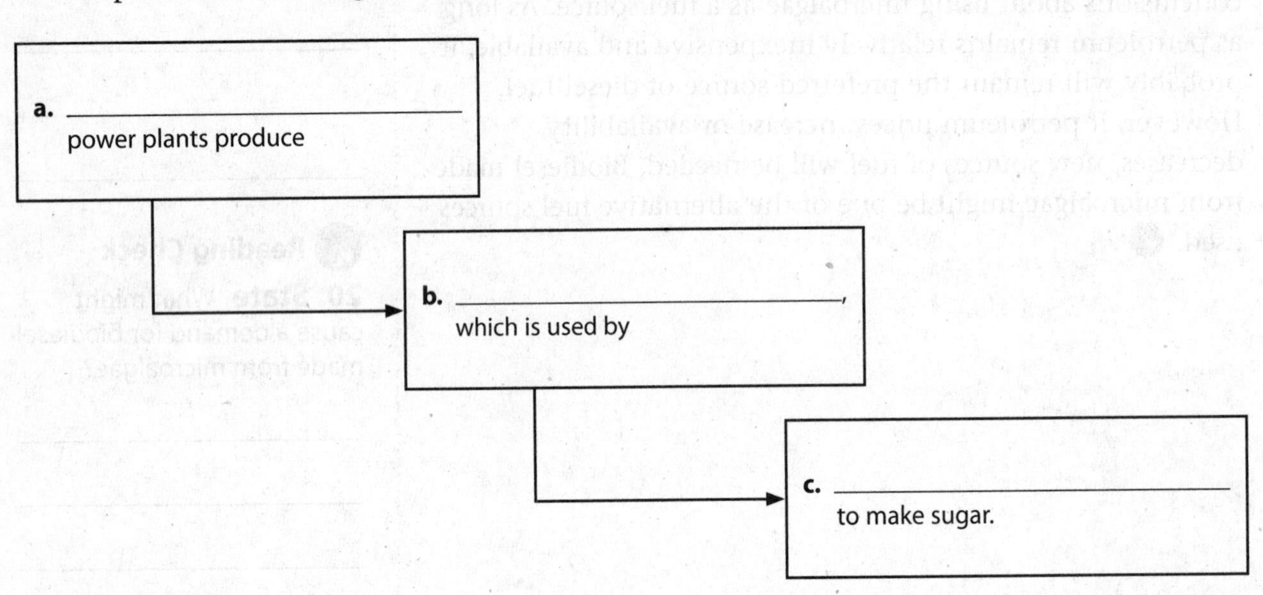

3. Explain how identifying the main idea of each paragraph helped you understand this lesson.

____

____

Log on to ConnectED.mcgraw-hill.com and access your textbook to find this lesson's resources.

Classifying and Exploring Life

Characteristics of Life

Before You Read

What do you think? Read the two statements below and decide whether you agree or disagree with them. Place an A in the Before column if you agree with the statement or a D if you disagree. After you've read this lesson, reread the statements to see if you have changed your mind.

Before	Statement	After
	1. All living things move.	
	2. The Sun provides energy for almost all organisms on Earth.	

Key Concept 
- What characteristics do all living things share?

Read to Learn

Characteristics of Life

Your classroom is full of nonliving things and living things. Desks, books, and lights are nonliving things. Your classmates, teacher, and plants are living things. What makes people and plants different from desks and lights?

People and plants, like all living things, have all the characteristics of life. All living things are organized. They grow and develop. All living things reproduce. They respond to their environment. All living things maintain certain internal conditions and use energy.

Nonliving things do not have all these characteristics. Books might be organized into chapters. Lights might use energy. But only those things that have all the characteristics of life are living. *Things that have all the characteristics of life are called* **organisms.**

Study Coach

Make Flash Cards Write each boldface word on one side of a flash card. Write the definition on the other side. Use the cards to quiz yourself.

Reading Check

1. Identify How do living things differ from nonliving things?

______

______

______

Organization

Your school has organization. The classrooms are for learning and the gym is for sports. Living things are also organized. Their organization involves cells. *A* **cell** *is the smallest unit of life.* An organism might be made of just one cell or of many cells. All organisms have structures with specific functions, or jobs.

Make a half-book and use it to organize your notes on the characteristics of living things.

Think it Over

2. Apply You are a multicellular organism. Name one function that groups of your cells carry out.

____

Unicellular Organisms *Living things that are made of only one cell are called* **unicellular** *organisms.* A unicellular organism has structures with specialized functions. Some structures control cell activities. Some take in nutrients. Other structures enable the organism to move.

Multicellular Organisms *Living things that are made of two or more cells are called* **multicellular** *organisms.* Some multicellular organisms only have a few cells, but others have trillions of cells. The cells of a multicellular organism usually do not all do the same things. Instead, groups of cells have specialized functions. These functions might include digestion or movement.

Growth and Development

Think about the tadpole in the figure below. The tadpole does not look like the frog it will become. The tadpole will lose its tail and grow legs. Like all organisms, the tadpole will grow and develop.

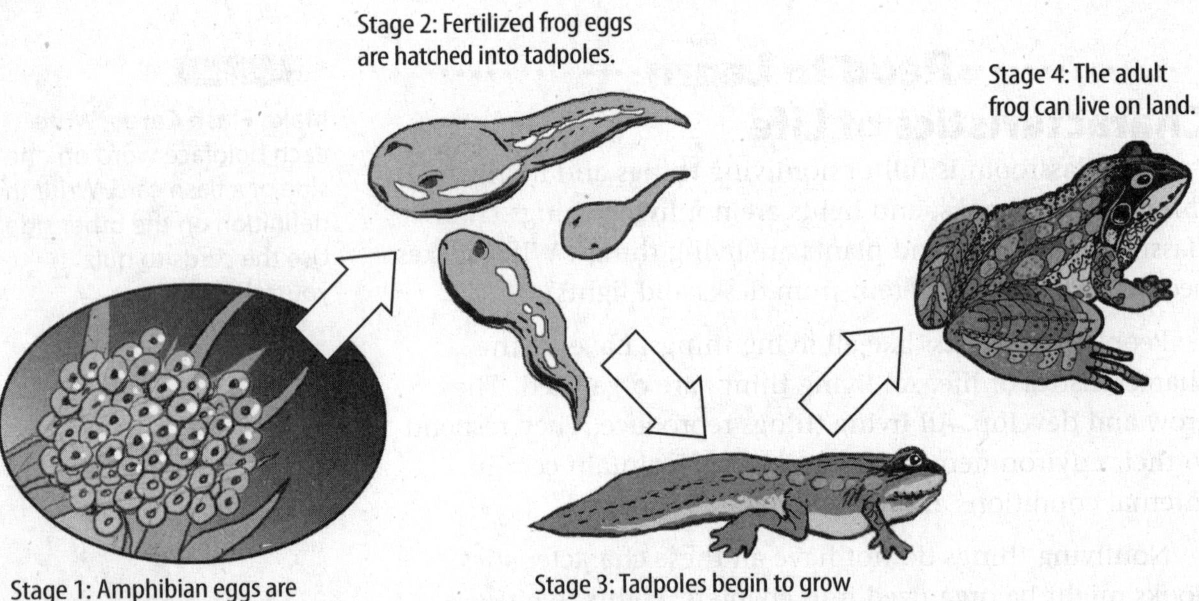

Visual Check

3. Identify the characteristics of life that you see in the figure.

____

____

____

____

How do organisms grow? When an organism grows, it increases in size. A unicellular organism grows as its one cell gets bigger. A multicellular organism grows when more cells are produced.

How do organisms develop? The changes in an organism during its lifetime are called development. A multicellular organism develops as cells become specialized into different cell types, such as skin and muscle cells. Some organisms have amazing developmental changes over their lifetimes. An example is a tadpole developing into a frog. ✓

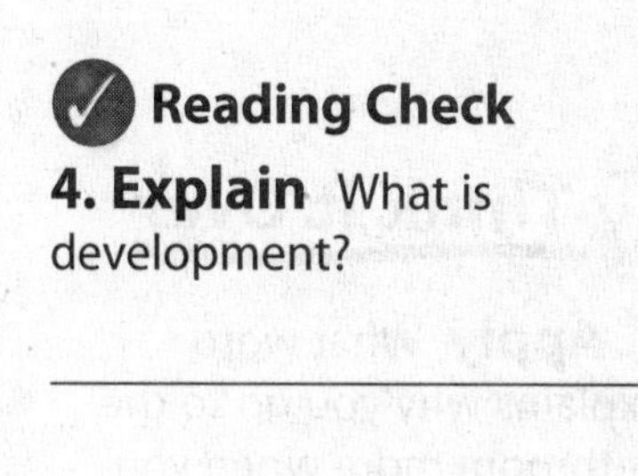

Reproduction

Reproduction is the process by which one organism makes one or more new organisms. Organisms must reproduce, or they will die out. Not all organisms reproduce, such as family pets. But if a type of organism is going to survive, some organisms of that type must reproduce.

Organisms reproduce in many ways. Some unicellular organisms divide and become two new organisms. Each new organism is just like the original cell. Some organisms must have a mate to reproduce. Other organisms can reproduce without a mate. Organisms produce different numbers of offspring. Humans usually produce only one or two offspring at a time. Other organisms, such as frogs, can produce hundreds of offspring at one time.

Responses to Stimuli

Organisms live in environments that change all the time. These changes are called stimuli (STIHM yuh li). One change is called a stimulus. All organisms respond to stimuli.

Internal Stimuli

Internal stimuli are changes inside an organism. They include hunger, thirst, and pain. If you feel hungry and look for food, you are responding to an internal stimulus—hunger. The feeling of thirst that causes you to look for water is another internal stimulus. ✓

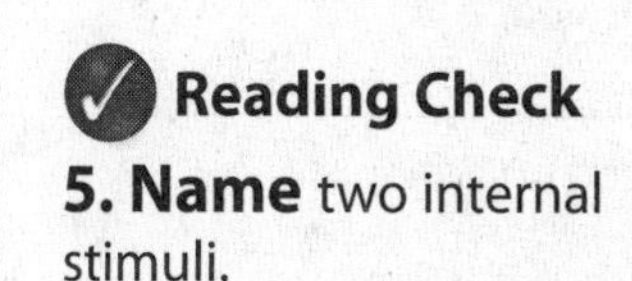

External Stimuli

External stimuli are changes outside an organism. They are usually changes in the environment that the organism lives in. Light and temperature are examples of external stimuli.

6. Contrast How do external stimuli differ from internal stimuli?

Light Many organisms respond to changes in light. Many plants will grow toward light. You respond to light, too. If you spend time in sunlight, your skin's response might be to darken, turn red, or freckle.

Temperature How does your body respond to changes in temperature? Like many animals, your body responds by increasing or decreasing the amount of blood flow to your skin. If the temperature gets warmer, your blood vessels respond by widening. Then more blood can flow to your skin. You feel cooler.

7. Apply What word explains why you go to the bathroom more when you drink a lot of water?

____

Homeostasis

All organisms are able to maintain some internal conditions. **Homeostasis** (hoh mee oh STAY sus) *is an organism's ability to maintain steady internal conditions when outside conditions change.* Have you ever noticed that if you drink more water than usual, you have to go to the bathroom more often than you usually do? Your body is keeping your internal conditions steady.

The Importance of Homeostasis

Cells need certain conditions to function the way they should. Homeostasis makes sure cells can function. If cells cannot function the way they should, an organism might get sick or die.

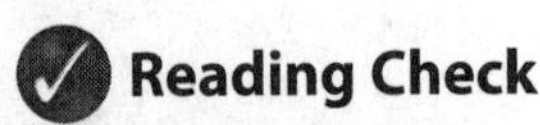

8. Explain Why is maintaining homeostasis important to organisms?

____

____

Methods of Regulation

Humans cannot survive if their body temperature changes more than a few degrees from 37°C. When your outside environment becomes too hot or too cold, your body responds. It sweats, shivers, or changes the flow of blood to maintain the body temperature of 37°C.

Both unicellular and multicellular organisms have ways to maintain homeostasis. Some unicellular organisms have a structure called a contractile vacuole (kun TRAK tul • VA kyuh wohl). It collects and pumps extra water out of the cell.

There is a limit to the amount of change that can occur inside an organism. For example, you could live for only a few hours in very cold water. Your body could not maintain steady internal conditions, or homeostasis, in this environment. Your cells could not function.

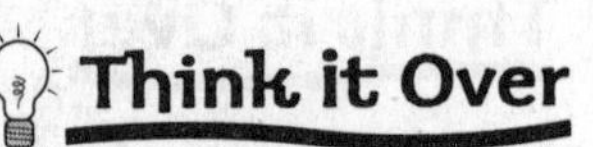

9. Generalize What conditions might be too harsh for an organism to maintain homeostasis?

____

____

Energy

All organisms use energy. Digesting food, thinking, reading, and sleeping use energy. Cells use energy to transport substances, make new cells, and perform chemical reactions. All of the characteristics of life use energy.

Energy's Origin Where does this energy come from? The energy that most organisms use originally came to Earth from the Sun, as shown below. The energy goes from one organism to another. Energy in the cactus comes from the Sun. The squirrel gets energy from the cactus that it eats. The coyote gets energy from eating the squirrel.

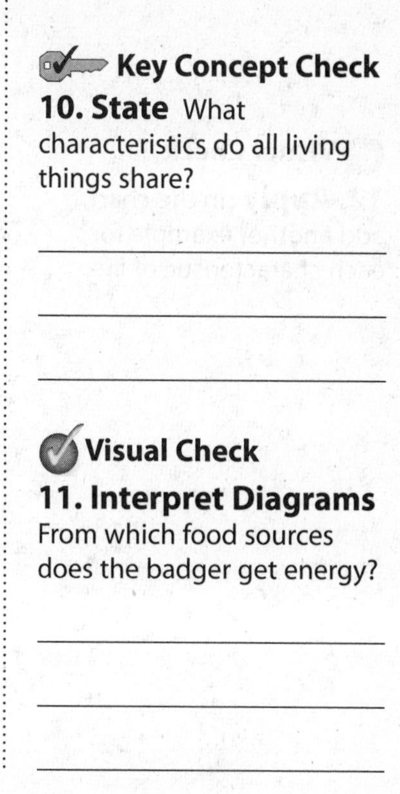

Key Concept Check

10. State What characteristics do all living things share?

____

____

____

Visual Check

11. Interpret Diagrams From which food sources does the badger get energy?

____

____

____

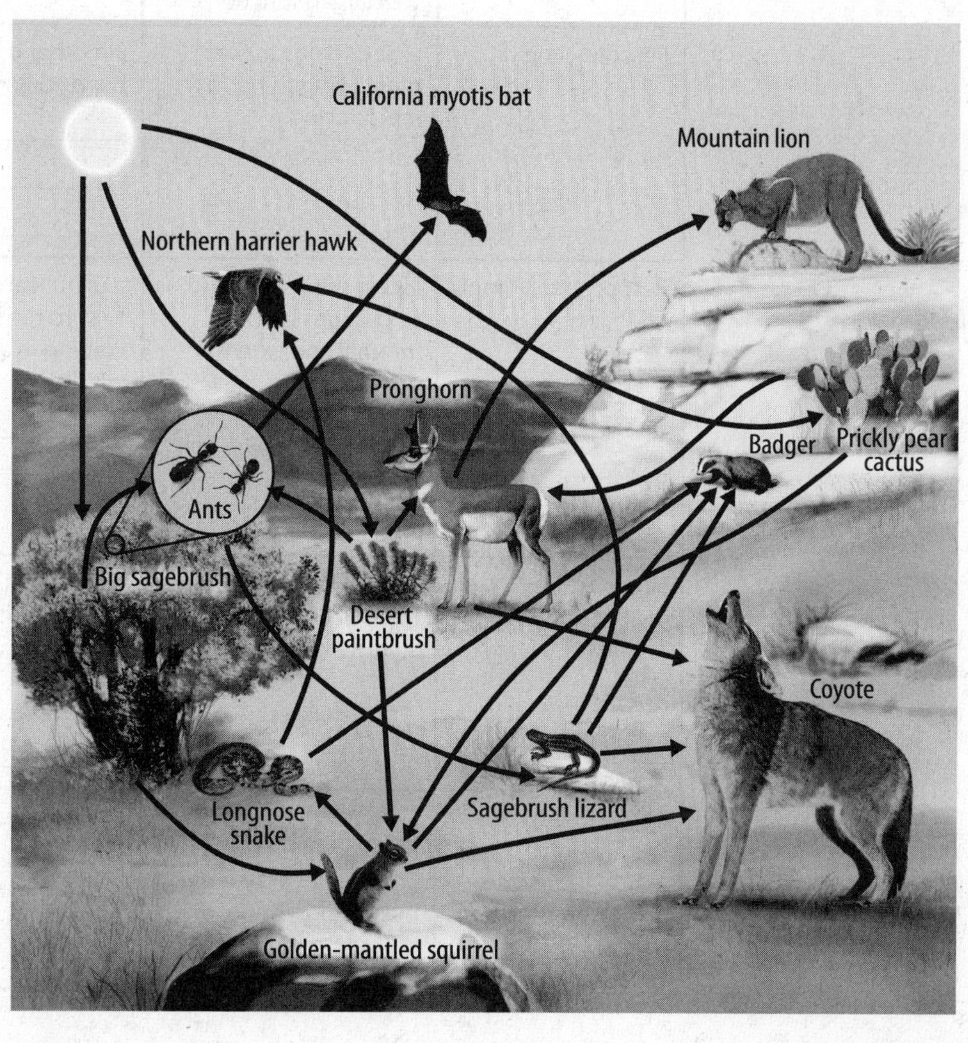

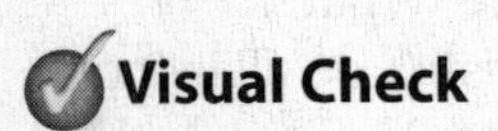

Visual Check

12. Apply In the chart, add another example for each characteristic of life.

Energy for Life There are six characteristics of life. From the chart below you will learn how each characteristic depends on energy from the Sun.

Characteristics of Life

Characteristic	Definition	Example
Organization	Living things have structures with their own functions or jobs. In living things with more than one cell, groups of cells work together. These living things have a higher level of organization than living things with only one cell do.	a leopard running ________ ________ ________
Growth and Development	Living things might grow by increasing cell size. They also might grow by making more cells. Living things develop when cells get specialized functions.	a tadpole changing into a frog ________ ________ ________
Reproduction	Living things make more living things by reproducing.	a mother duck and her ducklings ________ ________ ________
Response to Stimuli	Living things respond to changes in their internal and external environments.	a plant leaning toward the sunlight coming in a nearby window ________ ________ ________
Homeostasis	Living things keep their internal environment stable.	a girl drinking water after exercising ________ ________ ________
Use of Energy	Living things use energy for everything they do. They get their energy by making, eating, or absorbing food.	a squirrel eating a nut ________ ________ ________

Mini Glossary

cell: the smallest unit of life

homeostasis (hoh mee oh STAY sus): an organism's ability to maintain steady internal conditions when outside conditions change

multicellular: describes a living thing that is made of two or more cells

organism: a thing that has all the characteristics of life

unicellular: describes a living thing that is made of only one cell

1. Review the terms and their definitions in the Mini Glossary. Write one or two sentences to explain the difference between multicellular and unicellular organisms.

2. Fill in the table below to identify the six characteristics all living things share. Then add an example for each characteristic from your life or the life of someone you know.

Characteristic of Life	Personal Example

3. How did making flash cards help you learn the important terms in the lesson?

What do you think NOW?

Reread the statements at the beginning of the lesson. Fill in the After column with an A if you agree with the statement or a D if you disagree. Did you change your mind?

Log on to ConnectED.mcgraw-hill.com and access your textbook to find this lesson's resources.

Classifying and Exploring Life

Classifying Organisms

Key Concepts
- What methods are used to classify living things into groups?
- Why does every species have a scientific name?

Before You Read

What do you think? Read the two statements below and decide whether you agree or disagree with them. Place an A in the Before column if you agree with the statement or a D if you disagree. After you've read this lesson, reread the statements to see if you have changed your mind.

Before	Statement	After
	3. A dichotomous key can be used to identify an unknown organism.	
	4. Physical similarities are the only traits used to classify organisms.	

Mark the Text

Identify Main Ideas As you read this lesson, underline the main idea in each paragraph.

Read to Learn

Classifying Living Things

There have been many different ideas about how to organize, or classify, organisms. First you will learn about some early ideas for classifying organisms. Then you will learn about the system used today.

Greek philosopher Aristotle lived more than 2,000 years ago. He was one of the first people to classify organisms. He placed all organisms into two groups—plants and animals. Animals were classified based on whether or not the animal had "red blood," the animal's environment, and the shape and size of the animal. Plants were classified based on structure and size and on whether the plant was a tree, a shrub, or an herb.

Determining Kingdoms

In the 1700s, Carolus Linnaeus, a Swedish physician and botanist, classified organisms based on similar structures. Linneaus placed all organisms into two main groups, called kingdoms. For the next 200 years, people learned more about organisms and discovered new organisms. In 1969, Robert H. Whittaker, an American biologist, came up with a five-kingdom system for classifying organisms. Those kingdoms are Monera, Protista, Plantae, Fungi, and Animalia.

Science Use v. Common Use
kingdom
Science Use a classification category that ranks above phylum and below domain
Common Use a territory ruled by a king or queen

Determining Domains

The classification method used today is called systematics. It uses everything that is known about organisms to classify them. It looks at an organism's cell type, its habitat, the way it gets food and energy, the structure and function of its features, and the common ancestry of organisms. Systematics also uses molecular analysis—the study of molecules, such as DNA, within organisms.

Scientists using systematics found two distinct groups in Kingdom Monera. They added another classification level called domains. There are three domains—Bacteria, Archaea (ar KEE uh), and Eukarya (yew KER ee uh). They are shown below. All organisms are now classified into one of the three domains and then into one of the six kingdoms.

Key Concept Check
1. Summarize What evidence is used to classify living things into groups?

Domains and Kingdoms

Domain	Bacteria	Archaea	Eukarya			
Kingdom	**Bacteria**	**Archaea**	**Protista**	**Fungi**	**Plantae**	**Animalia**
Characteristics	Bacteria are simple, unicellular organisms.	Archaea are simple, unicellular organisms. They often live in very hot or salty environments.	Protists are unicellular and more complex than bacteria or archaea.	Fungi are unicellular or multicellular and absorb food.	Plants are multicellular and make their own food.	Animals are multicellular and take in their food.

Scientific Names

Suppose you did not have a name. What would people call you? All organisms, just like people, have names. We still use the naming system that Linnaeus created. It is called binomial nomenclature (bi NOH mee ul • NOH mun klay chur).

Visual Check
2. Name Why is a dog in Kingdom Animalia instead of Kingdom Fungi?

Binomial Nomenclature

Linneaus's naming system, **binomial nomenclature,** *gives each organism a two-word scientific name*. For example, the scientific name for the brown bear is *Ursus arctos*. This two-word scientific name is the name of an organism's species (SPEE sheez). *A* **species** *is a group of organisms that have similar traits and are able to produce fertile offspring*. In binomial nomenclature, the first word is the organism's genus (JEE nus) name, such as *Ursus*. *A* **genus** *is a group of similar species*. The second word might describe the way an organism looks or the way it acts.

How are organisms grouped? How do genus and species fit into kingdoms and domains? Similar species are grouped into one genus. (The term for more than one genus is *genera*.) Similar genera are grouped into families. Similar families are grouped into orders. Similar orders are grouped into classes. Similar classes are grouped into phyla. Similar phyla are grouped into kingdoms. And similar kingdoms are grouped into domains. The binomial nomenclature for the brown bear is shown below.

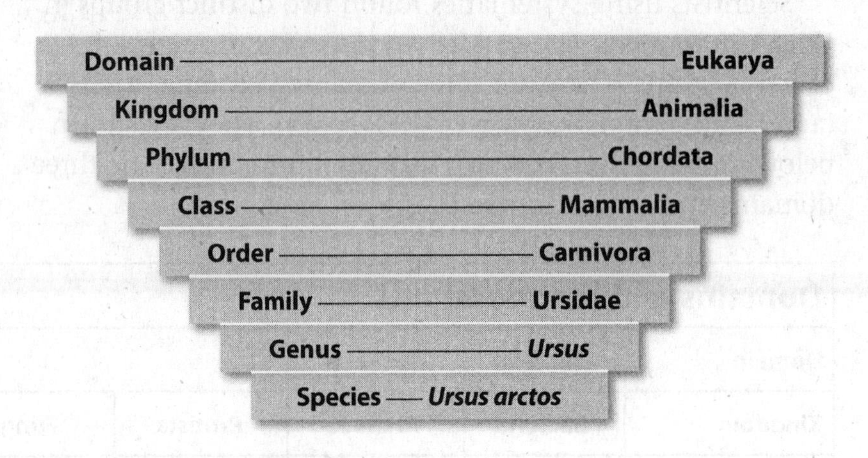

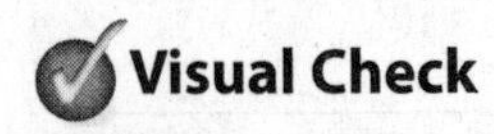

Visual Check

3. Name What domain does the brown bear belong to?

____

____

Uses of Scientific Names

Some people would call a large brown bear a brown bear. Others would call it a grizzly bear. But it has only one scientific name: *Ursus arctos*.

A common name might also refer to more than one type of organism. Imagine two different evergreen trees. Even though they are two different species, they have the same common name—pine trees. Scientific names are important for many reasons. Each species has its own scientific name. Scientific names are the same worldwide. This makes communication about organisms easier because everyone uses the same name for the same species.

Key Concept Check

4. Explain Why does every species have a scientific name?

____

____

____

Classification Tools

Imagine that you are fishing. You catch a fish that you have never seen before. How can you find out what type of fish you have caught? You can use several tools to identify organisms.

FOLDABLES®

Make a two-tab book to compare two of the tools scientists use to identify organisms—dichotomous keys and cladograms.

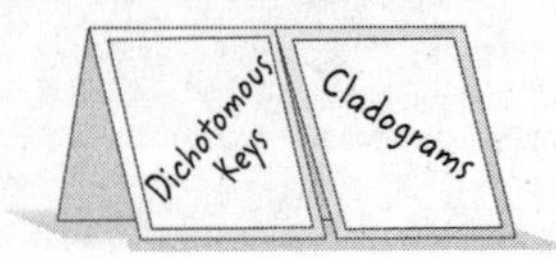

Dichotomous Keys

A **dichotomous key** *is a series of descriptions arranged in pairs that leads the user to the identification of an unknown organism.* Each chosen description leads to another description. You keep making choices until you reach the name of the organism. The dichotomous key below identifies some species of mice.

Key to Some Mice of North America	
1. Tail hair	**a.** no hair on tail; scales show plainly; house mouse, *Mus musculus* **b.** hair on tail, go to 2
2. Ear size	**a.** ears small and nearly hidden in fur, go to 3 **b.** ears large and not hidden in fur, go to 4
3. Tail length	**a.** less than 25 mm; woodland vole, *Microtus pinetorum* **b.** more than 25 mm; prairie vole, *Microtus ochrogaster*
4. Tail coloration	**a.** sharply bicolor, white beneath and dark above; deer mouse, *Peromyscus maniculatus* **b.** darker above than below but not sharply bicolor; white-footed mouse, *Peromyscus leucopus*

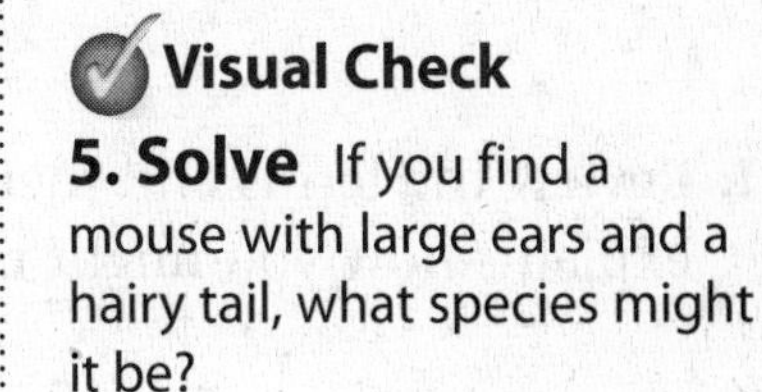

Visual Check

5. Solve If you find a mouse with large ears and a hairy tail, what species might it be?

Cladograms

Have any of your relatives made a family tree? Family trees are branching charts that show how family members are related. Biologists use a similar diagram to show how species are related. It is called a cladogram. *A* **cladogram** *is a branched diagram that shows the relationships among organisms, including common ancestors.*

The cladogram below has a series of branches. Each branch follows a new characteristic. Each characteristic can be seen in the species to its right. See what this cladogram tells you about the relationships among the living things that are shown. The salamander, lizard, hamster, and chimpanzee have lungs. The salmon does not have lungs. Therefore, the other animals are more closely related to each other than they are to the salmon.

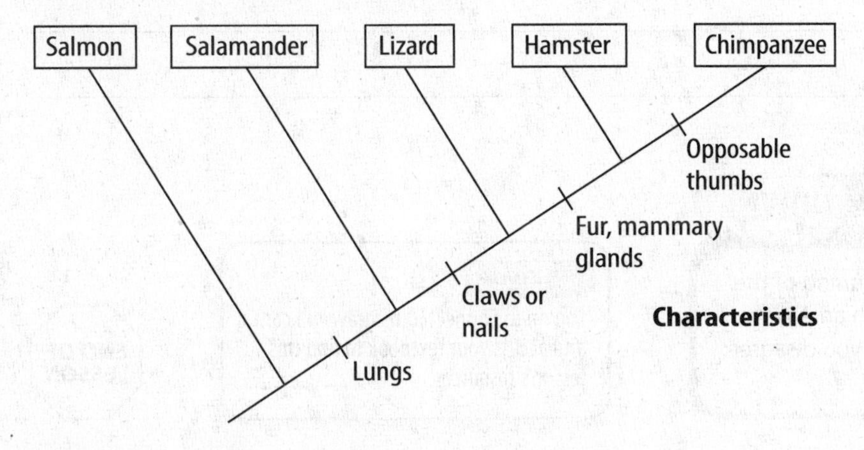

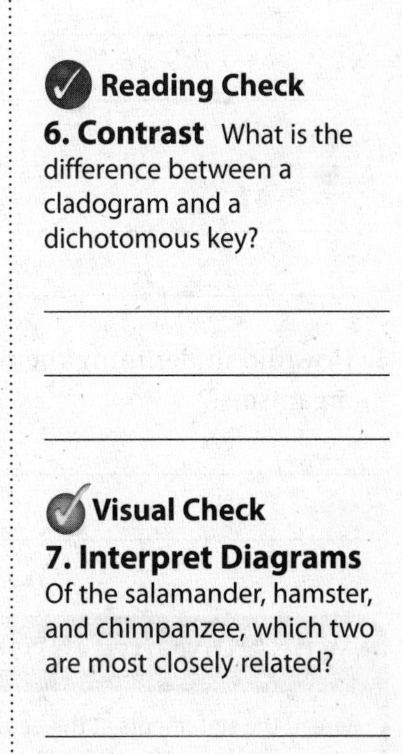

Reading Check

6. Contrast What is the difference between a cladogram and a dichotomous key?

Visual Check

7. Interpret Diagrams Of the salamander, hamster, and chimpanzee, which two are most closely related?

After You Read

Mini Glossary

binomial nomenclature (bi NOH mee ul • NOH mun klay chur): a naming system that gives each organism a two-word scientific name

cladogram: a branched diagram that shows the relationships among organisms, including common ancestors

dichotomous key: a series of descriptions arranged in pairs that lead the user to the identification of an unknown organism

genus (JEE nus): a group of similar species

species (SPEE sheez): a group of organisms that have similar traits and are able to produce fertile offspring

1. Review the terms and their definitions in the Mini Glossary. Write a sentence that explains how you might use a dichotomous key while birdwatching in the woods.

2. Fill in the upside-down pyramid below to show how living things are classified.

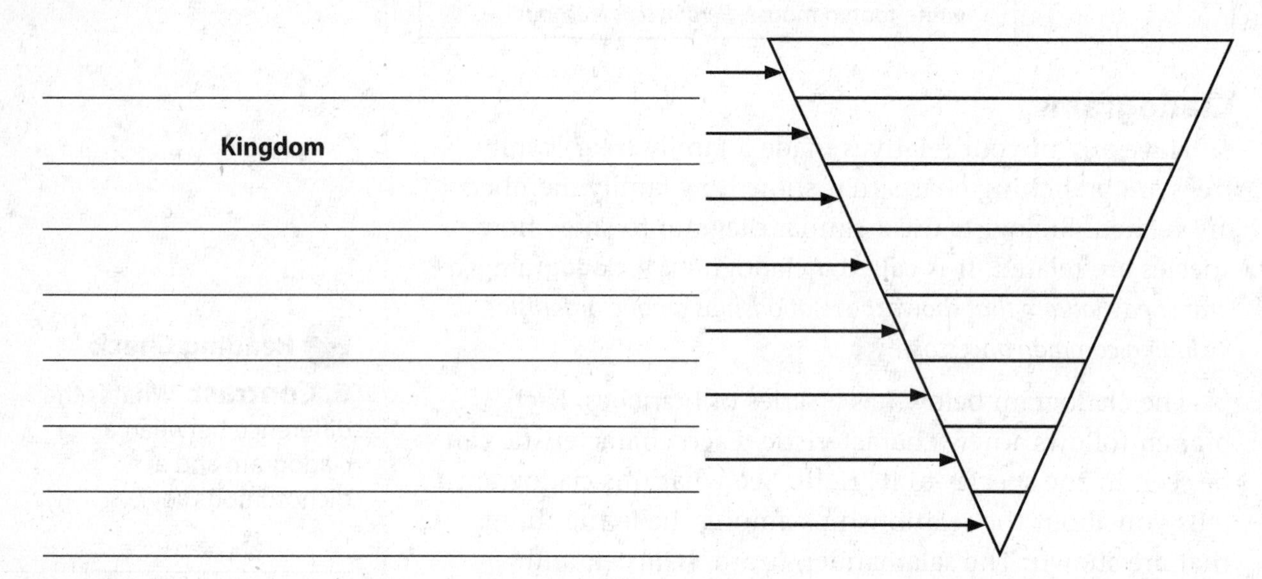

3. How did underlining the main idea in each paragraph help you learn about classifying organisms?

What do you think NOW?

Reread the statements at the beginning of the lesson. Fill in the After column with an A if you agree with the statement or a D if you disagree. Did you change your mind?

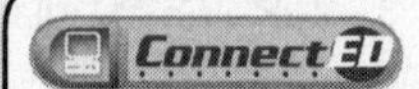

Log on to ConnectED.mcgraw-hill.com and access your textbook to find this lesson's resources.

Classifying and Exploring Life

Exploring Life

Before You Read

What do you think? Read the two statements below and decide whether you agree or disagree with them. Place an A in the Before column if you agree with the statement or a D if you disagree. After you've read this lesson, reread the statements to see if you have changed your mind.

Before	Statement	After
	5. Most cells are too small to be seen with the unaided eye.	
	6. Only scientists use microscopes.	

Key Concepts 

- How did microscopes change our ideas about living things?
- What are the types of microscopes, and how do they compare?

Read to Learn

The Development of Microscopes

Have you ever used a magnifying glass, or lens, to see an object in more detail? If so, you have used a tool similar to the first microscope. Using microscopes, people can see details that cannot be seen with the unaided eye. People have used microscopes to discover many things about organisms.

In the late 1600s, Anton van Leeuwenhoek (LAY vun hook) made one of the first microscopes. His microscope had one lens. It made things look about 270 times their original size. In the early 1700s, Robert Hooke observed and named cells while using a microscope. Before microscopes were invented, people did not know that living things are made of cells.

Study Coach

Record Questions As you read this lesson, write down any questions you have about microscopes and how they are used. Discuss these questions and their answers with another student or your teacher.

Types of Microscopes

One characteristic of all microscopes is that they magnify objects. Magnification makes an object look larger than it really is. Another characteristic of microscopes is resolution. Resolution is how clearly a magnified object can be seen.

The two main types of microscopes are light microscopes and electron microscopes. They are different in their magnification and resolution.

Key Concept Check

1. Describe How did microscopes change our ideas about living things?

____

____

____

Math Skills

The magnification of the ocular lens of a compound microscope is 10×. The magnification of the objective lens is also 10×. To determine a microscope's magnification, multiply the power of the ocular lens by the power of the objective lens. A microscope with a 10× ocular lens and a 10× objective lens magnifies an object 10 × 10, or 100 times.

2. Use Multiplication
What is the total magnification of a compound microscope with a 10× ocular lens and a 4× objective lens?

Reading Check

3. Name some ways an object can be examined under a light microscope.

FOLDABLES

Use a two-column folded chart to organize your notes about how different types of microscopes magnify images.

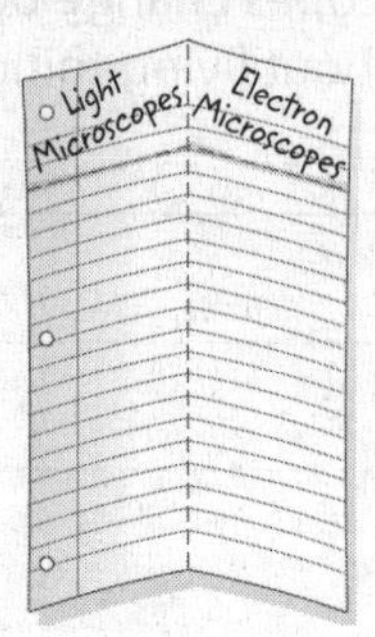

Light Microscopes

Your school might have a light microscope. **Light microscopes** *use light and lenses to enlarge an image of an object.* A simple light microscope has only one lens. *A light microscope that uses more than one lens to magnify an object is called a* **compound microscope.**

In a compound microscope, the objective lens magnifies the image. That image is further magnified by the ocular lens. The total magnification of an object is equal to the magnification of the ocular lens times the magnification of the objective lens.

Some light microscopes can enlarge images as much as 1,500 times their original size. A light microscope has a resolution of about 0.2 micrometers (µm), or two-millionths of a meter. A resolution of 2 µm means that you can clearly see points on an object that are at least 0.2 µm apart.

You can use light microscopes to view living or nonliving things. Sometimes the object is placed under the microscope. For other light microscopes, the object must be mounted on a slide. Some objects must be stained with a dye to see their details.

Electron Microscopes

You might recall that electrons are tiny particles inside atoms. *An* **electron microscope** *uses a magnetic field to focus a beam of electrons through an object or onto an object's surface.* An electron microscope can magnify an image 100,000 times or more. The resolution of an electron microscope can be as small as 0.2 nanometers (nm), or two-billionths of a meter. This resolution is up to 1,000 times greater than a light microscope's.

The two main types of electron microscopes are transmission electron microscopes (TEMs) and scanning electron microscopes (SEMs). Because objects must be mounted in plastic and then sliced very thinly, only dead organisms can be viewed with a TEM.

Transmission Electron Microscopes People use TEMs to study very small things, such as structures inside a cell. The electrons that pass through the object produce an image that can be viewed on a computer.

Scanning Electron Microscopes People use SEMs to study the surface of an object. Electrons bounce off the object, and a computer produces an image of the object in three dimensions.

Using Microscopes

Today's microscopes are useful tools in many fields. They are used in health care, police work, science research, and industry.

Health Care

People who work in health-care fields often use microscopes. Doctors and laboratory technicians find them useful. Surgeons use microscopes in cataract surgery and brain surgery. Microscopes enable doctors to view the surgical area in detail. The area can be shown on a TV screen so other people can watch and learn. Laboratory technicians use microscopes to study body fluids, such as blood and urine.

Other Uses

Health care is not the only field that uses microscopes. These tools are also useful in police work. Have you ever wondered how police figure out how and where a crime happened? Scientists use microscopes to study evidence from crime scenes. For example, a corpse might contain insects that could help identify when and where a crime happened. A microscope can help identify the type and age of the insects.

People who study fossils also use microscopes. They might examine a fossil and other materials from where the fossil was found.

There are many other ways that microscopes are used. People in the steel industry use microscopes to look for impurities in steel. Jewelers use microscopes to identify stones. They can also see markings and impurities in stones that they could not see with their unaided eyes.

Key Concept Check

4. Compare In what ways are TEMs different from SEMs?

Reading Check

5. Recall How might a health-care professional use a microscope?

ACADEMIC VOCABULARY

identify
(verb) to determine the characteristics of a person or thing

Reading Check

6. Summarize List some uses of microscopes.

After You Read

Mini Glossary

compound microscope: a light microscope that uses more than one lens to magnify an object

electron microscope: a microscope that uses a magnetic field to focus a beam of electrons through an object or onto an object's surface

light microscope: a tool that uses light and lenses to enlarge an image of an object

1. Review the terms and their definitions in the Mini Glossary. Write a sentence that describes how a compound microscope and a light microscope are related.

2. Complete the graphic organizer below by explaining how each kind of microscope differs from the previous one.

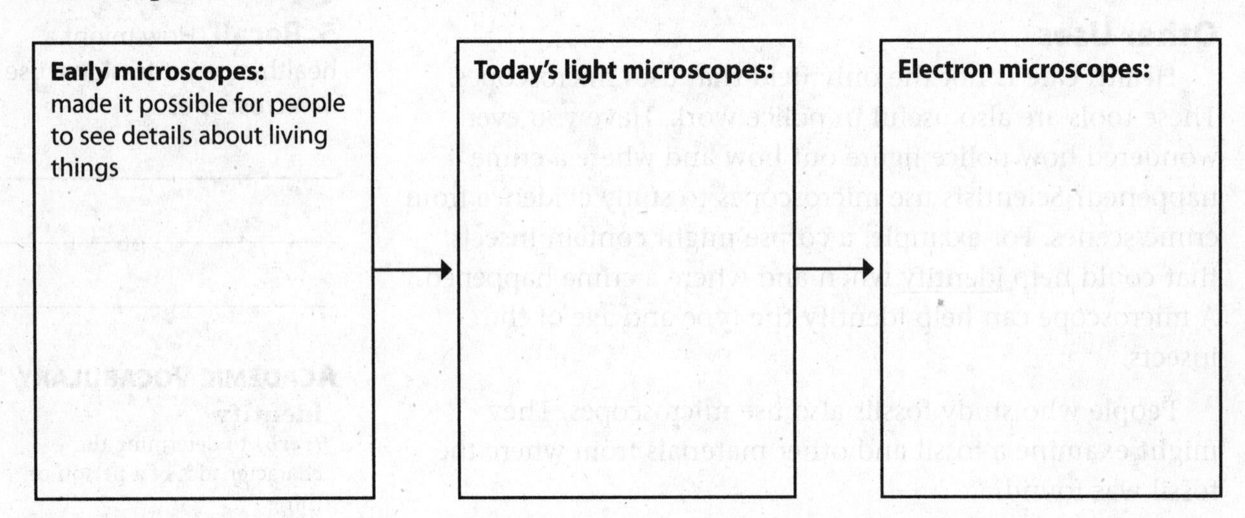

3. How did writing down questions about microscopes and then discussing their answers help you understand what you read?

What do you think NOW?

Reread the statements at the beginning of the lesson. Fill in the After column with an A if you agree with the statement or a D if you disagree. Did you change your mind?

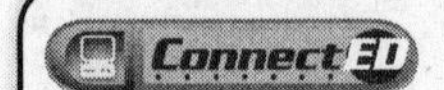

Log on to ConnectED.mcgraw-hill.com and access your textbook to find this lesson's resources.

Cell Structure and Function

Cells and Life

Before You Read

What do you think? Read the two statements below and decide whether you agree or disagree with them. Place an A in the Before column if you agree with the statement or a D if you disagree. After you've read this lesson, reread the statements to see if you have changed your mind.

Before	Statement	After
	1. Nonliving things have cells.	
	2. Cells are made mostly of water.	

Key Concepts 

- How did scientists' understanding of cells develop?
- What basic substances make up a cell?

Read to Learn

Understanding Cells

The cells that make up all living things are very small. Early scientists did not have the tools to see cells until the invention of the microscope. More than 300 years ago, Robert Hooke built a microscope. He used it to look at cork. He saw small openings in the cork similar to the honeycomb shown in the figure below. The openings reminded him of the small rooms, called cells, where monks lived. Hooke named these small structures cells.

Mark the Text

Identify Main Ideas As you read, highlight the main ideas under each heading. After you finish reading, review the main ideas of the lesson.

Visual Check

1. Identify The small openings of the honeycomb look most like which of the following? (Circle the correct answer.)

a. cells

b. plants

c. tiny animals

Think it Over

2. Define What are the three principles of the cell theory?

____

____

____

____

____

____

Key Concept Check

3. Explain how scientists' understanding of cells developed.

____

____

____

4. Identify Match the scientist with his part of the cell theory. Draw a line from the scientist to his observation.

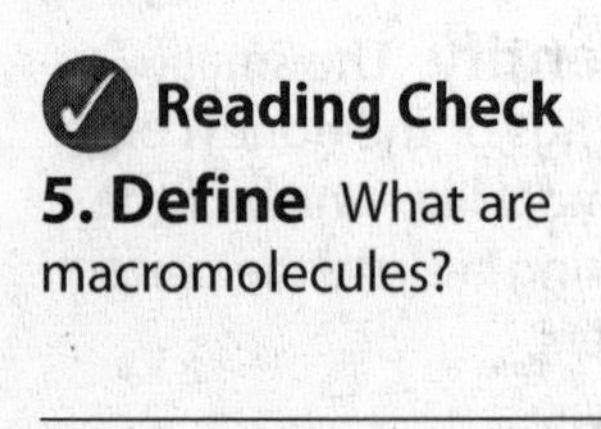

5. Define What are macromolecules?

____

____

____

The Cell Theory

Scientists made better microscopes. They looked for cells in places such as pond water and blood. The newer microscopes made it possible for scientists to see different structures inside cells. A scientist named Matthias Schleiden (SHLI dun) looked at plant cells. Another scientist, Theodore Schwann, studied animal cells. Later, Rudolf Virchow (VUR koh) said all cells come from cells that already exist. The observations made by these scientists, shown in the table below, became known as the cell theory. *The* **cell theory** *states that all living things are made of one or more cells, cells are the smallest unit of life, and all new cells come from cells that already exist.*

Cell Theory Matchup

Scientist	Observation
1. Theodore Schwann	**a.** By studying plants, he determined that all living things are made of one or more cells.
2. Rudolf Virchow	**b.** By studying animals, he determined that all living things are made of one or more cells.
3. Matthias Schleiden	**c.** All new cells come from cells that already exist.

Basic Cell Substances

The cell theory raised more questions for scientists. Scientists began to look into what cells are made of. *Cells are made of smaller parts called* **macromolecules** *that form when many small molecules join together.*

The Main Ingredient—Water

The main ingredient in every cell is water. Water makes up more than 75 percent of a cell. It is necessary for life. Water also surrounds cells. The water surrounding your cells helps to insulate your body. This helps your body maintain a stable internal environment, or homeostasis.

Water also is useful because it can dissolve other substances, such as salt (sodium chloride). For substances to move into and out of a cell, they must be dissolved in a liquid. In the figure below, the water molecules have a positive end and a negative end.

- The more negative end of a water molecule (–) can attract the positive part of another substance.
- The more positive end of a water molecule (+) can attract the negative part of another substance. With sodium chloride, the sodium (Na) ions and chloride (Cl) ions are more attracted by the water molecules. This attraction is similar to the attraction of magnets.

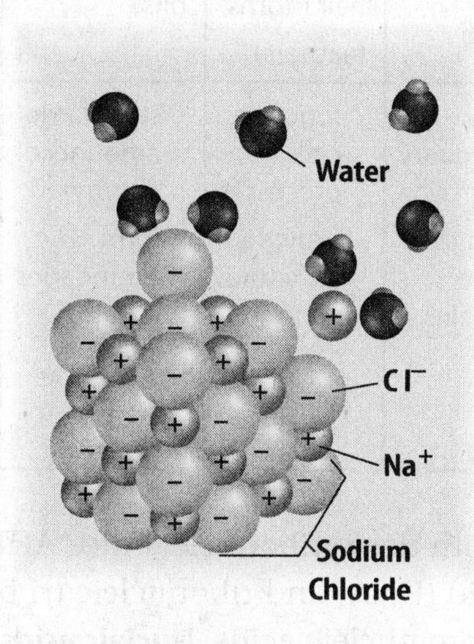

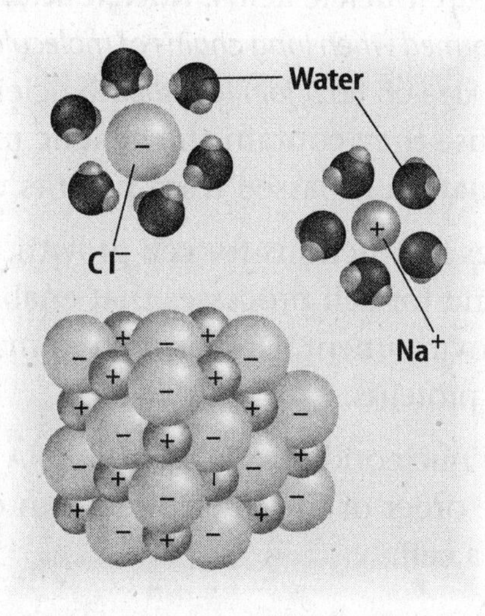

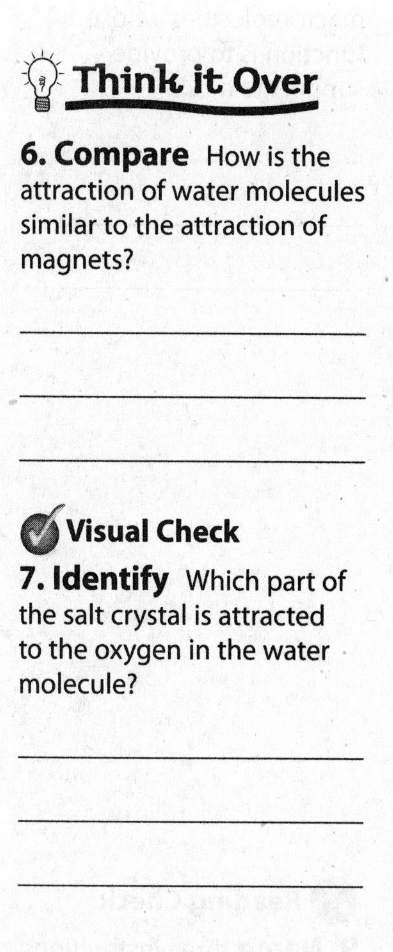

Think it Over

6. Compare How is the attraction of water molecules similar to the attraction of magnets?

____

____

____

Visual Check

7. Identify Which part of the salt crystal is attracted to the oxygen in the water molecule?

____

____

____

FOLDABLES®

Make a four-page book to organize information about macromolecules in a cell.

Nucleic acids
Proteins
Lipids
Carbohydrates

Macromolecules

All cells contain other substances besides water that help cells do what they do. There are four types of macromolecules in cells. They are nucleic acids, proteins, lipids, and carbohydrates. Each type of macromolecule has its own job, or function, in a cell. These functions range from growth and communication to movement and storage. The table below describes each macromolecule's function.

Macromolecules in Cells

	Nucleic Acids	Proteins	Lipids	Carbohydrates
Elements	carbon oxygen hydrogen nitrogen phosphorus	carbon oxygen hydrogen nitrogen sulfur	carbon oxygen hydrogen phosphorus	carbon hydrogen oxygen
Examples	DNA RNA	enzymes hair (horns, feathers)	fats oils	sugars starch cellulose
Function	• carry hereditary information • used to make proteins	• regulate cell processes • provide structural support	• store large amounts of energy • form boundaries around cells	• supply energy for cell processes • short-term energy storage • provide structural support

Visual Check

8. Identify two macromolecules whose function is to provide support to a cell.

Nucleic Acids Both deoxyribonucleic (dee AHK sih ri boh noo klee ihk) acid (DNA) and ribonucleic (ri boh noo KLEE ihk) acid (RNA) are nucleic acids. **Nucleic acids** *are macromolecules formed when long chains of molecules called nucleotides* (NEW klee uh tidz) *join together*. Nucleic acids are important because they contain the genetic material of a cell. This information is passed from parents to offspring.

DNA includes instructions for cell growth, for cell reproduction, and for cell processes that enable a cell to respond to its environment. DNA is used to make RNA. RNA is used to make proteins.

The order of nucleotides in DNA and RNA is important. A change in the order of the nucleotides can change the information in a cell.

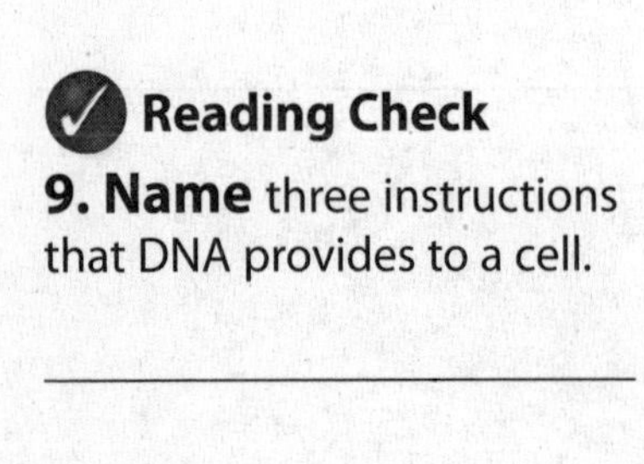

Reading Check

9. Name three instructions that DNA provides to a cell.

Proteins The macromolecules necessary for almost everything cells do are proteins. *A* **protein** *is a macromolecule made of long chains of amino acid molecules.* RNA contains instructions for joining amino acids together.

Cells have hundreds of proteins. Each protein has its own function. Some proteins help cells communicate with other cells. Other proteins move substances around inside cells. Some proteins help to break down nutrients in food. Other proteins, such as keratin (KER uh tun), which is found in hair, horns, and feathers, make up supporting structures. ✓

Lipids *A* **lipid** *is a large macromolecule that does not dissolve in water.* Because lipids do not dissolve in water, they protect cells. Lipids also are a large part of the cell membrane. Lipids store energy for cells and help with cell communication. Cholesterol (kuh LES tuh rawl), phospholipids (fahs foh LIH pids), and vitamin A are lipids. ✓

Carbohydrates *One sugar molecule, two sugar molecules, or a long chain of sugar molecules make up* **carbohydrates** (kar boh HI drayts). Carbohydrates store energy, provide structural support for cells, and help cells communicate. Sugars and starches are carbohydrates that store energy. Fruits contain sugars. Bread and pasta are mostly starch. The energy stored in sugars and starches can be released quickly through chemical reactions in cells. Cellulose is a carbohydrate in the cell walls of plants that provides support.

Reading Check

10. Identify three functions of proteins in cells.

Reading Check

11. Explain Why are lipids important to cells?

Key Concept Check

12. Name the basic substances that make up a cell.

After You Read

Mini Glossary

carbohydrate: (kar boh HI drayt): one sugar molecule, two sugar molecules, or a long chain of sugar molecules

cell theory: states that all living things are made of one or more cells, the cell is the smallest unit of life, and all new cells come from preexisting cells

lipid: a large macromolecule that does not dissolve in water

macromolecule: a substance that forms by joining many small molecules together

nucleic acid: a macromolecule that forms when long chains of molecules called nucleotides join together

protein: a long chain of amino acid molecules

1. Review the terms and their definitions in the Mini Glossary. Write a sentence that describes a lipid.

____

____

2. Fill in the chart below by identifying the different types of macromolecules and giving examples of each.

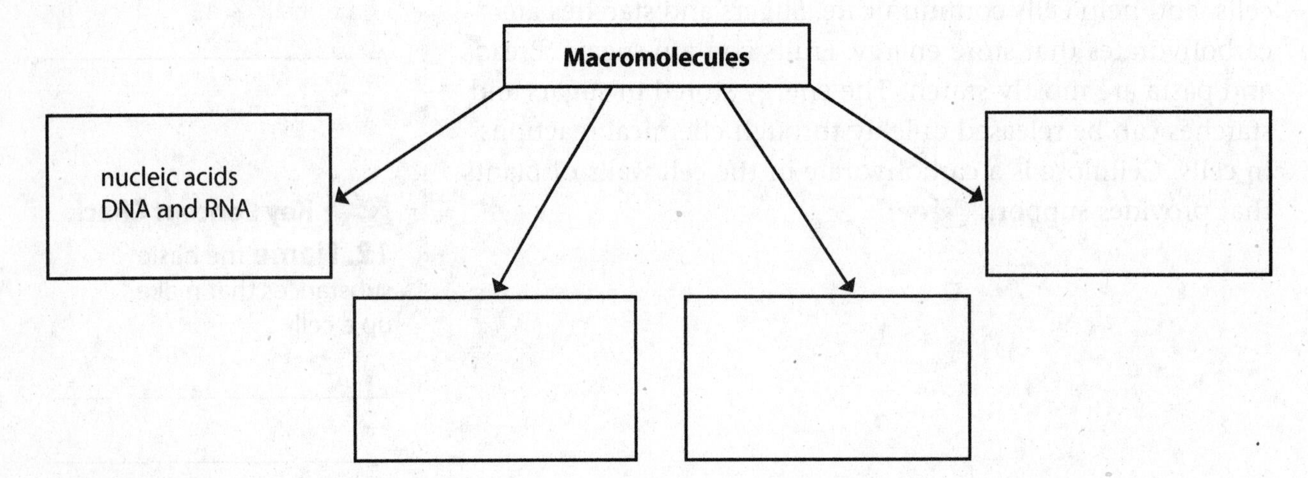

3. How did highlighting the main ideas in each section of this lesson improve your understanding of the cell?

____

____

What do you think NOW?

Reread the statements at the beginning of the lesson. Fill in the After column with an A if you agree with the statement or a D if you disagree. Did you change your mind?

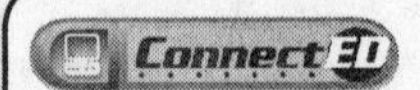

Log on to ConnectED.mcgraw-hill.com and access your textbook to find this lesson's resources.

Cell Structure and Function

The Cell

Before You Read

What do you think? Read the two statements below and decide whether you agree or disagree with them. Place an A in the Before column if you agree with the statement or a D if you disagree. After you've read this lesson, reread the statements to see if you have changed your mind.

Before	Statement	After
	3. Different organisms have cells with different structures.	
	4. All cells store genetic information in their nuclei.	

Key Concepts

- How are prokaryotic cells and eukaryotic cells similar, and how are they different?
- What do the structures in a cell do?

Read to Learn

Cell Shape and Movement

Cells come in many shapes and sizes. The size and shape of a cell is part of the function of the cell. Some cells, such as human red-blood cells, can be seen only by using a microscope. The cells can pass easily through small blood vessels because of their small size. Their disk shapes are important for carrying oxygen. Nerve cells have parts that jut out. These projections on nerve cells can send signals over long distances. Some plant cells are hollow. These hollow cells make up tubelike structures that can carry water and dissolved substances to parts of the plant.

The size and shape of a cell make it possible for the cell to carry out its functions. The parts that make up a cell have their own functions as well. A cell's parts are like the players on a football team who perform different tasks on the playing field. A cell is made up of different parts that perform different functions to keep the cell alive.

Study Coach

Use Prior Knowledge Before you read this lesson, look at the figures and headings to learn what the lesson is about. Write what you know about the cell on a piece of paper. As you read the lesson, fill in what you learned about the cell.

ACADEMIC VOCABULARY
function
(noun) the purpose for which something is used

Visual Check
1. Describe the location of the cell wall.

____

____

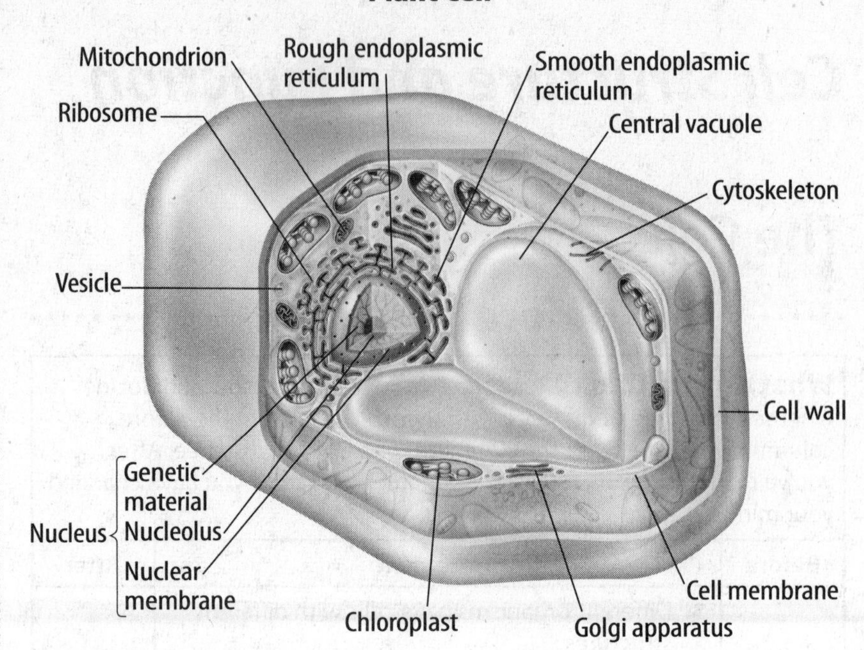

Cell Membrane

All cells have some parts, or structures, in common. One of these structures is a cell membrane. *A* **cell membrane** *is a flexible covering that protects the inside of a cell from the environment outside the cell.* You can see the cell membrane in both drawings on this page. Cell membranes are made of proteins and phospholipids.

Reading Check
2. Describe What are cell membranes made of?

____

____

Visual Check
3. Identify Circle the names of two parts in the animal cell that are also found in the plant cell.

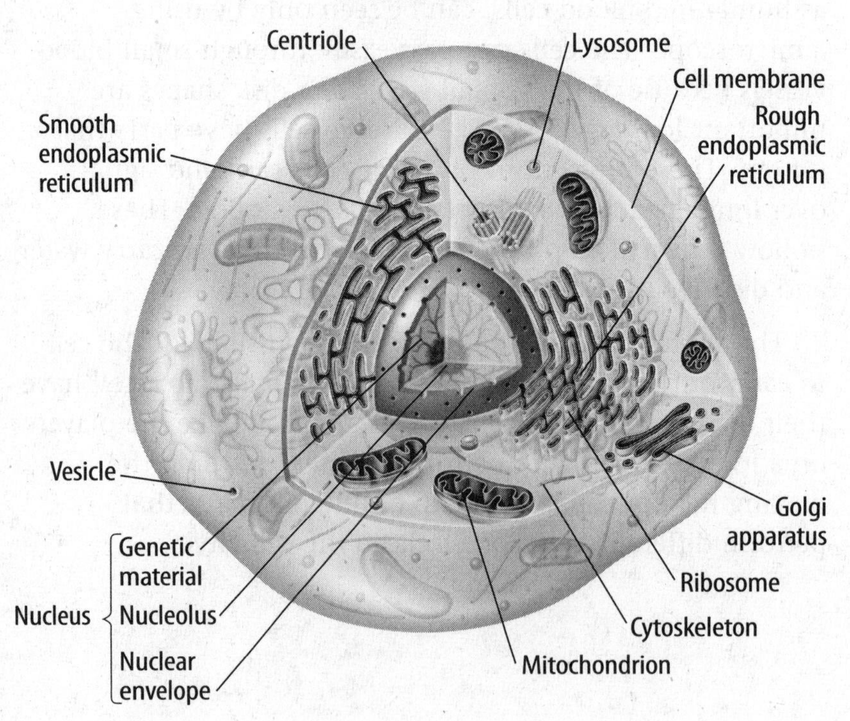

Cell Wall

Every cell has a cell membrane. But some cells also have a cell wall. Plant cells, fungal cells, bacterial cells, and some protists have cell walls. *A* **cell wall** *is a stiff structure outside the cell membrane*. A cell wall protects a cell from viruses and other harmful organisms. In some plant and fungal cells, the cell wall helps the cell keep its shape and gives it support.

Cell Appendages

If you look at a cell using a microscope, you might see structures on the outside of the cell. These appendages might look like hairs or long tails. They often help a cell move. Flagella (fluh JEH luh) (singular, flagellum) are long and tail-like. They whip back and forth to move the cell. Cilia (SIH lee uh) (singular, cilium) are short, hairlike structures. They can move a cell or move molecules away from a cell. The cilia in your windpipe move harmful particles away from your lungs.

Cytoplasm and the Cytoskeleton

The fluid inside a cell is made of water, salts, and other molecules and is called the **cytoplasm.** The cytoplasm contains a cell's cytoskeleton. *The* **cytoskeleton** *is made of threadlike proteins that are joined together*. The cytoskeleton is a framework that gives a cell its shape and helps it move. ✓

Reading Check

4. Describe the structure of the cytoskeleton.

Cell Types

Microscopes helped scientists discover that cells can be grouped into two types. There are prokaryotic (proh ka ree AH tihk) cells and eukaryotic (yew ker ee AH tihk) cells.

Prokaryotic Cells

The genetic material in a prokaryotic cell is not surrounded by a membrane. Look at the drawing below. Prokaryotic cells also do not have many of the cell parts other cells have. Most prokaryotic cells are unicellular organisms and are called prokaryotes.

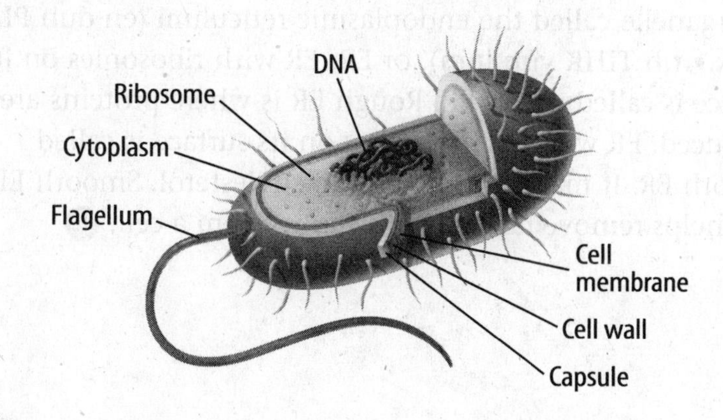

Visual Check

5. Name three parts of a prokaryotic cell.

Eukaryotic Cells

The cells of plants, animals, fungi, and protists are eukaryotic cells. The genetic material of eukaryotic cells is surrounded by a membrane. Every eukaryotic cell also has **organelles**—*other parts that are surrounded by a membrane and have specialized functions*. Eukaryotic cells are usually larger than prokaryotic cells.

Key Concept Check
6. Compare and Contrast How are prokaryotic cells and eukaryotic cells the same? How are they different?

Cell Organelles

The organelles of eukaryotic cells have different functions in the cell. Organelles help a cell carry out different functions at the same time. These functions include getting energy from food, storing information, and getting rid of waste material.

The Nucleus

The largest organelle inside most eukaryotic cells is the nucleus. *The **nucleus** is the part of a eukaryotic cell that directs cell activities and contains genetic information stored in DNA.*

DNA is in structures called chromosomes. The number of chromosomes in a nucleus is different for different species of organisms.

The nucleus also contains proteins and an organelle called the nucleolus (new KLEE uh lus). The nucleolus makes ribosomes, organelles that help produce proteins. Two membranes form the nuclear envelope that surrounds the nucleus. The nuclear envelope has many pores. Certain molecules, such as ribosomes and RNA, move into and out of the nucleus through these pores.

SCIENCE USE V. COMMON USE
envelope
Science Use an outer covering

Common Use a flat paper container for a letter

Manufacturing Molecules

You learned that proteins are important molecules in cells. Proteins are made of small organelles called ribosomes. A ribosome is not surrounded by a membrane. Ribosomes are in the cytoplasm of a cell. Ribosomes can be attached to an organelle called the endoplasmic reticulum (en duh PLAZ mihk • rih TIHK yuh lum), or ER. ER with ribosomes on its surface is called rough ER. Rough ER is where proteins are produced. ER without ribosomes on its surface is called smooth ER. It makes lipids such as cholesterol. Smooth ER also helps remove harmful substances from a cell.

Reading Check
7. Contrast smooth ER and rough ER.

Processing Energy

All living things must have energy to survive. Cells process some energy in specialized organelles called mitochondria (mi tuh KAHN dree uh) (singular, mitochondrion). Most eukaryotic cells contain hundreds of mitochondria. Some cells in a human heart can contain 1,000 mitochondria.

ATP A mitochondrion is surrounded by two membranes. Chemical reactions within mitochondria release energy. This energy is stored in high-energy molecules called ATP—adenosine triphosphate (uh DEH nuh seen • tri FAHS fayt). The energy in ATP molecules is used by the cell for growth, cell division, and transporting materials.

Chloroplasts The cells of some organisms, such as plants and algae, contain organelles called chloroplasts (KLOR uh plasts). **Chloroplasts** *are membrane-bound organelles that use light energy and make food, a sugar called glucose, from water and carbon dioxide in a process called photosynthesis* (foh toh SIHN thuh sus). The sugar has stored energy that can be used when the cells need it.

Processing, Transporting, and Storing Molecules

The Golgi (GAWL jee) apparatus is an organelle that looks like a stack of pancakes. It gets proteins ready for their specific jobs. It then packages the proteins into tiny membrane-bound, ball-like structures called vesicles. Vesicles are organelles that transport substances to other parts of the cell. Some vesicles in an animal cell are called lysosomes. Lysosomes help break down and recycle different parts of the cell.

Some cells also have structures called vacuoles (VA kyuh wohlz). Vacuoles are organelles that store food, water, and waste materials for a cell. A plant cell usually has one large vacuole. Some animal cells have many small vacuoles.

FOLDABLES®

Make a half-book to record information about cell organelles and their functions.

Reading Check

8. Identify the types of cells that contain chloroplasts.

Key Concept Check

9. Explain the function of the Golgi apparatus.

After You Read

Mini Glossary

cell membrane: a flexible covering that protects the inside of a cell from the environment outside a cell

cell wall: a stiff structure outside the cell membrane

chloroplast (KLOR uh plast): a membrane-bound organelle that uses light energy and makes food—a sugar called glucose—from water and carbon dioxide in a process known as photosynthesis

cytoplasm: a fluid inside a cell that contains salts and other molecules

cytoskeleton: a network of threadlike proteins that are joined together

nucleus: the part of a eukaryotic cell that directs cell activities and contains genetic information stored in DNA

organelle: a membrane-surrounded component of a cell that has specialized functions

1. Review the terms and their definitions in the Mini Glossary. Write a sentence that lists two functions of the nucleus of a eukaryotic cell.

____

____

2. Fill in the table below to identify the functions of each organelle.

Organelle	Function
Chloroplast	
Golgi apparatus	
Smooth ER	
Nucleus	

3. Name three tasks carried out by the organelles of eukaryotic cells.

____

____

What do you think NOW?

Reread the statements at the beginning of the lesson. Fill in the After column with an A if you agree with the statement or a D if you disagree. Did you change your mind?

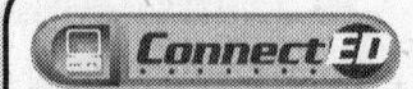

Log on to ConnectED.mcgraw-hill.com and access your textbook to find this lesson's resources.

Cell Structure and Function

Moving Cellular Material

Before You Read

What do you think? Read the two statements below and decide whether you agree or disagree with them. Place an A in the Before column if you agree with the statement or a D if you disagree. After you've read this lesson, reread the statements to see if you have changed your mind.

Before	Statement	After
	5. Diffusion and osmosis are the same process.	
	6. Cells with large surface areas can transport more than cells with smaller surface areas.	

Key Concepts 
- How do materials enter and leave cells?
- How does cell size affect the transport of materials?

Read to Learn

Passive Transport

The membranes of cells and organelles perform different functions. They form boundaries between cells. They also control the movement of substances into and out of cells.

Cell membranes are semipermeable. This means that only certain materials can enter or leave a cell. Substances can pass through a cell membrane by one of several different processes. The type of process depends on the physical and chemical properties of the substance that is passing through the membrane.

Small molecules, such as oxygen and carbon dioxide, pass through a cell's membrane by a process called passive transport. **Passive transport** *is the movement of substances through a cell membrane without using the cell's energy.* Passive transport depends on the amount of a substance on each side of the membrane. If there are more oxygen molecules outside a cell than there are inside a cell, oxygen molecules will move into the cell by passive transport. Oxygen molecules will move into a cell until the amount of oxygen outside the cell equals the amount of oxygen inside the cell. There are different types of passive transport.

Study Coach

Asking Questions Before you read the lesson, preview all the headings. Make a chart and write a *What* or *How* question for each heading. As you read, write the answers to your questions.

FOLDABLES®

Make a two-tab book to organize information about the different types of passive and active transport.

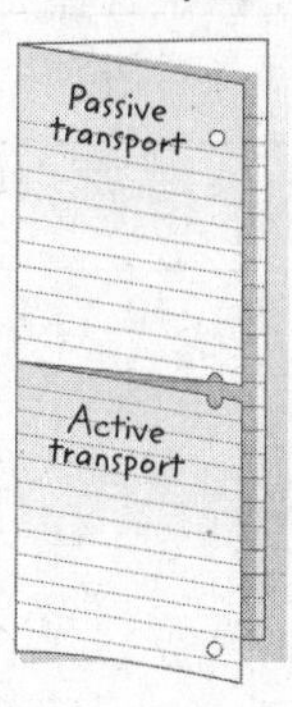

ACADEMIC VOCABULARY
concentration
(noun) the amount of a given substance in a certain area.

Diffusion

When the concentration, or amount per volume, of a substance is unequal on each side of a membrane, molecules will move from the side with a higher concentration of the substance to the side with the lower concentration. **Diffusion** *is the movement of substances from an area of higher concentration to an area of lower concentration.*

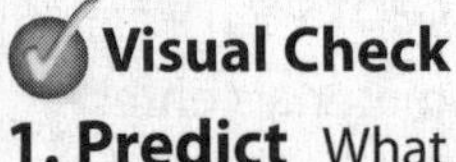

Visual Check

1. Predict What would the water in the beaker on the right look like if the membrane did not let anything through?

____

____

____

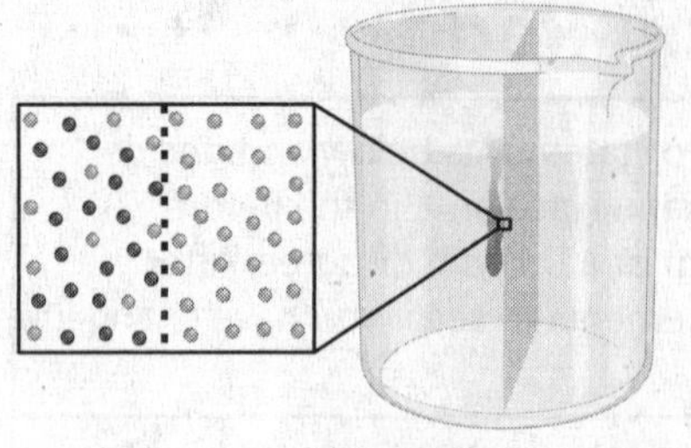

Dye added to water

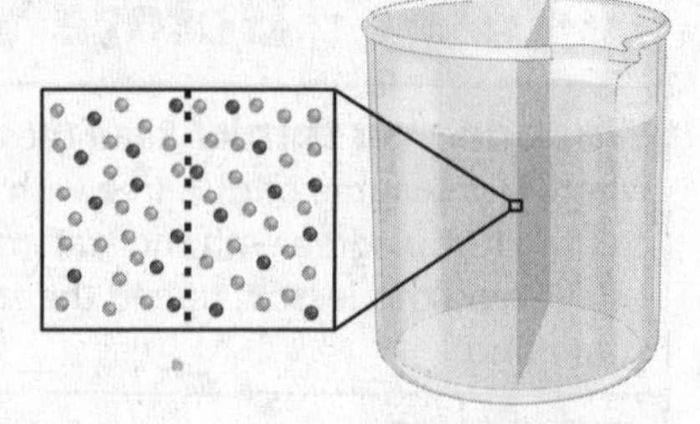

After 30 minutes

Diffusion will continue until the concentration on each side of the cell membrane is equal. The figure above shows how dye passed through the membrane into the clear water until there were equal concentrations of water and dye on both sides of the membrane.

Osmosis—The Diffusion of Water

Diffusion is the movement of any small molecules from areas of higher concentrations to areas of lower concentrations. **Osmosis** *is the diffusion of water molecules only through a membrane.* Water molecules pass through a semipermeable membrane from an area of high concentration to an area of low concentration. For example, plant cells lose water because of osmosis. The concentration of water in the air around a plant is less than the concentration of water in the cells of the plant. Water will leave plant cells and diffuse into the air. If the plant is not watered to replace the water lost by its cells, the plant will wilt and might die.

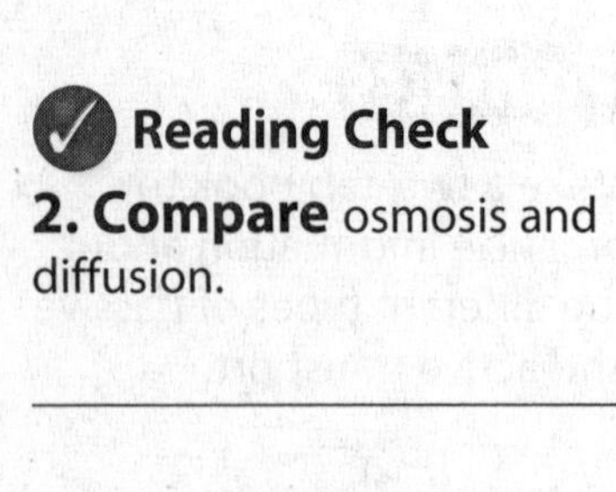

Reading Check

2. Compare osmosis and diffusion.

____

____

____

Facilitated Diffusion

Some molecules are too large or are chemically unable to move through a membrane by diffusion. **Facilitated diffusion** *is the movement of molecules through a cell membrane using special proteins called transport proteins.* Facilitated diffusion does not use the cell's energy to move the molecules. The transport proteins do the work. There are two types of transport proteins.

Carrier Proteins Carrier proteins are transport proteins. They carry large molecules, such as the sugar molecule glucose, through the cell membrane.

Channel Proteins Channel proteins are also transport proteins. They form pores through the cell membrane. Ions, such as sodium and potassium, pass through the cell membrane by channel proteins. Transport proteins are shown below.

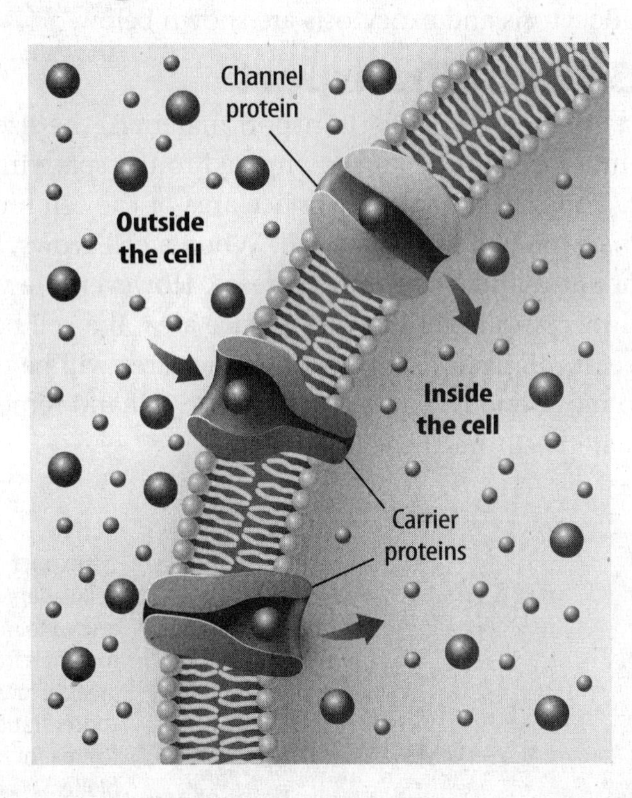

Active Transport

Sometimes a cell uses energy when a substance passes through its membrane. **Active transport** *is the movement of substances through a cell membrane only by using the cell's energy.*

Substances moving by active transport move from areas of lower concentration to areas of higher concentration. Active transport is important for cells and organelles. Cells can take in nutrients from the environment through carrier proteins by using active transport. Some molecules and waste materials leave cells by active transport.

Endocytosis and Exocytosis

Some substances are too large to enter a cell membrane by diffusion or by using a transport protein. There are other ways that substances can enter a cell.

Endocytosis *The process during which a cell takes in a substance by surrounding it with the cell membrane is called* **endocytosis** (en duh si TOH sus). Some cells take in bacteria and viruses using endocytosis.

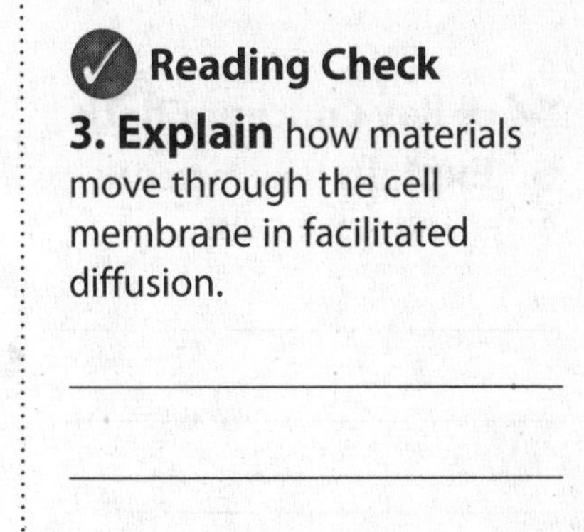

Reading Check

3. Explain how materials move through the cell membrane in facilitated diffusion.

____

____

____

Visual Check

4. Identify Circle the type of transport protein that carries large molecules through the cell membrane.

Reading Check

5. Summarize how a cell uses active transport.

____

____

____

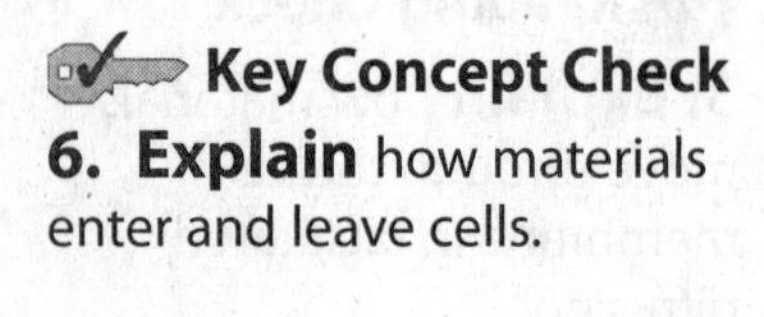

Key Concept Check

6. Explain how materials enter and leave cells.

Math Skills

A ratio is a comparison of two numbers, such as surface area and volume. If a cell were cube shaped, you would calculate surface area by multiplying its length *(ℓ)* by its width *(w)* by the number of sides (6).

Surface area: $\ell \times w \times 6$

You would calculate the volume of the cell by multiplying its length *(ℓ)* by its width *(w)* by its height *(h)*.

Volume: $\ell \times w \times h$

To find the surface-area-to-volume ratio of the cell, divide its surface area by its volume.

$$\frac{\text{Surface area}}{\text{Volume}}$$

7. Use Ratios What is the surface-area-to-volume ratio of a cube-shaped cell whose sides are 6 mm long?

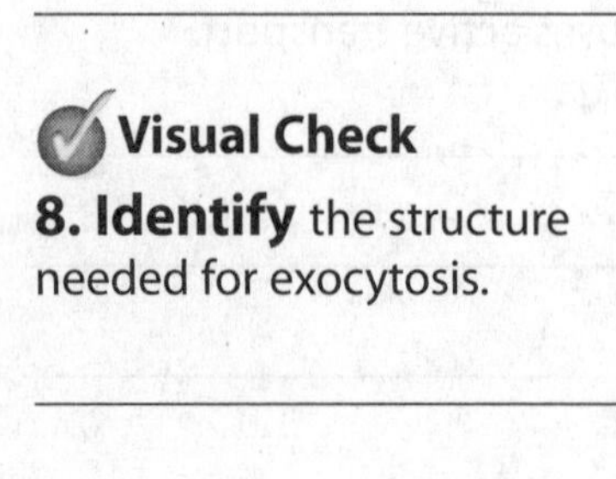

Visual Check

8. Identify the structure needed for exocytosis.

Exocytosis Some substances are too large to leave a cell by diffusion or by using a transport protein. They can leave using exocytosis (ek soh si TOH sus). **Exocytosis** *is the process during which a cell's vesicles release their contents outside the cell.* Proteins and other substances are removed from a cell through exocytosis. Both endocytosis and exocytosis are shown below.

Cell Size and Transport

For a cell to successfully transport materials, the size of the cell membrane must be large compared to the space inside of the cell. This means that the surface area of the cell must be larger than the volume of the cell. When a cell grows, both its surface area and its volume increase. However, the volume of a cell increases faster than its surface area. If a cell becomes too large, it might not survive. Its surface area will be too small to move enough nutrients into the cell and remove waste materials from the cell.

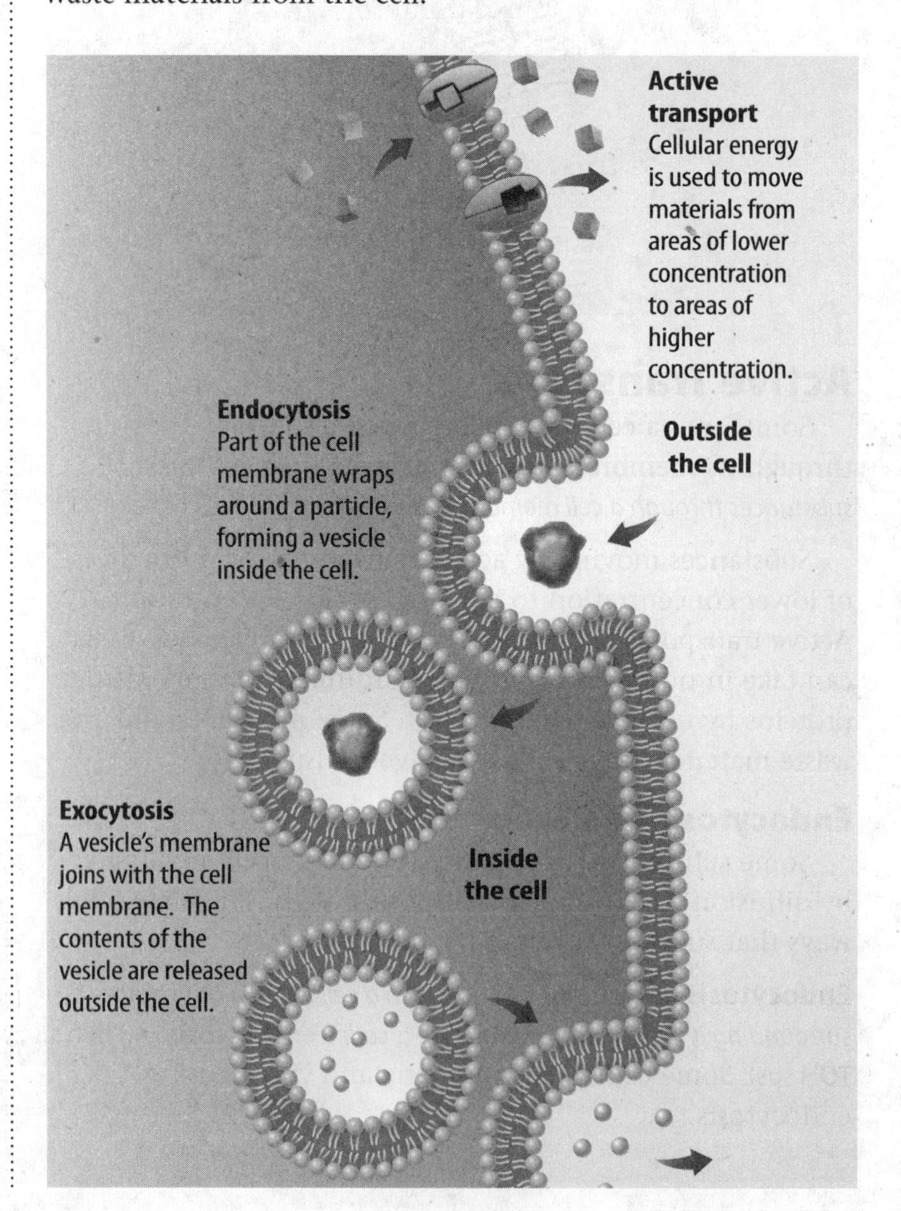

After You Read

Mini Glossary

active transport: the movement of substances through a cell membrane only by using the cell's energy

diffusion: the movement of substances from an area of higher concentration to an area of lower concentration

endocytosis (en duh si TOH sus): the process during which a cell takes in a substance by surrounding it with the cell membrane

exocytosis (ek soh si TOH sus): the process during which a cell's vesicles release their contents outside the cell

facilitated diffusion: when molecules pass through a cell membrane using special proteins called transport proteins

osmosis: the diffusion of water molecules only

passive transport: the movement of substances through a cell membrane without using the cell's energy

1. Review the terms and their definitions in the Mini Glossary. Write a sentence that compares passive and active transport.

2. Fill in the table below to compare active and passive transport.

	Energy needed?	Structures Involved	Examples
Active transport	yes/no		
Passive transport	yes/no		

What do you think NOW?

Reread the statements at the beginning of the lesson. Fill in the After column with an A if you agree with the statement or a D if you disagree. Did you change your mind?

Log on to ConnectED.mcgraw-hill.com and access your textbook to find this lesson's resources.

Cell Structure and Function

Cells and Energy

Key Concepts 

- How does a cell obtain energy?
- How do some cells make food molecules?

Before You Read

What do you think? Read the two statements below and decide whether you agree or disagree with them. Place an A in the Before column if you agree with the statement or a D if you disagree. After you've read this lesson, reread the statements to see if you have changed your mind.

Before	Statement	After
	7. ATP is the only form of energy found in cells.	
	8. Cellular respiration occurs only in lung cells.	

Study Coach

Use an Outline As you read, make an outline to summarize the information in the lesson. Use the main headings in the lesson as the main headings in the outline. Complete the outline with the information under each heading.

Read to Learn

Cellular Respiration

All living organisms need energy to survive. Cells use energy from food and make an energy-storing compound, ATP. **Cellular respiration** *is a series of chemical reactions that convert the energy in food into a usable form of energy called ATP.* Cellular respiration takes place in the cytoplasm and in the mitochondria of a cell.

Reactions in the Cytoplasm

The first step of cellular respiration is called glycolysis. It takes place in the cytoplasm of all cells. **Glycolysis** *is a process by which a sugar called glucose is broken down into smaller molecules.* Glycolysis produces some ATP molecules. It also uses energy from other ATP molecules.

More ATP is made during the second step of cellular respiration than during glycolysis.

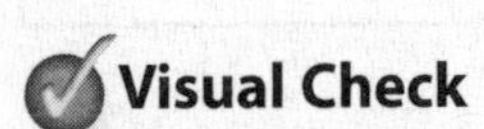

1. Locate Circle where sugar breaks down in the cell during glycolysis.

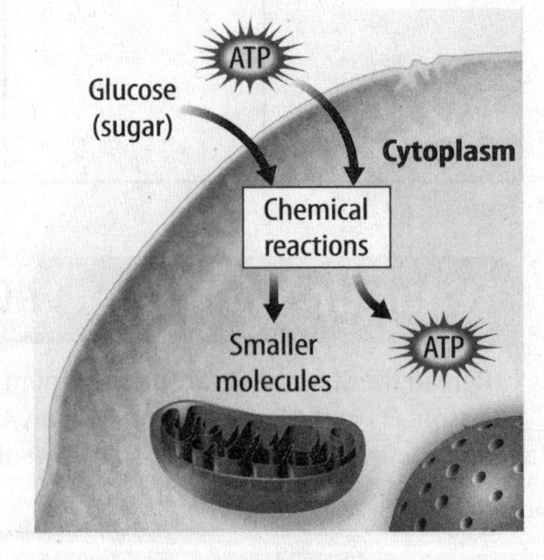

Reactions in the Mitochondria

The second step in cellular respiration, shown below, takes place in the mitochondria of eukaryotic cells. This step uses oxygen. The smaller molecules made during glycolysis are broken down. Many ATP molecules are made. Cells use ATP molecules to power all cellular processes. Two waste products, water and carbon dioxide (CO_2), are given off during this step of cellular respiration. The CO_2 released by cells as a waste product is used by plants and some unicellular organisms in a process called photosynthesis.

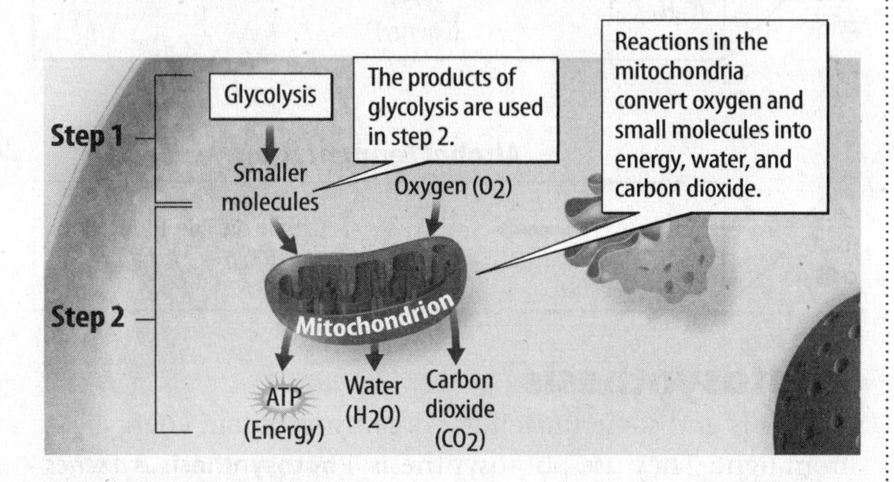

Fermentation

Sometimes, as you exercise, there is not enough oxygen in your cells to make ATP molecules through cellular respiration. When this happens, cells use a process called fermentation to obtain chemical energy. **Fermentation** *is a reaction that eukaryotic and prokaryotic cells use to obtain energy from food when oxygen levels are low.* Because no oxygen is used, fermentation makes less ATP than cellular respiration does. Fermentation takes place in a cell's cytoplasm, not in mitochondria.

Types of Fermentation

There are several types of fermentation. One type occurs when glucose is changed into ATP and a waste product called lactic acid.

Lactic-Acid Fermentation Some bacteria and fungi help produce cheese, yogurt, and sour cream using lactic-acid fermentation. The muscle cells in animals, including humans, can release energy during exercise using lactic-acid fermentation.

FOLDABLES

Make a half-book to record information about the different types of energy production.

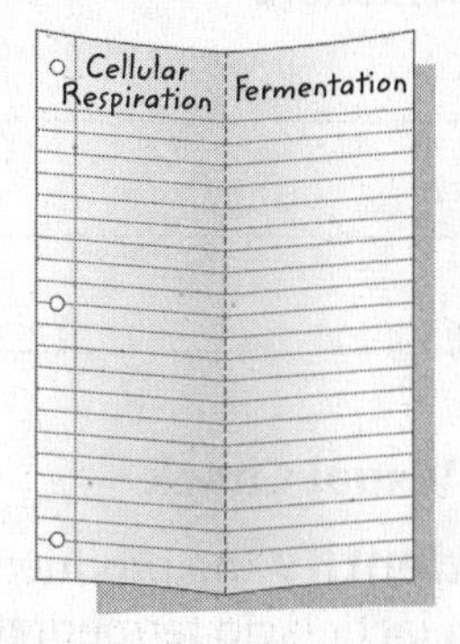

Visual Check

2. Compare the reactions in mitochondria with glycolysis.

____

____

____

Key Concept Check

3. Explain how a cell obtains energy.

____

____

____

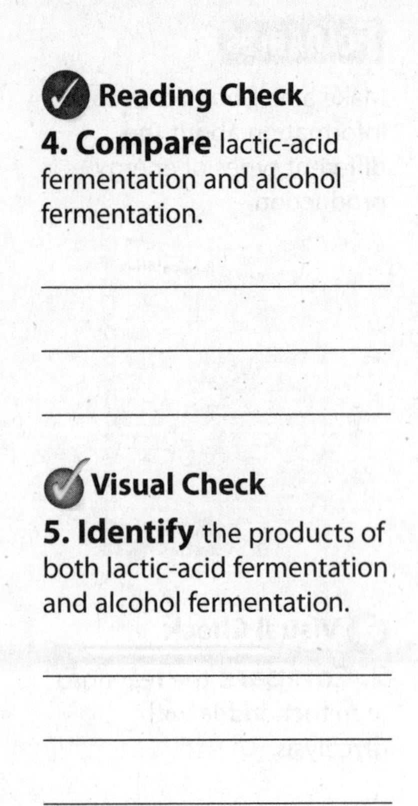

Reading Check

4. Compare lactic-acid fermentation and alcohol fermentation.

Visual Check

5. Identify the products of both lactic-acid fermentation and alcohol fermentation.

Alcohol Fermentation Some types of bacteria and yeast make ATP through a process called alcohol fermentation. Alcohol fermentation produces an alcohol, called ethanol, and carbon dioxide. Many types of bread are made using yeast. The carbon dioxide produced by yeast during alcohol fermentation makes bread dough rise. Lactic-acid fermentation and alcohol fermentation are shown below.

Lactic-Acid Fermentation

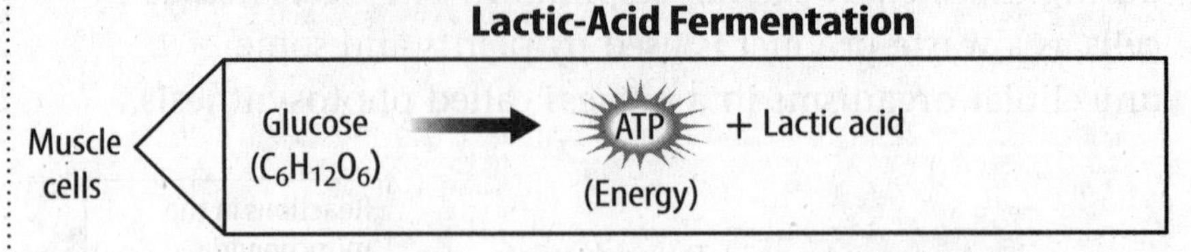

Alcohol Fermentation

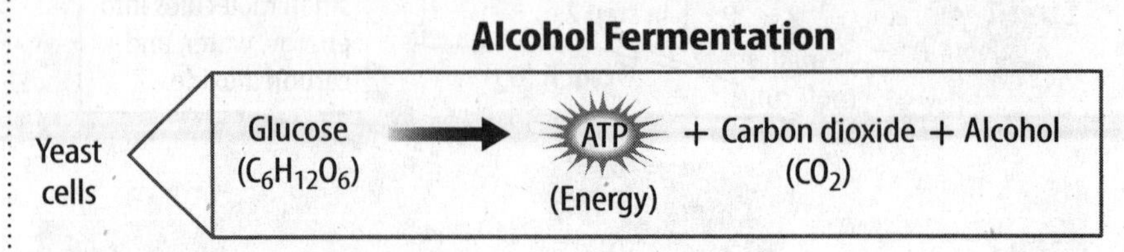

Photosynthesis

Plants and some unicellular organisms obtain energy from light. They use photosynthesis. **Photosynthesis** *is a series of chemical reactions that convert light energy, water, and carbon dioxide into the food-energy molecule glucose and the waste product oxygen.*

Light and Pigments

Photosynthesis uses light energy. In plants, pigments such as chlorophyll absorb light energy. As chlorophyll absorbs light, it absorbs all the colors in it except green.

The green light is reflected as the green color that you see in leaves and stems. Plants might also contain pigments that reflect other colors, such as red, yellow, or orange light.

Reactions in Chloroplasts

The chlorophyll that absorbs light energy for photosynthesis is in chloroplasts. Chloroplasts are organelles in plant cells that convert light energy to chemical energy in food. During photosynthesis, light energy, water, and carbon dioxide combine and make sugars. Photosynthesis also produces oxygen, which is released into the atmosphere.

Key Concept Check

6. Explain how some cells make food molecules.

Importance of Photosynthesis

Photosynthesis uses light energy and carbon dioxide to make food energy. Oxygen is released during this process. This food energy is stored as glucose. When an organism eats plant material, such as fruit, it takes in food energy. The cells of the organism will then go through cellular respiration. They will use the oxygen released during photosynthesis and convert the food energy into ATP. These organisms then release carbon dioxide into the atmosphere. The relationship between cellular respiration and photosynthesis is shown in the diagram below.

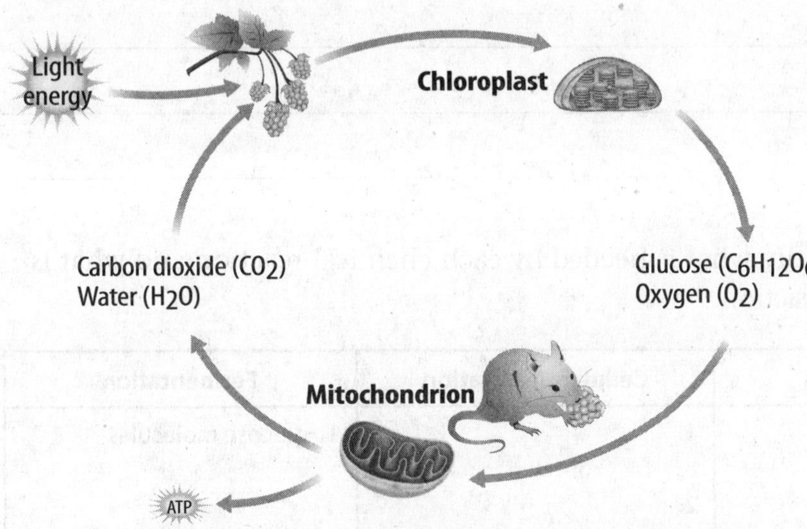

Light energy

Chloroplast

Carbon dioxide (CO_2)
Water (H_2O)

Glucose ($C_6H_{12}O_6$)
Oxygen (O_2)

Mitochondrion

ATP

$$C_6H_{12}O_6 + 6O_2 \rightarrow 6CO_2 + 6H_2O + \text{ATP (Energy)}$$

Cellular respiration

$$6CO_2 + 6H_2O \rightarrow C_6H_{12}O_6 + 6O_2$$

Photosynthesis

WORD ORIGIN

photosynthesis
from Greek *photo,* means "light," and *synthesis,* means "composition'

Visual Check

7. Explain the relationship between cellular respiration and photosynthesis.

____

____

____

After You Read

Mini Glossary

cellular respiration: a series of chemical reactions that convert the energy in food molecules into a usable form of energy called ATP

fermentation: a reaction that eukaryotic and prokaryotic cells use to obtain energy from food when oxygen levels are low

glycolosis: a process by which glucose, a sugar, is broken down into smaller molecules

photosynthesis: a series of chemical reactions that converts light energy, water, and carbon dioxide into the food-energy molecule glucose and gives off oxygen

1. Review the terms and their definitions in the Mini Glossary. Explain, using complete sentences, how photosynthesis and cellular respiration are related.

2. Fill in the table below to identify what is needed by each chemical reaction and what is produced by each chemical reaction.

	Photosynthesis	Cellular Respiration	Fermentation
What is needed?	**1.** **2.** **3.**	**1.** **2.**	**1.** glucose molecules
What is produced?	**1.** **2.**	**1.** **2.** **3.**	**1.** **2.** **3.**

3. As chlorophyll in plants absorbs light, it absorbs all the colors except one color. Which color is that?

What do you think NOW?

Reread the statements at the beginning of the lesson. Fill in the After column with an A if you agree with the statement or a D if you disagree. Did you change your mind?

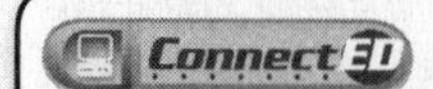

Log on to ConnectED.mcgraw-hill.com and access your textbook to find this lesson's resources.

From a Cell to an Organism

The Cell Cycle and Cell Division

Before You Read

What do you think? Read the three statements below and decide whether you agree or disagree with them. Place an A in the Before column if you agree with the statement or a D if you disagree. After you've read this lesson, reread the statements and see if you have changed your mind.

Before	Statement	After
	1. Cell division produces two identical cells.	
	2. Cell division is important for growth.	
	3. At the end of the cell cycle, the original cell no longer exists.	

Key Concepts 

- What are the phases of the cell cycle?
- Why is the result of the cell cycle important?

Read to Learn

The Cell Cycle

No matter where you live, you have probably noticed that the weather changes in a regular pattern each year. Some areas have four seasons—winter, spring, summer, and fall. As seasons change, temperature, precipitation, and the number of hours of sunlight change in a regular cycle.

Cells also go through cycles, just like the seasons. *Most cells in an organism go through a cycle of growth, development, and division called the* **cell cycle**. The cell cycle makes it possible for organisms

- to grow and develop,
- to replace cells that are old or damaged, and
- to produce new cells.

Phases of the Cell Cycle

There are two main phases in the cell cycle. These phases are interphase and the mitotic (mi TAH tihk) phase. **Interphase** *is the period of a cell's growth and development.* A cell spends most of its life in interphase.

Study Coach

Create a Quiz Write a question about the main idea under each heading. Exchange quizzes with another student. Together, discuss the answers to the quizzes.

Key Concept Check

1. Name What are the two main phases of the cell cycle?

______

______

______

FOLDABLES®

Make a folded book to illustrate the cell cycle.

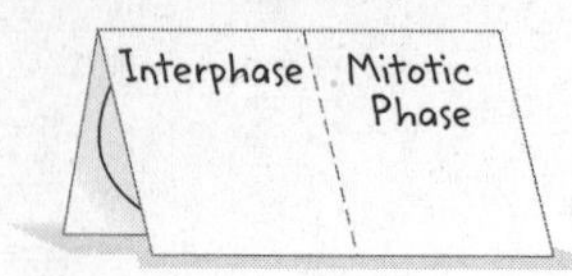

Visual Check

2. Identify Which stage of interphase is the longest?

REVIEW VOCABULARY

eukaryotic
related to a cell with membrane-bound structures

During Interphase Most cells go through three stages during interphase:

- rapid growth and replication, or copying, of the membrane-bound structures called organelles,
- copying of DNA, the genetic information in a cell, and
- preparation for cell division.

Interphase is followed by a shorter phase of the cell cycle called the mitotic phase.

During the Mitotic Phase A cell reproduces during the mitotic phase. The mitotic phase has two stages, as shown in the figure below. During the first stage, the contents of the nucleus divide. During the second stage, the cell's fluid, or cytoplasm, divides. The mitotic phase creates two new identical cells. The original cell no longer exists.

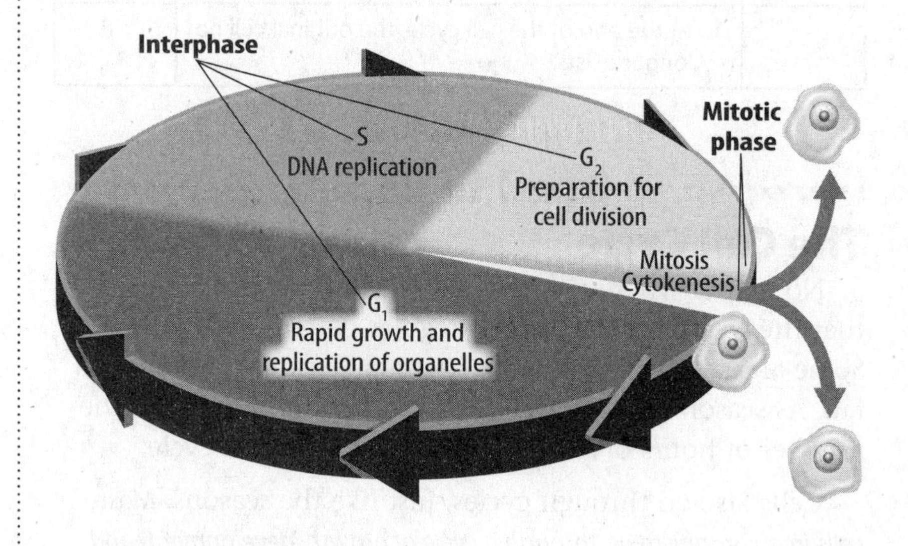

Length of a Cell Cycle

The time it takes a cell to complete the cell cycle depends on the type of cell that is dividing. Recall that a eukaryotic cell has membrane-bound organelles, including a nucleus. The cell cycle for some eukaryotic cells might only take eight minutes. The cell cycle for other eukaryotic cells might take up to one year. Most of the cells in the human body can complete the cell cycle in about 24 hours. The cells of some organisms divide very quickly. For example, the fertilized egg of the zebra fish divides into 256 cells in 2.5 hours.

Interphase

A new cell begins interphase with a period of rapid growth, in which the cell gets bigger. Cellular activities, such as making proteins, follow. Each cell that is actively dividing copies its DNA and prepares for cell division. A cell's DNA is called chromatin (KROH muh tun) during interphase. Chromatin is long, thin strands of DNA in the nucleus. If scientists add dye to a cell in interphase, the nucleus looks like a plate of spaghetti. This is because the nucleus contains strands of chromatin tangled together. ✓

✓ Reading Check

3. Identify What is the location of the chromatin in the cell?

______

______

______

Phases of Interphase

Interphase can be divided into three different stages, as shown in the table below.

The G_1 Stage The first stage of interphase is the G_1 stage. This is a period of rapid growth. G_1 is the longest stage of the cell cycle. During G_1, a cell grows and carries out its normal cell functions. For example, during G_1 the cells that line your stomach make enzymes that help you digest your food. Most cells continue the cell cycle. However, some cells stop the cell cycle at the G_1 stage. Mature nerve cells in your brain remain in G_1 and do not divide again.

Phases of the Cell Cycle

Phase	Stage	Description
Interphase	G_1	growth and cellular functions; organelle replication
	S	growth and chromosome replication; organelle replication
	G_2	growth and cellular functions; organelle replication
Mitotic Phase	mitosis	division of nucleus
	cytokinesis	division of cytoplasm

Interpreting Tables

4. Name What are the stages of interphase?

______

______

______

The S Stage The second stage of interphase is the S stage. During the S stage, a cell grows and copies its DNA. Strands of chromatin are copied, so there are now two identical strands of DNA. This is necessary because each new cell gets a copy of the genetic information. The new strands coil up and form chromosomes. A cell's DNA is arranged as pairs. Each pair is called a duplicated chromosome. *Two identical chromosomes called* ***sister chromatids*** *make up a duplicated chromosome. The sister chromatids are held together by a structure called a* ***centromere.*** ✓

✓ Reading Check

5. Explain What happens during the S stage?

______

______

______

Reading Check

6. Describe What happens during the G_2 stage?

Think it Over

7. Discuss Why is it important for a cell to make a copy of each organelle before division?

Think it Over

8. Calculate How many chromatids are in twenty duplicated chromosomes?

The G_2 Stage The last stage of interphase is the G_2 stage. This is another period of growth and the final preparation for mitosis. A cell uses energy to copy DNA during the S stage. During G_2, the cell stores energy that will be used during the mitotic phase of the cell cycle.

Organelle Replication

During cell division, the organelles are distributed between the two new cells. Before a cell divides, it makes a copy of each organelle. This way, the two new cells can function properly. Some organelles, such as the energy-processing mitochondria and chloroplasts, have their own DNA. These organelles can make copies of themselves on their own. A cell produces other organelles from materials such as proteins and lipids. A cell makes these materials using the information in the DNA inside the nucleus. Organelles are copied during all stages of interphase.

The Mitotic Phase

The mitotic phase of the cell cycle follows interphase. There are two stages of the mitotic phase: mitosis (mi TOH sus) and cytokinesis (si toh kuh NEE sus). *In* ***mitosis,*** *the nucleus and its contents divide. In* ***cytokinesis,*** *the cytoplasm and its contents divide.* ***Daughter cells*** *are the two new cells that result from mitosis and cytokinesis.*

During mitosis, the contents of the nucleus divide, forming two identical nuclei. The sister chromatids of the duplicated chromosomes separate from each other. This gives each daughter cell the same genetic information. For example, a cell that has ten duplicated chromosomes actually has 20 chromatids. When the cell divides, each daughter cell will have ten different chromatids. Chromatids are now called chromosomes.

During cytokinesis, the cytoplasm divides and two new daughter cells form. The organelles that were made during interphase are divided between the daughter cells.

Phases of Mitosis

Mitosis, like interphase, is a process that can be divided into different phases. Follow along with the diagrams on the next page as you read the descriptions in this section.

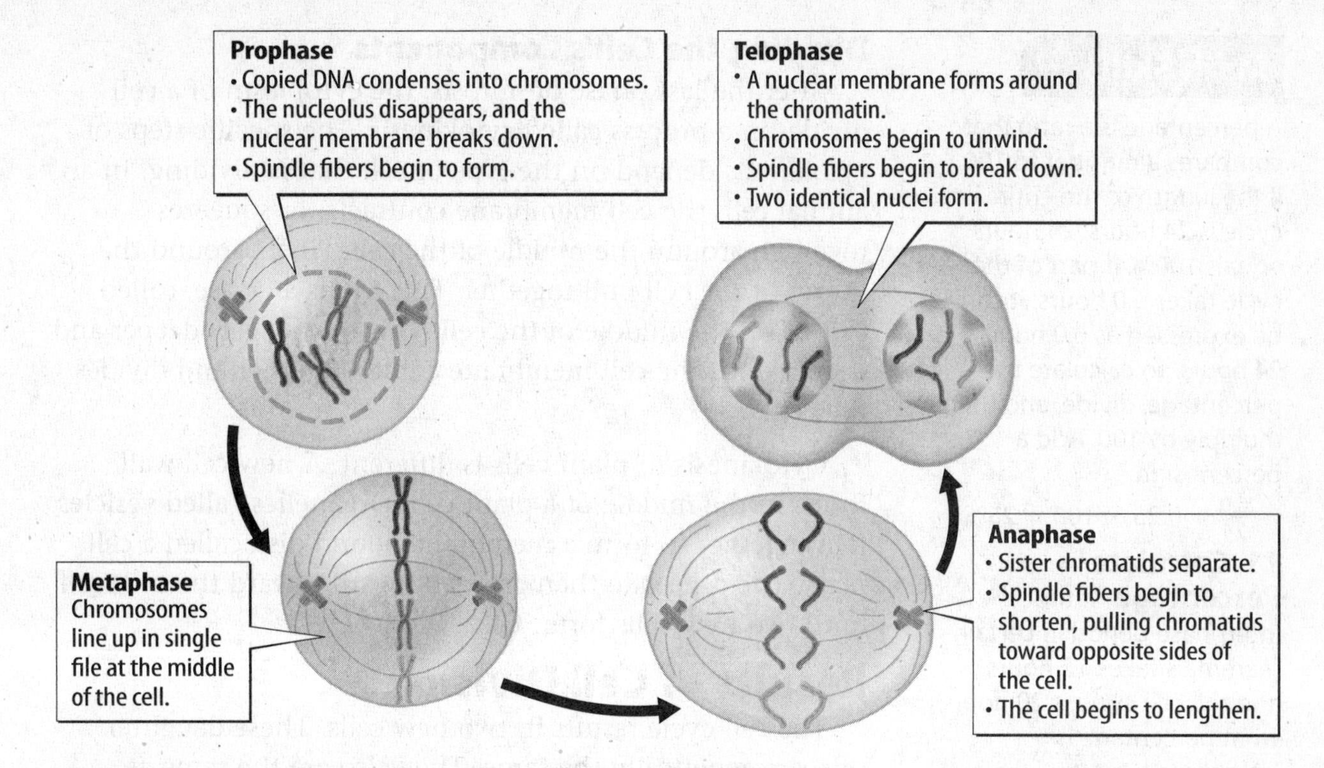

Prophase During the first phase of mitosis, called prophase, the copied chromatin coils together tightly. The coils form duplicated chromosomes that can be seen with a microscope. The nucleolus disappears and the nuclear membrane breaks down. Structures called spindle fibers form in the cytoplasm.

Metaphase The spindle fibers pull and push the duplicated chromosomes to the middle of the cell during metaphase. Notice in the figure above that the chromosomes line up along the middle of the cell. This makes sure that each new cell will receive one copy of each chromosome. Metaphase is the shortest phase in mitosis. It is an important phase because it makes the new cells the same.

Anaphase In anaphase, the two sister chromatids separate from each other and are pulled in opposite directions. Once they are separated, the chromatids are now two identical single-stranded chromosomes. As the single-stranded chromosomes move to opposite sides of the cell, the cell begins to get longer. Anaphase ends when the two sets of identical chromosomes reach opposite ends of the cell.

Telophase During telophase, the spindle fibers that helped divide chromosomes begin to disappear. The chromosomes begin to uncoil. A nuclear membrane grows around each set of chromosomes at either end of the cell. Two new identical nuclei form.

Visual Check

9. Identify During which phase of mitosis do the chromosomes line up in the center of a cell?

____

____

Reading Check

10. State What are the phases of mitosis?

____

____

____

____

Math Skills

A percentage is a ratio that compares a number to 100. If the length of the entire cell cycle is 24 hours, 24 hours equals 100%. If part of the cycle takes 6.0 hours, it can be expressed as 6.0 hours/24 hours. To calculate the percentage, divide, and then multiply by 100. Add a percent sign.

$\frac{6.0}{24} = 0.25 \times 100 = 25\%$

11. Calculate Percentage If the interphase period in certain mammals takes 12 hours, and the cell cycle is 20 hours, what percentage is interphase?

____

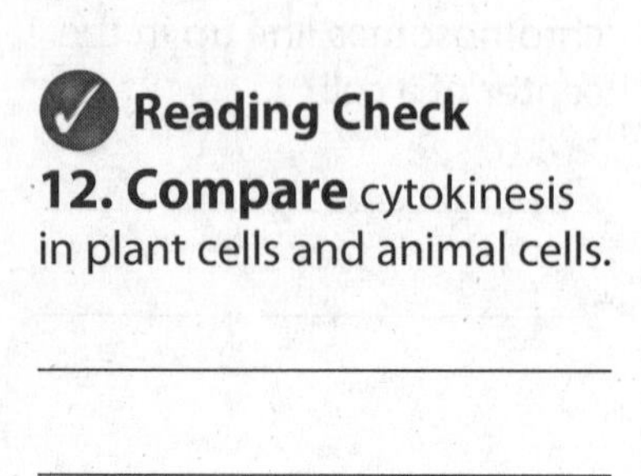

Reading Check

12. Compare cytokinesis in plant cells and animal cells.

____

____

____

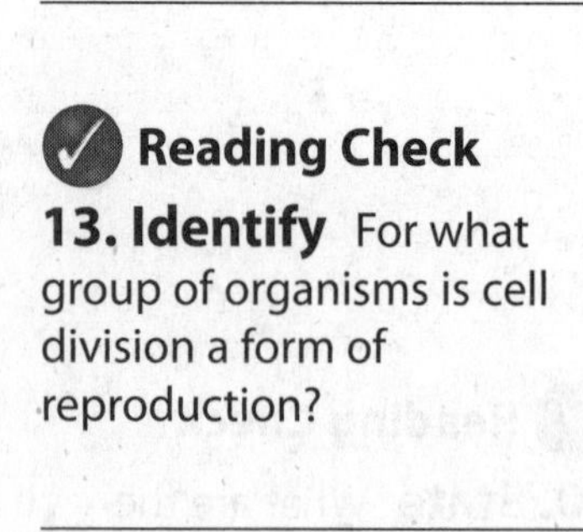

Reading Check

13. Identify For what group of organisms is cell division a form of reproduction?

____

____

Dividing the Cell's Components

After the last phase of mitosis, the cytoplasm of a cell divides in a process called cytokinesis. The specific steps of cytokinesis depend on the type of cell that is dividing. In an animal cell, the cell membrane contracts, or squeezes together, around the middle of the cell. Fibers around the center of the cell pull together. This forms a crease, called a furrow, in the middle of the cell. This furrow gets deeper and deeper until the cell membrane comes together and divides the cell.

Cytokinesis in plant cells is different. A new cell wall forms in the middle of a plant cell. Organelles called vesicles join together to form a membrane-bound disk called a cell plate. The cell plate then grows outward toward the cell wall until two new cells form.

Results of Cell Division

The cell cycle results in two new cells. These daughter cells are genetically the same. They also are the same as the original cell that no longer exists. A human cell has 46 chromosomes. When that cell divides, it produces two new cells, each with 46 chromosomes. The cell cycle is important for reproduction in some organisms. It is important for growth in multicellular organisms. The cell cycle also helps replace worn-out or damaged cells and repair damaged tissues.

Reproduction

Cell division is a form of reproduction for some unicellular organisms. For example, an organism called a paramecium reproduces by dividing into two new daughter cells, or two new paramecia. Cell division is also important in other methods of reproduction in which the offspring are identical to the parent organism.

Growth

Cell division allows multicellular organisms, such as humans, to grow and develop from one cell (a fertilized egg). In humans, cell division begins about 24 hours after fertilization. Cell division continues quickly for the first few years of life. During the next few years, you will probably go through a period of rapid growth and development. This happens because cells divide and increase in number as you grow and develop.

Replacement

Cell division continues even after an organism is fully grown. Cell division replaces cells that wear out or are damaged. The outermost layer of your skin is always rubbing or flaking off. A layer of cells below the skin's surface is constantly dividing. This produces millions of new cells each day to replace the ones that rub off.

Repair

Cell division is also important for repairing damage. When a bone breaks, cell division produces new bone cells. These new cells patch the broken pieces of bone back together.

Not all damage can be repaired. This is because not all cells continue to divide. Some nerve cells stop the cell cycle in interphase. Injuries to nerve cells often cause permanent damage.

Key Concept Check

14. Discuss Why is the result of the cell cycle important?

After You Read

Mini Glossary

cell cycle: a cycle of growth, development, and division that most cells in an organism go through

centromere: the structure that holds together two sister chromatids

cytokinesis (si toh kuh NEE sus): the stage in which the cytoplasm and its contents divide

daughter cells: the two new cells that result from mitosis and cytokinesis

interphase: the period of a cell's growth and development

mitosis (mi TOH sus): the phase in which the nucleus and its contents divide

sister chromatids: two identical chromosomes that make up a duplicated chromosome

1. Review the terms and their definitions in the Mini Glossary. Write a sentence that describes the cell cycle.

____

____

2. Complete the table below to explain what happens in each phase of mitosis.

Phases of Mitosis	Prophase	Metaphase	Anaphase	Telophase
What happens within the cell?		duplicated chromosomes line up in the middle of the cell		chromosomes uncoil, spindle fibers disappear, nuclear membrane grows around each set of chromosomes, and two new nuclei form

3. Choose one of your quiz questions and write it on the first line below. Then write your answer to that question on the line that follows.

____

____

What do you think NOW?

Reread the statements at the beginning of the lesson. Fill in the After column with an A if you agree with the statement or a D if you disagree. Did you change your mind?

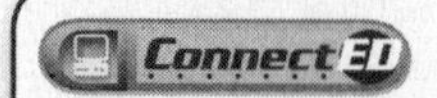

Log on to ConnectED.mcgraw-hill.com and access your textbook to find this lesson's resources.

From a Cell to an Organism

Levels of Organization

Before You Read

What do you think? Read the three statements below and decide whether you agree or disagree with them. Place an A in the Before column if you agree with the statement or a D if you disagree. After you've read this lesson, reread the statements and see if you have changed your mind.

Before	Statement	After
	4. Unicellular organisms do not have all the characteristics of life.	
	5. All the cells in a multicellular organism are the same.	
	6. Some organs work together as part of an organ system.	

Key Concepts 

- How do unicellular and multicellular organisms differ?
- How does cell differentiation lead to the organization within a multicellular organism?

Read to Learn

Life's Organization

All matter is made of atoms. Atoms combine and form molecules. Molecules make up cells. A large animal, such as a Komodo dragon, is not made of one cell. Instead, it is made of trillions of cells working together. The skin of the Komodo dragon is made of many cells that are specialized for protection. The Komodo dragon has other types of cells, such as blood cells and nerve cells, which perform other functions. Cells work together in the Komodo dragon and enable the whole organism to function. This is the same way that cells work together in you and in other multicellular organisms.

Recall that some organisms are made of only one cell. These unicellular organisms carry out all the activities necessary to survive, such as absorbing nutrients and getting rid of wastes. No matter their sizes, all organisms are made of cells. ✓

Study Coach

Make Flash Cards As you read, write each vocabulary word and key term from the text on one side of a flash card and its definition on the other side. Use your cards to review the material later.

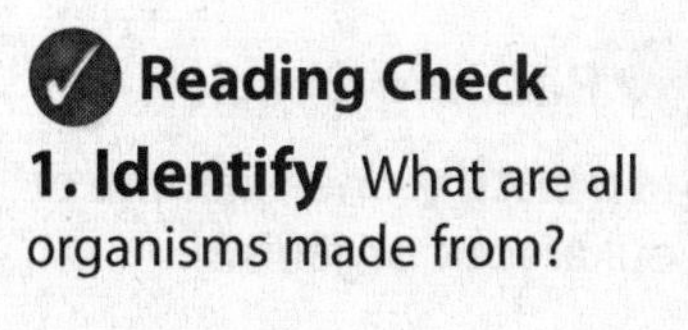

Reading Check

1. Identify What are all organisms made from?

Unicellular Organisms

Unicellular organisms have only one cell. These organisms do all the things needed for their survival within that one cell. An amoeba is a unicellular organism. It takes in, or ingests, other unicellular organisms for food to get energy. Unicellular organisms also respond to their environment, get rid of waste, grow, and reproduce. Unicellular organisms include both prokaryotes and some eukaryotes.

Prokaryotes

A cell without a membrane-bound nucleus is a prokaryotic cell. In general, prokaryotic cells are smaller than eukaryotic cells. As shown below on the left, prokaryotic cells also have fewer cell structures. A unicellular organism made of one prokaryotic cell is called a prokaryote. Some prokaryotes live in groups called colonies. Some can also live in extreme environments. The heat-loving bacteria that live in hot springs get their energy from sulfur instead of light.

2. Highlight each area where the hereditary material is located.

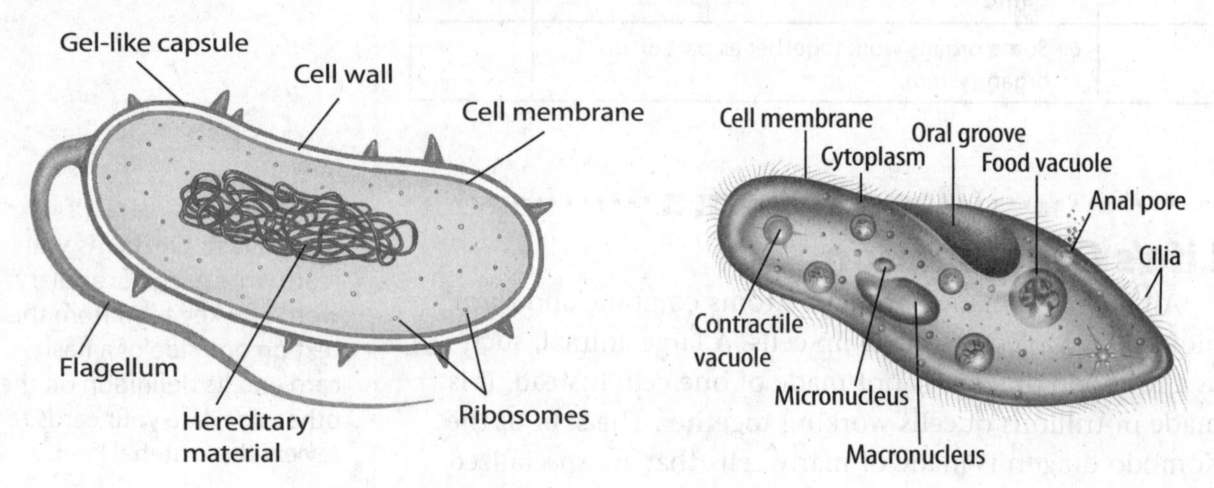

Eukaryotes

A eukaryotic cell has a nucleus surrounded by a membrane and many specialized organelles as shown above on the right. This paramecium has an organelle called a contractile vacuole. The contractile vacuole collects extra water from the paramecium's cytoplasm and pumps it out. The contractile vacuole keeps the paramecium from swelling and bursting.

A unicellular organism that is made of one eukaryotic cell is called a eukaryote. There are thousands of different unicellular eukaryotes. The alga that grows on the inside of an aquarium and the fungus that causes athlete's foot are unicellular eukaryotes.

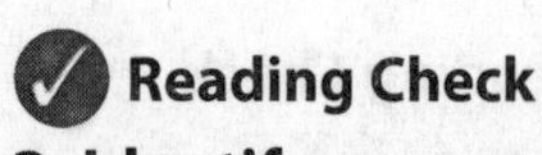

3. Identify one example of a eukaryotic organism.

____

____

Multicellular Organisms

A multicellular organism is made of many eukaryotic cells working together. Each type of cell in a multicellular organism has a specific job that is important to the survival of the organism. 

Cell Differentiation

Remember that all cells in a multicellular organism come from one cell, a fertilized egg. Cell division starts quickly after fertilization. The first cells made can become any type of cell, such as a muscle cell, a nerve cell, or a blood cell. *The process by which cells become different types of cells is called* **cell differentiation** (dihf uh ren shee AY shun).

A cell's instructions are contained in its chromosomes. Nearly all the cells in an organism have identical sets of chromosomes. If an organism's cells have identical sets of instructions, how can the cells be different? Different cell types use different parts of the instructions on the chromosomes. A few of the many different types of cells that can result from cell differentiation are shown in the figure below.

Animal Stem Cells Not all cells in a developing animal differentiate. **Stem cells** *are unspecified cells that are able to develop into many different cell types.* There are many stem cells in embryos but fewer in adult organisms. Adult stem cells are important for cell repair and replacement. For example, stem cells in your blood marrow can produce more than a dozen different types of blood cells. These replace the cells that are damaged or worn out. Stem cells in your muscles can produce new muscle cells. These can replace torn muscle fibers.

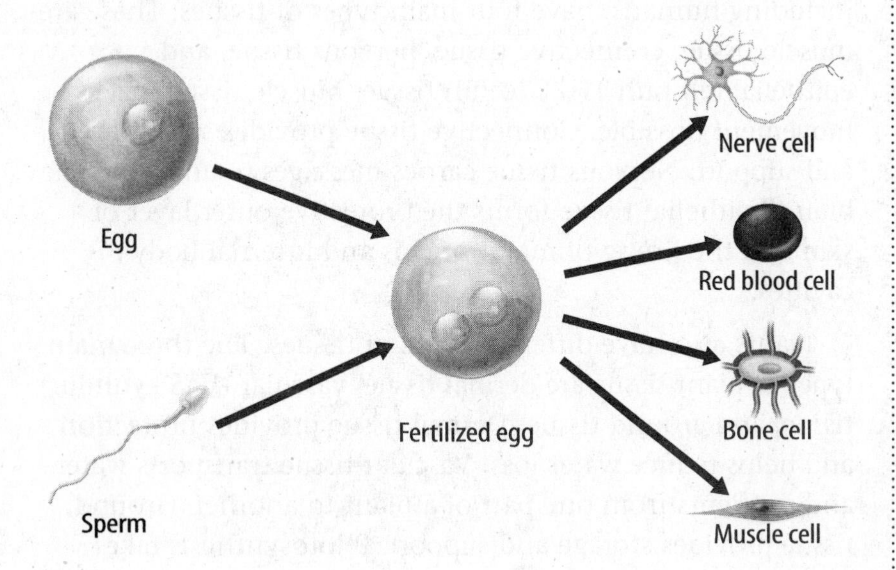

Key Concept Check
4. Describe How do unicellular and multicellular organisms differ?

FOLDABLES

Use a layered book to describe the levels of organization that make up organisms.

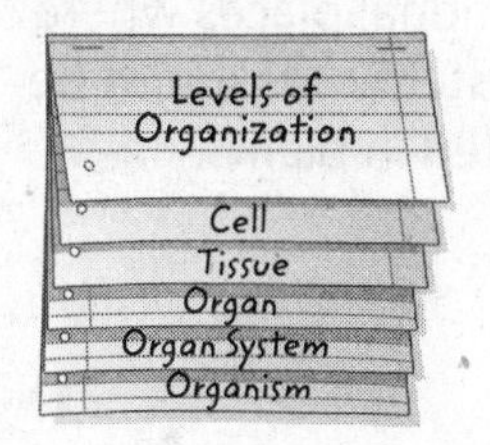

Visual Check
5. Name two types of cells that can result from cell differentiation.

Reading Check

6. Identify the three possible functions of meristems.

Visual Check

7. Identify Circle two additional places where meristem cells might be located in plants.

Reading Check

8. Compare animal and plant tissues.

Plant Cells Plants also have unspecialized cells, similar to the stem cells of animals. These cells are grouped in areas called meristems (MER uh stemz). Meristems are in different areas of a plant, including the tips of roots and stems. Cell division in meristems produces different types of plant cells with specialized structures and functions. These functions include transporting materials, making and storing food, or protecting the plant. Meristem cells might become part of stems, leaves, flowers, or roots. Meristems are shown in the figure below.

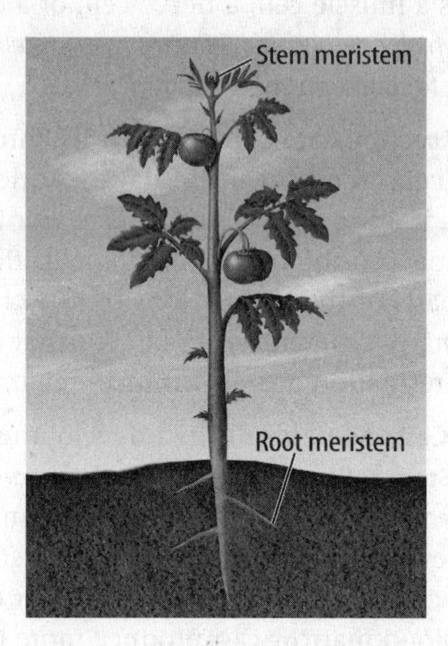

Tissues

In multicellular organisms, similar types of cells are organized into groups. **Tissues** *are groups of similar types of cells that work together to carry out specific tasks.* Most animals, including humans, have four main types of tissues. These are muscle tissue, connective tissue, nervous tissue, and epithelial (eh puh THEE lee ul) tissue. Muscle tissue makes movement possible. Connective tissue provides structure and support. Nervous tissue carries messages to and from the brain. Epithelial tissue forms the protective outer layer of skin and the lining of major organs and internal body cavities.

Plants also have different types of tissues. The three main types of plant tissue are dermal tissue, vascular (VAS kyuh lur) tissue, and ground tissue. Dermal tissue provides protection and helps reduce water loss. Vascular tissue transports water and nutrients from one part of a plant to another. Ground tissue provides storage and support. Photosynthesis takes place in ground tissue.

Organs

Complex jobs in organisms require more than one type of tissue. **Organs** *are groups of different tissues working together to perform a particular job*. Your stomach is an organ that breaks down food. It is made of all four types of tissue: muscle, epithelial, nervous, and connective. Each type of tissue performs a specific function necessary for the stomach to work properly and break down food. Muscle tissue contracts and breaks up food. Epithelial tissue lines the stomach. Nervous tissue signals when the stomach is full. Connective tissue supports the stomach wall.

Plants also have organs. A leaf is an organ specialized for photosynthesis. Each leaf is made of dermal tissue, ground tissue, and vascular tissue. Dermal tissue covers the outer surface of a leaf. The leaf is an important organ because it contains ground tissue that produces food for the rest of the plant. Ground tissue is where photosynthesis takes place. The ground tissue is tightly packed on the top half of the leaf. The vascular tissue moves both the food produced by photosynthesis and water throughout the leaf and plant.

Organ Systems

Most organs do not function alone. Instead, **organ systems** *are groups of different organs that work together to complete a series of tasks*. Human organ systems can be made of many different organs working together. For example, the digestive system is made of the stomach, the small intestine, the liver, and the large intestine. These organs all work together to break down food. Blood absorbs and transports nutrients from food to cells throughout the body.

Plants have two main organ systems—the shoot system and the root system. The shoot system includes leaves, stems, and flowers. The shoot system transports food and water throughout the plant. The root system anchors the plant and takes in water and nutrients.

Organisms

Multicellular organisms usually have many organ systems. The cells of these systems work together and carry out all the jobs needed for the organism to survive. There are many organ systems in the human body. Each organ system depends on the others and cannot work alone. For example, the respiratory system and circulatory system carry oxygen to the cells of the muscle tissue of the stomach. The oxygen aids in the survival of muscle tissue cells.

ACADEMIC VOCABULARY

complex
(adjective) made of two or more parts

Reading Check

9. Identify the major organ systems in plants.

____

____

Key Concept Check

10. Explain How does cell differentiation lead to the organization within a multicellular organism?

____

____

____

After You Read

Mini Glossary

cell differentiation (dihf uh ren shee AY shun): the process by which cells become different types of cells

organ: a group of tissues working together to perform a particular job

organ system: a group of organs that work together to complete a series of tasks

stem cell: an unspecified cell that is able to develop into many different cell types

tissue: a group of similar types of cells that work together to carry out specific tasks

1. Review the terms and their definitions in the Mini Glossary. Write two sentences describing some of the different types of cells within an organism.

2. Fill in the chart below to show the different levels of organization in a multicellular organism.

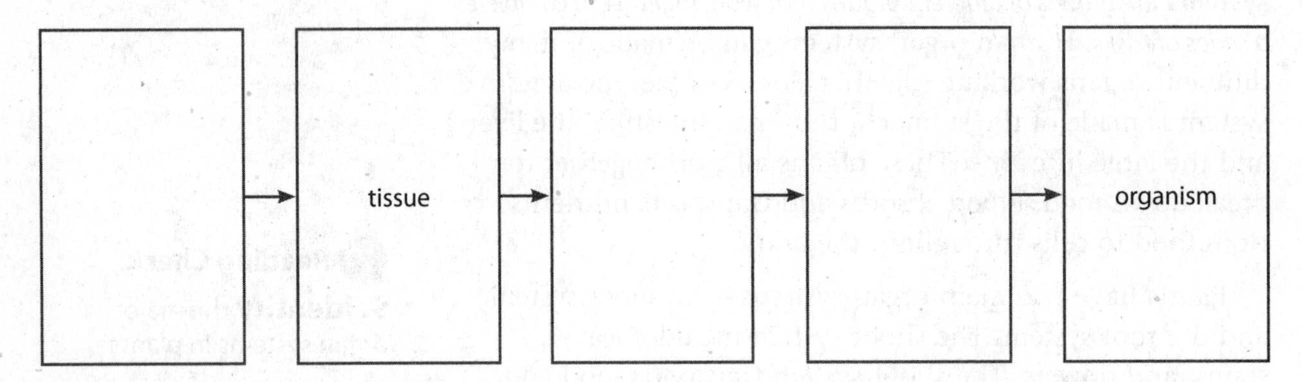

3. How did making flash cards of the important terms in the text help you review the material?

What do you think NOW?

Reread the statements at the beginning of the lesson. Fill in the After column with an A if you agree with the statement or a D if you disagree. Did you change your mind?

Log on to ConnectED.mcgraw-hill.com and access your textbook to find this lesson's resources.

Reproduction of Organisms

Sexual Reproduction and Meiosis

Before You Read

What do you think? Read the three statements below and decide whether you agree or disagree with them. Place an A in the Before column if you agree with the statement or a D if you disagree. After you've read this lesson, reread the statements to see if you have changed your mind.

Before	Statement	After
	1. Humans produce two types of cells: body cells and sex cells.	
	2. Environmental factors can cause variation among individuals.	
	3. Two parents always produce the best offspring.	

Key Concepts 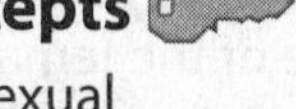

- What is sexual reproduction, and why is it beneficial?
- What is the order of the phases of meiosis, and what happens during each phase?
- Why is meiosis important?

Read to Learn

What is sexual reproduction?

Have you ever seen a litter of kittens? One kitten might have orange fur like its mother. A second kitten might have gray fur like its father. A third kitten might look like a combination of both parents. How does this happen?

The kittens look different because of sexual reproduction. **Sexual reproduction** *is a type of reproduction in which the genetic materials from two different cells combine, producing an offspring.* The cells that combine are called sex cells. Sex cells form in reproductive organs. There are two types of sex cells—eggs and sperm. *An* **egg** *is the female sex cell, which forms in an ovary. A* **sperm** *is the male sex cell, which forms in a testis.* **Fertilization** (fur tuh luh ZAY shun) *occurs when an egg cell and a sperm cell join together.* When an egg and a sperm join together, a new cell is formed. *The new cell that forms from fertilization is called a* **zygote.**

Study Coach

Vocabulary Quiz Write a question about each vocabulary term in this lesson. Exchange questions with another student. Together, discuss the answers to the questions.

Reading Check

1. Describe What is sexual reproduction?

2. Identify Circle the name of the female sex cell. Put a box around the name of the male sex cell.

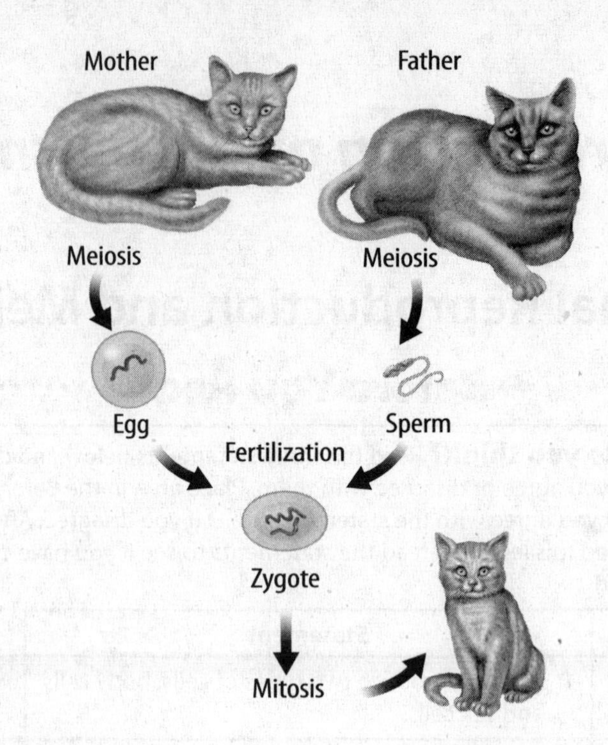

Diploid Cells

After fertilization, a zygote goes through mitosis and cell division, as shown above. Mitosis and cell division produce nearly all of the cells in a multicellular organism. The kitten in the picture above is a multicellular organism. Organisms that reproduce sexually form two kinds of cells—body cells and sex cells. In the body cells of most organisms, chromosomes occur in pairs. *Cells that have pairs of chromosomes are called* **diploid** *cells.*

Chromosomes

Pairs of chromosomes that have genes for the same traits arranged in the same order are called **homologous** (huh MAH luh gus) **chromosomes.** Because one chromosome is inherited from each parent, the chromosomes are not always identical. For example, the kittens you read about earlier inherited a gene for orange fur color from their mother. They also inherited a gene for gray fur color from their father. Some kittens might be orange, and some kittens might be gray. No matter what the color of a kitten's fur, both genes for fur color are found at the same place on homologous chromosomes. In this case, each gene codes for a different color.

FOLDABLES®

Make the following shutter-fold book, then use it to describe and illustrate the phases of meiosis.

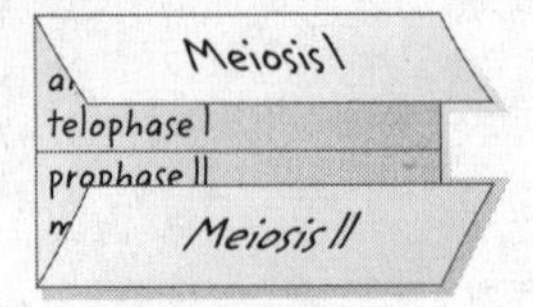

Chromosomes of a Human Cell

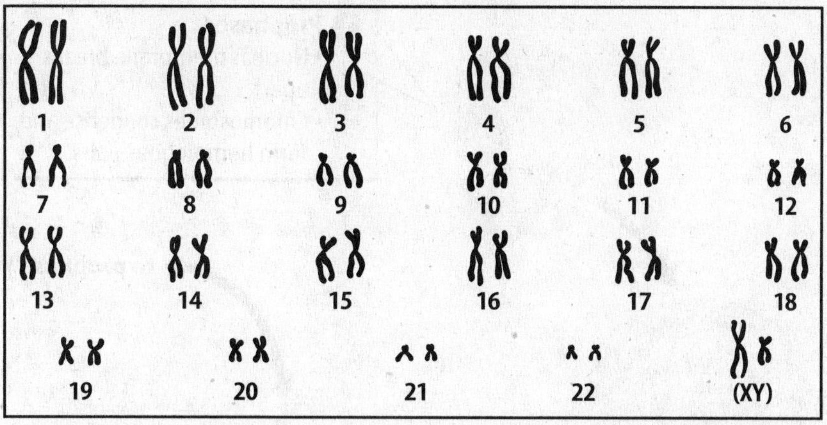

1	2	3	4	5	6
7	8	9	10	11	12
13	14	15	16	17	18
19	20	21	22	(XY)	

Different organisms have different numbers of chromosomes. Recall that diploid cells have pairs of chromosomes. Human diploid cells have 23 pairs of chromosomes, as shown in the picture above. This means that human diploid cells have a total of 46 chromosomes.

It is important to have the correct number of chromosomes. If a zygote has too many or too few chromosomes, it will not develop properly. The process of meiosis helps maintain the correct number of chromosomes. ✓

Haploid Cells

Organisms that reproduce sexually also form egg and sperm cells, or sex cells. Sex cells have only one chromosome from each pair of chromosomes. **Haploid** *cells are cells that have only one chromosome from each pair.*

Organisms produce sex cells using a special type of cell division called meiosis. *In* ***meiosis,*** *one diploid cell divides and makes four haploid sex cells.* Meiosis occurs only during the formation of sex cells. ✓

The Phases of Meiosis

Recall that mitosis and cytokinesis involve one division of the nucleus and cytoplasm. Meiosis involves two divisions of the nucleus and the cytoplasm. These two divisions are phases called meiosis I and meiosis II. Meiosis results in four haploid cells, each with half the number of chromosomes as the original cell.

Phases of Meiosis I

A reproductive cell goes through interphase before beginning meiosis I. During interphase, the reproductive cell grows and copies, or duplicates, its chromosomes. Each duplicated chromosome consists of two sister chromatids joined by a centromere.

✓ Visual Check

3. Identify How many chromosomes do human diploid cells have?

____

✓ Reading Check

4. Explain Why is it important for an organism to have the correct number of chromosomes?

____

____

____

✓ Reading Check

5. Contrast How do diploid cells differ from haploid cells?

____

____

____

Meiosis I

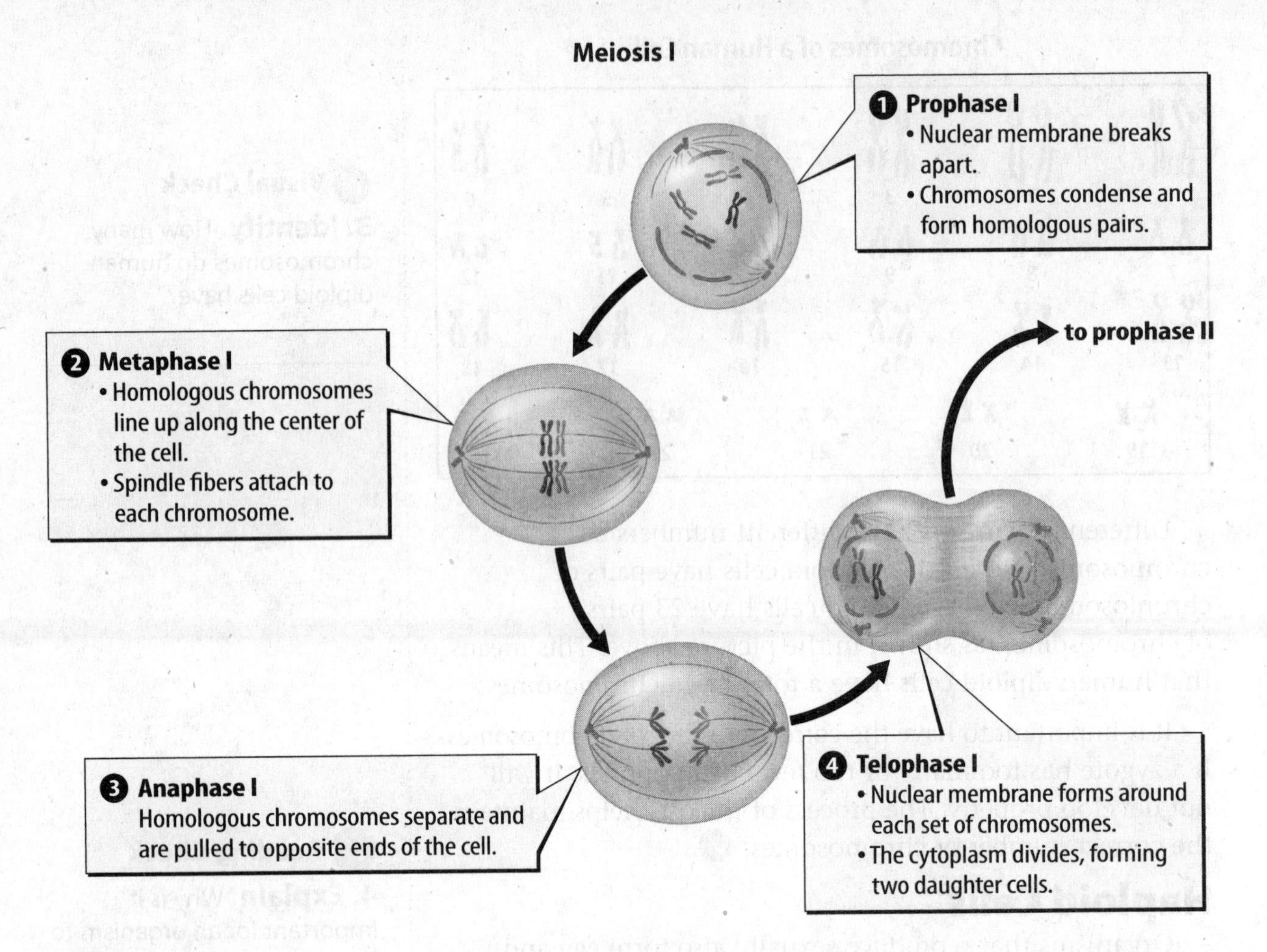

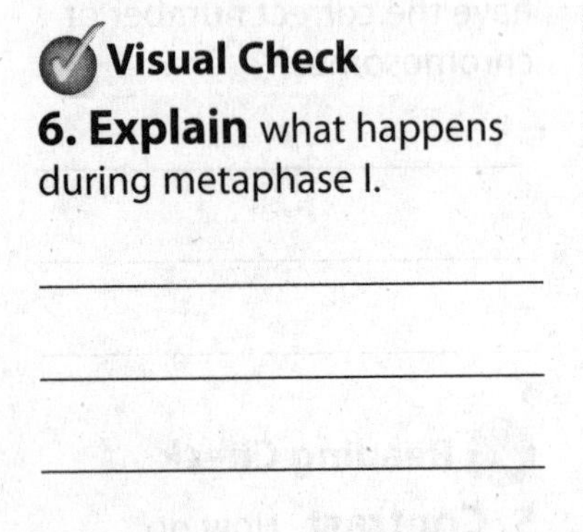

Visual Check

6. Explain what happens during metaphase I.

____

____

____

Reading Check

7. Describe what happens to the sister chromatids at the end of anaphase I.

____

____

____

As you read about the phases of meiosis I, refer to the figure above. Think about the process that produces cells with a reduced number of chromosomes.

1. Prophase I During prophase I, duplicated chromosomes condense, or shorten, and thicken. Homologous chromosomes come together and form pairs. The membrane around the nucleus breaks apart and the nucleolus disappears.

2. Metaphase I During metaphase I, homologous chromosome pairs line up along the middle of the cell, as shown in the figure above. A spindle fiber attaches to each chromosome.

3. Anaphase I During anaphase I, chromosome pairs separate and are pulled toward opposite ends of the cell. Notice in the figure above that the sister chromatids stay together. ✓

4. Telophase I During telophase I, a membrane forms around each group of duplicated chromosomes. The cytoplasm divides through cytokenesis, and two daughter cells form. Sister chromatids remain together.

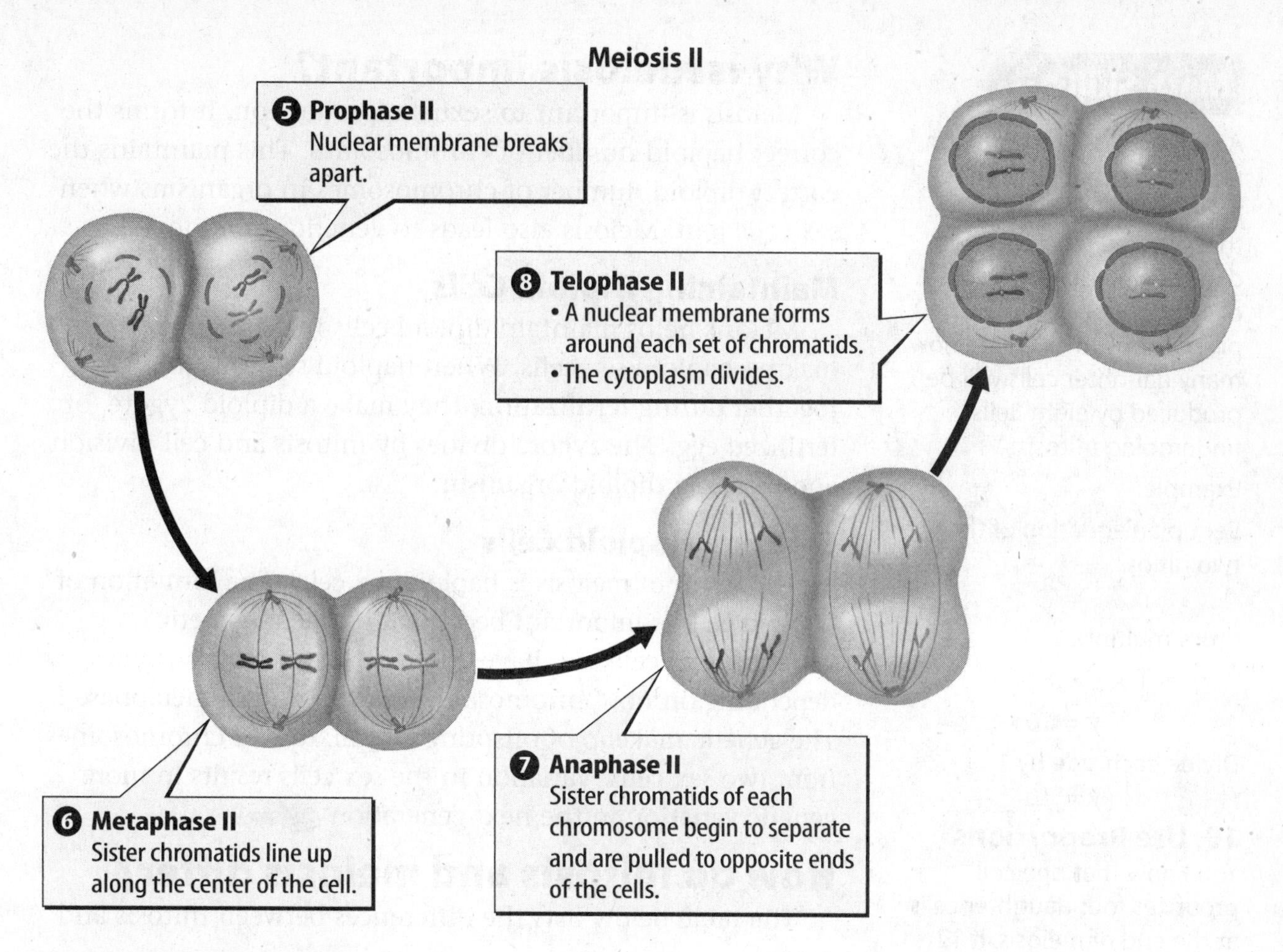

Phases of Meiosis II

After meiosis I, the two cells formed during this stage go through a second division of the nucleus and the cytoplasm. This process is called meiosis II. Meiosis II is shown in the figure above.

1. Prophase II Unlike prophase I, chromosomes are not copied again before prophase II. They remain short and thick sister chromatids. During prophase II, the membrane around the nucleus breaks apart, and the nucleolus disappears in each cell.

2. Metaphase II During metaphase II, the pairs of sister chromatids line up along the middle of the cell in single file.

3. Anaphase II During anaphase II, the sister chromatids of each duplicated chromosome are pulled apart. They then move toward opposite ends of the cells.

4. Telophase II The final phase of meiosis is telophase II. During telophase II, a nuclear membrane forms around each set of chromatids. The chromatids are again called chromosomes. The cytoplasm divides through cytokinesis, and four haploid cells form.

Visual Check
8. Differentiate How does telophase II differ from telophase I?

____

____

____

Key Concept Check
9. Name the phases of meiosis in order.

____

____

____

____

Math Skills

A proportion is an equation that shows that two ratios are equivalent. If you know that one cell produces two daughter cells at the end of mitosis, you can use proportions to calculate how many daughter cells will be produced by eight cells undergoing mitosis.

Example:

Set up an equation of the two ratios.

$$\frac{1}{2} = \frac{8}{y}$$

Cross-multiply.

$$1 \times y = 8 \times 2$$

$$1y = 16$$

Divide each side by 1.

$$y = 16$$

10. Use Proportions You know that one cell produces four daughter cells at the end of meiosis. If 12 sex cells undergo meiosis, how many daughter cells will be produced?

____

Key Concept Check

11. State why meiosis is important.

____

____

____

Visual Check

12. Compare How many cells are produced during mitosis? During meiosis?

____

____

Why is meiosis important?

Meiosis is important to sexual reproduction. It forms the correct haploid number of chromosomes. This maintains the correct diploid number of chromosomes in organisms when sex cells join. Meiosis also leads to genetic variation.

Maintaining Diploid Cells

Meiosis helps maintain diploid cells in offspring by making haploid sex cells. When haploid sex cells join together during fertilization, they make a diploid zygote, or fertilized egg. The zygote divides by mitosis and cell division and creates a diploid organism.

Creating Haploid Cells

The result of meiosis is haploid sex cells. The formation of haploid cells is important because it results in genetic variation. Sex cells can have different sets of chromosomes, depending on how chromosomes line up during metaphase I. The genetic makeup of offspring is a mixture of chromosomes from two sex cells. Variation in the sex cells results in more genetic variation in the next generation.

How do mitosis and meiosis differ?

The table below lists the differences between mitosis and meiosis.

Characteristic	Meiosis	Mitosis
Number of chromosomes in parent cell	diploid	diploid
Type of parent cell	reproductive	body
Number of divisions of the nucleus	2	1
Number of daughter cells produced	4	2
Chromosome number in daughter cells	haploid	diploid
Function in organism	forms sperm and egg cells	growth, cell repair, some types of reproduction

Advantages of Sexual Reproduction

The main advantage of sexual reproduction is that it results in genetic variation among offspring. Offspring inherit half their DNA from each parent. Inheriting different DNA means that each offspring has a different set of traits.

REVIEW VOCABULARY
DNA
the genetic information in a cell

Genetic Variation

Genetic variation exists among humans. You can look at your friends to see genetic variations. Some people have blue eyes; others have brown eyes. Some people have blonde hair; others have red hair. Genetic variation occurs in all organisms that reproduce sexually.

Because of genetic variation, individuals within a population have slight differences. These differences might be an advantage if the environment changes. Some individuals might have traits that make them able to survive harsh conditions. For example, some plants within a population might be able to survive long periods of dry weather. Sometimes the traits might help keep an organism from getting infected by a disease.

Key Concept Check
13. Identify Why is sexual reproduction beneficial?

Selective Breeding

Selective breeding is a process that involves breeding certain individuals within a population because of the traits they have. For example, a farmer might choose plants with the biggest flowers and stems. These plants would be allowed to reproduce and grow. Over time, the offspring of the plants would all have big flowers and stems. Selective breeding has been used to produce many types of plants and animals with certain traits.

Disadvantages of Sexual Reproduction

Sexual reproduction takes time and energy. Organisms have to grow and develop until they are mature enough to produce sex cells. Before they can reproduce, organisms have to find mates. Searching for a mate takes time and energy. The search might also expose individuals to predators, diseases, or harsh environmental conditions. Sexual reproduction can be limited by certain factors. For example, fertilization cannot take place during pregnancy, which can last as long as two years in some mammals.

Reading Check
14. State the disadvantages of sexual reproduction.

After You Read

Mini Glossary

diploid: cells that have pairs of chromosomes

egg: the female sex cell, which forms in an ovary

fertilization (fur tuh luh ZAY shun): the process in which an egg cell and a sperm cell join together

haploid: cells that have only one chromosome from each pair

homologous (huh MAH luh gus) chromosomes: pairs of chromosomes that have genes for the same traits arranged in the same order

meiosis: the process in which one diploid cell divides and makes four haploid cells

sexual reproduction: a type of reproduction in which the genetic materials from two different cells combine, producing an offspring

sperm: the male sex cell, which forms in a testis

zygote: the new cell that forms from fertilization

1. Review the terms and their definitions in the Mini Glossary. Use at least two words from the Mini Glossary in a sentence to describe the difference between the female and male sex cells.

____

____

2. In the table below, list the advantages and disadvantages of sexual reproduction.

Advantages	Disadvantages

3. How is genetic variation related to meiosis?

____

____

What do you think NOW?

Reread the statements at the beginning of the lesson. Fill in the After column with an A if you agree with the statement or a D if you disagree. Did you change your mind?

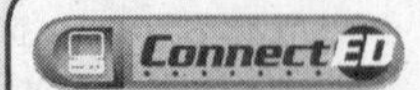

Log on to ConnectED.mcgraw-hill.com and access your textbook to find this lesson's resources.

Reproduction of Organisms

Asexual Reproduction

Before You Read

What do you think? Read the three statements below and decide whether you agree or disagree with them. Place an A in the Before column if you agree with the statement or a D if you disagree. After you've read this lesson, reread the statements to see if you have changed your mind.

Before	Statement	After
	4. Cloning produces identical individuals from one cell.	
	5. All organisms have two parents.	
	6. Asexual reproduction occurs only in microorganisms.	

Key Concepts 

- What is asexual reproduction, and why is it beneficial?
- How do the types of asexual reproduction differ?

Read to Learn

What is asexual reproduction?

In **asexual reproduction,** *one parent organism produces offspring without meiosis and fertilization.* Offspring produced by asexual reproduction inherit all of their DNA from one parent. Therefore, they are genetically the same as each other and their parent.

You have seen the results of asexual reproduction if you have ever seen mold on bread or fruit. Mold is a type of fungus (FUN gus) that can reproduce either sexually or asexually. Asexual reproduction is different from sexual reproduction.

Recall that sexual reproduction involves two parent organisms and the processes of meiosis and fertilization. Offspring inherit half of their DNA from each parent, resulting in genetic variation among the offspring.

Types Of Asexual Reproduction

There are many different types of organisms that reproduce asexually. Not only fungi, but also bacteria, protists, plants, and animals can reproduce asexually.

Study Coach

Discuss Read the first two paragraphs about asexual reproduction. Then take turns with a partner saying something about what you learned. Repeat this process with the other paragraphs in this lesson.

Key Concept Check

1. Describe What is asexual reproduction?

FOLDABLES®

Make the following six-celled chart, then use it to compare types of asexual reproduction.

Fission	Mitotic cell division	Budding
Animal regeneration	Vegetative reproduction	Cloning

Visual Check

2. Recognize What happens to the original cell's chromosome during fission?

Reading Check

3. Evaluate What advantage might asexual reproduction by fission have compared to sexual reproduction?

Fission

Recall that a prokaryotic cell, such as a bacterial cell, has a simpler cell structure than a eukaryotic cell. A prokaryote's DNA is not contained in a nucleus. For this reason, mitosis does not occur. Cell division in a prokaryote is a simpler process than in a eukaryote. *Cell division in prokaryotes that forms two genetically identical cells is known as* ***fission.***

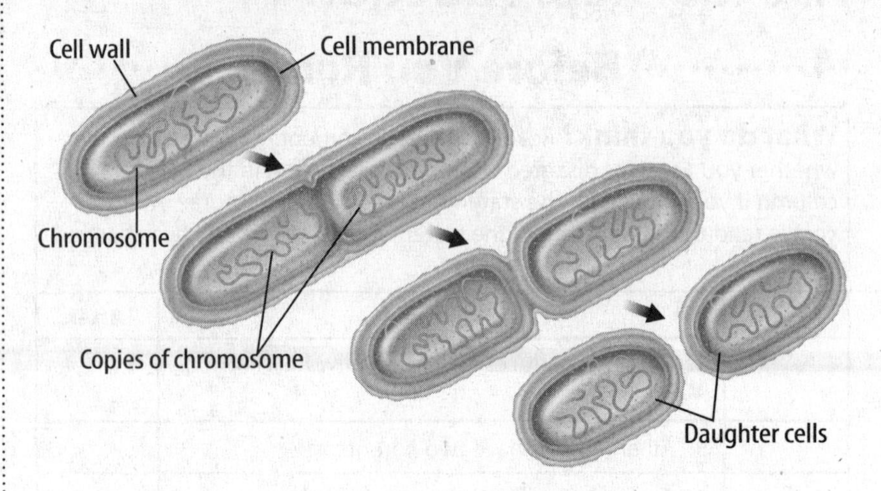

Fission begins when a prokaryote's DNA is copied, as shown in the figure above. Each copy attaches to the cell membrane. Then the cell begins to grow longer. The two copies of DNA are pulled apart. At the same time, the cell membrane starts to pinch inward along the middle of the cell. Finally the cell splits and forms two new identical offspring. The original cell no longer exists. Fission makes it possible for prokaryotes to divide rapidly.

Mitotic Cell Division

Many unicellular eukaryotes, such as amoebas, reproduce by mitotic cell division. In this type of asexual reproduction, an organism forms two offspring through mitosis and cell division. The nucleus of the cell divides by mitosis. Next, the cytoplasm and its contents divide through cytokinesis. Two new amoebas form.

Budding

In ***budding,*** *a new organism grows by mitosis and cell division on the body of its parent.* The bud, or offspring, is genetically identical to its parent. When the bud is large enough, it can break from the parent and live on its own. Organisms such as yeasts, which are fungi, reproduce through budding. Sometimes the bud stays attached to the parent and starts to form a colony. Corals are animals that form colonies through budding.

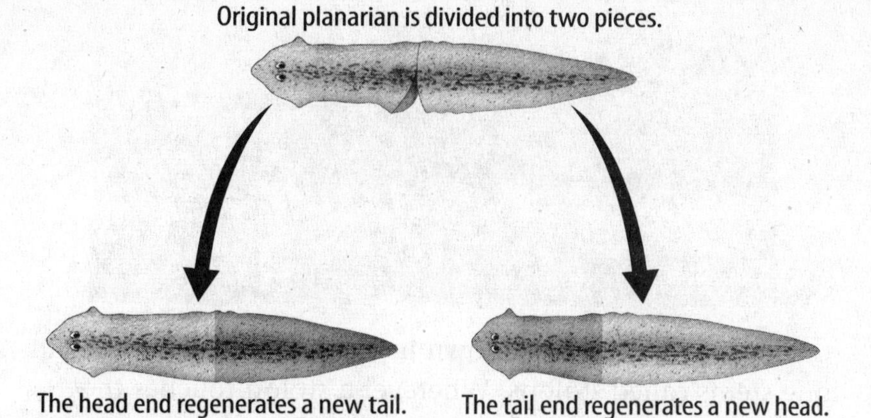

Animal Regeneration

Another type of asexual reproduction, **regeneration,** *occurs when an offspring grows from a piece of its parent.* Animals that can reproduce asexually through regeneration include sponges, sea stars, and planarians.

Producing New Organisms The figure above shows how a planarian reproduces through regeneration. If the planarian is cut into two pieces, each piece of the original planarian becomes a new organism.

If the arms are separated from the parent sea star, each of these arms has the potential to grow into a new organism. To regenerate a new sea star, the arm must have a part of the central disk of the parent. If conditions are right, one five-armed sea star can produce five new organisms. As with all types of asexual reproduction, the offspring are genetically the same as the parent.

Producing New Parts Some animals, such as newts, tadpoles, crabs, hydras, zebra fish, and salamanders, can regenerate a lost or damaged body part. Even humans are able to regenerate some damaged body parts, such as the skin and the liver. This type of regeneration is not considered asexual reproduction. It does not produce a new organism.

Vegetative Reproduction

Plants can also reproduce asexually in a process similar to regeneration. **Vegetative reproduction** *is a form of asexual reproduction in which offspring grow from part of a parent plant.* Strawberries, raspberries, potatoes, and geraniums are other plants that can reproduce this way

Visual Check

4. Describe What happens to a planarian when it is cut into two pieces?

____

____

____

ACADEMIC VOCABULARY
potential
(noun) possibility

Reading Check

5. Specify What is true of all cases of asexual reproduction?

____

____

____

Visual Check

6. Locate Circle the offspring of the strawberry plant in the figure.

The strawberry plant shown in the figure above sends out long stems called stolons. Wherever a stolon touches the ground, it can produce roots. Once a stolon grows roots, a new plant can grow, even if the stolon breaks off from the parent plant. Each new plant grown from a stolen is genetically identical to the parent plant. Roots, leaves, and stems are the structures that usually produce new plants.

Cloning

Cloning *is a type of asexual reproduction performed in laboratories. It produces identical individuals from a cell or from a cluster of cells taken from a multicellular organism.* Farmers and scientists often clone cells or organisms that have desirable traits.

Plant Cloning Some plants can be cloned from just a few cells using a method called a tissue culture. Tissue cultures make it possible for plant growers and scientists to make many copies of a plant with desirable traits. The new plants are genetically the same as the parent plant. Also, cloning produces plants more quickly than vegetative reproduction does.

SCIENCE USE V. COMMON USE
culture
Science Use the process of growing living tissue in a laboratory

Common Use the social customs of a group of people

A plant might be infected with a disease. To clone such a plant, a scientist can use cells from the meristem of the plant. Cells in meristems are disease-free. Therefore, if a plant becomes diseased, it can be cloned using meristem cells.

Think it Over

7. Synthesize Do all cloned plants have the same genetic makeup? Why or why not?

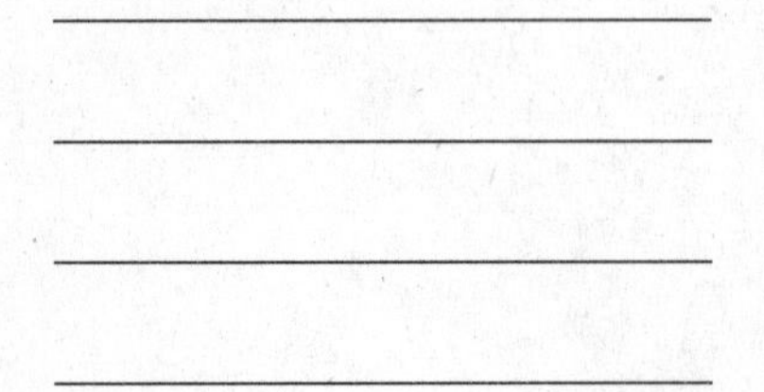

Animal Cloning In addition to cloning plants, scientists have been able to clone many animals. All of a clone's chromosomes come from one parent, the donor of the nucleus. This means that the clone is genetically the same as its parent. The first mammal cloned was a sheep named Dolly.

Steps in Cloning Dolly The first step in cloning Dolly was to remove cells from a sheep, as shown in the figure on the next page. DNA was then removed from an unfertilized egg of a second sheep. In a laboratory, the cells were fused, or combined, and the new cell had the DNA from the first sheep. The cell developed into an embryo. The embryo was then placed in a third sheep. The cloned sheep developed inside the third sheep and was later born.

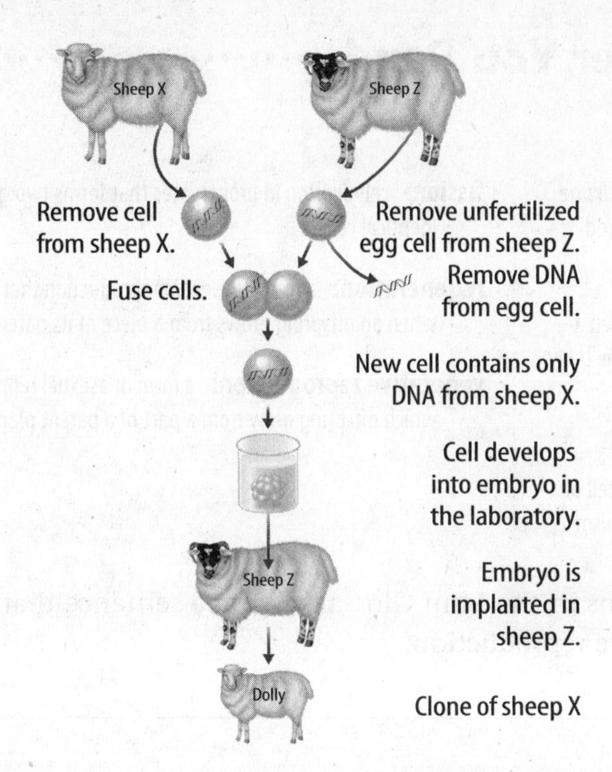

Visual Check

8. Identify Circle the two sheep that are genetically identical.

Cloning Issues Scientists are working to save some endangered species from extinction by cloning. Some people are concerned about the cost and ethical issues of cloning. Ethical issues include the possibility of human cloning.

Key Concept Check

9. Compare and contrast the different types of asexual reproduction.

Advantages of Asexual Reproduction

One advantage of asexual reproduction is that an organism can reproduce without a mate. Recall that finding a mate takes time and energy. Another advantage is that some organisms can quickly produce a large number of offspring. For example, crabgrass reproduces by underground stolons. This enables one plant to spread and colonize an area in a short period of time.

Key Concept Check

10. State the advantages of asexual reproduction.

Disadvantages of Asexual Reproduction

Asexual reproduction produces offspring that are genetically the same as the parent. This results in little genetic variation within a population. Genetic variation can give organisms a better chance of surviving if the environment changes. Imagine that all of the crabgrass plants in a lawn are genetically the same. If a weed killer can kill the parent plant, then it can kill all of the crabgrass plants in the lawn. This might be good for the lawn, but it is a disadvantage for the crabgrass. Another disadvantage involves genetic changes called mutations. A harmful mutation passed to asexually reproduced offspring could affect the offspring's ability to survive.

After You Read

Mini Glossary

asexual reproduction: a form of reproduction in which one parent organism produces offspring without meiosis and fertilization

budding: a form of asexual reproduction that occurs when a new organism grows by mitosis and cell division on the body of its parent

cloning: a type of asexual reproduction performed in a laboratory that produces identical individuals from a cell or from a cluster of cells taken from a multicellular organism

fission: cell division in prokaryotes that forms two genetically identical cells

regeneration: a form of asexual reproduction that occurs when an offspring grows from a piece of its parent

vegetative reproduction: a form of asexual reproduction in which offspring grow from a part of a parent plant

1. Review the terms and their definitions in the Mini Glossary. Write a sentence that compares regeneration and vegetative reproduction.

____

____

2. Fill in the spider map below with the different types of asexual reproduction. Use terms from the Mini Glossary.

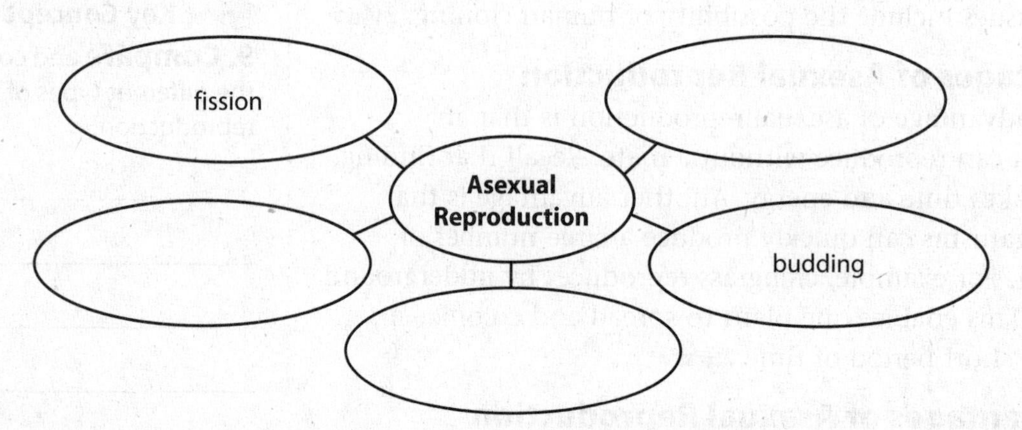

3. How did discussing what you learned from each paragraph with another student help you learn about asexual reproduction?

____

____

What do you think NOW?

Reread the statements at the beginning of the lesson. Fill in the After column with an A if you agree with the statement or a D if you disagree. Did you change your mind?

Log on to ConnectED.mcgraw-hill.com and access your textbook to find this lesson's resources.

Genetics

Mendel and His Peas

Before You Read

What do you think? Read the two statements below and decide whether you agree or disagree with them. Place an A in the Before column if you agree with the statement or a D if you disagree. After you've read this lesson, reread the statements to see if you have changed your mind.

Before	Statement	After
	1. Like mixing paints, parents' traits always blend in their offspring.	
	2. If you look more like your mother than you look like your father, then you received more traits from your mother.	

Key Concepts 

- Why did Mendel perform cross-pollination experiments?
- What did Mendel conclude about inherited traits?
- How do dominant and recessive factors interact?

Read to Learn

Early Ideas About Heredity

Have you ever mixed two different colors of paint to make a new color? Long ago, people thought that an organism's characteristics, or traits, were determined in the same way that paint colors can be mixed. People assumed this because offspring often resemble both parents. This is known as blending inheritance.

Today, scientists know that heredity (huh REH duh tee) is more complex. **Heredity** *is the passing of traits from parents to offspring.* For example, you and your brother might have blue eyes but both of your parents have brown eyes. How does this happen?

More than 150 years ago, Gregor Mendel, an Austrian monk, performed experiments that helped answer many questions about heredity. The results of his experiments also disproved the idea of blending inheritance.

Mendel's research into the questions of heredity gave scientists a basic understanding of genetics. **Genetics** (juh NE tihks) *is the study of how traits are passed from parents to offspring.* Because of his research, Mendel is known as the father of genetics. ✓

Study Coach

Vocabulary Quiz Write a question about each vocabulary term in this lesson. Exchange quizzes with a partner. After completing the quizzes, discuss the answers with your partner.

✓ **Reading Check**

1. Define What is genetics?

______

______

______

Mendel's Experimental Methods

During the 1850s, Mendel studied genetics by doing controlled breeding experiments with pea plants. Pea plants were ideal for genetics studies because

- they reproduce quickly. Mendel was able to grow many plants and collect a lot of data.
- they have easily observed traits, such as flower color and pea shape. Mendel was able to observe whether or not a trait was passed from one generation to the next.
- Mendel could control which pairs of plants reproduced. He was able to find out which traits came from which plant pairs.

Think it Over

2. Explain In his breeding experiments, how did Mendel know which traits came from which pair of plants?

Pollination in Pea Plants

To observe how a trait was inherited, Mendel controlled which plants pollinated other plants. Pollination occurs when pollen lands on the pistil of a flower. Sperm cells from the pollen then fertilize egg cells in the pistil.

REVIEW VOCABULARY

sperm
a haploid sex cell formed in the male reproductive organs

egg
a haploid sex cell formed in the female reproductive organs

Self-pollination occurs when pollen from one plant lands on the pistil of a flower on the same plant. Cross-pollination occurs when pollen from one plant reaches the pistil of a flower on a different plant. Mendel allowed one group of flowers to self-pollinate. With another group, he cross-pollinated the plants himself.

True-Breeding Plants

Mendel began his experiments with plants that were true-breeding for the trait that he would test. When a true-breeding plant self-pollinates, it always produces offspring with traits that match the parent. For example, when a true-breeding pea plant with wrinkled seeds self-pollinates, it produces only plants with wrinkled seeds. In fact, it will produce wrinkled seeds generation after generation.

Mendel's Cross-Pollination

By cross-pollinating plants himself, Mendel was able to select which plants pollinated other plants. Mendel cross-pollinated hundreds of plants for each set of traits he wanted to learn more about. The traits included flower color (purple or white), seed color (green or yellow), and seed shape (round or wrinkled).

Key Concept Check

3. Explain Why did Mendel perform cross-pollination experiments?

With each cross-pollination, Mendel recorded the traits that appeared in the offspring. By testing such a large number of plants, Mendel was able to predict which crosses would produce which traits.

Mendel's Results

Once Mendel had enough true-breeding plants for a trait that he wanted to test, he cross-pollinated selected plants. His results are described below.

First-Generation Crosses

Crosses between true-breeding plants with purple flowers produced true-breeding plants with only purple flowers. Crosses between true-breeding plants with white flowers produced true-breeding plants with only white flowers. However, when Mendel crossed true-breeding plants with purple flowers and true-breeding plants with white flowers, all of the offspring had purple flowers.

New Questions Raised

Why did crossing plants with purple flowers and plants with white flowers always produce offspring with purple flowers? Why were there no white flowers? Why didn't the cross produce offspring with pink flowers—a combination of white and purple? Mendel carried out more experiments to answer these questions. ✓

Second-Generation (Hybrid) Crosses

Mendel's first-generation purple-flowering plants are called hybrid plants. They came from true-breeding parent plants with different forms of the same trait. When Mendel cross-pollinated two purple-flowering hybrid plants, some of the offspring had white flowers. The trait that had disappeared in the first-generation always reappeared in the second-generation.

Mendel got similar results each time he cross-pollinated hybrid plants. For example, a true-breeding yellow-seeded pea plant crossed with a true-breeding green-seeded pea plant always produced yellow-seeded hybrids. A second-generation cross of two yellow-seeded hybrids always produced plants with yellow seeds and plants with green seeds.

More Hybrid Crosses

Mendel cross-pollinated many hybrid plants. He counted and recorded the traits of offspring. He analyzed these data and noticed patterns. In crosses between hybrid plants with purple flowers, the ratio of purple flowers to white flowers was about 3:1. This means that purple-flowering pea plants grew from this cross three times more often than white-flowering pea plants grew from the cross. Mendel calculated similar ratios for all seven traits that he tested.

FOLDABLES®

Make a two-tab book and organize your notes on dominant and recessive factors.

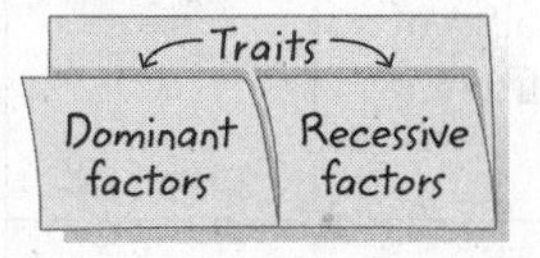

✓ Reading Check

4. Predict the offspring of a cross between two true-breeding pea plants with smooth seeds.

Math Skills

A ratio is a comparison of two numbers or quantities by division. For example, the ratio comparing 6,022 yellow seeds to 2,001 green seeds can be written as follows:

6,022 to 2,001 or
6,022:2,001 or
$\frac{6,022}{2,001}$

To simplify the ratio, divide the first number by the second number.

$$\frac{6,022}{2,001} = \frac{3}{1} = 3:1$$

5. Use Ratios A science class has 14 girls and 7 boys. Simplify the ratio.

Results of Hybrid Crosses

Characteristics	Flower Color	Flower Position	Seed Color	Seed Shape	Pod Shape	Pod Color	Stem Length
Dominant Trait; # of Offspring	Purple; 705	Axial; 651	Yellow; 6022	Round; 5474	Smooth; 882	Green; 428	Long; 781
Recessive Trait: # of Offspring	White; 224	Terminal; 207	Green; 2001	Wrinkled; 1850	Bumpy; 299	Yellow; 152	Short; 277
Ratio	3.15 : 1	3.14 : 1	3.01 : 1	2.96 : 1	2.95 : 1	2.82 : 1	2.84 : 1

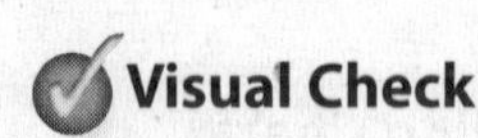
Visual Check

6. Predict If a cross between two hybrid plants with purple flowers produced 12 offspring, how many offspring would you expect to have purple flowers?

Key Concept Check
7. Summarize What did Mendel conclude about inherited traits?

Key Concept Check
8. Describe How do dominant and recessive factors interact?

Mendel's Conclusions

After analyzing the results of his experiments, Mendel concluded that two factors control each inherited trait. He also proposed that when organisms reproduce, the sperm and the egg each contribute one factor for each trait. Mendel's results are shown in the table above.

Dominant and Recessive Traits

Recall that when Mendel cross-pollinated a true-breeding plant with purple flowers and a true-breeding plant with white flowers, the hybrid offspring had only purple flowers. He hypothesized that the hybrid offspring had one genetic factor for purple flowers and one genetic factor for white flowers. But why were there no white flowers? Mendel also hypothesized that the purple factor was dominant, blocking the white factor. *A genetic factor that blocks another genetic factor is called a* **dominant** (DAH muh nunt) **trait.** A dominant trait, such as purple pea flowers, is seen when offspring have either one or two dominant factors. *A genetic factor that is blocked by the presence of a dominant factor is called a* **recessive** (rih SE sihv) **trait.** A recessive trait, such as white pea flowers, is seen only when two recessive genetic factors are present in offspring.

From Parents to Second Generation

For the second generation, Mendel cross-pollinated two hybrids that had purple flowers. About 75 percent of the second-generation plants had purple flowers. These plants had at least one dominant factor. Twenty-five percent of the second-generation plants had white flowers. These plants had the same two recessive factors.

After You Read

Mini Glossary

dominant (DAH muh nunt) trait: a genetic factor that blocks another genetic factor

genetics (juh NE tihks): the study of how traits are passed from parents to offspring

heredity (huh REH duh tee): the passing of traits from parents to offspring

recessive (rih SE sihv) trait: a genetic factor that is blocked by the presence of a dominant factor

1. Review the terms and their definitions in the Mini Glossary. Write one or two sentences that compare and contrast dominant traits and recessive traits.

2. The tables below show a sequence of crosses for the trait of pod color in a type of plant. Study the tables and fill in the trait or traits that the second-generation cross would produce in the offspring.

First-Generation Cross	
Plants Crossed	**Offspring**
true-breeding green-pod × true-breeding yellow-pod	all green-pod hybrids

Second-Generation (Hybrid) Cross	
Plants Crossed	**Offspring**
green-pod hybrid × green-pod hybrid	

3. Why were the pea plants that Mendel used in his experiments a good choice for genetics studies?

What do you think NOW?

Reread the statements at the beginning of the lesson. Fill in the After column with an A if you agree with the statement or a D if you disagree. Did you change your mind?

Log on to ConnectED.mcgraw-hill.com and access your textbook to find this lesson's resources.

Genetics

Understanding Inheritance

Key Concepts 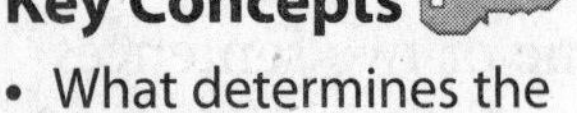

- What determines the expression of traits?
- How can inheritance be modeled?
- How do some patterns of inheritance differ from Mendel's model?

Before You Read

What do you think? Read the two statements below and decide whether you agree or disagree with them. Place an A in the Before column if you agree with the statement or a D if you disagree. After you've read this lesson, reread the statements to see if you have changed your mind.

Before	Statement	After
	3. All inherited traits follow Mendel's patterns of inheritance.	
	4. Scientists have tools to predict the form of a trait an offspring might inherit.	

Study Coach

Build Vocabulary Skim this lesson and circle any words you do not know. After you've read the lesson, review the circled words. Look up the definitions in the dictionary for any words you cannot define.

Read to Learn

What controls traits?

Mendel concluded that two factors control each trait. One factor comes from the egg cell and one factor comes from the sperm cell. What are these factors? How are they passed from parents to offspring?

Chromosomes

Inside each cell is a nucleus that has threadlike structures called chromosomes. Chromosomes contain genetic information that controls traits. What Mendel called "factors" are parts of chromosomes. Each cell in an offspring contains chromosomes from both parents. These chromosomes exist in pairs—one chromosome from each parent.

Genes and Alleles

Each chromosome can have information about hundreds or thousands of traits. *A* **gene** (JEEN) *is a section on a chromosome that has genetic information for one trait.* For example, a gene of a pea plant might have information about flower color. An offspring inherits two genes (factors) for each trait, one from each parent. The genes can be the same or different, such as purple or white for pea flower color. *The different forms of a gene are called* **alleles** (uh LEELS). Pea plants can have two purple alleles, two white alleles, or one of each allele. ✓

 **Reading Check**

1. Identify How many alleles controlled flower color in Mendel's experiments?

Genotype and Phenotype

Geneticists call how a trait appears, or is expressed, the trait's **phenotype** (FEE nuh tipe). A person's eye color is an example of phenotype. The trait of eye color can be expressed as blue, brown, green, or other colors.

Mendel concluded that two alleles control the expression or phenotype of each trait. *The two alleles that control the phenotype of a trait are called the trait's* **genotype** (JEE nuh tipe). You cannot see an organism's genotype. But you can make guesses about a genotype based on its phenotype. For example, you have already learned that a pea plant with white flowers has two recessive alleles for that trait. These two alleles are its genotype. The white flower is its phenotype.

Symbols for Genotype Scientists use symbols to represent the alleles in a genotype. The table below shows the possible genotypes for both round and wrinkled seed phenotypes. Uppercase letters represent dominant alleles and lowercase letters represent recessive alleles. The dominant allele, if present, is written first.

The round pea seed can have either of these two genotypes—*RR* or *Rr*. Both genotypes have a round phenotype. *Rr* results in round seeds because the round allele *(R)* is dominant to the wrinkled allele *(r)*.

The wrinkled pea can have only one genotype—*rr*. The wrinkled phenotype is possible only when two recessive alleles *(rr)* are present in the genotype.

Homozygous and Heterozygous *When the two alleles of a gene are the same, its genotype is* **homozygous** (hoh muh ZI gus). Both *RR* and *rr* are homozygous genotypes. The *RR* genotype has two dominant alleles. The *rr* genotype has two recessive alleles.

If the two alleles of a gene are different, its genotype is **heterozygous** (he tuh roh ZI gus). *Rr* is a heterozygous genotype. It has one dominant and one recessive allele.

Phenotype and Genotype

Phenotype (observed traits)	Genotype (alleles of a gene)
Round	Homozygous dominant (*RR*)
	Heterozygous (*Rr*)
Wrinkled	Homozygous recessive (*rr*)

Think it Over

2. Draw Conclusions In a pea plant, an allele for a tall stem is dominant to an allele for a short stem. You see a pea plant with a short stem. What can you conclude about the genotype of this plant?

Key Concept Check

3. Explain How do alleles determine the expression of traits?

Interpreting Tables

4. Identify the genotype of a plant that produces a wrinkled pea.

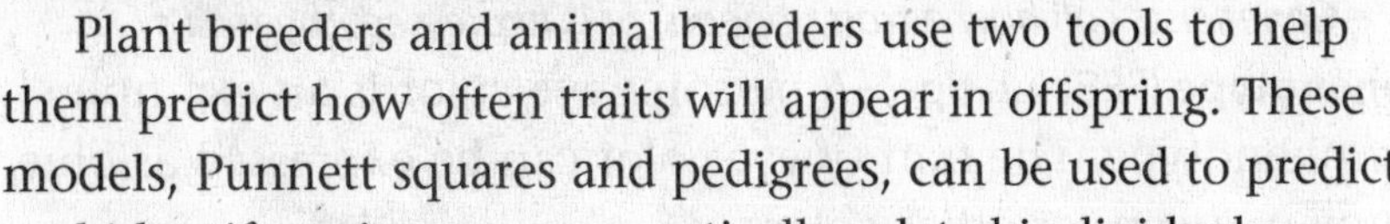

Modeling Inheritance

Plant breeders and animal breeders use two tools to help them predict how often traits will appear in offspring. These models, Punnett squares and pedigrees, can be used to predict and identify traits among genetically related individuals.

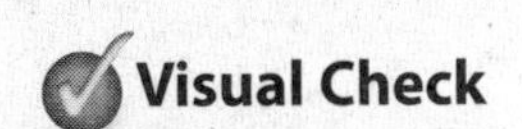

Visual Check

5. Specify What phenotypes are possible for pea offspring of this cross?

Punnett Squares

If the genotypes of the parents are known, then the different genotypes and phenotypes of the offspring can be predicted. *A* **Punnett square** *is a model used to predict possible genotypes and phenotypes of offspring.* Follow the steps shown below to learn how to make a Punnett square.

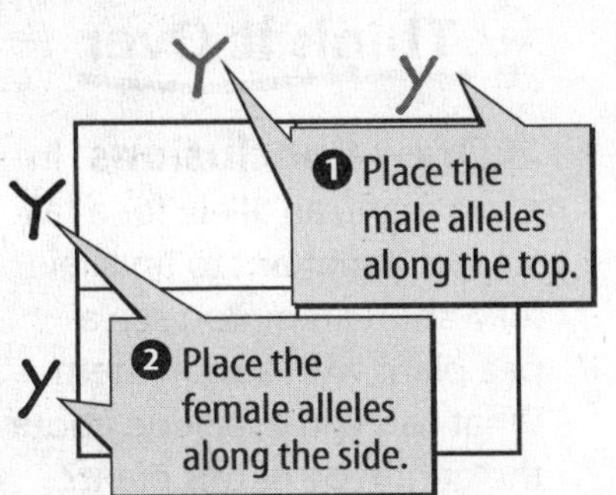

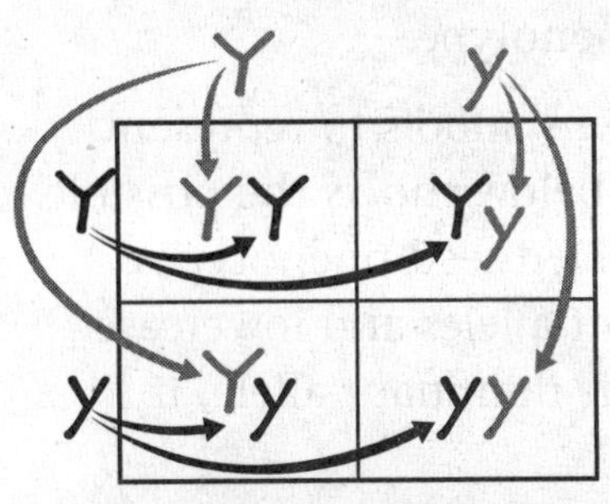

3 Copy female alleles across each row. Copy male alleles down each column.

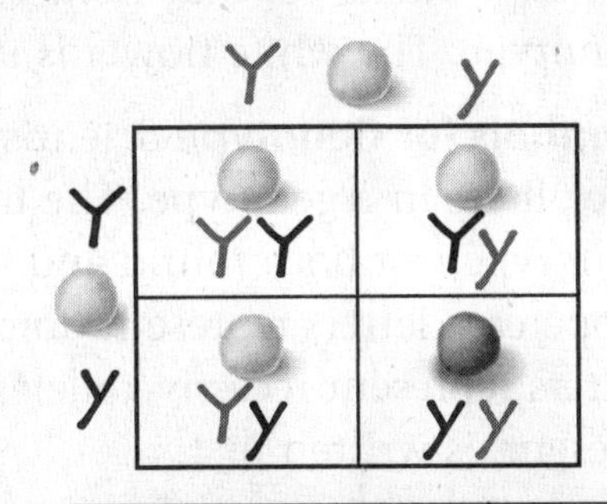

Analysis The ratio of phenotypes is 3:1, yellow:green. The ratio of genotypes is 1:2:1, *YY:Yy:yy.*

Analyzing a Punnett Square

The figure shows a cross between two pea plants that are heterozygous for seed color—*Yy* and *Yy*. Yellow is the dominant allele—*Y*. Green is the recessive allele—*y*. The offspring can have one of three genotypes—*YY*, *Yy*, or *yy*. The ratio of genotypes is written as 1:2:1.

Because *YY* and *Yy* represent the same phenotype (yellow), the offspring can have one of only two phenotypes—yellow or green. The ratio of phenotypes is written 3:1. About 75 percent of the offspring of the cross between two heterozygous plants will produce yellow seeds. About 25 percent of the plants will produce green seeds.

Think it Over

6. Apply If the cross shown in the figure produced 100 offspring, about how many offspring would have green peas?

Using Ratios to Predict

A 3:1 ratio means that an offspring of heterozygous parents has a 3:1 chance of having yellow seeds. It does not mean that any group of four offspring will have three plants with yellow seeds and one with green seeds. This is because one offspring does not affect the phenotype of other offspring. But if you examine large numbers of offspring from a particular cross, as Mendel did, the overall ratio will be close to the ratio predicted by a Punnett square.

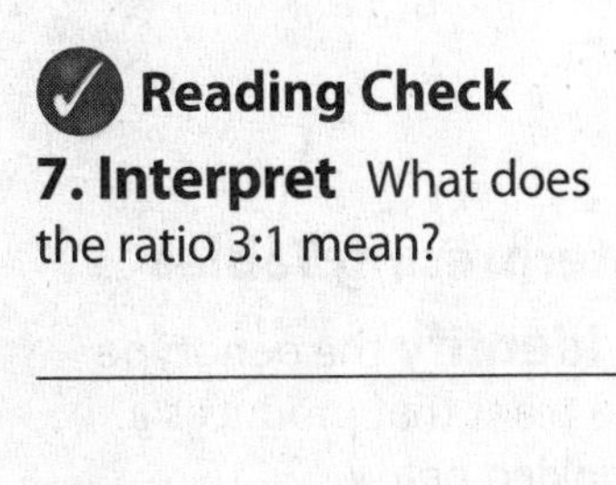

Reading Check

7. Interpret What does the ratio 3:1 mean?

Pedigrees

Another model that can show inherited traits is a pedigree. A pedigree shows phenotypes of genetically related family members. It can also help determine genotypes. In the pedigree shown below, three offspring have a trait—attached earlobes—that the parents do not have. If these offspring received one allele for this trait from each parent but neither parent displays this trait, the offspring must have received two recessive alleles. If either allele was dominant, the offspring would have the dominant phenotype—unattached earlobes.

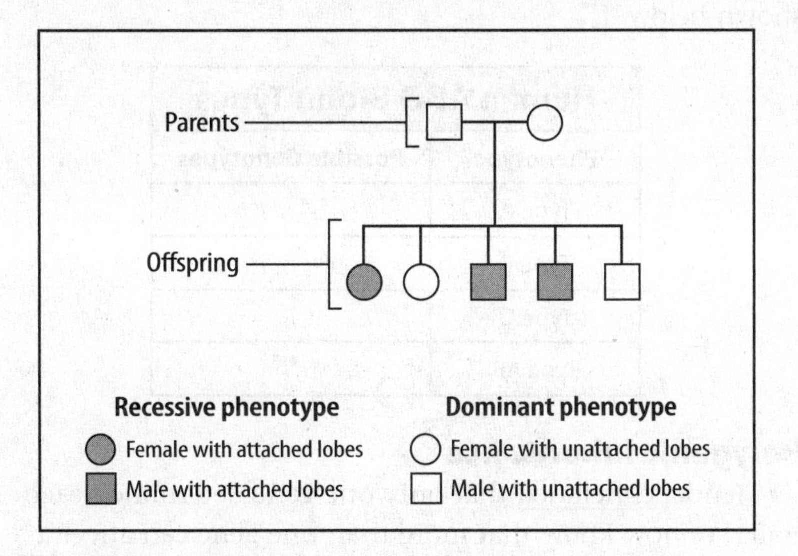

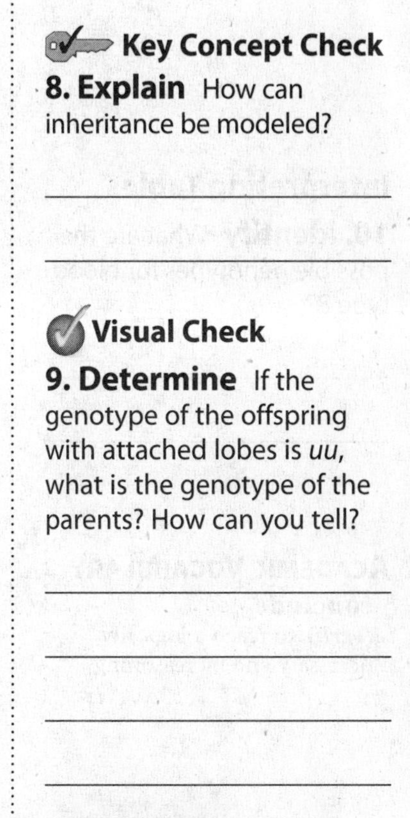

Key Concept Check

8. Explain How can inheritance be modeled?

Visual Check

9. Determine If the genotype of the offspring with attached lobes is *uu*, what is the genotype of the parents? How can you tell?

Complex Patterns of Inheritance

Mendel studied traits influenced by only one gene with two alleles. We know now that not all traits are inherited this way. Some traits have more complex inheritance patterns.

Types of Dominance

Recall that in pea plants, the presence of one dominant allele produces a dominant phenotype. However, not all allele pairs have a dominant-recessive interaction.

Incomplete Dominance Sometimes traits appear to be blends of alleles. *Alleles show* **incomplete dominance** *when the offspring's phenotype is a blend of the parents' phenotypes*. For example, a pink camellia flower results from incomplete dominance. A cross between a camellia plant with white flowers and a camellia plant with red flowers produces only camellia plants with pink flowers.

Codominance *When both alleles can be observed in a phenotype, this type of interaction is called* **codominance.** For example, if a cow inherits the allele for white coat color from one parent and the allele for red coat color from the other parent, the cow will have both red and white hairs.

FOLDABLES®

Make a layered book and organize your notes on inheritance patterns.

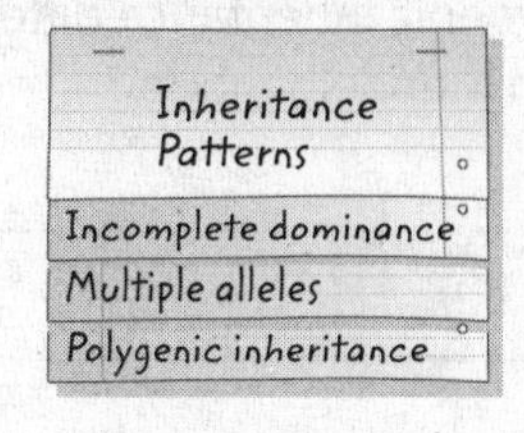

Multiple Alleles

Unlike the genes in Mendel's pea plants, some genes have more than two alleles, or multiple alleles. Human ABO blood type is an example of a trait that is determined by multiple alleles. There are three alleles for the ABO blood type—I^A, I^B, and *i*. The way the alleles combine results in one of four blood types—A, B, AB, or O. The I^A and I^B alleles are codominant to each other, but they both are dominant to the *i* allele. Even though there are multiple alleles, a person inherits only two of these alleles, one from each parent, as shown below.

Interpreting Tables

10. Identify What are the possible genotypes for blood type B?

Human ABO Blood Types	
Phenotype	**Possible Genotypes**
Type A	I^AI^A or I^Ai
Type B	I^BI^B or I^Bi
Type O	ii
Type AB	I^AI^B

Polygenic Inheritance

Mendel concluded that only one gene determined each trait. We now know that more than one gene can affect a trait. **Polygenic inheritance** *occurs when multiple genes determine the phenotype of a trait.* Because several genes determine a trait, many alleles affect the phenotype even though each gene has only two alleles. Therefore, polygenic inheritance has many possible phenotypes. Eye color in humans is an example of polygenic inheritance. Polygenic inheritance also determines the human characteristics of height, weight, and skin color.

ACADEMIC VOCABULARY
conclude
(verb) to reach a logically necessary end by reasoning

Key Concept Check

11. Explain How does polygenic inheritance differ from Mendel's model?

Genes and the Environment

Recall that an organism's genotype determines its phenotype. However, genes are not the only factors that can affect phenotypes. An organism's environment can also affect its phenotype. For example, the flower color of one type of hydrangea is determined by the soil in which the hydrangea plant grows. Acidic soil produces blue flowers. Basic, or alkaline, soil produces pink flowers.

For humans, healthful choices can also affect phenotype. Many genes affect a person's chances of having heart disease. However, what a person eats and the amount of exercise he or she gets can influence whether heart disease will develop.

After You Read

Mini Glossary

allele (uh LEEL): any of the different forms of a gene

codominance: occurs when both alleles can be observed in a phenotype

gene (JEEN): a section on a chromosome that has genetic information for one trait

genotype (JEE nuh tipe): the two alleles that control the phenotype of a trait

heterozygous (he tuh roh ZI gus): an organism's genotype when the two alleles of a gene are different

homozygous (hoh muh ZI gus): an organism's genotype when the two alleles of a gene are the same

incomplete dominance: occurs when an offspring's phenotype is a blend of the parents' phenotypes

phenotype (FEE nuh tipe): how a trait appears or is expressed

polygenic inheritance: occurs when multiple genes determine the phenotype of a trait

Punnett square: a model used to predict possible genotypes and phenotypes of offspring

1. Review the terms and their definitions in the Mini Glossary. Compare and contrast genotype and phenotype.

____

____

2. Complete the Punnett square below. Predict the genotypes of offspring produced by crossing a heterozygous pea plant with round seeds and a homozygous pea plant with wrinkled seeds. Round (*R*) is dominant to wrinkled (*r*).

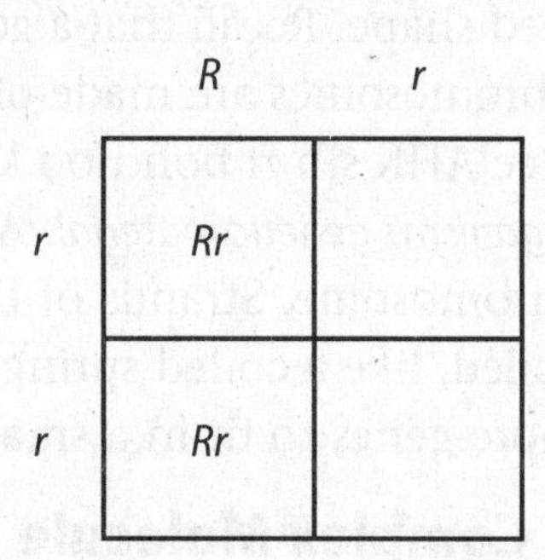

3. Predict the ratio of round phenotype to wrinkled phenotype in the offspring from the cross in Question 2.

____

What do you think NOW?

Reread the statements at the beginning of the lesson. Fill in the After column with an A if you agree with the statement or a D if you disagree. Did you change your mind?

Log on to ConnectED.mcgraw-hill.com and access your textbook to find this lesson's resources.

Genetics

DNA and Genetics

Key Concepts

- What is DNA?
- What is the role of RNA in protein production?
- How do changes in the sequence of DNA affect traits?

Before You Read

What do you think? Read the two statements below and decide whether you agree or disagree with them. Place an A in the Before column if you agree with the statement or a D if you disagree. After you've read this lesson, reread the statements to see if you have changed your mind.

Before	Statement	After
	5. Any condition present at birth is genetic.	
	6. A change in the sequence of an organism's DNA always changes the organism's traits.	

Study Coach

Asking Questions As you read the lesson, write a question for each paragraph. Answer the question using information from the paragraph.

Read to Learn

The Structure of DNA

Cells put molecules together by following a set of directions. Genes provide the directions for a cell to put together molecules that express traits, such as eye color or seed shape. Recall that a gene is a section of a chromosome. Chromosomes are made of proteins and deoxyribonucleic (dee AHK sih ri boh noo klee ihk) acid, or DNA. **DNA** *is an organism's genetic material.* A gene is a segment of DNA on a chromosome. Strands of DNA in a chromosome are tightly coiled, like a coiled spring. This coiling makes it possible for more genes to fit in a small space.

Key Concept Check

1. Explain What is DNA?

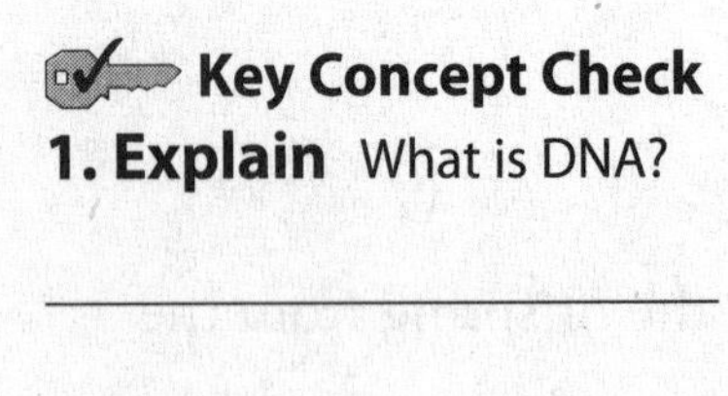

A Complex Molecule

The shape of DNA is like a twisted ladder. It is called a double helix. You can see a double helix in the figure on the next page. How did scientists discover the shape of DNA? Rosalind Franklin and Maurice Wilkins used X-rays to study DNA. Some of the X-rays showed that DNA has a helix shape. Another scientist, James Watson, saw one of the DNA X-rays. Watson worked with Francis Crick to build a model of DNA. They used information from the X-rays and chemical information about DNA discovered by another scientist, Erwin Chargaff. Eventually, Watson and Crick were able to build a model that showed how smaller molecules of DNA bond together and form a double helix.

Four Nucleotides Shape DNA

DNA has a twisted-ladder shape that is caused by molecules called nucleotides. *A* **nucleotide** *is a molecule made of a nitrogen base, a sugar, and a phosphate group.* Sugar-phosphate groups form the sides of the DNA ladder. The nitrogen bases bond and form the rungs of the ladder. There are four nitrogen bases: adenine (A), cytosine (C), thymine (T), and guanine (G). A and T always bond together, and C and G always bond together. The figure below shows how the sugar-phosphate groups and the nitrogen bases form the twisted DNA shape. ✓

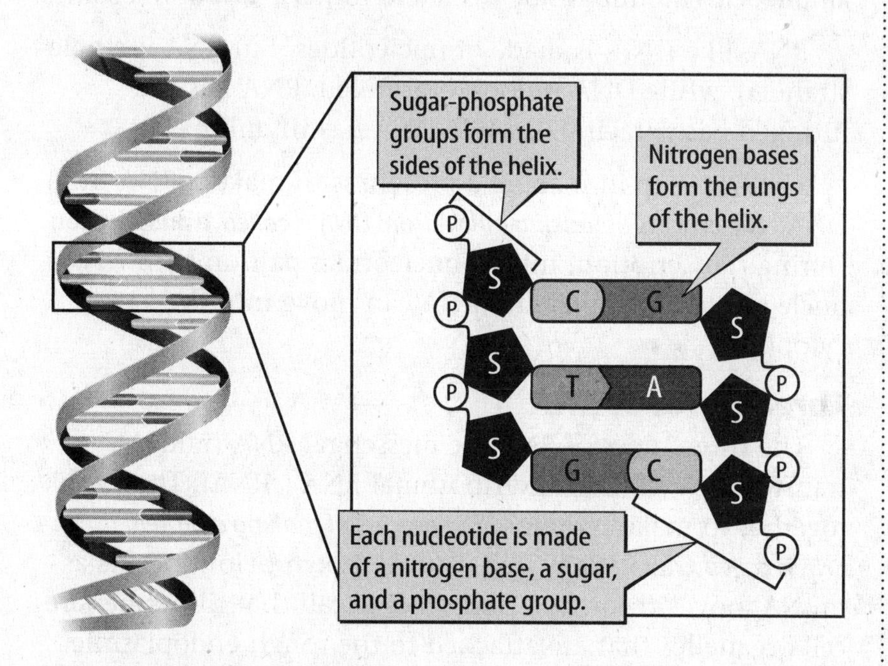

✓ Reading Check

2. Define What is a nucleotide?

✓ Visual Check

3. Identify What forms the rungs of the DNA double helix?

How DNA Replicates

Cells contain DNA in chromosomes. So, every time a cell divides, all chromosomes must be copied for the new cell. The new DNA is identical to existing DNA. **Replication** *is the process of copying a DNA molecule to make another DNA molecule.*

In the first part of replication, the strands separate in many places and the nitrogen bases are exposed. Nucleotides move into place and form new nitrogen base pairs. This produces two identical strands of DNA. ✓

✓ Reading Check

4. Describe What is replication?

Making Proteins

Proteins are important for every cellular process. The DNA of each cell carries a complete set of genes that provides instructions for making all the proteins a cell needs. Most genes contain instructions for making proteins. Some genes contain instructions for when and how quickly proteins are made.

Make a vertical three-tab book and record information about the three types of RNA and their functions.

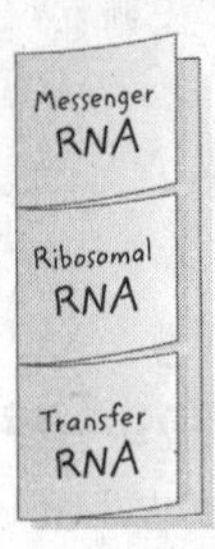

Junk DNA

All genes are segments of DNA on a chromosome. However, about 97 percent of DNA on human chromosomes is not part of any gene. Segments of DNA that are not parts of genes are often called junk DNA. It is not known whether junk DNA has functions that are important to cells.

The Role of RNA in Making Proteins

Proteins are made with the help of ribonucleic acid. **Ribonucleic acid (RNA)** *is a type of nucleic acid that carries the code for making proteins from the nucleus to the cytoplasm.* RNA also carries amino acids around inside a cell and forms a part of ribosomes.

RNA, like DNA, is made of nucleotides. But RNA is single-stranded, while DNA is double-stranded. RNA has the nitrogen base uracil (U), while DNA has thymine (T).

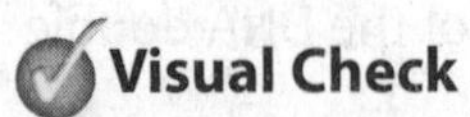

Key Concept Check

5. Explain What is the role of RNA in protein production?

The first step in making a protein is to make mRNA from DNA. *The process of making mRNA from DNA is called* **transcription.** During transcription, mRNA nucleotides pair up with DNA nucleodtides. Completed mRNA can move into the cytoplasm.

Three Types of RNA

Visual Check

6. Explain What is the role of rRNA during translation?

The three types of RNA are messenger RNA (mRNA), transfer RNA (tRNA), and ribosomal RNA (rRNA). They work together to make proteins. *The process of making a protein from RNA is called* **translation.** Translation, shown below, occurs as mRNA moves through a ribosome. Recall that ribosomes are cell organelles that are attached to the rough endoplasmic reticulum (rough ER).

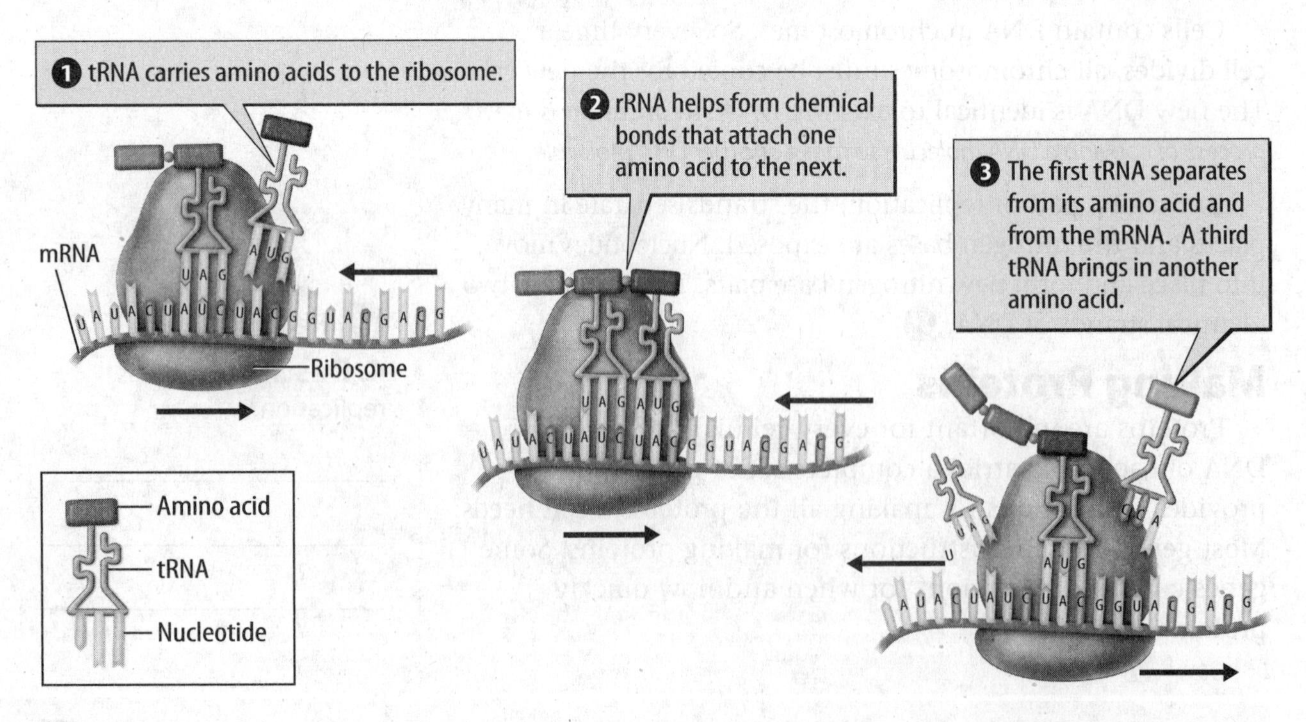

Translating the RNA Code

Making a protein from mRNA is like using a secret code. Proteins are made of amino acids. The order of the nitrogen bases in mRNA determines the order of the amino acids in a protein. Three nitrogen bases on mRNA form the code for one amino acid.

Each series of three nitrogen bases on mRNA is called a codon. There are 64 codons, but only 20 amino acids. Some of the codons code for the same amino acid. One of the codons codes for an amino acid that is the beginning of a protein. This codon signals that translation should start. Three of the codons do not code for any amino acid. Instead, they code for the end of a protein. They signal that translation should stop. ✓

Mutations

A change in the nucleotide sequence of a gene is called a **mutation.** Sometimes, mistakes happen during replication. Most mistakes are corrected before replication is finished. An uncorrected mistake can result in a mutation. Mutations can be caused by exposure to X-rays, ultraviolet light, radioactive materials, and some kinds of chemicals.

Types of Mutations

There are several types of DNA mutations. In a deletion mutation, one or more nitrogen bases are left out of the DNA sequence. In an insertion mutation, one or more nitrogen bases are added to the DNA. In a substitution mutation, one nitrogen base is replaced by a different nitrogen base.

Each type of mutation changes the sequence of nitrogen base pairs. A change can cause a mutated gene to code for a protein that is different from a normal gene. Some mutated genes do not code for any protein. For example, a cell might lose the ability to make one of the proteins it needs.

Results of a Mutation

The effects of a mutation depend on where in the DNA sequence the mutation happens and the type of mutation. Proteins express traits. Because mutations can change proteins, they can cause traits to change. Some mutations in human DNA cause genetic disorders. With more research, scientists hope to find cures and treatments for genetic disorders.

Not all mutations have negative effects. Some mutations do not change proteins, so they do not affect traits. Other mutations can cause a trait to change in a way that benefits an organism.

Reading Check

7. Define What is a codon?

Think it Over

8. Specify What are some causes of mutations?

Key Concept Check

9. Explain How do changes in the sequence of DNA affect traits?

After You Read

Mini Glossary

DNA: an organism's genetic material

mutation: a change in the nucleotide sequence of a gene

nucleotide: a molecule made of a nitrogen base, a sugar, and a phosphate group

replication: the process of copying a DNA molecule to make another DNA molecule

RNA: a type of nucleic acid that carries the code for making proteins from the nucleus to the cytoplasm

transcription: the process of making mRNA from DNA

translation: the process of making a protein from RNA

1. Review the terms and their definitions in the Mini Glossary. Write as many sentences as you need to explain how DNA and RNA are related.

2. Under each heading in the chart below, explain what happens during each type of mutation.

Deletion	Insertion	Substitution
	one or more nitrogen bases are added to the DNA sequence	

3. How did writing a question for each paragraph and then finding the answer in the paragraph help you learn about DNA and genetics?

What do you think NOW?

Reread the statements at the beginning of the lesson. Fill in the After column with an A if you agree with the statement or a D if you disagree. Did you change your mind?

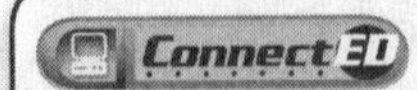

Log on to ConnectED.mcgraw-hill.com and access your textbook to find this lesson's resources.

The Environment and Change Over Time

Fossil Evidence of Evolution

Before You Read

What do you think? Read the two statements below and decide whether you agree or disagree with them. Place an A in the Before column if you agree with the statement or a D if you disagree. After you've read this lesson, reread the statements to see if you have changed your mind.

Before	Statement	After
	1. Original tissues can be preserved as fossils.	
	2. Organisms become extinct only in mass extinction events.	

Key Concepts 

- How do fossils form?
- How do scientists date fossils?
- How are fossils evidence of biological evolution?

Read to Learn

The Fossil Record

An oak tree changes a little when it loses its leaves. A robin changes when it loses some of its feathers. Yet, these living organisms change little from day to day. It might seem as if oak trees and robins have been on Earth forever. If you were to go back a few million years in time, you would not see oak trees or robins. You would see different species of trees and birds. That is because species change over time.

You might already know that fossils are the remains or evidence of once-living organisms. *The* **fossil record** *is made up of all the fossils ever discovered on Earth.* It has millions of fossils that come from many thousands of species. Most of these species are no longer alive on Earth.

The fossil record provides evidence that species have changed over time. Fossils help scientists picture what species looked like. Based on fossil evidence, scientists can re-create the physical appearance of species that are no longer alive.

The fossil record is huge, but it still has many missing parts. Scientists are still looking for more fossils to fill these missing parts. Scientists hypothesize that the fossil record represents only a small fraction of all the organisms that ever lived on Earth.

Study Coach

Vocabulary Quiz Write a question about each vocabulary term in this lesson. Exchange quizzes with a partner. After completing the quizzes, discuss the answers with your partner.

Think it Over

1. Explain Why don't scientists know the exact number of species that have lived on Earth?

____

____

____

Science Use v. Common Use . . .

tissue

Science Use similar cells that work together and perform a function

Common Use a piece of soft, absorbent paper

Reading Check

2. Analyze Why is it rare for soft tissue to become a fossil?

Fossil Formation

When animals eat a dead animal, they usually leave little behind. Any soft tissues that are not eaten are broken down by bacteria. Only the hard parts—bones, shells, and teeth—remain. Usually, these parts also break down over time. Sometimes they become fossils. Very rarely, the soft tissues of animals or plants—skin, muscles, or leaves—can also become fossils.

Mineralization

After an organism dies, its body might be buried under mud or sand in a river. Minerals in the water can take the place of the organism's original material and harden into rock. When this happens, a fossil forms. This process is called mineralization. Minerals in water also can fill the small spaces of a dead organism's tissues and become rock. Shells and bones are the most common mineralized fossils. Wood can also become a fossil in this way.

Carbonization

In carbonization, a fossil forms when a dead plant or animal is subjected to pressure over time. Pressure drives off the organism's liquids and gases. Only the carbon outline, or film, of the organism is left behind.

Molds and Casts

Sometimes the shell or bone of a dead animal makes an impression, the outline of its shape, in mud or sand. When the mud or sand hardens, so does the impression. *The impression of an organism in a rock is called a* ***mold.*** Sand or mud can later fill in the mold and harden to form a cast. *A* ***cast*** *is a fossil copy of an organism in a rock.* Molds and casts show only the outside parts of living organisms.

Trace Fossils

Fossils can give clues about the movement or behavior of once-living organisms. *A* ***trace fossil*** *is the preserved evidence of the activity of an organism.* For example, an organism might walk across mud and leave tracks. The tracks can become trace fossils if they fill with mud or sand that later hardens.

Key Concept Check

3. List the different ways fossils can form.

Original Material

In rare cases, the actual parts of an organism can be preserved. Some original-material fossils include mammoths frozen in ice and saber-toothed cats preserved in tar pits. Even the bodies of ancient humans have been found in bogs. Some insects were preserved when they got stuck in sap that hardened into amber.

Determining a Fossil's Age

Scientists cannot date most fossils directly. Instead, they find the age of the rocks in which the fossils are found.

Relative-Age Dating

You might be younger than a brother but older than a sister. This is your relative age. In the same way, a rock is either older or younger than rocks near it. In relative-age dating, scientists find the order in which rock layers formed. Some layers of rock have not moved since they formed. In these layers, scientists know that the bottom layers are older than the top layers, as shown in the figure below. Knowing the order in which the rocks formed helps scientists date the fossils in them. In this way, they can find the relative order in which species have appeared on Earth over time.

Dating Fossils

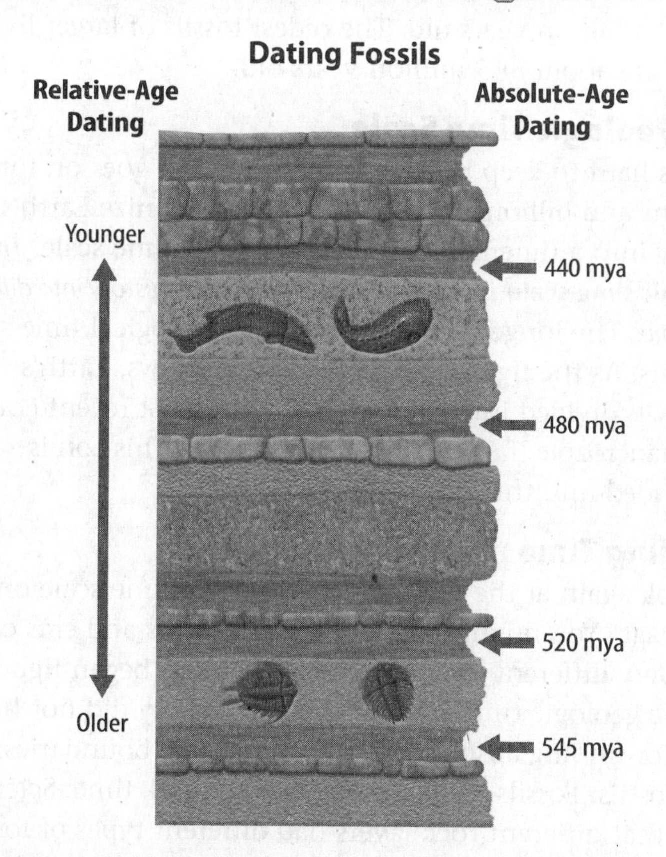

Absolute-Age Dating

Absolute-age dating is more exact than relative-age dating. A rock's absolute age is its age in years. To find absolute age, scientists use radioactive decay, which is a natural process that happens at a known rate. In radioactive decay, unstable isotopes in rocks change into stable isotopes over time. Scientists measure the ratio of unstable isotopes to stable isotopes to find the age of a rock. This ratio is best measured in igneous rocks, as shown above.

FOLDABLES®

Make a small shutterfold book. Label it as shown. Under the left tab, describe relative-age dating. Under the right tab, describe absolute-age dating.

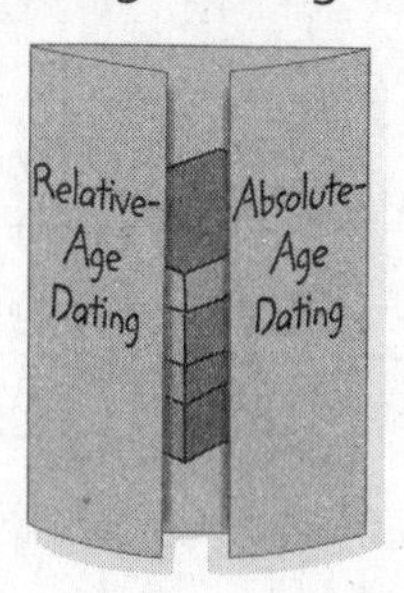

Key Concept Check

4. Explain How does relative-age dating help scientists learn about fossils?

Visual Check

5. Estimate What is the estimated age of the trilobite fossils (bottom layer of fossils)?

REVIEW VOCABULARY

isotopes
atoms of the same element that have different numbers of neutrons

Think it Over

6. Analyze Why is it difficult for scientists to absolute-age date sedimentary rock formations using isotopes?

____

____

____

Dating Igneous Rock Absolute age is easiest to determine in igneous rocks. Igneous rocks form from volcanic magma. Magma is so hot that it is rare for parts of organisms in it to remain and form fossils.

Dating Sedimentary Rock Most fossils form in mud and sand, which become sedimentary rock. To measure the age of a sedimentary rock layer, scientists find the ages of igneous layers above and below it. They can estimate an age in between these ages for the fossils found in the sedimentary layer. Absolute-age dating is illustrated in the figure on the previous page.

Fossils over Time

How old do you think Earth's oldest fossils are? Evidence of microscopic, unicellular organisms has been found in rocks 3.4 billion years old. The oldest fossils of larger living things are about 565 million years old.

The Geologic Time Scale

It is hard to keep track of something that goes on for millions and billions of years. Scientists organize Earth's history into a time line called the geologic time scale. *The* **geologic time scale** *is a chart that divides Earth's history into different time units.* The longest time units in the geological time scale are eons. As the figure on the next page shows, Earth's history is divided into four eons. Earth's most recent eon is the Phanerozoic (fa nuh ruh ZOH ihk) eon. This eon is subdivided into three eras.

Reading Check

7. Describe What is the geologic time scale?

____

____

____

____

Dividing Time

Look again at the figure of the geologic time scale on the next page. You might have noticed that eons and eras can have very different lengths. When scientists began figuring out the geologic time scale in the 1800s, they did not have ways for finding absolute age. To mark time boundaries, they used fossils. Fossils were an easy way to mark time. Scientists knew that different rock layers had different types of fossils. Some of the fossils scientists use to mark the time boundaries are shown in the figure.

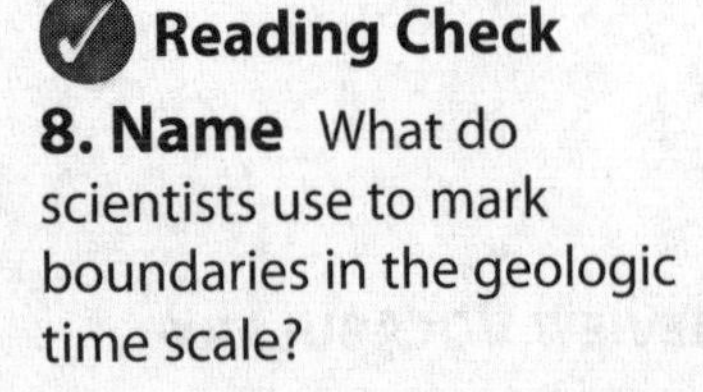

Reading Check

8. Name What do scientists use to mark boundaries in the geologic time scale?

____

Often, a type of fossil found in one rock layer was not in layers above it. Even more surprising, entire groups of fossils found in one layer were sometimes missing from layers above them. It seemed as if whole communities of living organisms had suddenly disappeared. What could have caused them to disappear?

The Geologic Time Scale

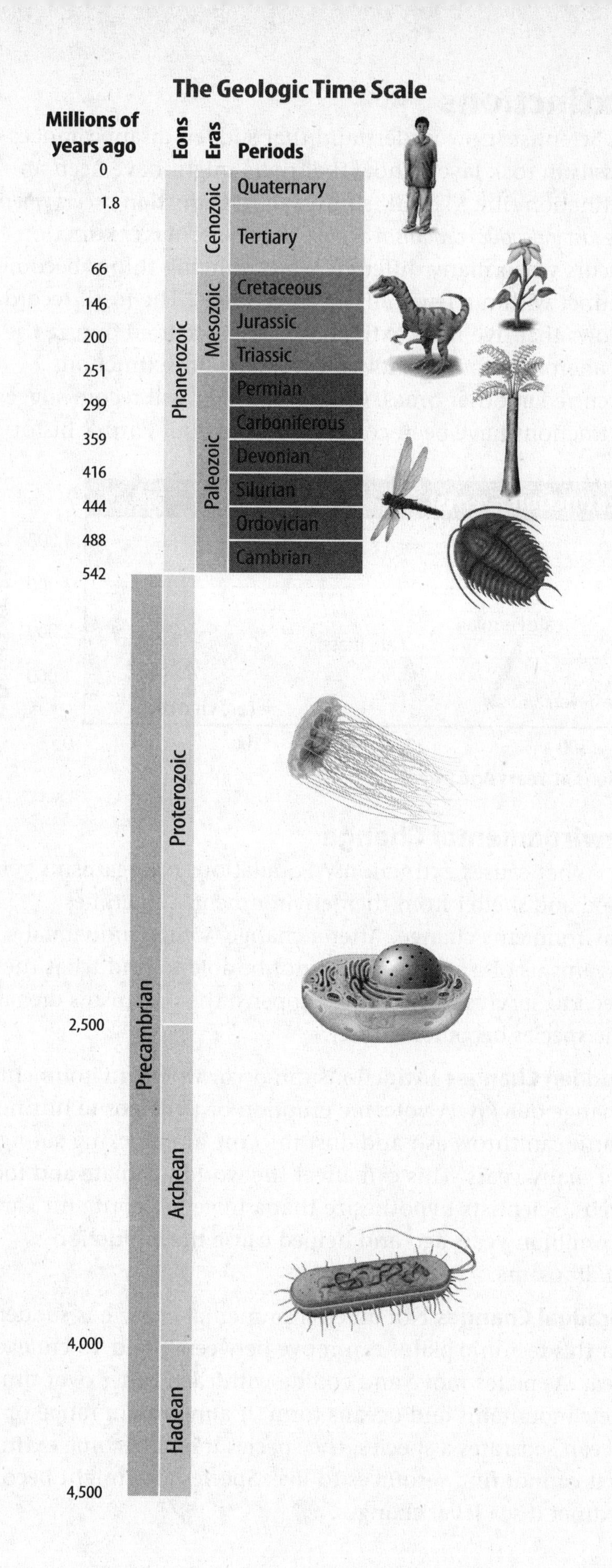

Visual Check

9. Sequence List the following living organisms in the order in which they first appeared on Earth: humans, insects, single cells, dinosaurs.

____

____

____

____

Math Skills

Numbers that refer to the ages of Earth's fossils are very large, so scientists use scientific notation to work with them. For example, mammals appeared on Earth about 200 mya or 200,000,000 years ago. Change this number to scientific notation using the following process.

Move the decimal point until only one nonzero digit remains on the left.

$200{,}000{,}000 = 2.00000000$

Count the number of places you moved the decimal point (8) and use that number as a power of ten.

$200{,}000{,}000 = 2.0 \times 10^8$ years.

10. Use Scientific Notation The first vertebrates appeared on Earth about 490,000,000 years ago. Express this time in scientific notation.

____

Visual Check

11. Analyze When did the most recent mass extinction happen? Circle the name of the time period on the graph. About how long ago did it happen?

Extinctions

Scientists now understand that sudden disappearances of fossils in rock layers show that there might have been an extinction (ihk STINGK shun) event. **Extinction** *occurs when the last individual organism of a species dies.* A mass extinction occurs when many different kinds of living things become extinct within a few million years or less. The fossil record shows that five mass extinctions have occurred during the Phanerozoic eon, as shown below. Smaller extinctions occurred at other times. Clues from the fossil record suggest extinctions have been common throughout Earth's history.

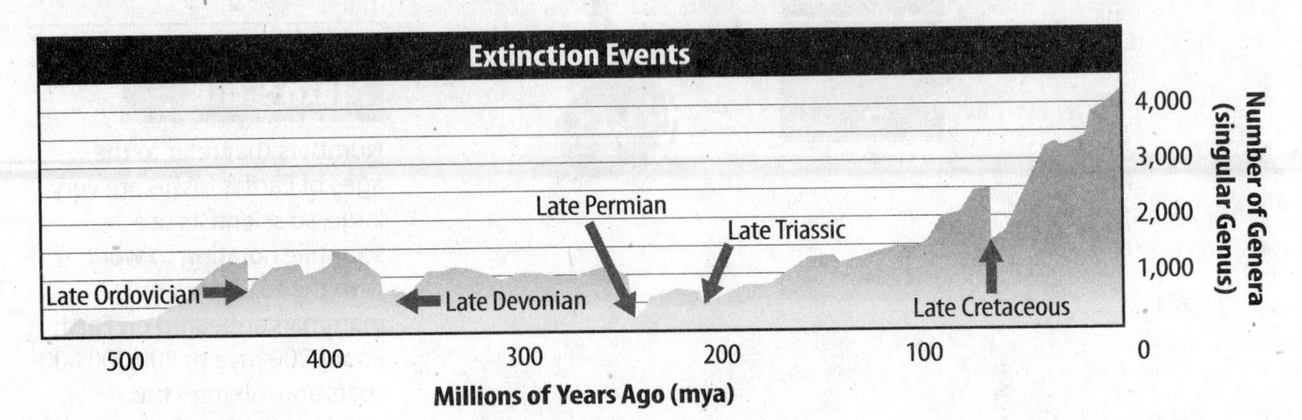

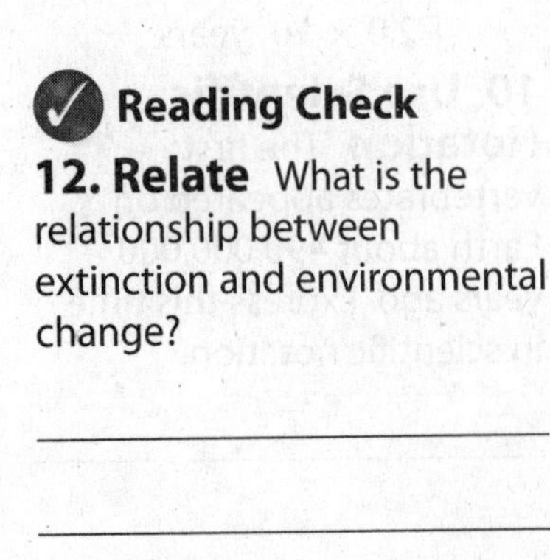

Reading Check

12. Relate What is the relationship between extinction and environmental change?

Environmental Change

What causes extinctions? Populations of organisms get food and shelter from their environment. Sometimes environments change. After a change occurs, individual organisms of a species might not be able to find what they need to survive. When this happens, the organisms die, and the species becomes extinct.

Sudden Changes Extinctions can occur when environments change quickly. A volcanic eruption or a meteorite hitting Earth can throw ash and dust into the air, blocking sunlight for many years. This can affect the world's climate and food webs. Scientists hypothesize that a huge meteorite hit Earth 65 million years ago and helped cause the extinction of dinosaurs.

Gradual Changes Not all environmental change is sudden. Earth's tectonic plates can move between 1 and 15 cm each year. As plates move and collide with each other over time, new mountains and oceans form. If a mountain range or an ocean separates a species, the species might become extinct if it cannot find resources to live. Species also might become extinct if sea level changes.

Extinctions and Evolution

The fossil record has obvious clues about the extinction of species over time. But it also has clues about the appearance of many new species. How do new species form?

Many early scientists thought that each species appeared on Earth independently of every other species. However, as scientists found more fossils, they began to see patterns in the fossil record. Many fossil species in nearby rock layers had similar body plans and similar body parts. These similar species seemed to be related to each other. For example, the series of horse fossils in the figure below suggests that the modern horse is related to other extinct species.

These species changed over time in what appeared to be a sequence. Change over time is evolution. **Biological evolution** *is the change over time in populations of related organisms.* Charles Darwin developed a theory about how species evolve from other species. You will read about Darwin's theory in the next lesson.

Horse Fossils

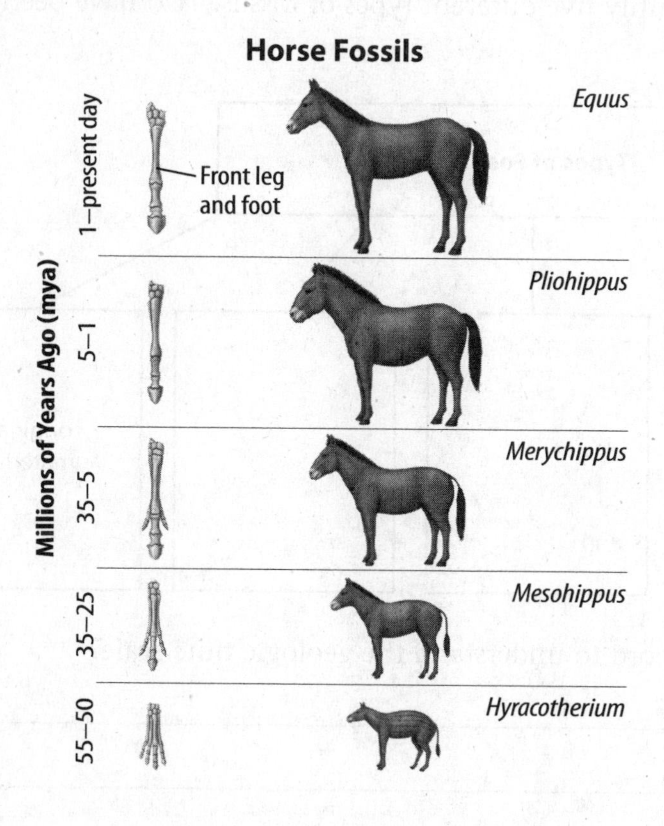

Reading Check

13. State What patterns did scientists find in the fossil record?

Key Concept Check

14. Analyze How are fossils evidence of biological evolution?

Visual Check

15. Evaluate In what ways did the horse change over time?

After You Read

Mini Glossary

biological evolution: the change over time in populations of related organisms

cast: a fossil copy of an organism in a rock

extinction (ihk STINGK shun): when the last individual organism of a species dies

fossil record: all the fossils ever discovered on Earth

geologic time scale: a chart that divides Earth's history into different time units

mold: the impression of an organism in a rock

trace fossil: the preserved evidence of the activity of an organism

1. Review the terms and their definitions in the Mini Glossary. Write a sentence that explains the difference between a cast and a mold.

2. Fill in the graphic organizer to identify five different types of fossils. Two have been done for you.

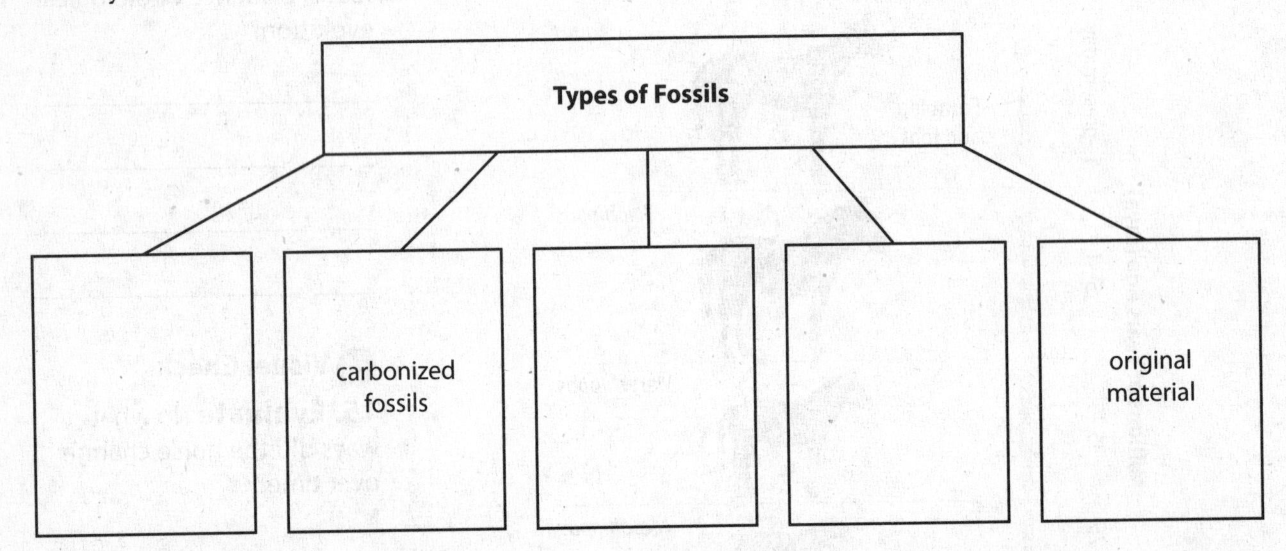

3. How do scientists use the fossil record to understand the geologic time scale?

What do you think NOW?

Reread the statements at the beginning of the lesson. Fill in the After column with an A if you agree with the statement or a D if you disagree. Did you change your mind?

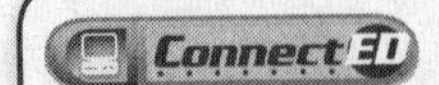

Log on to ConnectED.mcgraw-hill.com and access your textbook to find this lesson's resources.

The Environment and Change Over Time

Theory of Evolution by Natural Selection

Before You Read

What do you think? Read the two statements below and decide whether you agree or disagree with them. Place an A in the Before column if you agree with the statement or a D if you disagree. After you've read this lesson, reread the statements to see if you have changed your mind.

Before	Statement	After
	3. Environmental change causes variations in populations.	
	4. Variations can lead to adaptations.	

Key Concepts 

- Who was Charles Darwin?
- How does Darwin's theory of evolution by natural selection explain how species change over time?
- How are adaptations evidence of natural selection?

Read to Learn

Charles Darwin

How many species of birds can you name? Robins, penguins, and chickens are a few. There are about 10,000 species of birds on Earth today. Each species has wings, feathers, and beaks. Scientists hypothesize that all birds evolved from an earlier, or ancestral, group of birdlike organisms. As this group evolved into different species, birds developed different sizes, colors, songs, and ways of eating. Yet, they kept their key bird traits.

How do species evolve? Charles Darwin, a scientist, worked to answer this question. Darwin was an English naturalist who, in the mid-1800s, developed a theory of how evolution works. *A* **naturalist** *is a person who studies plants and animals by observing them.* Darwin spent years studying plants and animals in nature before developing his theory. Recall that a theory is an explanation of the natural world that is well supported by evidence. Darwin's theory of evolution was not the first, but his theory is the one best supported by evidence today.

Study Coach

Make Flash Cards Think of a quiz question for each paragraph. Write the question on one side of a flash card. Write the answer on the other side. Work with a partner to quiz each other using your flash cards.

Key Concept Check

1. Describe Who was Charles Darwin?

______

______

______

______

Voyage of the *Beagle*

Darwin worked as a naturalist on the HMS *Beagle*, a ship of the British navy. During his trip around the world, Darwin observed and collected many plants and animals.

FOLDABLES®

Make a small four-door shutterfold book. Use it to investigate the who, what, when, and where of Charles Darwin, the Galápagos Islands, and the theory of evolution by natural selection.

The Galápagos Islands

Darwin was interested in the organisms he saw on the Galápagos (guh LAH puh gus) Islands. These islands are 1,000 km off the South American coast in the Pacific Ocean. Darwin saw that each island had a slightly different environment. Some were dry. Some were more humid. Others had mixed environments.

Tortoises Darwin saw that the giant tortoises on each island looked different. On one island, tortoises had shells that came close to their necks. They could eat only short plants. On other islands, tortoises had more space between the shell and neck. They could eat taller plants.

Mockingbirds and Finches Darwin was also curious about the different mockingbirds and finches he saw. Like the tortoises, different types of mockingbirds and finches lived in different island environments. Later, he was surprised to find that many were different enough to be separate species. ✓

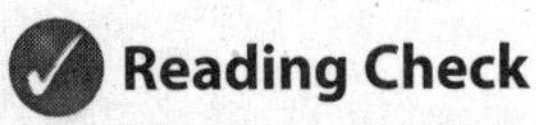

Reading Check

2. Explain What made Darwin become curious about the organisms that lived on the Galápagos Islands?

______

______

______

______

Darwin's Theory

Darwin discovered a relationship between each species and the food found on the island where it lived. Tortoises with long necks lived on islands that had tall cacti. Their long necks made it possible for them to reach high to eat the cacti. The tortoises with short necks lived on islands that had plenty of short grass.

Common Ancestors

Darwin became convinced that all the tortoise species were related. He thought they all shared a common ancestor. He suggested that millions of years before, a storm had carried a group of tortoises to one of the islands from South America. In time, the tortoises spread to the other islands. Their neck lengths and shell shapes changed to match their islands' food sources. How did this happen?

ACADEMIC VOCABULARY

convince
(verb) to overcome by argument

Variations

Darwin knew that individual members of a species have slight differences, or variations. *A* **variation** *is a slight difference in the appearance of individual members of a species.* Variations arise naturally in populations. They occur in the offspring as a result of sexual reproduction. You might recall that variations are caused by random mutations, or changes, in genes. Mutations can lead to changes in phenotype. Recall that an organism's phenotype is all of the observable traits and characteristics of the organism. Genetic changes to phenotype can be passed on to future generations.

Natural Selection

Darwin did not know about genes. But he saw that variations were the key to how evolution worked. He knew that there was not enough food on each island to feed every tortoise that was born. Tortoises had to compete for food. As the tortoises spread to the different islands, some were born with random variations in neck length. If a variation helped a tortoise compete for food, the tortoise lived longer than other tortoises without the variation. Because it lived longer, it reproduced more. It passed on the helpful variation to its offspring.

This is Darwin's theory of evolution by natural selection. **Natural selection** *is the process by which populations of organisms with variations that help them survive in their environments live longer, compete better, and reproduce more than those that do not have the variations.* Natural selection explains how Galápagos tortoises became matched to their food sources, as shown below. It also explains why there were so many different kinds of Galápagos finches and mockingbirds. Birds with beak variations that helped them compete for food lived longer and reproduced more.

Key Concept Check

3. Analyze What role do variations have in the theory of evolution by natural selection?

____

____

____

Visual Check

4. Illustrate Mark all the tortoises in the figure that have short necks with the letter *S*. Mark those that have long necks with the letter *L*. What trend do you see over time?

____

____

Natural Selection

1 Reproduction A population of tortoises produces many offspring that inherit its characteristics.

2 Variation A tortoise is born with a variation that makes its neck slightly longer.

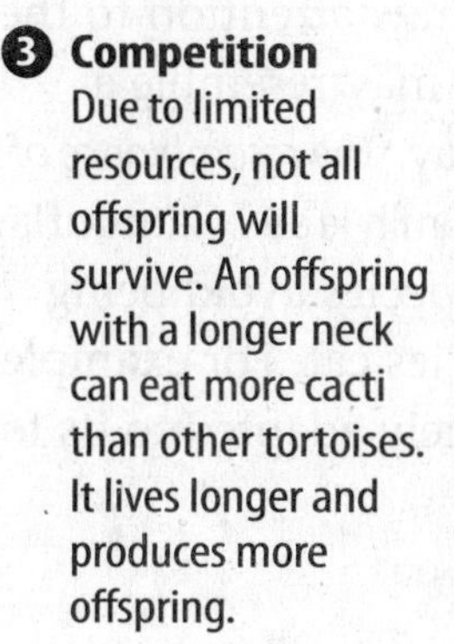

3 Competition Due to limited resources, not all offspring will survive. An offspring with a longer neck can eat more cacti than other tortoises. It lives longer and produces more offspring.

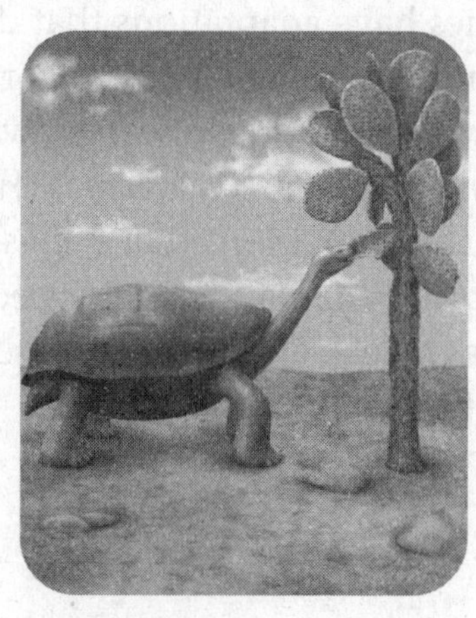

4 Selection Over time, the variation is inherited by more and more offspring. Eventually, all tortoises have longer necks.

Adaptations

Natural selection explains how all species change over time as their environments change. Through natural selection, a helpful variation in one individual can eventually pass to future members of a population.

As time passes, more variations come about. The buildup of many similar variations can lead to an adaptation (a dap TAY shun). *An* ***adaptation*** *is a characteristic of a species that enables the species to survive in its environment.* The long neck of certain species of tortoises is an adaptation to an environment with tall cacti.

Key Concept Check
5. Explain How do variations lead to adaptations?

___

___

___

___

Types of Adaptations

Every species has many adaptations. Scientists classify adaptations into three categories: structural, behavioral, and functional.

Structural Adaptations These adaptations involve color, shape, and other physical characteristics. The shape of a tortoise's neck is a structural adaptation.

Behavioral Adaptations The way an organism behaves or acts is a behavioral adaptation. Hunting at night and moving in herds are behavioral adaptations.

Functional Adaptations The last category is functional adaptations. These adaptations involve chemical changes in body systems. A drop in body temperature during hibernation is a functional adaptation.

Think it Over
6. Apply An opossum will play dead when a predator frightens it. That way the predator might think it is not good food and will leave it alone. What kind of adaptation is this? (Circle the correct answer.)

a. structural
b. behavioral
c. functional

Environmental Interactions

Many species have evolved adaptations that make them nearly invisible. For example, a seahorse may be the same color as and similar in texture to the coral it rests on. This is a structural adaptation called camouflage (KAM uh flahj). ***Camouflage*** *is an adaptation that enables species to blend in with their environments.*

Some species have adaptations that draw attention to them or make them more visible. A caterpillar may resemble a snake. Predators see it and are scared away. *The resemblance of one species to another species is* ***mimicry*** (MIH mih kree). Camouflage and mimicry are adaptations that help species avoid being eaten. Many other adaptations help species eat. For example, the pelican has a beak and mouth uniquely adapted to its food source—fish.

Reading Check
7. Contrast How do camouflage and mimicry differ?

___

___

___

___

Role of Environment Environments are complex. Species must adapt to an environment's living parts as well as to its nonliving parts. Some nonliving things are temperature, water, nutrients in soil, and climate. Deciduous trees shed their leaves due to changes in climate. Camouflage, mimicry, and mouth shape are adaptations mostly to an environment's living parts.

Extinct Species Living and nonliving factors are always changing. Even slight environmental changes affect how species adapt. If a species is unable to adapt, it becomes extinct. The fossil record contains many fossils of species unable to adapt to change.

Artificial Selection

Adaptations show how closely Earth's species match their environments. This is exactly what Darwin's theory of evolution by natural selection predicted. Darwin gave many examples of adaptation in *On the Origin of Species,* the book he wrote to explain his theory. Darwin wrote his book 20 years after he developed his theory. He spent those years collecting more evidence for his theory.

Darwin also had a hobby of breeding pigeons. He bred pigeons of different colors and shapes. In this way, he produced new, fancy varieties. *The breeding of organisms for desired characteristics is called* **selective breeding.** Like many plants and animals produced from selective breeding, pigeons look different from their ancestors.

Darwin saw that changes caused by selective breeding were much like changes caused by natural selection. Instead of nature selecting variations, humans selected them. Darwin called this process artificial selection. ✓

Artificial selection explains and supports Darwin's theory. In Lesson 3, you will read about other evidence that supports the idea that species evolve from other species.

Think it Over

8. Synthesize How do you think some fur-bearing species might adapt to a gradual change in climate in which global temperature increased?

Reading Check

9. Compare How are artificial selection and natural selection alike?

After You Read

Mini Glossary

adaptation (a dap TAY shun): a characteristic of a species that enables the species to survive in its environment

camouflage (KAM uh flahj): an adaptation that enables species to blend in with their environments

mimicry (MIH mih kree): the resemblance of one species to another species

naturalist: a person who studies plants and animals by observing them

natural selection: the process by which populations of organisms with variations that help them survive in their environments live longer, compete better, and reproduce more than those that do not have the variations

selective breeding: the breeding of organisms for desired characteristics

variation: a slight difference in the appearance of individual members of a species

1. Review the terms and their definitions in the Mini Glossary. Describe a living organism that depends on camouflage or mimicry to survive.

2. Write a letter in each box to show the correct sequence that demonstrates the process of natural selection.
   - **a.** Birds eat more light green beetles. Dark green beetles live longer and reproduce more.
   - **b.** A beetle is born with a variation in its color: It is dark green.
   - **c.** Over time, all beetles in the environment are dark green.
   - **d.** A population of beetles is light green. They stand out against dark green leaves.

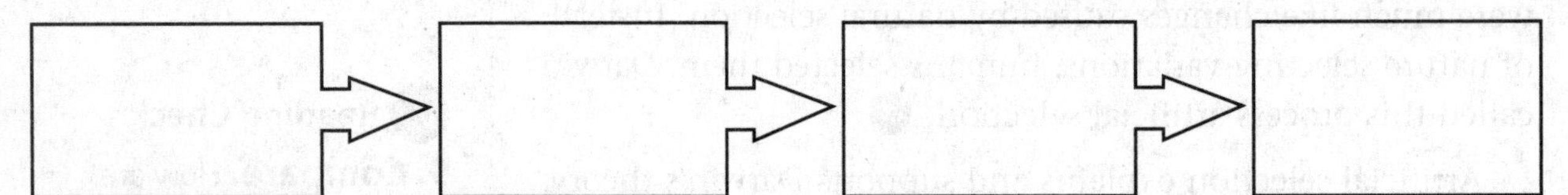

3. Compare selective breeding and evolution.

What do you think NOW?

Reread the statements at the beginning of the lesson. Fill in the After column with an A if you agree with the statement or a D if you disagree. Did you change your mind?

Log on to ConnectED.mcgraw-hill.com and access your textbook to find this lesson's resources.

The Environment and Change Over Time

Biological Evidence of Evolution

Before You Read

What do you think? Read the two statements below and decide whether you agree or disagree with them. Place an A in the Before column if you agree with the statement or a D if you disagree. After you've read this lesson, reread the statements to see if you have changed your mind.

Before	Statement	After
	5. Living species contain no evidence that they are related to each other.	
	6. Plants and animals share similar genes.	

Key Concepts

- What evidence from living species supports the theory that species descended from other species over time?
- How are Earth's organisms related?

Read to Learn

Evidence for Evolution

The pictures of horse fossils in Lesson 1 seem to show that horses evolved in a straight line. That is, one species replaced another in a series of orderly steps. Evolution does not occur this way. Different horse species were sometimes alive at the same time. They are related to one another because each descended from a common ancestor.

Living species that are closely related share a close common ancestor. How closely they are related depends on how closely in time they diverged, or split, from that ancestor. Evidence of common ancestors can be found in the fossil record and in living organisms.

Comparative Anatomy

It is easy to see that some species evolved from a common ancestor. For example, robins, finches, and hawks have similar body parts. They all have feathers, wings, and beaks. The same is true for tigers, leopards, and house cats. But how are hawks related to cats?

Studying the structural and functional similarities and differences in species that do not look alike can show the relationships. *The study of similarities and differences among structures of living species is called* **comparative anatomy.**

Mark the Text

Identify Main Ideas
Highlight the main idea of each paragraph. Highlight two details that support each main idea with a different color. Use your highlighted copy to review what you studied in this lesson.

FOLDABLES®

Make a table with five rows and three columns. Label the rows and columns of the table as shown below. Give your table a title.

	Explanation	Example
Comparative Anatomy		
Vestigial Structures		
Developmental Biology		
Molecular Biology		

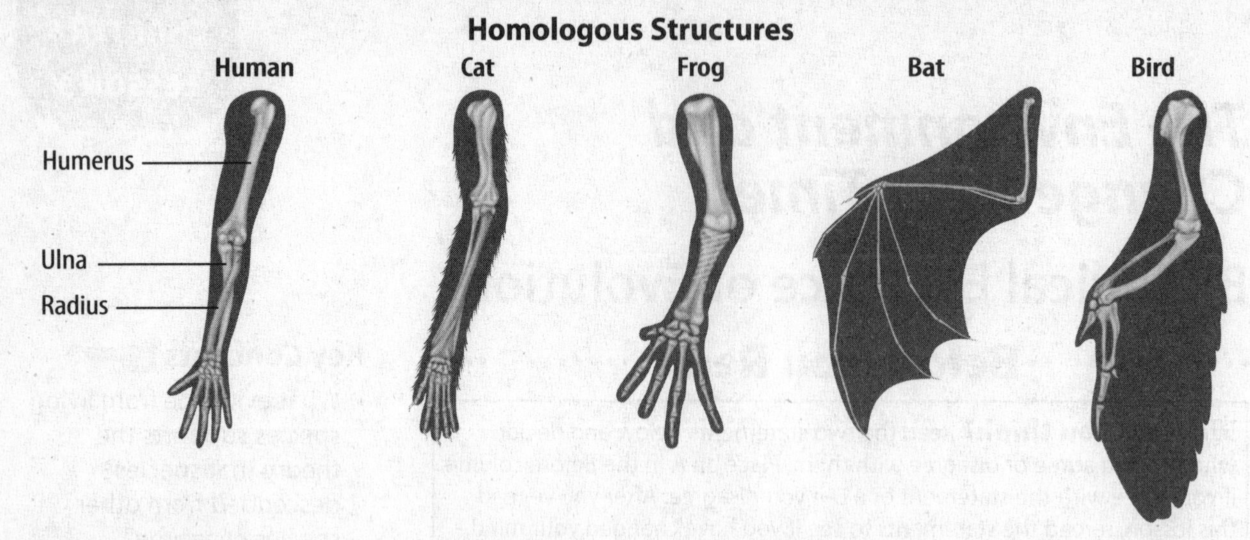

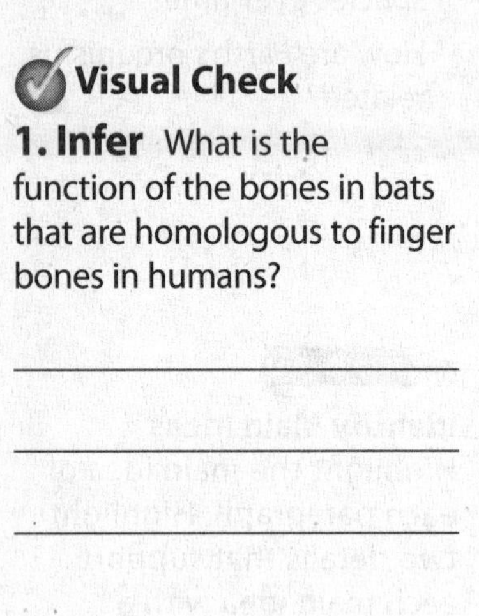

Visual Check

1. Infer What is the function of the bones in bats that are homologous to finger bones in humans?

Key Concept Check

2. Explain How do homologous structures provide evidence for evolution?

Homologous Structures Humans, cats, frogs, bats, and birds look different and move in different ways. Humans use their arms for balance and their hands to grasp objects. Cats use their forelimbs to walk, run, and jump. Frogs use their forelimbs to jump. The forelimbs of bats and birds are wings and are used for flying. However, the forelimb bones of all these species show similar patterns, as shown in the figure above. The forelimbs of the species in the figure are different sizes, but their placement and structure suggest common ancestry.

Homologous (huh MAH luh gus) **structures** *are body parts of organisms that are similar in structure and position but different in function.* Homologous structures, such as the forelimbs of humans, cats, frogs, bats, and birds, suggest that these species are related. The more alike two structures are, the more likely it is that the species have evolved from a recent common ancestor.

Analogous Structures Can you think of a body part in two species that does the same job but differs in structure? How about the wings of birds and flies? The wings in both species are used for flight. But bird wings are covered with feathers. Fly wings are covered with tiny hairs. Though used for the same function—flight—the wings of birds and insects are too different in structure to suggest close common ancestry.

Bird wings and fly wings are analogous (uh NAH luh gus) structures. **Analogous structures** *are body parts that perform a similar function but differ in structure.* The differences in wing structure show that birds and flies are not closely related.

Vestigial Structures

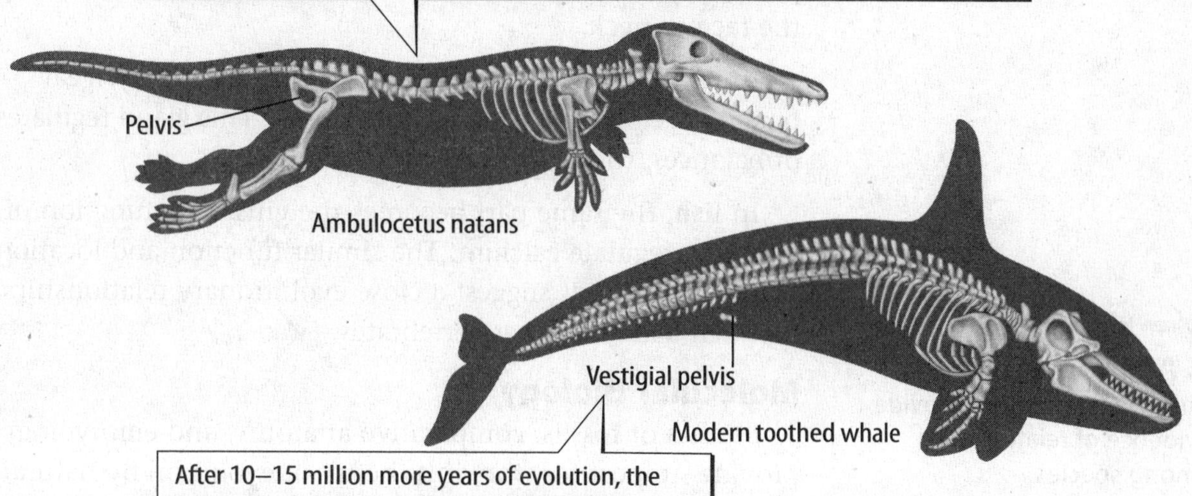

Vestigial Structures

Ostriches have wings. Yet they cannot fly. An ostrich's wings are an example of vestigial structures. **Vestigial** (veh STIH jee ul) **structures** *are body parts that have lost their original function through evolution.* The best explanation for vestigial structures is that the species with a vestigial structure is related to an ancestral species that used the structure for a specific purpose.

The whale shown in the figure above has tiny pelvic bones inside its body. Pelvic bones are hip bones, which in many species attach the leg bones to the body. Modern whales do not have legs. The pelvic bones in whales suggest that whales came from ancestors that used legs for walking on land. The fossil evidence supports this conclusion. Many fossils of whale ancestors show a slow loss of legs over millions of years. They also show, at the same time, that whale ancestors became better adapted to their watery environments.

Developmental Biology

Studying the internal structures of living organisms is not the only way that scientists learn about common ancestors. Studying how embryos develop can also show how species are related. *The science of the development of embryos from fertilization to birth is called* **embryology** (em bree AH luh jee).

Visual Check

3. Infer Why does a vestigial pelvis show that the ancestors of the modern whale once had legs?

___

___

___

Key Concept Check

4. Explain How are vestigial structures evidence of descent from ancestral species?

___

___

___

Pharyngeal Pouches Embryos of different species often look like each other at different stages of their growth. For example, all vertebrate embryos have pharyngeal (fuh rihn JEE ul) pouches at one stage. These pouches become different body parts in each vertebrate. Yet, in all vertebrates, each part is in the face or neck.

In reptiles, birds, and humans, part of the pharyngeal pouch develops into a gland in the neck. This gland regulates, or balances, the body's calcium levels.

In fish, the same part becomes the gills. One function of gills is to regulate calcium. The similar function and location of gills and glands suggest a close evolutionary relationship between fish and other vertebrates.

Key Concept Check
5. Analyze How do pharyngeal pouches provide evidence of relationships among species?

Molecular Biology

Studies of fossils, comparative anatomy, and embryology provide support for Darwin's theory of evolution by natural selection. Molecular biology is the study of gene structure and function.

Discoveries in molecular biology have confirmed and extended much of the data already collected about the theory of evolution. Darwin did not know about genes, but scientists today know that mutations in genes are the source of variations upon which natural selection acts. Genes provide powerful support for evolution.

Reading Check
6. Describe What is molecular biology?

Comparing Sequences All living organisms have genes. All genes are made of DNA, and all genes work in similar ways. This supports the idea that all living organisms are related.

Scientists can study how living organisms are related by comparing their genes. For example, nearly all organisms have a gene for cytochrome *c*, a protein required for cellular respiration. Some species, such as humans and rhesus monkeys, have nearly identical cytochrome *c*. The more closely related two species are, the more similar their genes and proteins are.

Key Concept Check
7. Explain How is molecular biology used to determine relationships among species?

Divergence Scientists have found that some stretches of shared DNA mutate at regular, predictable rates. Scientists use this "molecular clock" to estimate when in the past living species split from common ancestors. This is how scientists have shown that whales and porpoises are more closely related to hippopotamuses than they are to other living things. Whales and hippopotamuses share an ancestor that lived 50–60 million years ago.

The Study of Evolution Today

The theory of evolution by natural selection is the cornerstone of modern biology. Since Darwin published his theory, scientists have confirmed, refined, and extended his work. They have observed natural selection in hundreds of living species. Their studies of fossils, anatomy, embryology, and molecular biology have shown relationships among living and extinct species. ✓

How New Species Form

New evidence supporting the theory of evolution by natural selection is discovered nearly every day. But scientists debate some of the details. The figure below shows how scientists have different ideas about the rate at which natural selection produces new species. Some say it works slowly and gradually. Others say it works quickly, in bursts. How different species first came about is difficult to study on human time scales. It is also difficult to study with the incomplete fossil record. Yet, new fossils that fill in the holes are discovered all the time. Further fossil discoveries will help scientists study more details about the origin of new species. ✓

Diversity

Evolution has produced Earth's wide diversity of living things using the same basic building blocks called genes. This is an active area of study in evolutionary biology. Scientists are finding that genes can be reorganized in simple ways and give rise to dramatic changes in organisms. Scientists now study evolution by looking at molecules. Yet, they still use the same basic ideas that Darwin came up with over 150 years ago.

Rates of Evolution

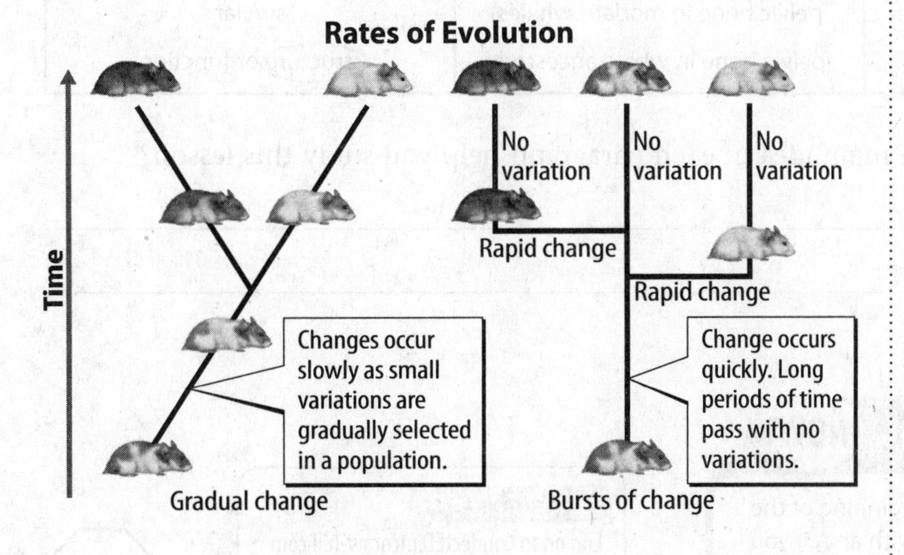

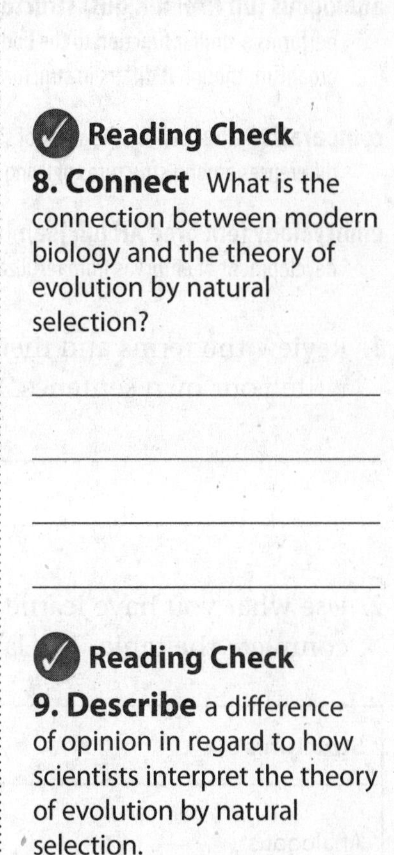

✓ Reading Check

8. Connect What is the connection between modern biology and the theory of evolution by natural selection?

✓ Reading Check

9. Describe a difference of opinion in regard to how scientists interpret the theory of evolution by natural selection.

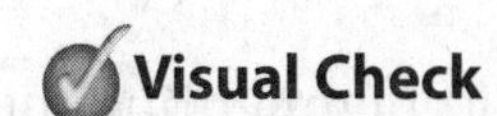

✓ Visual Check

10. Analyze What does a flat (horizontal) line mean in the figure? (Circle the correct answer.)

a. gradual change

b. no variation

c. rapid change

After You Read

Mini Glossary

analogous (uh NAH luh gus) structure: a body part that performs a similar function to the body part of another organism, though it differs in structure

comparative anatomy: the study of similarities and differences among structures of living species

embryology (em bree AH luh jee): the science of the development of embryos from fertilization to birth

homologous (huh MAH luh gus) structure: a body part that is similar in structure and position to the body part of another organism, though it has a different function

vestigial (veh STIH jee ul) structure: a body part that has lost its original function through evolution

1. Review the terms and their definitions in the Mini Glossary. Use one of the terms to write your own sentence.

____

____

2. Use what you have learned about analogous, homologous, and vestigial structures to complete the table. The last row has been completed for you.

Structures	Example Pair of Structures	Similar Structure or Function (circle one)
Analogous		similar structure or function
Homologous		similar structure or function
Vestigial	pelvic bone in modern whales pelvic bone in whale ancestors	similar (structure) or function

3. How did highlighting the main idea in each paragraph help you study this lesson?

____

____

What do you think NOW?

Reread the statements at the beginning of the lesson. Fill in the After column with an A if you agree with the statement or a D if you disagree. Did you change your mind?

Log on to ConnectED.mcgraw-hill.com and access your textbook to find this lesson's resources.

Bacteria and Viruses

What are bacteria?

Before You Read

What do you think? Read the two statements below and decide whether you agree or disagree with them. Place an A in the Before column if you agree with the statement or a D if you disagree. After you've read this lesson, reread the statements to see if you have changed your mind.

Before	Statement	After
	1. A bacterium does not have a nucleus.	
	2. Bacteria cannot move.	

Key Concept
- What are bacteria?

Read to Learn

Characteristics of Bacteria

You are surrounded by billions of tiny organisms too small to be seen. They even live inside your body. These organisms are called bacteria. **Bacteria** (singular, bacterium) *are microscopic prokaryotes.* As you learned, a prokaryote is a unicellular organism that has no nucleus or other membrane-bound organelles.

Bacteria live almost everywhere. Their habitats include the air, glaciers, the ocean floor, and soil. Bacteria also live in or on almost every organism, both living and dead. Bacteria live both inside you and on your skin. In fact, your body has more bacterial cells than human cells! The bacteria in your body outnumber human cells by 10 to 1.

Bacteria have many of the same traits as archaea (ar KEE uh; singular, archaean). Bacteria and archaea are prokaryotes. Both lack membrane-bound organelles. Archaea can live where most living things cannot survive, such as in very warm areas or areas with little oxygen. Both bacteria and archaea are important to life on Earth.

Mark the Text

Identify the Main Idea As you read, highlight the main idea of each paragraph. Use a different color to highlight a detail or example that might help you understand that main idea.

Key Concept Check
1. Define What are bacteria?

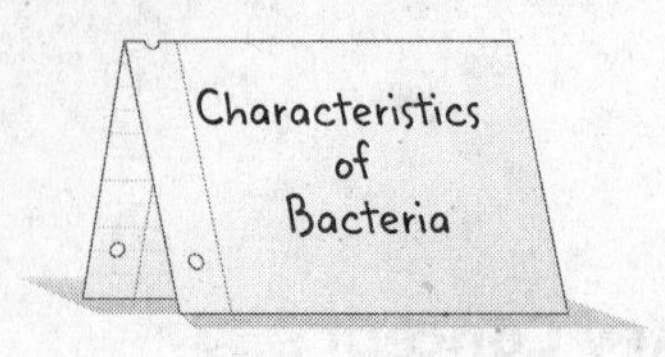

Make a folded book. Use it to organize your notes on the characteristics of bacteria.

2. Identify Circle the names of the two structures that surround the bacterium's DNA and cytoplasm.

3. State the three basic shapes of bacteria.

Structure of Bacteria

Look at the typical bacterium shown below. The bacterium is made up of cytoplasm and DNA. These parts are surrounded by a cell membrane and a cell wall. The cytoplasm also contains ribosomes. Most bacteria have DNA that is one coiled, circular chromosome. Many bacteria also have one or more plasmids. Plasmids are small circular pieces of DNA that are separate from other DNA.

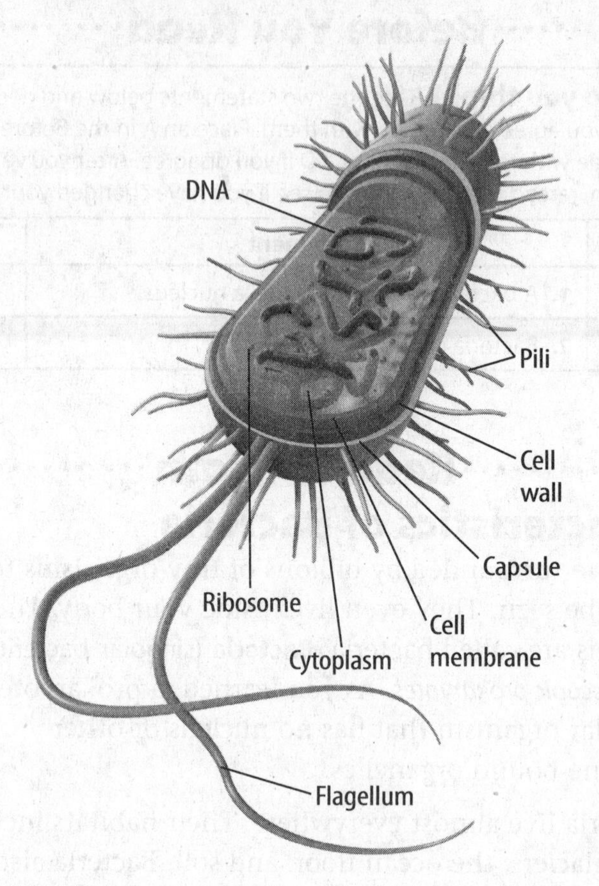

Some bacteria have specialized structures. The bacterium that causes pneumonia (noo MOH nyuh) has a thick covering around its cell wall. This structure, called a capsule, helps the bacterium survive. The capsule keeps the bacterium from drying out. It also keeps things from getting into the bacterium. Many bacteria have capsules with pili (PI li). These hairlike structures help bacteria stick to surfaces.

Size and Shapes of Bacteria

Bacteria are much smaller than plant or animal cells. Bacteria are generally only 1–5 micrometers (μm) (1 m = 1 million μm) wide. An average eukaryotic cell is 10–100 μm wide. Scientists estimate that as many as 100 bacteria could be lined up across the head of a pin. Bacteria have three basic shapes. Bacteria can be shaped like spheres, rods, or spirals.

Obtaining Food and Energy

Bacteria live in many different places. Because their environments are different, bacteria get food in different ways. Some bacteria take in food and break it down to get energy. Many of these bacteria feed on dead organisms or organic waste. Other bacteria take in nutrients from living hosts. Bacteria that cause tooth decay, for example, live in the plaque on your teeth. These bacteria feed on sugars in the things you eat and drink.

Some bacteria make their own food. They use light energy and make food, like most green plants do. These bacteria live where there is much light, such as on the surface of a lake. Other bacteria use energy from chemical reactions and make their food. These bacteria live where there is no sunlight, such as on the dark ocean floor.

Most organisms, including humans, cannot survive without oxygen. However, not all bacteria need oxygen to survive. Bacteria that can live without oxygen are called anaerobic (a nuh ROH bihk) bacteria. Bacteria that need oxygen are called aerobic (er OH bihk) bacteria. Most bacteria in the environment are aerobic.

Movement

Some bacteria can move around to find the resources they need to survive. These bacteria move using special structures. *Many bacteria have long whiplike structures called* **flagella** (fluh JEH luh; singular, flagellum). If you look again at the diagram of the bacterium, you can see a flagellum. Some bacteria twist or spiral as they move. Still others use their pili as hooks as they move. Some bacteria make threadlike structures that they use to push away from a surface.

Reproduction

You learned that organisms reproduce asexually or sexually. Bacteria reproduce asexually by a process called fission. **Fission** *is cell division that forms two genetically identical cells.* Fission can occur quickly. A cell might divide as often as every 20 minutes if conditions are right.

Genetic Diversity Bacteria produced by fission are identical to the parent cell. However, genetic variation can be increased by a process called conjugation.

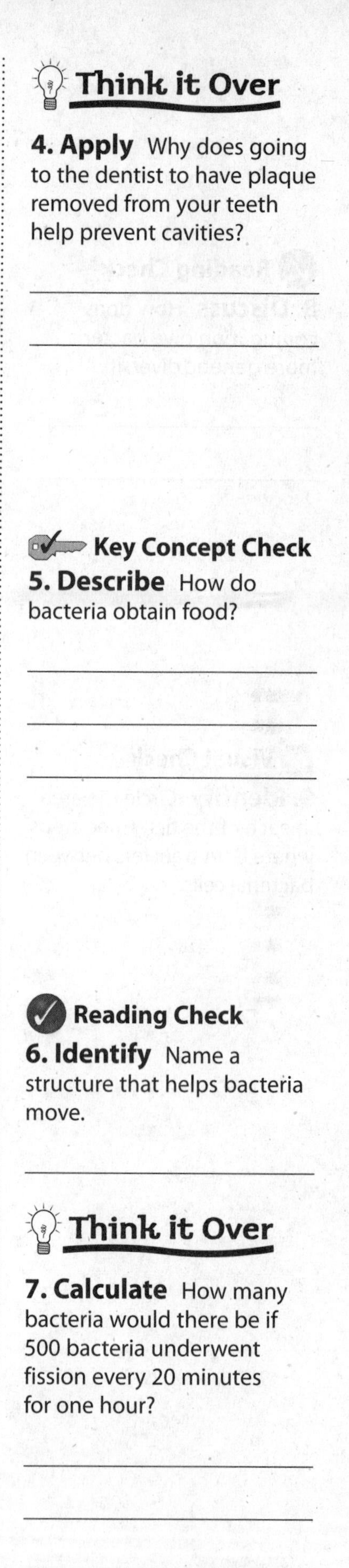

Conjugation During **conjugation** (kahn juh GAY shun), *two bacteria of the same species attach to each other and combine their genetic material*. As shown below, DNA is transferred between the bacteria. This results in new combinations of genes and an increase in genetic diversity. Conjugation does not produce new organisms, so the process is not considered reproduction.

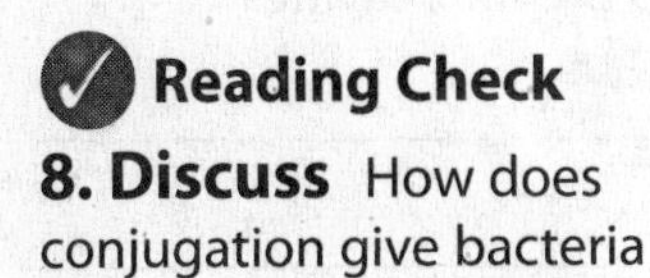

Reading Check

8. Discuss How does conjugation give bacteria more genetic diversity?

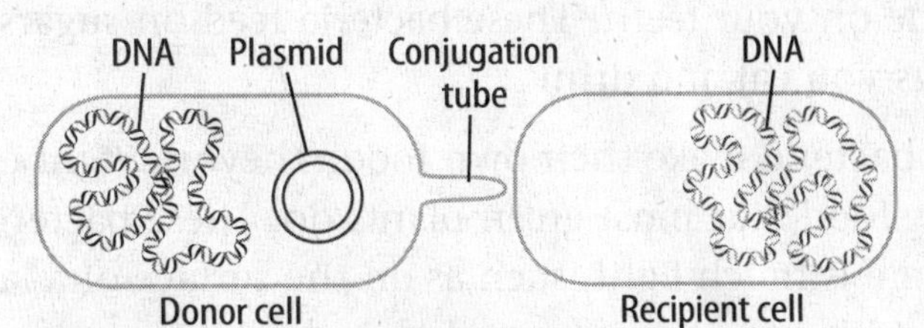

1 The donor cell and recipient cell both have circular chromosomal DNA. The donor cell also has DNA as a plasmid. The donor cell forms a conjugation tube and connects to the recipient cell.

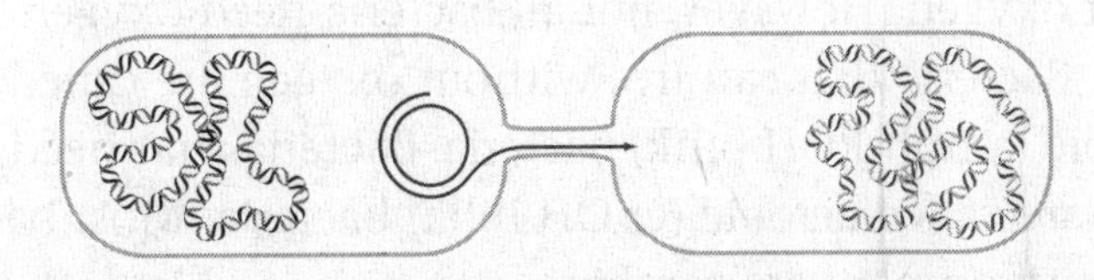

2 The conjugation tube connects both cells. The plasmid splits in two and one plasmid strand moves through the conjugation tube into the recipient cell.

Visual Check

9. Identify Circle the area in each of the first three steps where DNA transfers between bacterial cells.

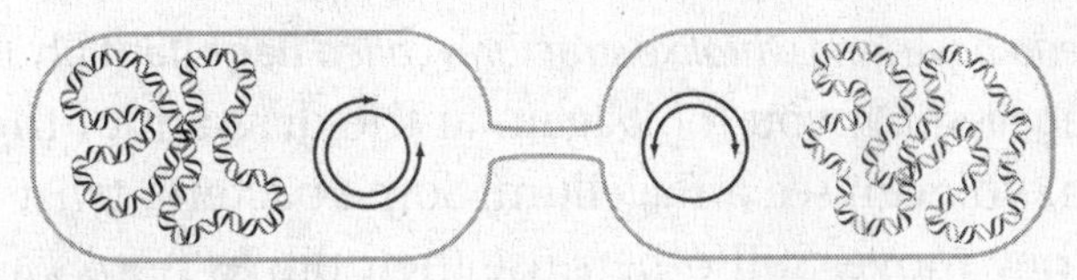

3 The complimentary strands of the plasmids are completed in both bacteria.

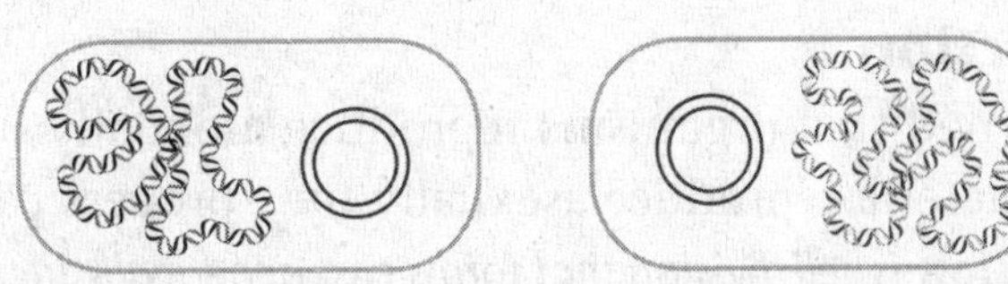

4 With the new plasmids complete, the bacteria separate from each other. The recipient cell now contains plasmid DNA from the donor cell as well as its own chromosomal DNA.

Endospore Formation

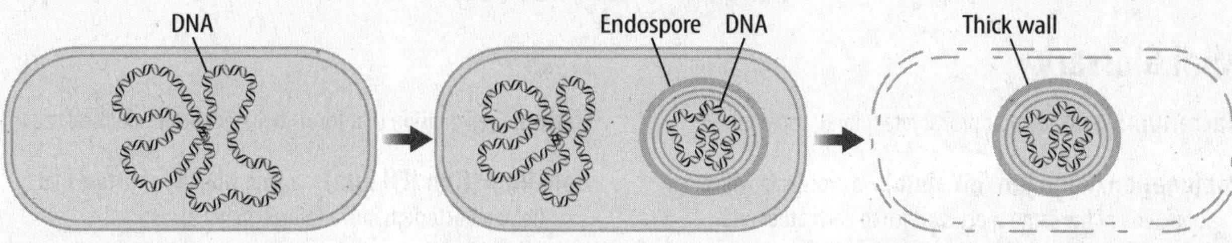

❶ Bacterial cells in favorable conditions form without endospores.

❷ As conditions become unfavorable, the cell forms an endospore around some of its DNA.

❸ The cell dissolves, leaving the endospore-protected DNA to survive in the harsh conditions.

Endospores

Sometimes environmental conditions are unfavorable for the survival of bacteria. Some of these conditions are extreme heat, cold, and drought. In these cases, some bacteria can form endospores. *An* **endospore** (EN duh spor) *forms when a bacterium builds a thick internal wall around its chromosome and part of the cytoplasm.* Endospores protect bacteria. A bacterium can stay dormant for months or even hundreds of years until conditions are better. Endospores enable bacteria to survive extreme conditions. The diagram at the top of the page shows how an endospore forms.

Archaea

Prokaryotes called archaea were once considered bacteria. Like a bacterium, an archaean has a cell wall and no nucleus or membrane-bound organelles. Its chromosome is also circular, like those in bacteria. However, archaea and bacteria differ in important ways. The ribosomes of archaea resemble the ribosomes of eukaryotes, rather than those of bacteria. Archaea also contain molecules in their plasma membranes that are not in any other organisms. Archaea often live in environments where other organisms cannot. Some archea live in hot springs, while others live in salty lakes. Some scientists refer to archaea as extremophiles (ik STREE muh filez), which is a word that means "those that love extremes."

Visual Check

10. Locate Color the DNA in each of the three stages of endospore formation.

Math Skills

Each time bacteria undergo fission, the population doubles in size. To calculate how many bacteria there will be, use this formula:

$$n = x \times 2^f$$

Example: 100 bacteria undergo fission 3 times.

$f = 3$, so 2^f is 2 multiplied by itself 3 times. $(2 \times 2 \times 2 = 8)$

$n = 100 \times 8 = 800$ bacteria

11. Use a Formula How many bacteria would there be if a population of 10 bacteria underwent fission 5 times?

After You Read

Mini Glossary

bacterium: a microscopic prokaryote (plural, bacteria)

conjugation (kahn juh GAY shun): a process in which two bacteria of the same species attach to each other and combine their genetic material

endospore (EN duh spor): what forms when a bacterium builds a thick internal wall around its chromosome and part of the cytoplasm

fission: cell division that forms two genetically identical cells

flagellum (fluh JEH lum): a long whiplike structure that helps a bacterium move (plural, flagella)

1. Review the terms and their definitions in the Mini Glossary. Write a sentence that explains what happens to a bacterium's cell during fission.

2. Fill in the diagram below with another way bacteria move.

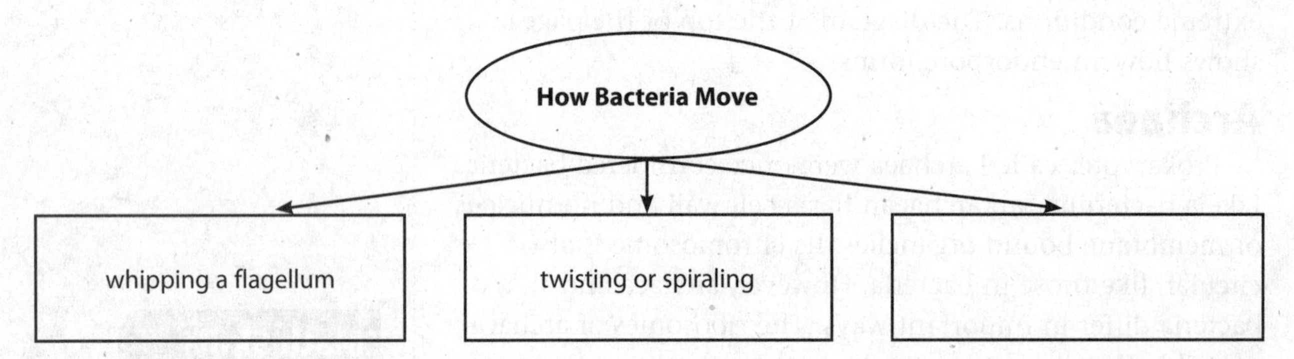

3. Write a paragraph that explains what might happen to a bacterium when environmental conditions become too dry for survival.

What do you think NOW?

Reread the statements at the beginning of the lesson. Fill in the After column with an A if you agree with the statement or a D if you disagree. Did you change your mind?

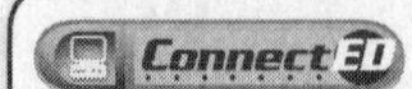

Log on to ConnectED.mcgraw-hill.com and access your textbook to find this lesson's resources.

Bacteria and Viruses

Bacteria in Nature

Before You Read

What do you think? Read the two statements below and decide whether you agree or disagree with them. Place an A in the Before column if you agree with the statement or a D if you disagree. After you've read this lesson, reread the statements to see if you have changed your mind.

Before	Statement	After
	3. All bacteria cause diseases.	
	4. Bacteria are important for making many types of food.	

Key Concepts 

- How can bacteria affect the environment?
- How can bacteria affect health?

Read to Learn

Beneficial Bacteria

When you hear the word *bacteria*, you might think about illness. Not all bacteria cause disease, though. In fact, most bacteria are beneficial, or helpful. Many organisms, including humans, depend on bacteria to survive.

Some types of bacteria live in your intestines. They help you digest your food. One type of bacteria makes vitamin K, which helps your blood to clot. Other beneficial bacteria in your intestines prevent harmful bacteria from growing.

Animals benefit from bacteria as well. For example, bacteria live in a cow's stomach. The bacteria help break down substances in the plants that the cow eats.

Decomposition

What do you think would happen if organic wastes such as food scraps and dead leaves never decayed? **Decomposition** *is the breaking down of dead organisms and organic waste.* It is an important process in nature. When a tree dies, bacteria and other organisms called decomposers feed on the dead organic matter. As decomposers break down the tree, they release molecules such as carbon and phosphorus into the soil. Other organisms can take in these molecules and use them for life processes.

Mark the Text

Underline Unfamiliar Words As you read, underline any words that you do not know. Look up and write the meaning of the underlined words in the margin. Refer to the words and their meanings as you study this lesson.

FOLDABLES®

Make a four-door book to summarize the ways in which bacteria are beneficial.

Decomposition	Nitrogen Fixation
Bioremediation	Bacteria and Food

Nitrogen Fixation

Organisms use nitrogen to make proteins. About 78 percent of the atmosphere is nitrogen gas. However, it is in a form that plants and animals cannot use. How do living things get the nitrogen they need? Some plants get nitrogen from bacteria in a process called nitrogen fixation.

Nitrogen fixation *is the conversion of atmospheric nitrogen into nitrogen compounds that are usable by living things.* Nitrogen fixation takes place in nodules, special structures on the roots of some plants. These nodules contain bacteria that change nitrogen from the atmosphere into a form that plants can use. Beans and peas have nodules that contain nitrogen-fixing bacteria.

Key Concept Check
1. State List some ways that bacteria can benefit the environment.

____

____

____

Bioremediation

Can you imagine an organism that eats pollution? Some bacteria do just that. *The use of organisms, such as bacteria, to clean up environmental pollution is called* **bioremediation** (bi oh rih mee dee AY shun). These organisms often break down sewage and other harmful substances into less-harmful materials. These materials then can be used as landfill or fertilizers. Bacteria are often used to clean contaminated areas. Some bacteria can help clean up radioactive waste, such as uranium. Without the bacteria, these substances would take centuries to break down.

Reading Check
2. Generalize Why might using bacteria to clean up environmental spills be a good choice?

____

____

____

____

Bacteria and Food

Certain foods are made with the help of bacteria. For example, some pickles are made when bacteria change the sugar in cucumbers into an acid. Bacteria are used to make foods such as yogurt, cheese, buttermilk, vinegar, and soy sauce. Bacteria also are used to make chocolate. The bacteria help break down the covering of the cocoa bean. Bacteria give chocolate some of its flavor.

Harmful Bacteria

Of the 5,000 species of bacteria, very few are considered **pathogens** (PA thuh junz)—*agents that cause disease.* Some pathogens live in your body but make you sick only when your immune system becomes weak. For example, the bacterium *Steptococcus pneumonia* lives in your throat. It can cause pneumonia if your immune system is weakened. Other pathogens can enter your body through a cut, in the air you breathe, or in the food you eat. Once they are inside your body, they can reproduce and cause disease.

Key Concept Check
3. Describe one way that bacteria can be harmful to your health.

____

____

Bacterial Diseases

Bacteria can harm your body and cause disease in two ways. Some bacteria make you sick by damaging tissues in your body. Tuberculosis is caused by a bacterium that invades lung tissue and breaks it down for food. Other bacteria can make you sick by producing toxins. One type of bacteria grows and produces toxins in foods that have been canned improperly. If the contaminated foods are eaten, the toxins can cause food poisoning, resulting in paralysis or death.

Treating Bacterial Diseases Most bacterial diseases in humans can be treated with antibiotics. **Antibiotics** (an ti bi AH tihks) *are medicines that stop the growth and reproduction of bacteria.* Many antibiotics work by keeping bacteria from building cell walls. Others affect ribosomes in bacteria, making it hard for them to produce proteins.

Many types of bacteria have developed resistance to antibiotics. This means that antibiotics no longer stop the growth and reproduction of these bacteria. Some diseases, such as tuberculosis, pneumonia, and meningitis, are now harder to treat.

Bacterial Resistance The figure below shows how bacteria can become resistant to antibiotics. The DNA in a bacterium undergoes random mutations.

SCIENCE USE V. COMMON USE

resistance

Science Use the capacity of an organism to defend itself against disease

Common Use the act of opposing something

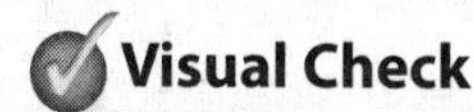

Visual Check

4. Identify What could be used to kill the bacteria in step 4? (Circle the correct answer.)

a. the nonresistant bacteria

b. a different antibiotic

c. a virus

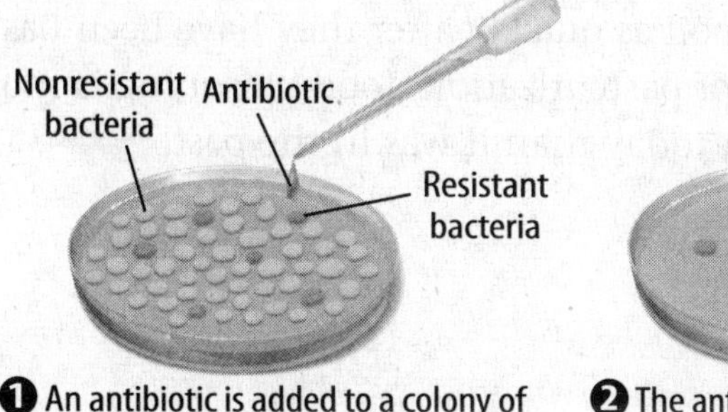

❶ An antibiotic is added to a colony of bacteria. A few of the bacteria have mutations that enable them to resist the antibiotic.

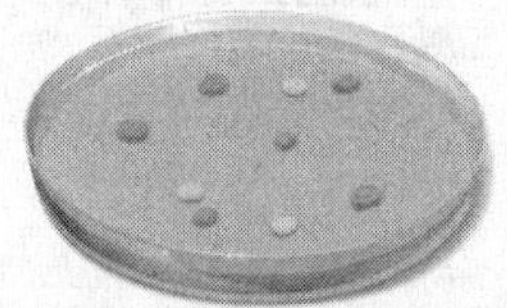

❷ The antibiotic kills most of the nonresistant bacteria. The resistant bacteria survive and reproduce, creating a growing colony of bacteria.

❸ Surviving bacteria are added to another plate containing more of the same antibiotic.

❹ The antibiotic now affects only a small percentage of the bacteria. The surviving bacteria continue to reproduce. Most of the bacteria are resistant to the antibiotic.

Bacterial Mutations Some bacterial mutations enable a bacterium to survive when a particular antibiotic is present. The bacterium is said to "resist" that antibiotic. When that antibiotic is given as a treatment, it will kill all of the bacteria except the one with the mutation.

Over time, the resistant bacterium will reproduce and become more common. The antibiotic will no longer be effective in killing that bacterium. A different antibiotic must be used to fight the disease. Scientists are always working to develop more effective antibiotics to which bacteria are not resistant.

Reading Check
5. Discuss How do bacteria develop resistance to antibiotics?

Food Poisoning

All untreated or unprocessed food contains bacteria. Over time, these bacteria reproduce and break down the food, causing it to spoil. As you read earlier, eating food that is contaminated by some bacteria can cause food poisoning. Proper treatment or processing of food before the food is stored or eaten kills most bacteria. This makes it easier to avoid food poisoning and other illnesses.

Pasteurization *The process of heating food to a temperature that kills most harmful bacteria is called* **pasteurization** (pas chuh ruh ZAY shun). Foods such as milk, ice cream, yogurt, and fruit juice are pasteurized in factories. Then they are taken to stores and sold. Pasteurization makes foods much safer to eat. The foods also do not spoil as quickly after they have been pasteurized. Because of pasteurization, food poisoning is much less common today than it was in the past.

Key Concept Check
6. Explain How does pasteurization affect human health?

After You Read

Mini Glossary

antibiotic (an ti bi AH tihk): a medicine that stops the growth and reproduction of bacteria

bioremediation (bi oh rih mee dee AY shun): the use of organisms, such as bacteria, to clean up environmental pollution

decomposition: the breaking down of dead organisms and organic waste

nitrogen fixation: the conversion of atmospheric nitrogen into nitrogen compounds that are usable by living things

pasteurization (pas chuh ruh ZAY shun): the process of heating food to a temperature that kills most harmful bacteria

pathogen (PA thuh jun): an agent that causes disease

1. Review the terms and their definitions in the Mini Glossary. Write a sentence that describes how antibiotics affect pathogens.

____

____

2. Draw a line from each role that bacteria play in the environment to the description at the right.

decomposition	
food poisoning	**Beneficial Bacteria**
disease	
nitrogen fixation	**Harmful Bacteria**
bioremediation	

3. How did researching the meaning of unfamiliar words as you read the lesson help you understand the roles of bacteria in your environment?

____

____

What do you think NOW?

Reread the statements at the beginning of the lesson. Fill in the After column with an A if you agree with the statement or a D if you disagree. Did you change your mind?

Log on to ConnectED.mcgraw-hill.com and access your textbook to find this lesson's resources.

CHAPTER 7
LESSON 3

Bacteria and Viruses

What are viruses?

Key Concepts
- What are viruses?
- How do viruses affect human health?

Before You Read

What do you think? Read the two statements below and decide whether you agree or disagree with them. Place an A in the Before column if you agree with the statement or a D if you disagree. After you've read this lesson, reread the statements to see if you have changed your mind.

Before	Statement	After
	5. Viruses are the smallest living organisms.	
	6. Viruses can replicate only inside an organism.	

Study Coach

Think-Pair-Share Work with a partner. As you read the text, discuss what you learn about bacteria for each topic. Then compare viruses with bacteria.

Visual Check

1. Name two shapes that viruses can have.

____

____

Read to Learn

Characteristics of Viruses

Have you heard of chicken pox, mumps, measles, or polio? You might have had shots to protect you from these diseases. You might have also received a shot to protect you from influenza, which is called the flu. What do these diseases have in common? They are caused by different viruses. *A* ***virus*** *is a strand of DNA or RNA surrounded by a layer of protein that can infect and replicate in a host cell.* If you have ever had a cold, you have been infected by a virus.

A virus does not have a cell wall, a nucleus, or any other organelles present in cells. Viruses are 20 to 100 times smaller than most bacteria. Viruses can have different shapes, as shown below.

Cylinder

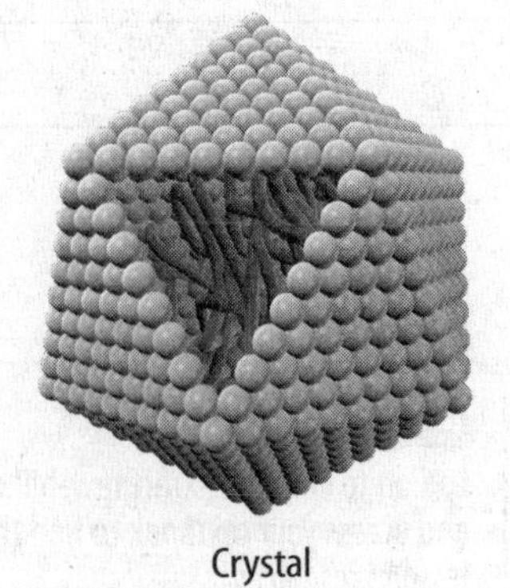
Crystal

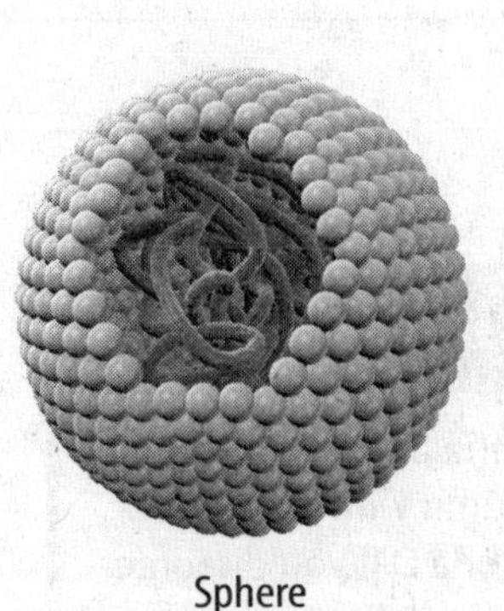
Sphere

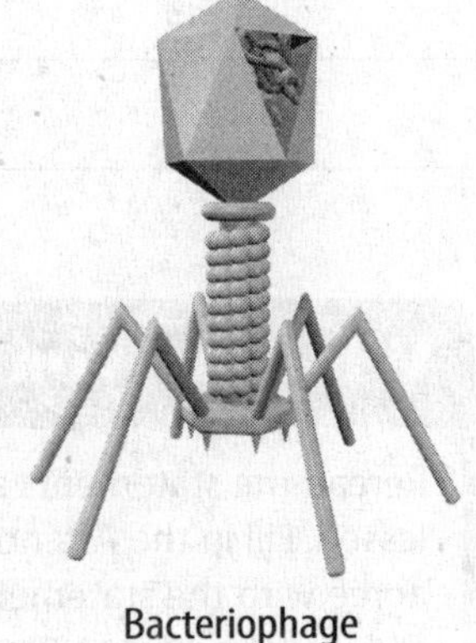
Bacteriophage

Dead or Alive?

Scientists do not consider viruses to be alive because they do not have all the characteristics of a living organism. Recall that living things are organized, respond to stimuli, use energy, grow, and reproduce. Viruses cannot do any of these things. A virus can make copies of itself in a process called replication.

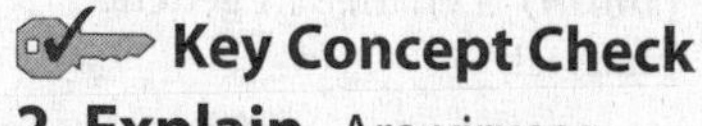

Key Concept Check

2. Explain Are viruses alive? Why or why not?

Viruses and Organisms

Viruses must use organisms to carry out their processes. Viruses have no organelles, so they cannot take in nutrients or use energy. A virus also uses the cellular parts of another organism to copy itself. Viruses must be inside a cell to replicate. The living cell that a virus infects is called a host cell. When a virus enters a cell, it can be either active or latent. The inactive stage of a virus is called the latent stage. At this stage, the virus does not replicate into more viruses.

FOLDABLES®

Make a folded book from a sheet of paper to organize your notes on viral replication.

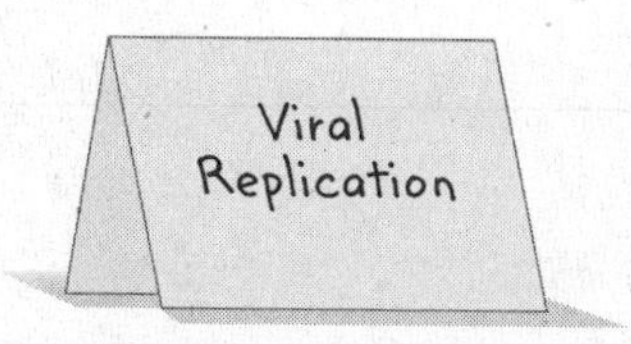

Replication

A virus cannot infect every kind of cell. A virus can only attach to a host cell with specific molecules on its cell wall or cell membrane. The diagram below shows what happens when a virus infects a host cell. After the virus attaches to the host cell, its DNA or RNA enters the host cell. Once inside, the virus does one of two things. It either becomes latent, or it starts to replicate. After a virus becomes active and replicates, the virus destroys the host cell. Copies of the virus are then released into the host organism, where they can infect other cells. When latent viruses become active again, they take control of the host cells and replicate.

Visual Check

3. Describe What occurs when a virus becomes latent?

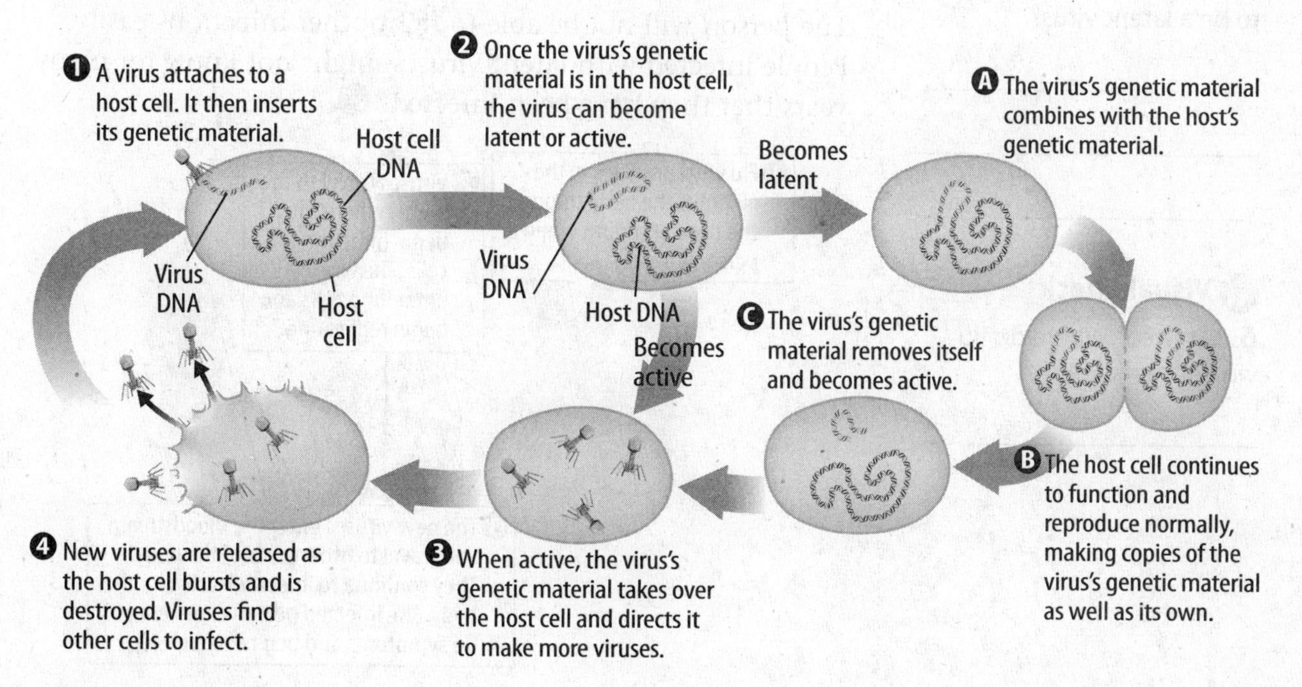

REVIEW VOCABULARY
mutation
(noun) a change in genetic material

Mutations

When viruses replicate, their DNA or RNA often mutates, or changes. These mutations are helpful to the viruses. The mutations make it possible for viruses to adjust to changes in their host cells. For example, the molecules on the outside of the host cells might change so that the viruses can no longer attach to the cell. Mutations in the viruses produce new ways for the viruses to attach to host cells. Viruses change quickly. Curing or preventing viral diseases can be difficult because the viruses can mutate again.

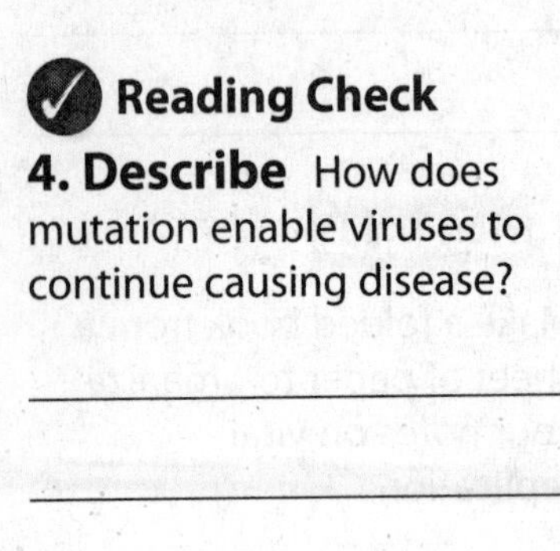

Reading Check
4. Describe How does mutation enable viruses to continue causing disease?

Viral Diseases

You have learned that viruses cause human diseases such as chicken pox, influenza, HIV (human immunodeficiency virus), and the common cold. Viruses also infect animals and plants. Most viruses attack and destroy specific cells. This cell destruction causes the symptoms of the disease.

Some viruses cause symptoms soon after infection. As shown below, flu viruses begin to replicate immediately. Flu symptoms, such as a runny nose and scratchy throat, appear within 2–3 days.

Other viruses might not cause symptoms for years. These viruses are sometimes called latent viruses. Latent viruses continue replicating without damaging the host cell. HIV is one example of a latent virus that might not cause immediate symptoms. This virus infects white blood cells, which are part of the body's immune system. At first, the infected cells function normally, so the person does not seem sick. If the virus becomes active, it destroys cells in the immune system. The person will not be able to fight other infections easily. People infected with latent viruses might not know for many years that they have been infected.

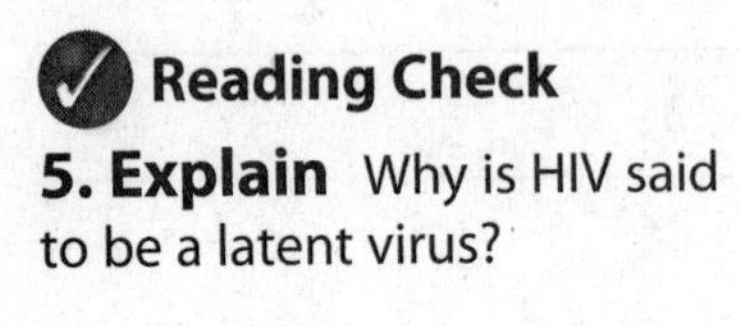

Reading Check
5. Explain Why is HIV said to be a latent virus?

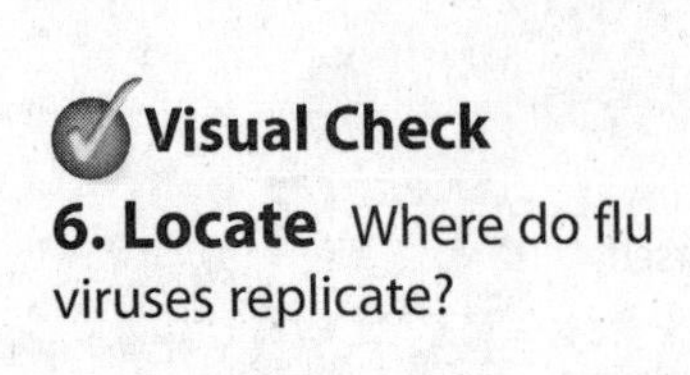

Visual Check
6. Locate Where do flu viruses replicate?

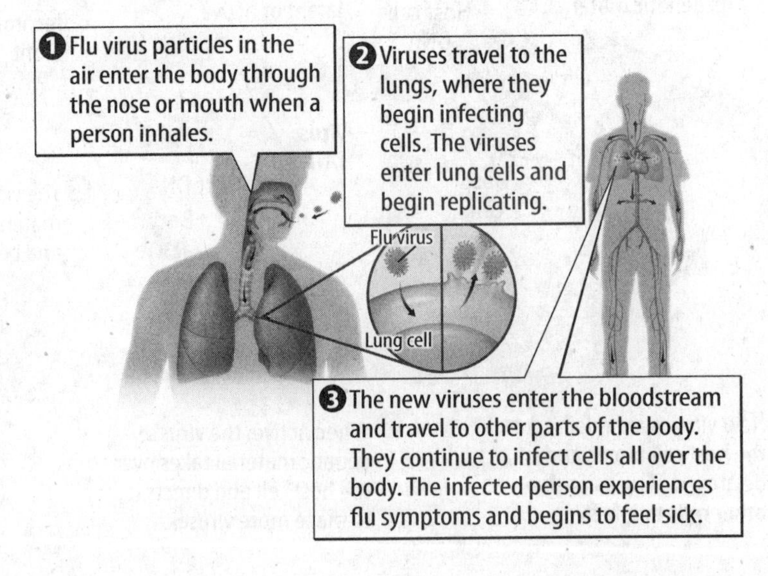

Treating and Preventing Viral Diseases

Viruses are always changing, so it can be hard to treat viral diseases. Antibiotics work only against bacteria, not against viruses. Antiviral medicines can be used to treat some viral diseases or prevent infection. These medicines keep the virus from entering cells or stop the virus from replicating. Like bacteria, viruses can change quickly and become resistant to medicines.

One of the best ways to prevent a viral infection is to limit contact with an infected human or animal. The most important way is to practice good hygiene, such as washing your hands.

Immunity

Has anyone you know ever had chicken pox? Did the person get it only once? Most people who become infected with chicken pox develop an immunity to the disease. This is an example of acquired immunity.

Acquired Immunity When a virus infects a person's body, the body begins to make proteins called antibodies. *An* ***antibody*** *is a protein that can attach to a pathogen and make it useless.* The figure below shows how antibodies fight infection. They bind to viruses and other pathogens, keeping them from attaching to a host cell. Antibodies also target viruses and signal the body to destroy them. These antibodies can multiply quickly if the same pathogen enters the body again.

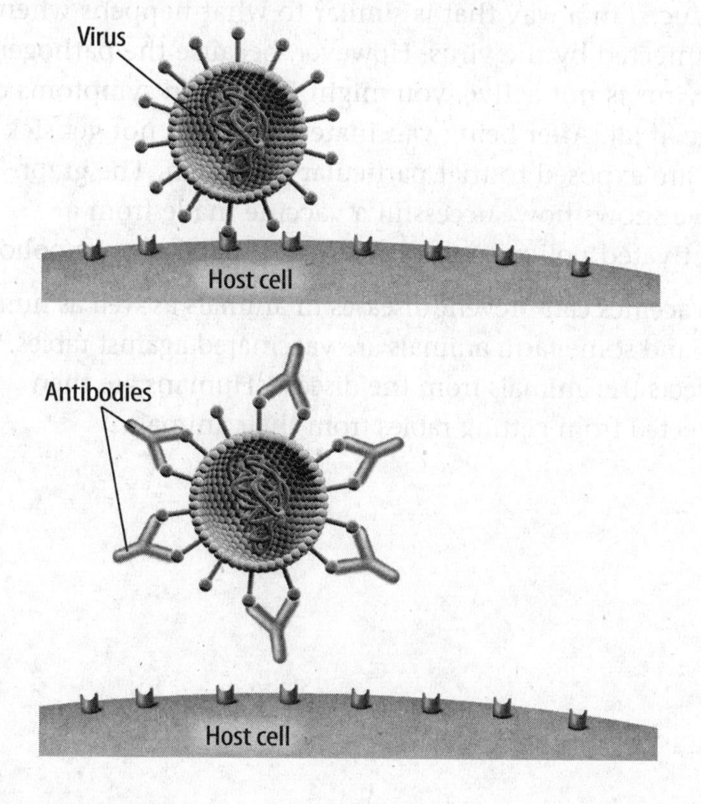

Think it Over

7. Specify If someone in your family gets a cold, how could you limit contact with him or her?

Reading Check

8. State What prevents a person from getting chicken pox more than once? (Circle the correct answer.)

a. mutation
b. replication
c. acquired immunity

Visual Check

9. Explain How does the antibody prevent the virus from attaching to the host cell?

Natural Immunity A mother can pass antibodies on to her unborn baby. This condition is called natural immunity. Like acquired immunity, natural immunity helps the body fight infection.

Reading Check

10. Define What is the condition called when a pregnant woman passes along her antibodies to her unborn baby?

____

Vaccines

You have read that people can avoid getting viral diseases through natural immunity or acquired immunity. Another way to prevent viral diseases is through vaccination. *A* **vaccine** *is a mixture containing material from one or more deactivated pathogens, such as viruses.* A vaccine is given in a vaccination.

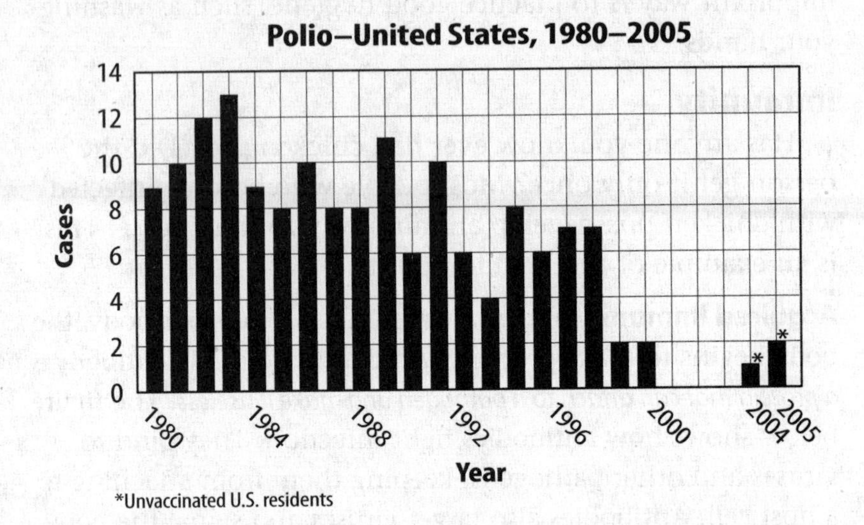

Interpreting a Graph

11. Explain There were no new cases of polio in the United States during what period? (Circle the correct answer.)

a. 2000 to 2005
b. 2000 to 2003
c. 1980 to 1999

You might get a vaccine for a viral disease. When you do, it causes your body to make antibodies. The antibodies are produced in a way that is similar to what happens when you are infected by the virus. However, because the pathogen in the virus is not active, you might have mild symptoms or none at all. After being vaccinated, you will not get sick if you are exposed to that particular pathogen. The graph above shows how successful a vaccine made from a deactivated poliovirus was against the viral disease polio.

Reading Check

12. Summarize Why don't you get the flu when you get a flu shot?

____

____

____

____

Vaccines can prevent diseases in animals as well as humans. Pets and some farm animals are vaccinated against rabies. This protects the animals from the disease. Humans are then protected from getting rabies from their animals.

Gene Transfer

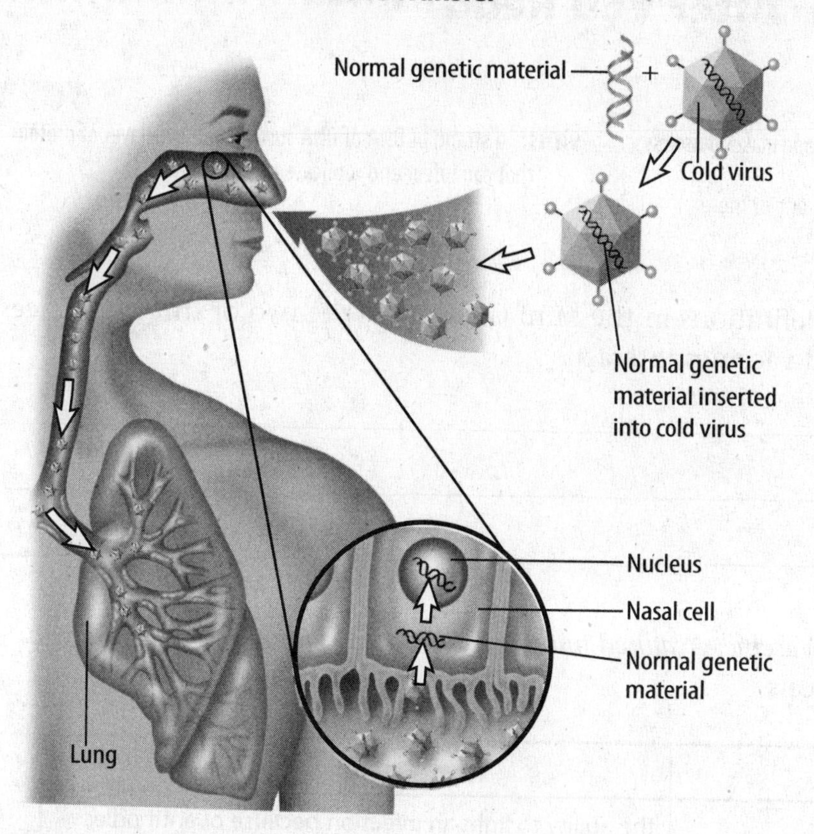

Research with Viruses

Scientists study viruses. They look for new ways to treat and prevent viral diseases in humans, animals, and plants. They also study the link between viruses and cancer. Viruses can cause changes in a host's DNA or RNA that result in tumors or abnormal growth. Because viruses can change quickly, scientists must always be working on new ways to treat and prevent viral diseases.

Did you know that viruses also can be helpful? They can be used to treat genetic disorders and cancer in a process called gene transfer. In this process, shown above, scientists use viruses to insert normal genetic information into a specific cell. Someday, gene transfer might be used to treat genetic disorders that are caused by one gene. Two of these disorders are cystic fibrosis and hemophilia.

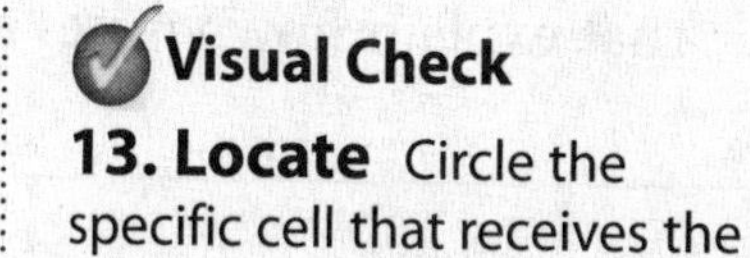

Visual Check

13. Locate Circle the specific cell that receives the normal genetic information.

Key Concept Check

14. Identify Name two ways in which viruses can affect human health.

____

____

After You Read

Mini Glossary

antibody: protein that attaches to a pathogen and makes it useless

vaccine: a mixture containing material from one or more deactivated pathogens, such as viruses

virus: a strand of DNA or RNA surrounded by a layer of protein that can infect and replicate in a host cell

1. Review the terms and their definitions in the Mini Glossary. Write two or three sentences that explain how viruses and vaccines interact.

2. Fill in the table below with *Vaccine, Acquired immunity,* and *Natural immunity* to identify ways to prevent viral infections.

Prevention Process	Definition
	the ability to fight an infection because of antibodies in our bodies from a previous infection
	the ability to fight an infection because of antibodies passed on by your mother before you were born
	the ability to fight an infection because of the injection of deactivated pathogens that cause the body to make antibodies

3. State three things you learned about bacteria and viruses.

What do you think NOW?

Reread the statements at the beginning of the lesson. Fill in the After column with an A if you agree with the statement or a D if you disagree. Did you change your mind?

Log on to ConnectED.mcgraw-hill.com and access your textbook to find this lesson's resources.

Protists and Fungi

What are protists?

Before You Read

What do you think? Read the three statements below and decide whether you agree or disagree with them. Place an A in the Before column if you agree with the statement or a D if you disagree. After you've read this lesson, reread the statements and see if you have changed your mind.

Before	Statement	After
	1. Protists are grouped together because they all look similar.	
	2. Some protists cause harm to other organisms.	
	3. Many protists make their own food.	

Key Concepts 
- What are the different types of protists and how do they compare?
- How are protists beneficial?

Read to Learn

What are protists?

When you see a living thing, you might think about whether it is a plant or an animal. You might know that a dog is an animal because of its fur and that a flower is a plant because of its green leaves. If you looked only at the cells of a plant or an animal, you might still know what it is. A plant cell has a cell wall and often contains chloroplasts. An animal cell has a cellular membrane but does not have chloroplasts. But some organisms, such as protists, are not as easy to identify and classify.

A **protist** *is a eukaryotic organism that can be plantlike, animal-like, or funguslike.* Protists share some characteristics with plants, animals, and fungi. But they do not share enough characteristics to be classified within any of those groups. Although all protists belong to one group, they are very different from one another. They have many different methods of movement and finding food. ✓

Study Coach

Use an Outline As you read, make an outline to summarize the information in the lesson. Use the main headings in the lesson as the main headings in your outline. Complete the outline with the information under each heading.

✓ Reading Check

1. Identify What is a protist?

REVIEW VOCABULARY

asexual reproduction
a type of reproduction in which one parent reproduces without a sperm and egg joining

Reproduction of Protists

Most protists reproduce asexually. The offspring are exact copies of the parent. Asexual reproduction creates new organisms quickly.

Many protists can also reproduce sexually. The offspring are genetically different from the parents. Sexual reproduction takes more time, but it creates new organisms with a variety of characteristics.

Classification of Protists

Most organisms are classified according to their similarities. But scientists must use a different approach for classifying protists. Usually, any eukaryote that is not a plant, an animal, or a fungus is classified as a protist. Protists, however, might still look and act like plants, animals, or fungi. Therefore, scientists classify protists in categories based on which group they are most similar to. The table below introduces the plantlike, animal-like, and funguslike protists.

Key Concept Check
2. Identify What are the different types of protists?

Interpreting Tables
3. Identify List two types of protists that can be multicellular.

Classification	Plantlike	Animal-Like	Funguslike
Example	algae	paramecia	slime molds
Characteristics	make their own food; unicellular or multicellular	eat other organisms for food; mostly microscopic and unicellular	break down organic matter for food; mostly multicellular

Plantlike Protists

You might have seen brown, green, or red seaweed at the beach or in an aquarium. These seaweeds are algae (AL jee). **Algae** *are plantlike protists that use light energy and carbon dioxide to produce food through photosynthesis*. There are several kinds of plantlike protists. Most plantlike protists are much smaller than funguslike multicellular organisms.

Diatom

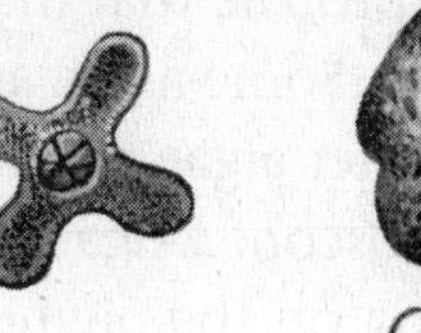

Diatoms

A **diatom** (DI uh tahm) *is a type of microscopic, plantlike protist with a hard outer wall.* Diatoms are so common that a cup of lake or pond water might hold thousands of diatoms. The cell walls of diatoms contain a large amount of silica, the main mineral in glass. Diatoms often look like colored glass. An example of a diatom is shown above.

Dinoflagellates

A dinoflagellate (di noh FLA juh lat) is a unicellular, plantlike protist that has flagella. Flagella are whiplike parts that help some protists move. The flagella whip back and forth, causing the dinoflagellate to spin. Some dinoflagellates glow in the dark when they are disturbed. This is caused by a chemical reaction in their cells. An example of a dinoflagellate is shown above. ✓

Euglenoids

A euglenoid (yew GLEE noyd) is a unicellular, plantlike protist with a flagellum at one end of its body. Euglenoids do not have cell walls. Instead, they are covered by a rigid, rubbery cell coat called a pellicle (PEL ih kul). Euglenoids have eyespots that detect light and determine where they move. A euglenoid is shown above. Euglenoids swim quickly in water or move slowly along moist surfaces. Euglenoids have chloroplasts and make their own food. If there is not enough light for photosynthesis to occur, they can absorb nutrients from decaying matter in the water. Animals such as tadpoles and small fish eat euglenoids. ✓

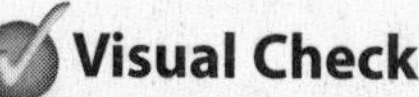

4. Identify Which plantlike protists do not have flagella? (Circle the correct answer.)

a. diatoms

b. dinoflagellates

c. euglenoids

Reading Check

5. Explain What is the purpose of flagella?

____

____

Reading Check

6. Name one characteristic that euglenoids share with plants.

____

____

____

Algae

Recall that algae are plantlike protists that make their own food through photosynthesis. Some algae are large and multicellular, such as seaweed. Other algae are unicellular and can be seen only with a microscope. Algae are classified as red, green, or brown, depending on the pigments they contain. Some types of red and brown algae look similar to plants. But they do not have the ability to transport water and nutrients, as plants do. They do not have roots. They have holdfasts, structures with a gluelike substance that helps them stick to rocks.

One unusual green algae is volvox. Volvox cells form a large sphere held together with strands of cytoplasm. These cells move as a group and beat their flagella at the same time. Some cells have the parts necessary for sexual reproduction. Volvox cells in the front of the group have larger eyespots that sense light for photosynthesis.

Reading Check

7. Describe How do red and brown multicellular algae differ from plants?

The Importance of Algae

You might be surprised by all of the materials you use that contain algae. Algae are in many ice creams, marshmallows, and puddings. Algae are also in products such as toothpaste, lotions, fertilizers, and some swimming-pool filters.

Think it Over

8. Discuss How are algae beneficial to an ecosystem?

Algae and Ecosystems

Algae are food for animals and animal-like protists. They also give shelter to many organisms that live in water. Groups of tall brown algae are called kelp forests.

Algae and other photosynthetic protists can help remove pollution from water. But this pollution can be food for the algae, causing the population of algae to increase quickly. The algae produce waste products that can poison other organisms. When large numbers of algae are present, the water can look red or brown. This is called red tide, which is a harmful algal bloom.

Reading Check

9. Explain What causes a red tide?

Animal-Like Protists

Some protists are more like animals than plants. **Protozoans** (proh tuh ZOH unz) *are protists that resemble tiny animals*. These animal-like protists share several characteristics. They do not have chloroplasts or make their own food. Protozoans are usually microscopic and unicellular. Most are found in wet environments.

Ciliates

Many protists have cilia (SIH lee uh). **Cilia** *are short, hairlike structures that grow on the surface of some protists.* Protists with cilia are called ciliates. As shown in the figure below, the entire surface of the cell is covered by cilia. The beating cilia help move the animal-like protist through water.

A **paramecium** (pa ruh MEE see um) *is a protist with cilia and two types of nuclei.* A paramecium, like most ciliates, gets its food by forcing water though a groove in its side. The groove closes and a food vacuole, or storage area, forms within the cell. The food particles are digested and the extra water is forced back out of the cell. Ciliates reproduce asexually. They can exchange some genetic material through a process called conjugation (kahn juh GAY shun).

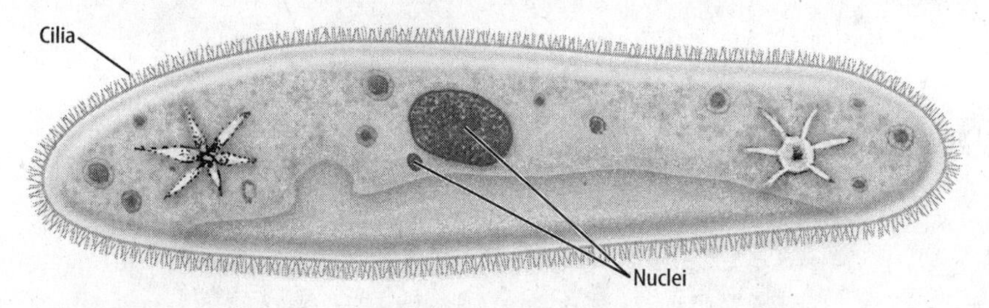

Flagellates

Recall that dinoflagellates are a type of plantlike protist that uses one or more flagella to move. Flagellates are a type of protozoan that also has one or more flagellum. A flagellate does not always spin when it moves.

Flagellates eat decaying matter, including plants, animals, and other protists. Many flagellates live in the digestive systems of animals and absorb nutrients from food moving through the systems.

Reading Check

10. Identify What function do cilia perform?

Visual Check

11. Explain What are the characteristics of a paramecium?

Reading Check

12. Identify two different sources of food for flagellates.

Sarcodines

Animal-like protists called sarcodines (SAR kuh dinez) have no specific shape. At rest, they look like a glob of cytoplasm, or cellular material. These animal-like protists can ooze into almost any shape. *An* **amoeba** (uh MEE buh) *is one common sarcodine with an unusual adaptation for movement and getting nutrients.*

An amoeba moves with a pseudopod, as shown below. *A* **pseudopod** *is a temporary "foot" that forms as the organism pushes part of its body outward.* An amoeba moves by stretching out a pseudopod, then oozing the rest of its body up into the pseudopod.

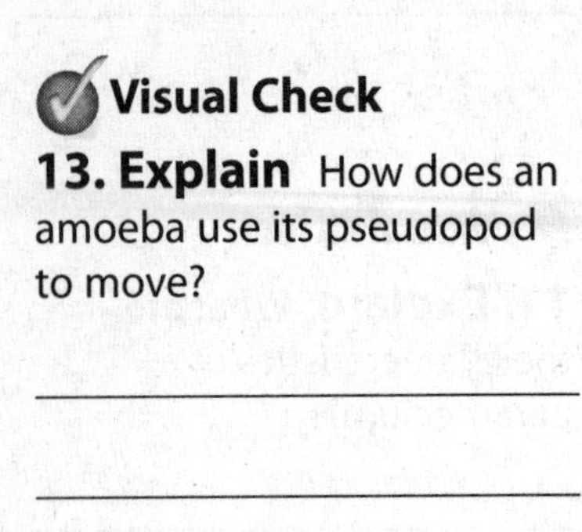

Visual Check

13. Explain How does an amoeba use its pseudopod to move?

____

____

____

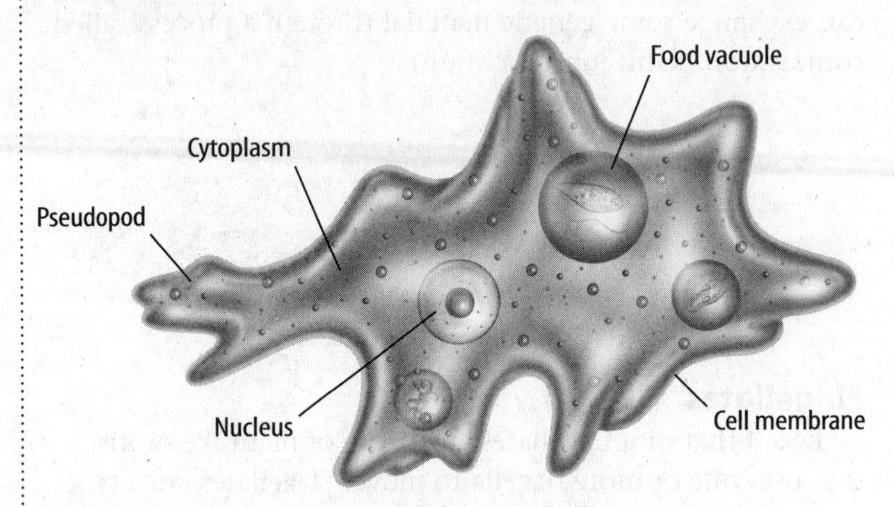

Amoebas also use pseudopods to get nutrients. An amoeba will surround a smaller organism or food particle with its pseudopod and then ooze around it. A food vacuole forms inside the pseudopod, where the food is digested.

Most sarcodines get nutrients and energy by eating other organisms. Some make their own food. Others live in human digestive systems, where they get nutrients and energy.

FOLDABLES®

Use a three-tab book to organize your notes about protozoans and how they move.

How Protozoans Move

Flagella | Cilia | Pseudopod

The Importance of Protozoans

Protozoans are important members of ecosystems. They break down dead plant and animal matter. This decomposed matter is then recycled into the environment and used by other living organisms.

Some protozoans can cause diseases. They act as parasites and live inside a host organism. Inside the host organism, the protozoan feeds off the host and can kill it.

Malaria is one disease caused by protists. Malaria is spread to humans by mosquitoes, as shown in the figure below. Protozoan parasites called plasmodia live and reproduce in red blood cells. Malaria kills more than a million people yearly.

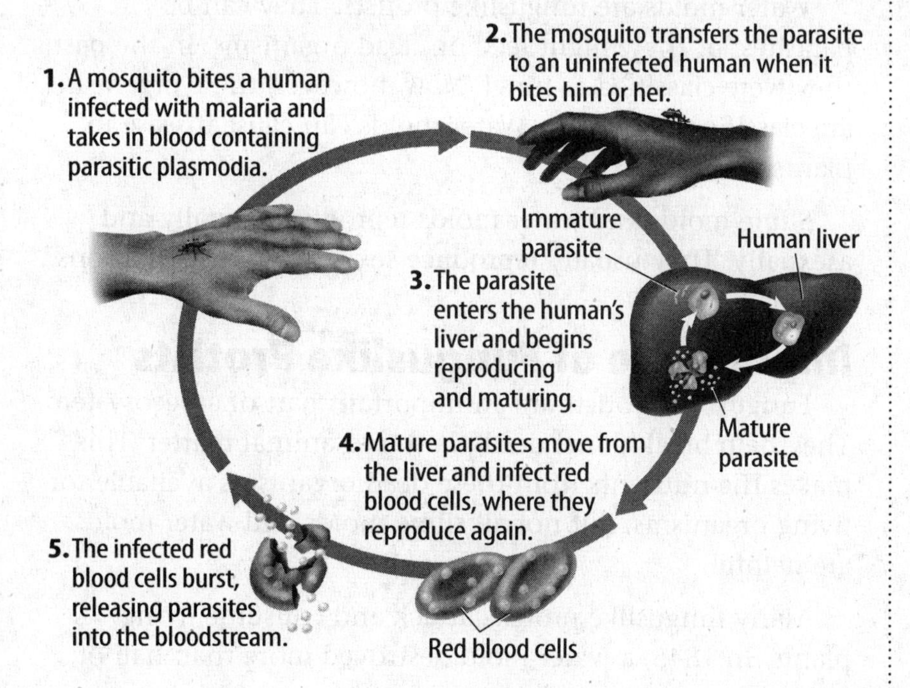

Key Concept Check

14. Discuss How are protists helpful and harmful to humans?

____

____

Visual Check

15. Identify Circle the stage when the parasite transfers to a healthy human.

Funguslike Protists

There is another category of protists, the funguslike protists. Funguslike protists share some characteristics with fungi, but they have some differences that cause them to be classified as protists.

Slime and Water Molds

Slime molds are funguslike protists. They come in a variety of colors and forms. Slime molds often live on the surface of plants. The body of a slime mold is made of cell material and nuclei floating in a slimy mass. Most slime molds absorb nutrients from organic materials in the environment. ✓

Water molds are funguslike protists. They can be parasites, or they might feed on dead organisms. In the past, they were classified as fungi. Now, however, these organisms are classified as protists. Water molds can cause disease in plants.

Slime molds and water molds reproduce sexually and asexually. They usually reproduce sexually when conditions are harsh.

✓ Reading Check

16. Explain Where do slime molds get their nutrients?

____

____

Importance of Funguslike Protists

Funguslike protists are an important part of an ecosystem. They help break down dead plant and animal matter. This makes the nutrients from these dead organisms available for living organisms. But not all slime molds and water molds are helpful.

Many funguslike protists attack and consume living plants. In 1845, a water mold destroyed more than half of Ireland's potato crop. This was known as the Irish Potato Famine. More than one million people starved as a result of this famine.

Think it Over

17. Discuss How are funguslike protists beneficial to an environment?

____

____

After You Read

Mini Glossary

algae (AL jee): plantlike protists that produce food through photosynthesis using light energy and carbon dioxide

amoeba (uh MEE buh): one common sarcodine with an unusual adaptation for movement and getting nutrients

cilia (SIH lee uh): short, hairlike structures that grow on the surface of some protists

diatom (DI uh tahm): a type of microscopic plantlike protist with a hard outer wall

paramecium (pa ruh MEE see um): a protist with cilia and two types of nuclei

protist: a eukaryote organism that can be plantlike, animal-like, or funguslike

protozoans (proh tuh ZOH unz): protists that resemble tiny animals

pseudopod: a temporary "foot" that forms as an organism pushes part of its body outward

1. Review the terms and their definitions in the Mini Glossary. Write a sentence comparing two different types of protists.

2. Fill in the chart below about the different types of protists.

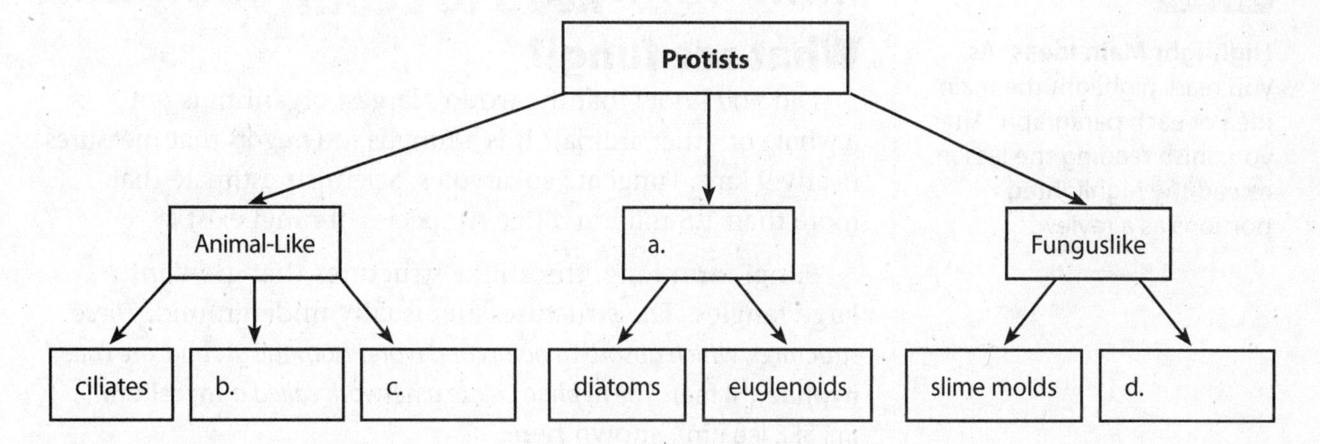

3. How did the use of an outline help you understand what you read?

What do you think NOW?

Reread the statements at the beginning of the lesson. Fill in the After column with an A if you agree with the statement or a D if you disagree. Did you change your mind?

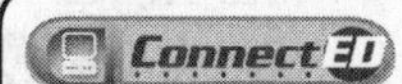

Log on to ConnectED.mcgraw-hill.com and access your textbook to find this lesson's resources.

Protists and Fungi

What are fungi?

Key Concepts
- What are the different types of fungi and how do they compare?
- Why are fungi important?
- What are lichens?

Before You Read

What do you think? Read the three statements below and decide whether you agree or disagree with them. Place an A in the Before column if you agree with the statement or a D if you disagree. After you've read this lesson, reread the statements and see if you have changed your mind.

Before	Statement	After
	4. Mushrooms and yeasts are two types of fungi.	
	5. Fungi are always helpful to plants.	
	6. Some fungi can be made into foods or medicines.	

Mark the Text

Highlight Main Ideas As you read, highlight the main idea of each paragraph. After you finish reading the lesson, reread the highlighted portions as a review.

Read to Learn

What are fungi?

Did you know that the world's largest organism is not a whale or other animal? It is a fungus in Oregon that measures nearly 9 km^2. Fungi are eukaryotes. Scientists estimate that more than 1.5 million different species of fungi exist.

Fungi form long, threadlike structures that grow into large tangles. The structures are usually underground. *These structures, which absorb minerals and water from the ground, are called* **hyphae** (HI fee). *The hyphae create a network called a* **mycelium** (mi SEE lee um), shown here. Fungi are heterotrophs, which means they cannot make their own food. Some fungi are parasites. These fungi get their nutrients from living organisms. Most fungi get their food by releasing chemicals that break down dead matter. The fungi then absorb the nutrients.

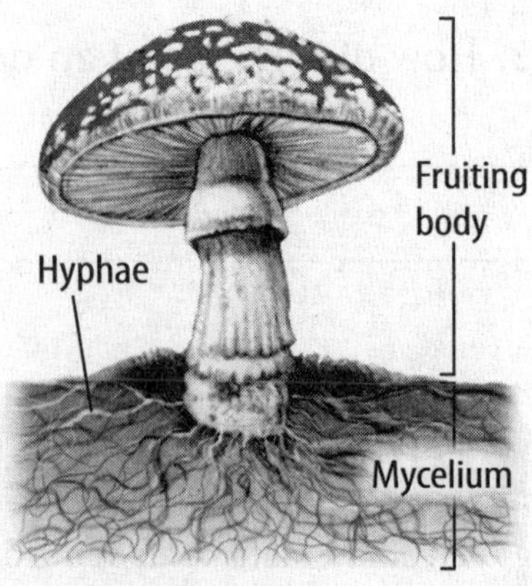

Reading Check

1. Describe How are hyphae and mycelium related?

Types of Fungi

Scientists group fungi based on how they look and how they reproduce. Fungi can reproduce sexually or asexually. Almost all fungi reproduce asexually by producing spores. Spores are small reproductive cells with a strong outer covering. The spores can grow into new fungi.

There are four groups of fungi:

- club fungi
- sac fungi
- zygote fungi
- imperfect fungi

The classification of fungi and funguslike protists often changes as scientists learn more about them. Technology and discoveries might lead to more changes in the classification of some imperfect fungi.

Club Fungi

When you read the title of this lesson, you might have thought about mushrooms. Mushrooms are a type of club fungi. Their reproductive structures have a clublike shape. The part of a mushroom that grows above ground is called a basidiocarp (bus SIH dee oh karp). Inside the basidiocarp are the **basidia** (buh SIH dee uh; singular, basidium), *reproductive structures that produce sexual spores*.

Most of a club fungus is a network of hyphae that grows underground and absorbs nutrients. Many club fungi are named for their shapes and characteristics. Puff-balls, stinkhorns, and birds' nest fungi belong to this group. There is even a club fungus that glows in the dark because of a chemical reaction in its basidiocarp.

Sac Fungi

Did you know that bread and diaper rash have something in common? A type of sac fungus causes bread dough to rise. A different sac fungus causes a rash to develop on damp skin. Sac fungi can cause diseases in plants and animals. Humans eat some sac fungi, such as truffles and morels.

Sac fungi are named for their reproductive structures. *The* **ascus** (AS kuhs) *is the reproductive structure where spores develop on sac fungi.* The ascus often looks like the bottom of a tiny bag or sack. The spores of a sac fungus are called ascospores (AS kuh sporz). Sac fungi can reproduce sexually or asexually. Many yeasts are sac fungi. The gas released by the yeast during cellular respiration causes bread dough to rise.

Key Concept Check

2. Name the four groups of fungi.

____

____

____

____

FOLDABLES®

Make a four-door book and use it to organize information about the characteristics of the different classifications of fungi.

Zygote fungi | Sac fungi | Club fungi | Imperfect fungi

Think it Over

3. Predict How do sac fungi cause bread to rise?

____

____

____

Zygote Fungi

Bread mold is caused by another type of fungus called a zygote fungus. Zygote fungi grow in moist areas, such as a damp basement or a bathroom shower.

The hyphae of a zygote fungus grow over the surface of a material, such as bread. There, the hyphae dissolve the material and adsorb the nutrients. *When this fungus undergoes sexual reproduction, tiny stalks called* ***zygosporangia*** (ZI guh spor AN jee uh) *form.* The zygosporangia release spores called zygospores. New zygote fungi can grow from zygospores.

Reading Check

4. Contrast How are sac fungi and zygote fungi different?

Imperfect Fungi

Athlete's foot and blue cheese are two things that can be caused by imperfect fungi. The moist environments near a shower and in a sweaty shoe are perfect places for the imperfect fungus that causes athlete's foot to grow. The blue color in blue cheese comes from colonies of imperfect fungi that are added during the cheese-making process.

These fungi are called imperfect because scientists have not observed a sexual, or "perfect," reproductive stage in their life cycle. Because fungi are classified by the shape of their reproductive structures, these types of fungi do not fit other categories.

Reading Check

5. Explain Why are imperfect fungi classified that way?

The Importance of Fungi

Fungi are important to humans. Chocolate, carbonated soda, cheese, bread, and medicines are made using fungi. Some fungi are used as a substitute for meat because they are high in protein and low in cholesterol. Other fungi are used in making antibiotics.

Decomposers

Fungi are important because they are eaten by people. Fungi are also important because of the things they eat. Fungi are necessary to the environment because they break down dead plant and animal matter. Without fungi and other organisms, dead plants and animals would pile up year after year. Fungi also help break down pollution in soil. Without fungi, this pollution would build up in the environment.

When fungi break down dead matter, they help put nutrients back into soil. Plants need these nutrients to grow. Fungi help keep soil fertile.

Think it Over

6. Identify three products you have used recently that were made using fungi.

Fungi and Plant Roots

Many fungi and plants help each other as they grow together. Recall that the hyphae of a fungus take in minerals and water underground. *The roots of a plant and the hyphae of a fungus weave together to form a structure called a* **mycorrhiza** (mi kuh RI zuh), as shown below.

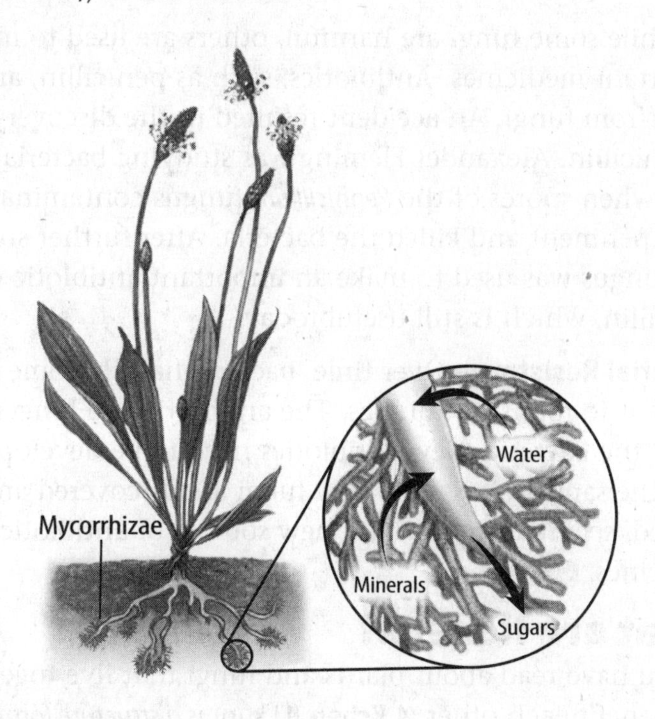

Mychorrhizae can exchange molecules. As fungi break down decaying matter in soil, they make nutrients more available to plants. They also increase the amount of water that plants can absorb because they increase the surface area of the roots.

Fungi do not use light energy to make their own food. The fungi in mycorrhiza take in some of the sugars from plants. Plant produce these sugars through photosynthesis. The plants receive more nutrients and water because of the fungi. The fungi continue to grow using the sugars from the plants. So, both a plant and a fungus benefit in this relationship. Scientists suspect that most plants can benefit from mychorrhizae. 

Health and Medicine

Fungi, like protists, can be harmful to humans. A small number of people die every year from eating poisonous mushrooms or spoiled food containing harmful fungi.

Math Skills

Under certain conditions, 100 percent of the cells in fungus A reproduce in 24 hours. The number of cells of fungus A doubles once each day.

Day 1 = 10,000 cells
Day 2 = 20,000 cells
Day 3 = 40,000 cells
Day 4 = 80,000 cells

When an antibiotic is added to the fungus, the growth is reduced by 50 percent. Only half the cells reproduce each day.

Day 2 = 15,000 cells
Day 3 = 22,500 cells
Day 4 = 33,750 cells

7. Calculate Without an antibiotic, how many cells of fungus A would there be on day 6?

____

Reading Check

8. Describe How do mychorrhizae benefit both a plant and a fungus?

____

____

Effects of Fungi You do not have to eat fungi for them to make you sick or uncomfortable. Some fungi can cause allergies, athlete's foot rashes, diaper rashes, pneumonia, and thrush. Thrush is a yeast infection that grows in the mouth. It is most common in infants and in people with weak immune systems.

While some fungi are harmful, others are used to make important medicines. Antibiotics, such as penicillin, are made from fungi. An accident resulted in the discovery of penicillin. Alexander Fleming was studying bacteria in 1928 when spores of the *Pennicillium* fungus contaminated his experiment and killed the bacteria. After further study, this fungus was used to make an important antibiotic called penicillin, which is still useful today.

Bacterial Resistance Over time, bacteria have become resistant to many antibiotics. The antibiotics no longer work to kill the bacteria. New antibiotics need to be developed to treat the same diseases. As new fungi are discovered and studied, scientists might find new sources of antibiotics and medicines.

Key Concept Check
9. Describe two ways that fungi are important to humans.

____

____

What are lichens?

You have read about plants and fungi that live together and benefit each other. *A* **lichen** (LI kun) *is a structure formed when fungi and some other photosynthetic organisms grow together.* Usually, a lichen consists of a sac fungus or a club fungus that lives in a partnership with either a green alga or a photosynthetic bacterium. The fungus hyphae grow in a layer around the algal cells.

WORD ORIGIN
lichen
from Greek *leichen,* means "what eats around itself"

Green algae and photosynthetic bacteria are autotrophs. This means that they can make their own food using photosynthesis. Lichens are beneficial to both organisms, just as mycorrhizae are. In lichens, the fungus provides water and minerals, while the bacterium or alga provide the sugars and oxygen made from photosynthesis.

The Importance of Lichens

Lichens are found in many harsh environments. They might be found on a steep, rocky cliff near the ocean. The fungi can absorb water, help break down the rocks, and get minerals for the algae or bacterium. The algae can make food for the fungi through photosynthesis. The diagram below shows how lichens form.

Once lichens have settled into an area, it becomes a better environment for other organisms. Many animals that live in harsh environments survive by eating lichens. Plants benefit from lichens because the fungi help break down the rocks and form soil. Plants can grow in the soil, providing a source of food for other organisms.

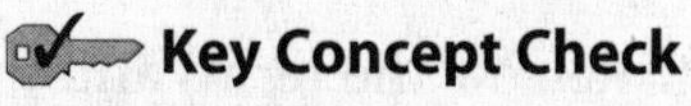

Key Concept Check

10. Identify Algae and fungi help form which of the following? (Circle the correct answer.)

a. plants
b. air
c. lichens

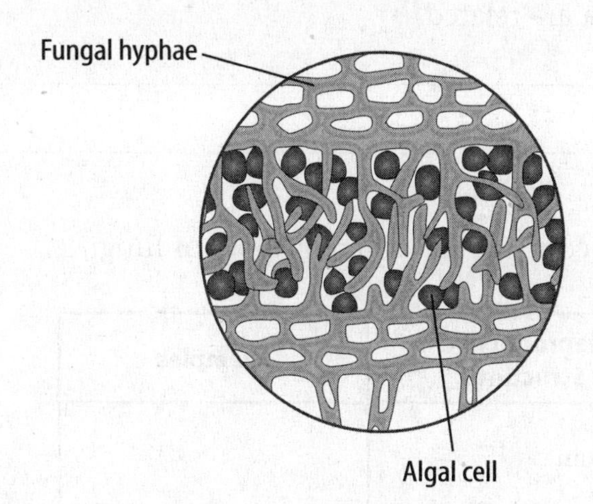

Visual Check

11. Circle the two organisms that form a lichen.

After You Read

Mini Glossary

ascus (AS kuhs): the reproductive structure where spores develop on sac fungi

basidium (buh SIH dee um): a reproductive structure of club fungi that produces sexual spores

hyphae (HI fee): long, threadlike structures that grow into large tangles, usually underground, that absorb minerals and water

lichen (LI kun): a structure formed when fungi and certain other photosynthetic organisms grow together

mycelium (mi SEE lee um): a network of hyphae

mycorrhiza (mi kuh RI zuh): a structure formed when the roots of plants and the hyphae of fungi weave together

zygosporangia (zi guh spor AN jee uh): tiny stalks that form when a zygote fungus undergoes sexual reproduction

1. Review the terms and their definitions in the Mini Glossary. Write a sentence that explains how hyphae and a mycorrhiza are related.

____

____

2. Fill in the table below to compare and contrast the different types of fungi.

Types of Fungi	Reproductive Structures	Examples
Club	basidium	
Sac		yeasts, morels
Zygote	zygosporangia	
Imperfect		athlete's foot

3. Why are fungi an important part of the environment?

____

____

What do you think NOW?

Reread the statements at the beginning of the lesson. Fill in the After column with an A if you agree with the statement or a D if you disagree. Did you change your mind?

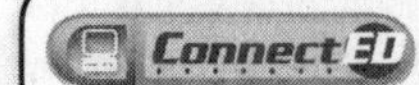

Log on to ConnectED.mcgraw-hill.com and access your textbook to find this lesson's resources.

Diversity

is a plant?

Before You Read

t do you think? Read the two statements below and decide
ther you agree or disagree with them. Place an A in the Before column
u agree with the statement or a D if you disagree. After you've read
lesson, reread the statements to see if you have changed your mind.

re	Statement	After
	1. All plants produce flowers and seeds.	
	2. Humans depend on plants for their survival.	

Key Concepts 
- What characteristics are common to all plants?
- What adaptations have enabled plant species to survive Earth's changing environments?
- How are plants classified?

Read to Learn

eristics of Plants

n important part of life on Earth. As you read
k for the characteristics that make plants so
her organisms.

e of eukaryotic cells, which have membrane-
Some of a plant cell's organelles are shown
v. A plant cell differs from an animal cell
oroplasts and a cell wall. Chloroplasts
gy to chemical energy. The cell wall
and protection. A mature plant cell also has
oles that store a watery liquid called sap.

Study Coach

Preview Headings Before you read the lesson, preview all the headings. Make a chart and write a question for each heading beginning with *What* or *How*. As you read, write the answers to your questions.

Reading Check

1. Describe the structure of a plant cell.

Plant Cell Structure

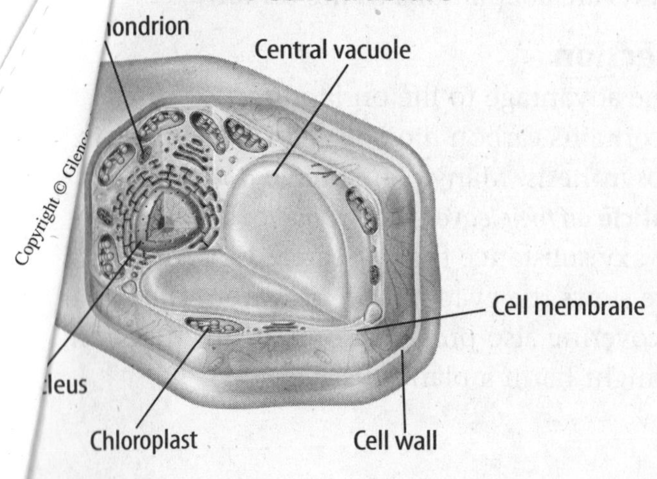

Visual Check

2. Illustrate Highlight the cell wall in the figure.

Multicellular

Plants are multicellular. This means they are made of many cells. The cells carry out specialized functions and work together to keep the plant alive. Some plants, such as the reproductive stage of some ferns, are microscopic. Other plants, such as redwood trees, are some of the largest organisms on Earth.

Producers

Organisms that use an outside energy source, such as the Sun, to make their own food are called **producers.** Plants are producers. Plants make their own food, a simple sugar called glucose, during a process called photosynthesis. All other organisms rely on producers, either directly or indirectly, for their sources of food.

Key Concept Check
3. Name What characteristics are common to all plants?

____

____

____

Plant Adaptations

Millions of years ago, no land plants existed. Scientists hypothesize that present-day land plants and green algae evolved from a common ancestor. They base their hypothesis on chemical similarities between green algae and plants. Green algae and land plants have some of the same kind of pigments. DNA similarities are also found between these two groups of organisms.

The first land plants probably lived in moist areas. Life o[n] land would have provided some advantages to plants. Plent[y] of sunlight would have been available for photosynthesis t[o] occur. The air that surrounded those plants would have bee[n] a mixture of gases, including carbon dioxide. Carbon dioxi[de] is needed for photosynthesis. As land plants became more abundant, the amount of oxygen in the atmosphere increased because oxygen is a product of photosynthesis.

Plant species also had to adapt to survive without bein[g] surrounded by water. Many of the characteristics we now [see] in plants are adaptations to life on land.

FOLDABLES
Make a vertical two-tab book to organize your notes on common plant characteristics and more-specific plant adaptations.

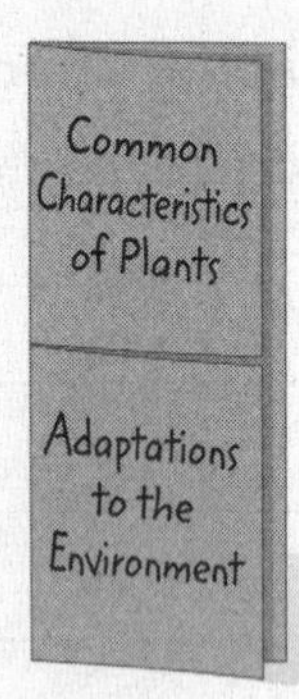

Protection

One advantage to life on land is a constant supply of [air] that contains carbon dioxide. Carbon dioxide is needed f[or] photosynthesis. Many plants have *a waxy, protective layer c[alled] the* ***cuticle*** *on their leaves, stems, and flowers.* The cuticle is m[ade] of a waxy substance that is secreted by the cells. Its waxy nature slows the evaporation of water from a plant's sur[face]. This covering also provides some protection from insect[s] that might harm a plant's tissues.

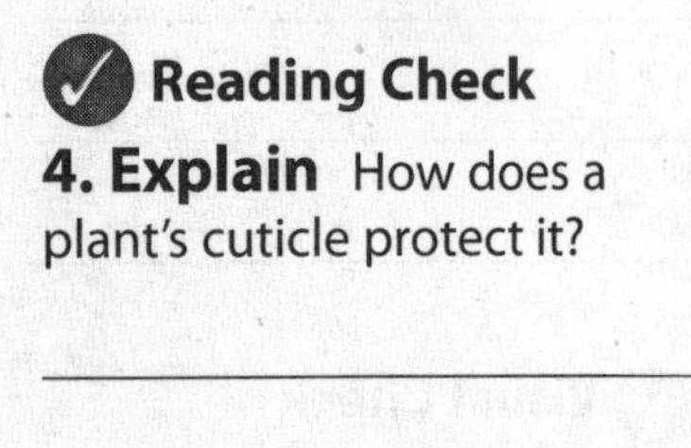

Reading Check
4. Explain How does a plant's cuticle protect it?

____

____

____

Support

The water that surrounds aquatic plants supports them. Land plants support themselves. Like all cells, a plant cell has a cell membrane. Recall that a rigid cell wall surrounds the cell membrane in a plant cell. The cell wall provides support and is made of cellulose. **Cellulose** *is an organic compound made of chains of glucose molecules.*

Many land plants also produce a chemical compound called lignin (LIG nun). Lignin strengthens cellulose in the cell walls and makes the walls more rigid. The combined strength of all of a plant's cell walls provides support for the plant. Wood is mostly made of cellulose and lignin.

Transporting Materials

In order for a plant to survive, water and nutrients must move throughout its tissues. In some plants, such as mosses, these materials move from cell to cell by the processes of osmosis and diffusion. This means that water and other materials dissolved in water move from areas of a plant where they are more concentrated to areas where they are less concentrated.

Other plants, such as grasses and trees, have specialized tissues called vascular tissue. **Vascular tissue** *is composed of tubelike cells that transport water and nutrients in some plants.* Vascular tissue can carry materials over great distances throughout a plant. Material can travel up to hundreds of meters.

Reproduction

Water carries the reproductive cells of aquatic plants from plant to plant. Adaptations in land plants enable them to reproduce in other ways.

Some plants have water-resistant seeds or spores that are part of their reproductive process. Seeds and spores move throughout environments in different ways. Animals transport seeds, and environmental factors such as wind and water move seeds from place to place. For example, burrs containing seeds may cling to a dog's fur, the wind carries milkweed seeds, and coconut seeds float in water.

Plant Classification

Recall that kingdoms such as the animal kingdom consist of smaller groups called phyla. Members of the plant kingdom are organized into groups called divisions instead of phyla. Like all organisms, each plant has a two-word scientific name. For example, the scientific name for a red oak is *Quercus rubra.*

ACADEMIC VOCABULARY
transport
(verb) to carry somebody or something

Key Concept Check
5. Identify What adaptations of plants have enabled them to survive Earth's changing environments?

Key Concept Check
6. State How are plants classified?

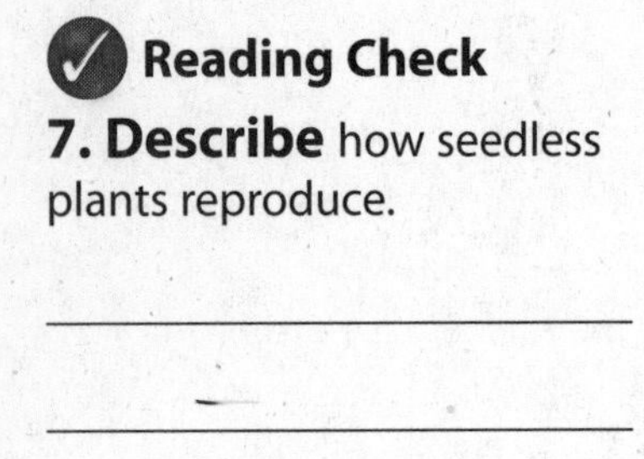

Seedless Plants

Liverworts and mosses reproduce by structures called spores. Plants that reproduce by spores often are called seedless plants.

Seedless plants do not have flowers. Some seedless plants do not have vascular tissue and are called nonvascular plants. Other seedless plants, such as ferns, have vascular tissue and are called vascular plants. Seedless plants are classified into several divisions.

Seed Plants

Most of the plants around you—such as pine trees, grasses, petunias, and oak trees—are seed plants. Almost all the plants we use for food are seed plants.

Some seed plants have flowers that produce fruit with one or more seeds. Others, such as pine trees, produce their seeds in cones. Each seed has tissues that surround, nourish, and protect the tiny plant embryo inside it. It is thought that all present-day land plants originated from a common ancestor, an ancient green algae.

Think it Over

8. Compare How are pine trees and petunias alike?

After You Read

Mini Glossary

cellulose: an organic compound made of chains of glucose molecules

cuticle: a waxy, protective layer on a plant's leaves, stems, and flowers

producer: an organism that uses an outside energy source, such as the Sun, to make its own food

vascular tissue: composed of tubelike cells that transport water and nutrients in some plants

1. Review the terms and their definitions in the Mini Glossary. Use one of the terms in an original sentence.

____

____

2. Use what you have learned about plant adaptations to complete the table.

Adaptation	Description
cuticle	• •
	• helps make plant rigid to provide support
seeds or spores	• •
	• transports water and nutrients

3. In the space below, record a concept you now understand by answering one of the questions from your chart.

____

____

____

What do you think NOW?

Reread the statements at the beginning of the lesson. Fill in the After column with an A if you agree with the statement or a D if you disagree. Did you change your mind?

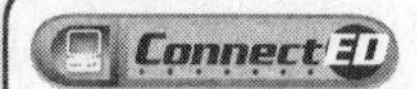

Log on to ConnectED.mcgraw-hill.com and access your textbook to find this lesson's resources.

Plant Diversity

Seedless Plants

Key Concept 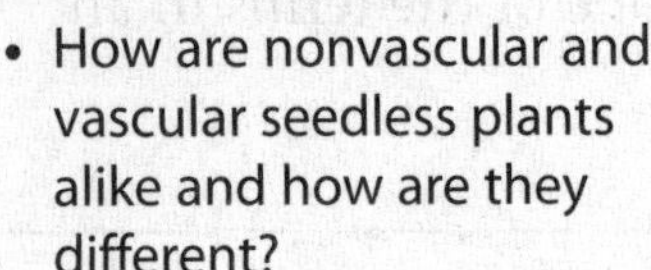
- How are nonvascular and vascular seedless plants alike and how are they different?

Before You Read

What do you think? Read the two statements below and decide whether you agree or disagree with them. Place an A in the Before column if you agree with the statement or a D if you disagree. After you've read this lesson, reread the statements to see if you have changed your mind.

Before	Statement	After
	3. Some plants move water only by diffusion.	
	4. Mosses can grow only in moist, shady places.	

Read to Learn

Nonvascular Seedless Plants

Mark the Text

Identify the Main Ideas Write a phrase beside each paragraph that summarizes the main point of the paragraph. Use the phrases to review the lesson.

If someone asked you to make a list of plants, your list might include plants such as your favorite flowers or trees that grow near your home. Your list probably would not include any nonvascular seedless plants.

Many scientists refer to all nonvascular seedless plants as bryophytes (BRI uh fites). Bryophytes usually are small. These plants lack vascular tissue, the tubelike structures that transport water and nutrients. In bryophytes, materials move from cell to cell by diffusion and osmosis. Bryophytes usually grow in moist environments.

REVIEW VOCABULARY
osmosis
the diffusion of water molecules

Because bryophytes do not have vascular tissue, they do not have roots, stems, or leaves. They have rootlike structures called rhizoids. **Rhizoids** *are structures that anchor a nonvascular seedless plant to a surface*. Rhizoids can be unicellular—consisting of only one cell—or they can be multicellular. Rhizoids are shown in the figure at the top of the next page.

 **Reading Check**

1. State the characteristics that are common in bryophytes.

____

____

____

____

____

The photosynthetic tissue of bryophytes is often only one cell layer thick. This layer does not have a cuticle, which most other plants have.

Bryophytes reproduce by spores. This requires water. Mosses, liverworts, and hornworts are bryophytes.

Rhizoids

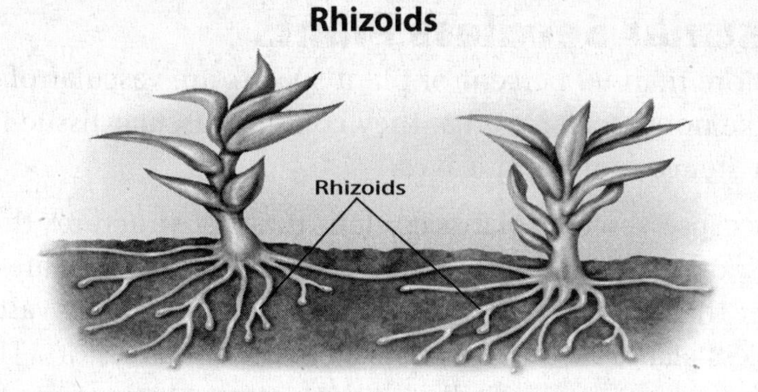

Mosses

The most common bryophytes are mosses. These small, green plants grow in forests, in parks, and sometimes in the cracks of sidewalks. Mosses usually grow in shady, damp environments, but they are able to survive periods of dryness. ✓

Mosses have leaflike structures that grow on a stemlike structure called a stalk. They have multicellular rhizoids.

Mosses play an important role in the ecosystem. They are often the first plants to grow in barren areas or after a natural disturbance such as a fire or a mudslide.

Peat moss is formed from partially decomposed mosses. It is a useful additive for potting soil because it is able to retain large amounts of water. This moss has been used to enrich soil and as a fuel source.

Liverworts

Hundreds of years ago, people thought this plant could be used to treat liver diseases. Liverwort also gets its name from its appearance—it resembles the flattened lobes of a liver.

The rhizoids of liverworts are unicellular. The two common forms of liverworts are leafy and thallose (THA los) liverworts.

Hornworts

The long, hornlike reproductive structures of these bryophytes give this group of plants its name. These reproductive structures produce spores.

Hornworts are only about 2.5 cm in diameter. One unusual characteristic of hornworts is that each of their photosynthetic cells has only one chloroplast. Before they produce their long, hornlike reproductive structures, hornworts resemble liverworts.

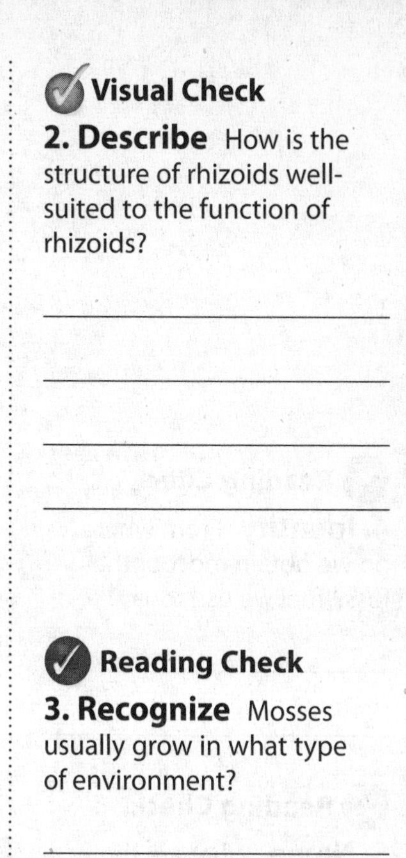

Visual Check

2. Describe How is the structure of rhizoids well-suited to the function of rhizoids?

____

____

____

____

Reading Check

3. Recognize Mosses usually grow in what type of environment?

____

____

____

FOLDABLES®

Make a vertical three-tab Venn book to compare and contrast vascular and nonvascular seedless plants.

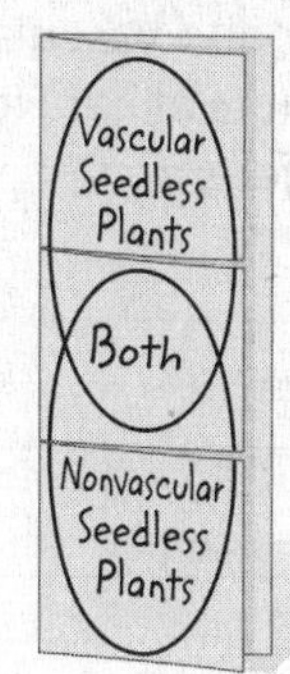

Vascular Seedless Plants

More than 90 percent of plant species are vascular plants. Unlike nonvascular plants, they contain vascular tissue in their stems, roots, and leaves.

Because vascular plants contain tubelike structures that transport water and nutrients, these plants generally are larger than nonvascular plants. However, present-day vascular seedless plants are smaller than their ancient ancestors. These ancient plants grew as tall as today's trees. Much of the fossil fuel that we use today formed from the remains of these ancient plants. ✓

Reading Check

4. Identify From where do we obtain most of the fossil fuel we use today?

____

____

Ferns

The **fronds,** *or leaves of ferns*, make up most of a fern. Ferns range in size from a few centimeters to several meters tall. Ferns grow in a variety of habitats, including damp, swampy areas and dry, rocky cliffs. Ferns often are grown as houseplants. They are also used as landscape plants. Some people consider young fronds, also called fiddleheads, a gourmet treat. ✓

Reading Check

5. Name What are the leaves of ferns called?

____

Club Mosses

Unlike mosses, club mosses have roots, stems, and leaves. Club mosses are small plants that rarely grow taller than 50 cm. The stems often grow along the ground. The leaves are scalelike and resemble small evergreen needles. Some club mosses reproduce by producing spores in two or three cylindrical, yellow-green–colored cones. The spores of club mosses are flammable. In fact, the powdery spores have been used to make fireworks.

Horsetails

Horsetails have small, feathery branches or leaves growing in circles around a hollow stem. The very small leaves can be difficult to see. Horsetail stems are the main photosynthetic structure of the plant. The tissues contain silica, a mineral found in sand. Because silica makes the plants abrasive, horsetails once were used for scrubbing pots and pans. Horsetails can be grown in water gardens but can crowd out other plants because they tend to spread rapidly.

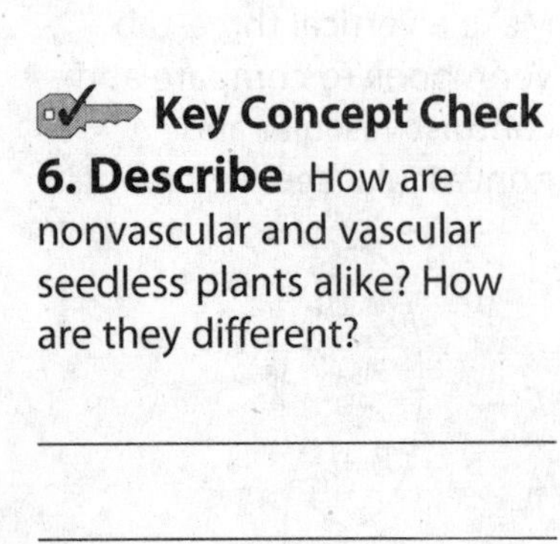

Key Concept Check

6. Describe How are nonvascular and vascular seedless plants alike? How are they different?

____

____

____

____

After You Read

Mini Glossary

frond: the leaf of a fern

rhizoid: a structure that anchors a nonvascular seedless plant to a surface

1. Review the terms and their definitions in the Mini Glossary. Write a sentence explaining why mosses are important in the ecosystem.

2. Fill in the graphic organizer to categorize the following seedless plants as nonvascular or vascular: club mosses, ferns, hornworts, horsetails, liverworts, mosses.

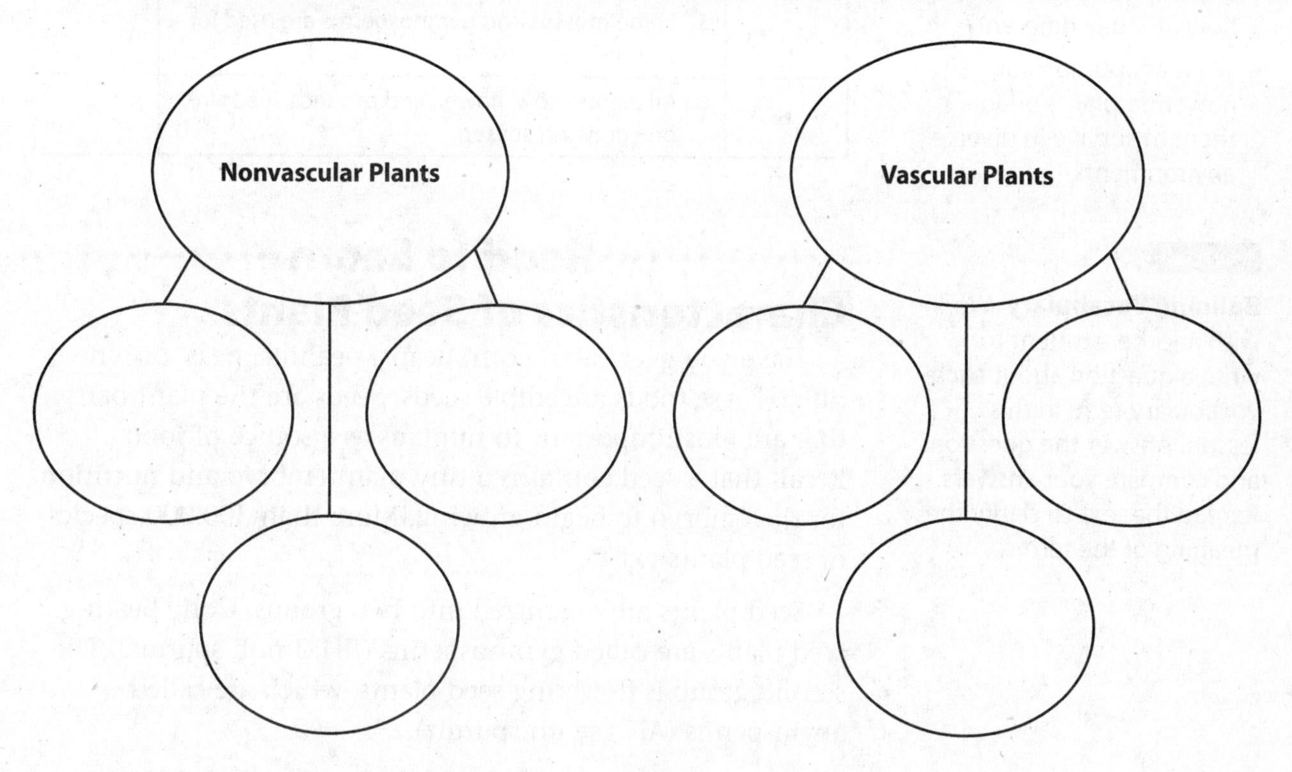

3. Why are vascular seedless plants generally larger than nonvascular plants?

What do you think NOW?

Reread the statements at the beginning of the lesson. Fill in the After column with an A if you agree with the statement or a D if you disagree. Did you change your mind?

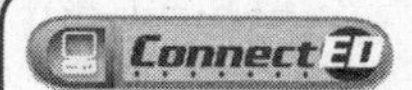

Log on to ConnectED.mcgraw-hill.com and access your textbook to find this lesson's resources.

CHAPTER 9
LESSON 3

Plant Diversity

Seed Plants

Key Concepts

- What characteristics are common to seed plants?
- How do other organisms depend on seed plants?
- How are gymnosperms and angiosperms alike, and how are they different?
- What adaptations of flowering plants enable them to survive in diverse environments?

Study Coach

Building Vocabulary Work with another student to write a question about each vocabulary term in this lesson. Answer the questions and compare your answers. Reread the text to clarify the meaning of the terms.

Before You Read

What do you think? Read the two statements below and decide whether you agree or disagree with them. Place an A in the Before column if you agree with the statement or a D if you disagree. After you've read this lesson, reread the statements to see if you have changed your mind.

Before	Statement	After
	5. Some mosses and gymnosperms are used for commercial purposes.	
	6. All plants grow, flower, and produce seeds in one growing season.	

Read to Learn

Characteristics of Seed Plants

Have you ever eaten corn, beans, peanuts, peas, or pine nuts? These foods are edible seeds. Seeds are the plant parts that are most important to humans as a source of food. Recall that a seed contains a tiny plant embryo and nutrition for the embryo to begin growing. More than 300,000 species of seed plants exist.

Seed plants are organized into two groups. Cone-bearing seed plants are called gymnosperms (JIHM nuh spurmz). The second group is flowering seed plants, which are called angiosperms (AN gee uh spurmz).

All seed plants have vascular tissue that transports water and nutrients throughout the plant. This means they also have roots, stems, and leaves. You will read more about the characteristics of seed plants in this lesson.

Key Concept Check

1. Identify What characteristics do all seed plants have in common?

Vascular Tissue

All seed plants contain vascular tissues in their roots, stems, and leaves. This tissue transports water and nutrients throughout a plant.

The two types of vascular tissue are xylem (ZI lum) and phloem (FLOH em). *The* **cambium** *is a layer of tissue that produces new vascular tissue and grows between xylem and phloem.*

Xylem *One type of vascular tissue—**xylem**—carries water and dissolved nutrients from the roots to the stem and the leaves.* Some xylem cells have thickened cell walls that help support the plant.

Two kinds of xylem cells are tracheids (TRAY kee udz) and vessel elements. The xylem of all vascular plants is made of tracheid cells. ✓

Xylem Cells

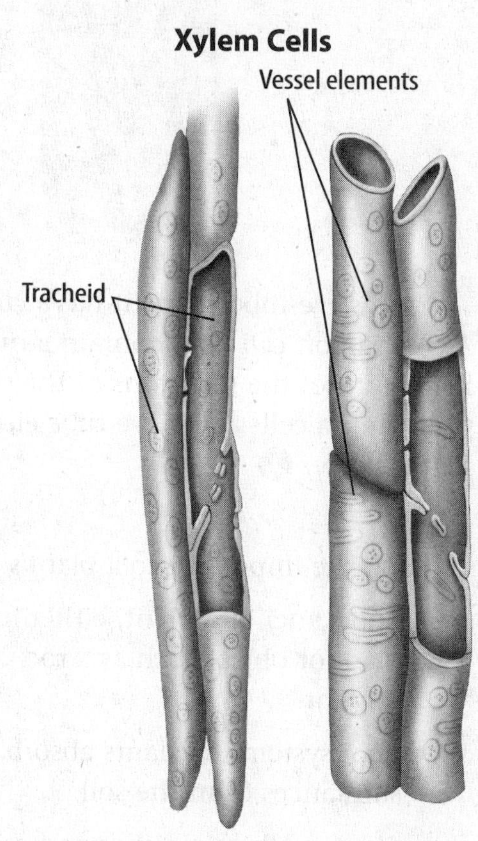

As shown in the figure to the right, tracheid cells are long and narrow with tapered ends. The cells grow end-to-end and form a strawlike tube. Water passes through openings in the end wall of each cell. Tracheid cells die at maturity, leaving a hollow tube. This enables water to flow freely through them.

The xylem in flowering plants also includes a type of cell called a vessel element. The diameter of a vessel element is larger than that of a tracheid. The end walls of vessel elements have large openings where water can pass through, as shown in the figure above. In some vessel elements, the end walls are completely open. Because of these large openings, vessel elements are more efficient at transporting water than are tracheids. ✓

Phloem *Another type of vascular tissue—**phloem**—carries dissolved sugars throughout a plant.* Phloem is made up of two types of cells. These are sieve-tube elements and companion cells.

Sieve-tube elements are specialized phloem cells. These long, thin cells are stacked end-to-end and form long tubes. The end walls have holes in them. A sieve-tube element contains cytoplasm but does not have a nucleus, mitochondria, or ribosomes.

✓ **Reading Check**

2. Point Out What are the two types of xylem cells?

____

____

✓ **Visual Check**

3. Show Circle the type of cell that has the larger opening.

✓ **Reading Check**

4. Explain why water and dissolved nutrients flow freely through xylem.

____

____

____

____

____

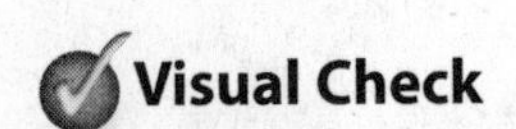

Visual Check

5. Identify Add a line to show one more companion cell and a line to show another sieve-tube element. Add a label to the figure that describes their function.

Reading Check

6. Describe List and describe the function of each of the two types of vascular tissue in plants.

Reading Check

7. Restate Why is a stem important to a plant?

Phloem

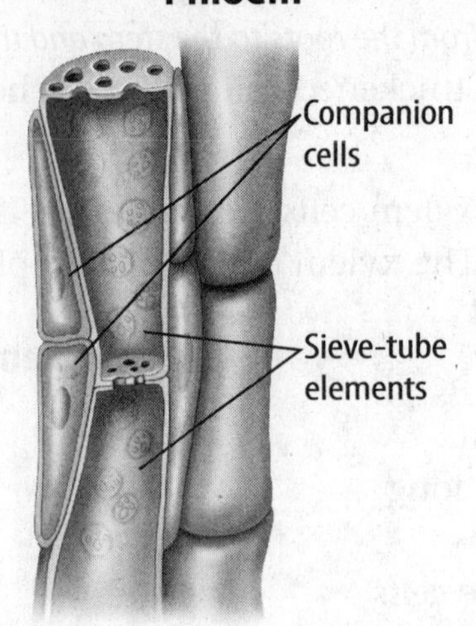

Each sieve-tube element has a companion cell next to it. A companion cell does contain a nucleus. A companion cell helps control the functions of the sieve-tube element. Companion cells and sieve-tube elements are shown in the figure above.

Roots

Roots are important to a plant's survival.

- Roots anchor a plant, either in soil or onto another plant or object such as a rock. Roots help the plant stay upright.
- Root systems of plants absorb water and other substances from the soil.
- Plants such as radishes and carrots store food in their roots. This food can be used to grow new plant tissues after a dry period or a cold season.

Stems

The stem connects a plant's roots to its leaves. Some plants, like trees, have stems that are easy to see. Other plants, like the potato and iris, have underground stems that are often mistaken for roots. Stems support branches and leaves. Their vascular tissues transport water, minerals, and food. Xylem carries water and minerals from the roots to the leaves. The sugar that a plant produces during photosynthesis flows through a stem's phloem to all parts of the plant. Another important function of stems is the production of new cells for growth. Only certain regions of a stem produce new cells.

Plant stems are classified as either woody or herbaceous. Woody stems are stiff and typically not green, like those of trees. Herbaceous stems usually are soft and green.

Leaves

Leaves come in many shapes and sizes. Most leaves have an important function in common—they are the main location for photosynthesis in the plant. They capture light energy and convert it to chemical energy. This is how leaves make the plant's food. ✓

Most leaves are made of layers of cells, as shown in the figure below. The top and bottom layers of a leaf are made of epidermal (eh puh DUR mul) tissue. Epidermal cell walls are transparent, and light passes through them easily. These cells produce a waxy outer layer called the cuticle. The cuticle helps reduce the amount of water that evaporates from a leaf.

Most leaves have small openings in the epidermis called **stomata** (STOH muh tuh; singular, stoma). When the stomata open, carbon dioxide, oxygen, and water vapor can pass through them. Two guard cells surround each stoma and control its size.

Rows of palisade (pa luh SAYD) mesophyll (MEH zuh fil) cells are below the upper epidermis. These cells are tightly packed. Photosynthesis mainly occurs in these cells.

Under the palisade mesophyll cells is the spongy mesophyll layer. The arrangement of these cells enables gases to diffuse throughout a leaf. A leaf's xylem and phloem transport materials throughout the leaf.

Angiosperm and gymnosperm leaves each have some unique characteristics. An angiosperm leaf tends to be flat with a broad surface area. A gymnosperm leaf is usually needlelike or scalelike and often has a thick cuticle. This characteristic benefits gymnosperms that grow in drier areas. The thick cuticle helps conserve water.

Reading Check

8. Locate What is the main location of photosynthesis in plants?

____

FOLDABLES®

Make a three-tab Venn book to compare and contrast seed plants.

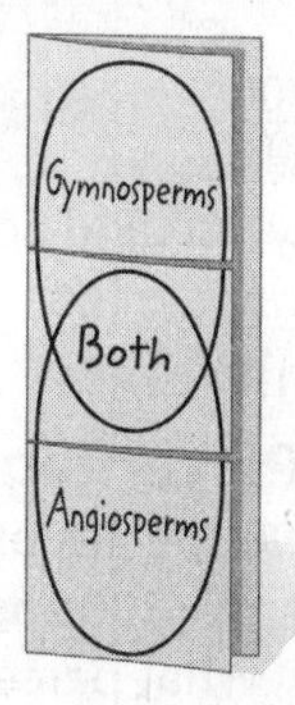

Visual Check

9. Consider Describe the location of the stomata. What role do they play in photosynthesis?

____

____

____

Leaf Cross Section

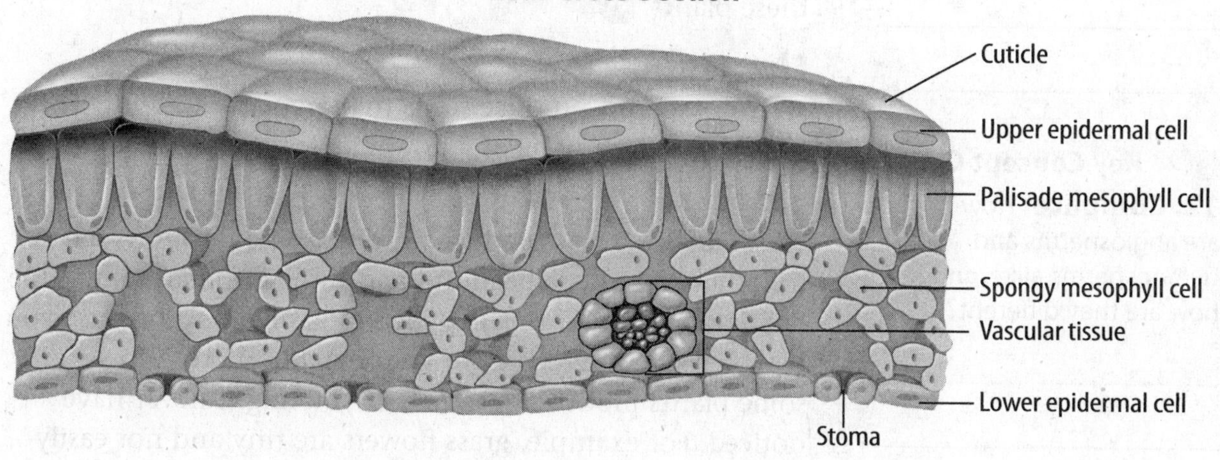

Gymnosperms

In a gymnosperm, the seeds are produced in a cone. Gymnosperms include the oldest plant (the bristlecone pine at 4,900 years old); the tallest plant (the coast redwood that can grow to 115 m); and perhaps Earth's largest organism (the sequoia). Gymnosperms include some familiar conifers such as spruces, pines, and redwoods. You might not be as familiar with some other types of gymnosperms, such as cycads, ginkgoes, and gnetophytes.

Conifers grow on all continents except Antarctica. Cycads usually grow in tropical regions. Although cycads might look like ferns, they are seed plants. DNA evidence has shown that they are closely related to other gymnosperms. One gymnosperm group has only one species—ginkgo. Ginkgoes have broad leaves and are popular as ornamental trees in urban areas. The gnetophytes (NEE tuh fites) are an unusual and diverse group of gymnosperms.

Humans use gymnosperms in a variety of ways, including as building materials; in paper production; as medicines; and as ornamental plants in gardens, along streets, and in parks.

Angiosperms

More than 260,000 species of flowering plants, or angiosperms, exist. Angiosperms began to thrive about 80 million years ago. Angiosperms grow in a variety of habitats, from deserts to the tundra.

Almost all of the food eaten by humans comes from angiosperms or from animals that eat angiosperms. Grains, vegetables, herbs, and spices are just a few examples of foods that come from angiosperms. Many other items, such as clothing, medicines, and building materials, also come from these plants.

Flowers

Angiosperms produce seeds that are part of a fruit. This fruit grows from parts of a flower after pollination and fertilization.

All angiosperms produce flowers. Some flowers, such as tulips and roses, are beautiful and showy. You also might be familiar with other flowers, such as dandelions, because you have seen them growing in your neighborhood. However, some plants produce flowers that you might never have noticed. For example, grass flowers are tiny and not easily seen.

Math Skills

A percentage compares a part to a whole. For example, out of about 1,090 species of gymnosperms, 700 species are conifers. What percentage of gymnosperms are conifers?

Express the information as a fraction.

$$\frac{700 \text{ conifers}}{1{,}090 \text{ gymnosperms}}$$

Change the fraction to a decimal.

$$\frac{700}{1{,}090} = 0.64$$

Multiply by 100 and add a percent sign.

$$0.64 \times 100 = 64\%$$

10. Use Percentages Out of 1,090 species of gymnosperms, 300 species are cycads. What percentage of gymnosperm species are cycads?

____

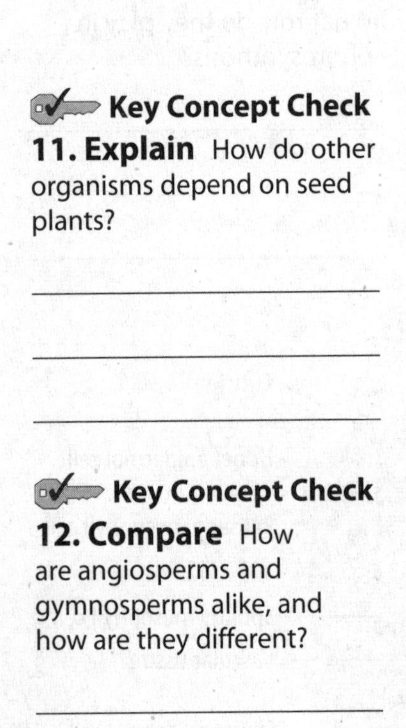

Key Concept Check

11. Explain How do other organisms depend on seed plants?

____

____

____

Key Concept Check

12. Compare How are angiosperms and gymnosperms alike, and how are they different?

____

____

____

Annuals, Biennials, and Perennials

Annuals Plants that grow, flower, and produce seeds in one growing season are called annuals. After one growing season, the plants die. Tomatoes, beans, pansies, and many common weeds are annuals.

Biennials Plants that complete their life cycles in two growing seasons are called biennials. During the first year, the plant grows roots, stems, and leaves. The part of the plant that is above ground might become dormant during the winter months. In the second growing season, the plant produces new stems and leaves. It flowers and produces seeds during this second growing season. After flowering and producing seeds, the plant dies. Carrots, beets, and foxglove are biennials.

Perennials Plants that can live for more than two growing seasons are called perennials. Trees and shrubs are perennials. The leaves and stems of some herbaceous perennial plants die in the winter. Stored food in the roots is used each spring to produce new growth.

Monocots and Dicots

Flowering plants traditionally have been organized into two groups—monocots and dicots. These groups are based on the number of leaves in early development, or cotyledons (kah tuh LEE dunz), in a seed. Researchers have learned that dicots can be organized further into two groups based on the structure of their pollen. However, because these two groups of dicots share many characteristics, we will continue to refer to them all as dicots. The table below shows some of the differences between monocots and dicots.

Monocots and Dicots

Structure	Monocots	Dicots
Leaves	narrow with parallel veins	branched veins
Flowers	flower parts in multiples of three	flower parts in multiples of four or five
Stems	vascular tissue in bundles scattered throughout the stem	vascular tissue in bundles in rings
Seeds	one cotyledon	two cotyledons

Reading Check

13. Describe how the growing seasons of annuals, biennials, and perennials differ.

Key Concept Check

14. Explain What adaptations of flowering plants enable them to survive in diverse environments?

Visual Check

15. Differentiate How are the leaves, flowers, stems, and seeds of monocots different from those of dicots?

After You Read

Mini Glossary

cambium: a layer of tissue that produces new vascular tissue and grows between xylem and phloem

phloem (FLOH em): a type of vascular tissue that carries dissolved sugars throughout a plant

stoma (STOH muh): a small opening in the epidermis of most leaves

xylem (ZI lum): one type of vascular tissue that carries water and dissolved nutrients from the roots to the stem and leaves

1. Review the terms and their definitions in the Mini Glossary. Write a sentence to compare and contrast xylem and phloem.

____

____

2. Compare and contrast gymnosperms with angiosperms by writing the letter of each item under one of the headings in the graphic organizer.

a. have roots, stems, and leaves

b. may produce beautiful flowers

c. produce almost all food eaten by humans

d. Seeds are part of a fruit.

e. include the oldest and tallest plants

f. have vascular tissue

g. conifers, cycads, ginkgoes, gnetophytes

h. Seeds are produced in a cone.

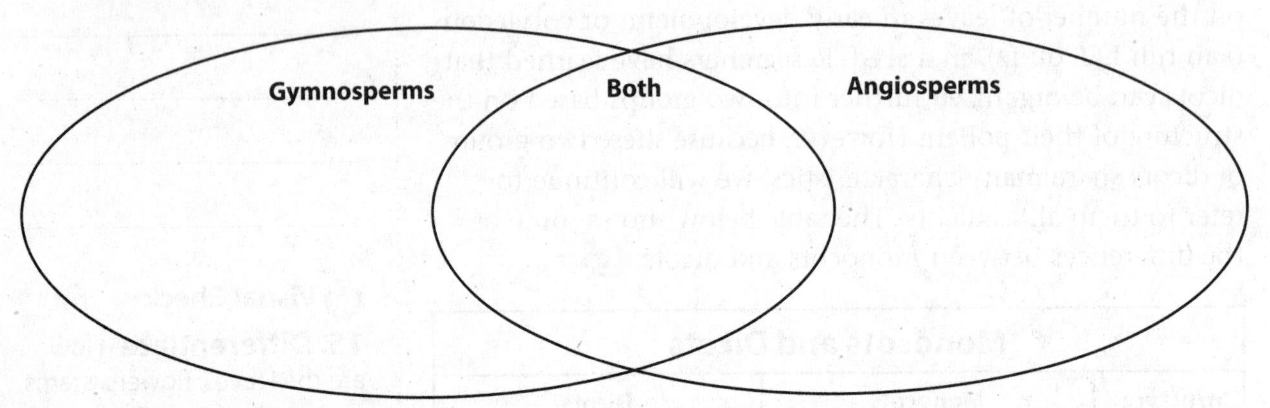

3. Compare the leaves of gymnosperms and angiosperms. What is the benefit of a gymnosperm's leaf structure?

____

____

What do you think NOW?

Reread the statements at the beginning of the lesson. Fill in the After column with an A if you agree with the statement or a D if you disagree. Did you change your mind?

Log on to ConnectED.mcgraw-hill.com and access your textbook to find this lesson's resources.

Plant Processes and Reproduction

Energy Processing in Plants

Before You Read

What do you think? Read the three statements below and decide whether you agree or disagree with them. Place an A in the Before column if you agree with the statement or a D if you disagree. After you've read this lesson, reread the statements to see if you have changed your mind.

Before	Statement	After
	1. Plants do not carry on cellular respiration.	
	2. Plants are the only organisms that carry on photosynthesis.	
	3. Plants make food in their underground roots.	

Key Concepts 

- How do materials move inside plants?
- How do plants perform photosynthesis?
- What is cellular respiration?
- How are photosynthesis and cellular respiration alike, and how are they different?

Read to Learn

Materials for Plant Processes

Food, water, and oxygen are three things you need to survive. Some of your organ systems process these materials. Other systems transport them throughout your body. Like you, plants need food, water, and oxygen to survive. Unlike you, plants do not take in food. Most plants make their own food.

Study Coach

Make an Outline Summarize the information in the lesson in an outline. Use the main headings in the lesson as the main headings in your outline. Use your outline to review the lesson.

Moving Materials Inside Plants

Xylem (ZI lum) and phloem (FLOH em) are the vascular tissue in most plants. These tissues transport materials throughout a plant.

After water enters a plant's roots, it moves into xylem. Water then flows inside xylem to all parts of a plant. Without enough water, plant cells wilt.

Most plants make their own food—a liquid sugar. The liquid sugar moves out of food-making cells, enters phloem, and flows to all plant cells. Cells break down the sugar and release energy. Some plant cells can store food.

Plants require oxygen and carbon dioxide to make food. Like you, plants produce water vapor as a waste product. Carbon dioxide, oxygen, and water vapor pass into and out of a plant through tiny openings in leaves.

Key Concept Check

1. Determine How do materials move through plants?

______

______

______

______

Photosynthesis

WORD ORIGIN
photosynthesis
from Greek *photo–*, means "light"; and *synthesis*, means "composition"

Plants make their own food through a process called photosynthesis (foh toh SIHN thuh sus). **Photosynthesis** *is a series of chemical reactions that convert light energy, water, and carbon dioxide into the food-energy molecule glucose and give off oxygen.*

Leaves are the major food-producing organs of plants. This means that photosynthesis takes place in the plant's leaves. The structure of a leaf is well-suited to its role in photosynthesis.

Leaves and Photosynthesis

FOLDABLES®

Make a shutterfold book and label it as shown to use as a diagram of leaf structure.

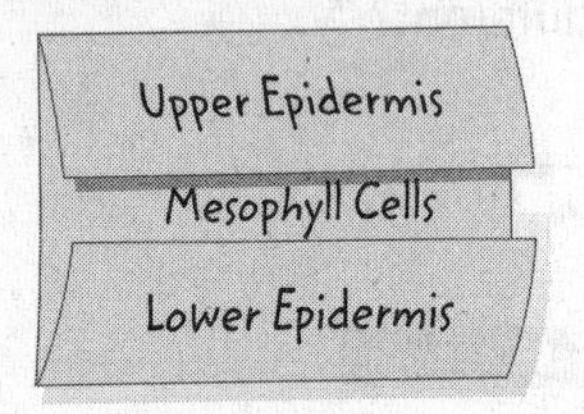

The figure below shows the many types of cells in a leaf. The epidermal (eh puh DUR mul) cells make up the upper and lower layers of the leaf. Epidermal cells are flat and irregularly shaped. The bottom epidermal layer of most leaves has small openings called stomata (STOH muh tuh). Carbon dioxide, water vapor, and oxygen pass through stomata. Epidermal cells can produce a waxy covering called the cuticle.

Most photosynthesis occurs in two types of mesophyll (ME zuh fil) cells inside a leaf. Mesophyll cells contain chloroplasts, which are the organelles where photosynthesis occurs. The palisade mesophyll cells are near the top surface of the leaf. They are packed close together. This arrangement exposes the most cells to light. Spongy mesophyll cells are below the palisade mesophyll cells. They have open spaces between them. Gases needed for photosynthesis flow through the spaces between the spongy mesophyll cells.

Visual Check
2. Locate Which layer of cells contains vascular tissue?

Cross Section of a Leaf

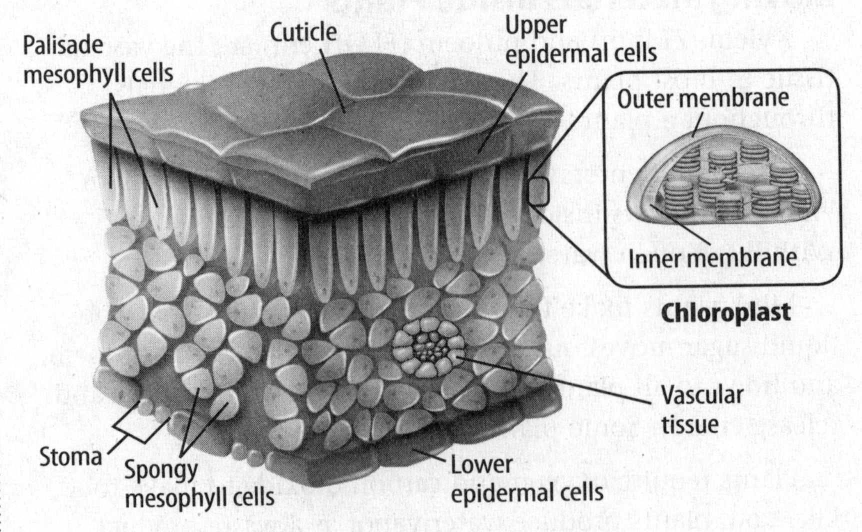

Capturing Light Energy

Photosynthesis is a complex chemical process. It consists of two basic steps: capturing light energy and using that energy to make sugars. Look at the figure below as you read about these steps to help you understand the process.

In the first step, chloroplasts capture the energy in light. Chloroplasts contain plant pigments. Pigments are chemicals that can absorb and reflect light. Chlorophyll is the most common plant pigment. It is necessary for photosynthesis. Most plants appear green because chlorophyll reflects green light. Chlorophyll absorbs other colors of light. This light energy is used during photosynthesis.

Chlorophyll traps and stores light energy. Then this energy can be transferred to other molecules. During photosynthesis, water molecules are split apart. The oxygen from the water molecules is released into the atmosphere, as shown below. The hydrogen atoms in the water are used to make sugars in the second step of photosynthesis.

Making Sugars

In the second step of photosynthesis, sugars are made from the light energy. In chloroplasts, carbon dioxide from the air is converted into sugars by using the energy stored and trapped by chlorophyll. Carbon dioxide combines with hydrogen atoms from the splitting of water molecules and forms sugar molecules. Plants can use this sugar as an energy source. Plants can also store the sugar for later use. Potatoes and carrots are examples of plant structures where plants store excess sugar.

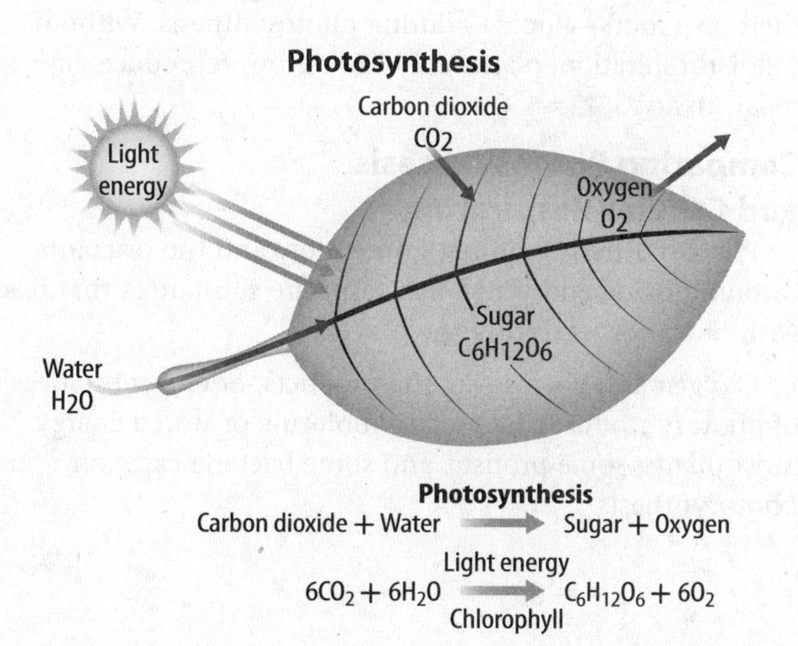

Reading Check

3. Identify How do plants capture light energy?

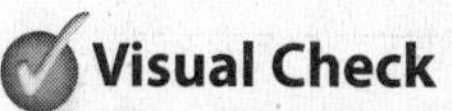

Key Concept Check

4. Name What are the two steps of photosynthesis?

Visual Check

5. Describe Use the figure to explain to a partner the first step of photosynthesis. Then have your partner use the figure to explain the second step of photosynthesis.

Why is photosynthesis important?

Try to imagine a world without plants. The world would certainly look different, and its atmosphere would also be different. How would humans or other animals get the oxygen they need?

Plants help maintain the atmosphere you breathe. Photosynthesis produces as much as 90 percent of the oxygen in the atmosphere. Without green plants, humans would not have enough oxygen to breathe.

Academic Vocabulary
energy
(noun) usable power

Cellular Respiration

All organisms require energy to survive. Energy is in the chemical bonds in food molecules. A process called cellular respiration releases energy. **Cellular respiration** *is a series of chemical reactions that convert the energy in food molecules into a usable form of energy called ATP.*

Reading Check
6. Name Which cellular process converts food energy into usable energy?

Releasing Energy from Sugars

Glucose is the sugar produced by photosynthesis. Glucose molecules break down during cellular respiration. Much of the energy released during this process is used to make ATP. ATP is an energy storage molecule.

Cellular respiration occurs in the cytoplasm and mitochondria of cells. This process requires oxygen and produces water and carbon dioxide as waste products.

Why is cellular respiration important?

If your body did not break down the food you eat, you would not have energy to do anything. All organisms must break down their food to produce energy. Plants produce their own food—glucose—during photosynthesis. Without cellular respiration, plants could not grow, reproduce, or repair tissues.

Key Concept Check
7. Explain What is cellular respiration?

Comparing Photosynthesis and Cellular Respiration

Photosynthesis requires light energy and the reactants carbon dioxide and water. Reactants are substances that react with one another during the process.

Oxygen and glucose are the products, or end substances, of photosynthesis. Glucose is a molecule of stored energy. Most plants, some protists, and some bacteria carry on photosynthesis.

Photosynthesis and Cellular Respiration Work Together

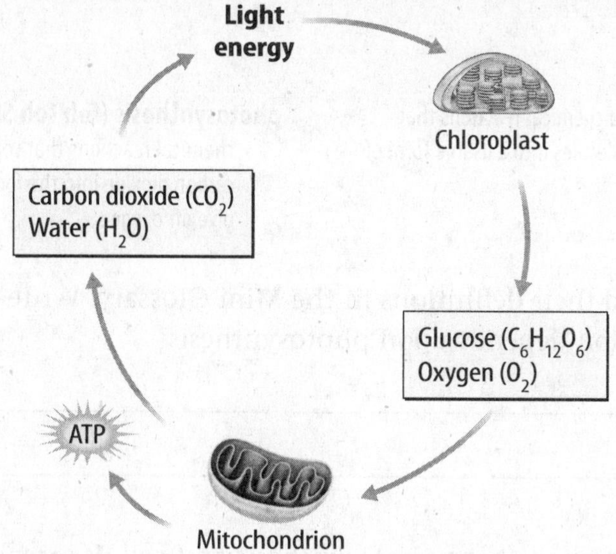

$$6CO_2 + 6H_2O \xrightarrow[\text{Chlorophyll}]{\text{Light energy}} C_6H_{12}O_6 + 6O_2$$

Photosynthesis

$$C_6H_{12}O_6 + 6O_2 \longrightarrow 6CO_2 + 6H_2O + \text{ATP (Energy)}$$

Cellular respiration

Chemical Equation for Photosynthesis Look at the chemical equation for photosynthesis on the right in the figure above. Notice that photosynthesis requires carbon dioxide (CO_2) and water (H_2O) molecules. These molecules react with light energy and produce glucose ($C_6H_{12}O_6$) and oxygen (O_2).

Chemical Equation for Cellular Respiration The chemical equation for cellular respiration is on the left in the figure above. The reactants are glucose ($C_6H_{12}O_6$) and oxygen (O_2). It produces carbon dioxide (CO_2) and water (H_2O) molecules. Cellular respiration releases energy in the form of ATP.

Most organisms carry on cellular respiration. The connection between photosynthesis and cellular respiration is shown in the table below. Life on Earth depends on a balance of these two processes.

Comparing Photosynthesis and Cellular Respiration

Process	Photosynthesis	Cellular Respiration
Reactants	light energy, CO_2, H_2O	glucose (sugar), O_2
Products	glucose, O_2	CO_2, H_2O, ATP
Organelle in which it occurs	chloroplasts	mitochondria
Type of organism	photosynthetic organisms including plants and algae	most organisms, including plants and animals

Visual Check

8. Distinguish Refer to the figure above and the table below. What are the reactants of cellular respiration? What are the products?

Key Concept Check

9. Compare How are photosynthesis and cellular respiration alike, and how are they different?

After You Read

Mini Glossary

cellular respiration: a series of chemical reactions that convert the energy in food molecules into a usable form of energy called ATP

photosynthesis (foh toh SIHN thuh sus): a series of chemical reactions that convert light energy, water, and carbon dioxide into the food-energy molecule glucose and give off oxygen

1. Review the terms and their definitions in the Mini Glossary. Write a sentence describing how cellular respiration depends upon photosynthesis.

____

____

2. Use what you have learned about photosynthesis and cellular respiration to complete the graphic organizer below.

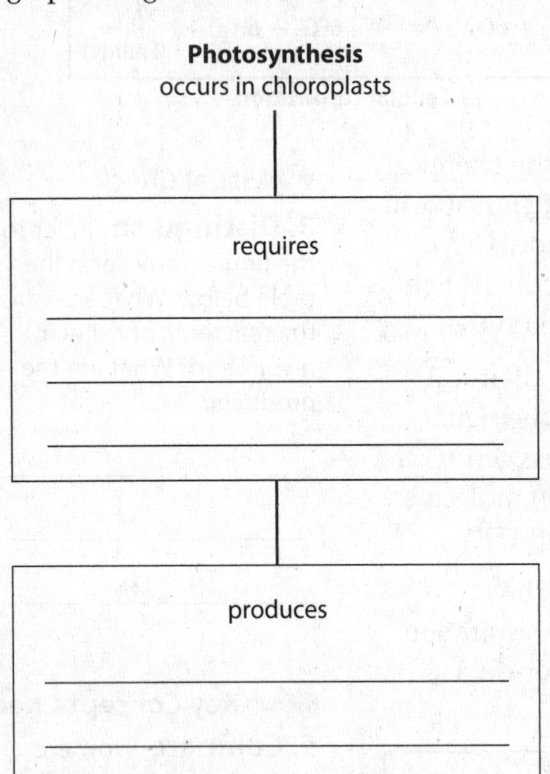

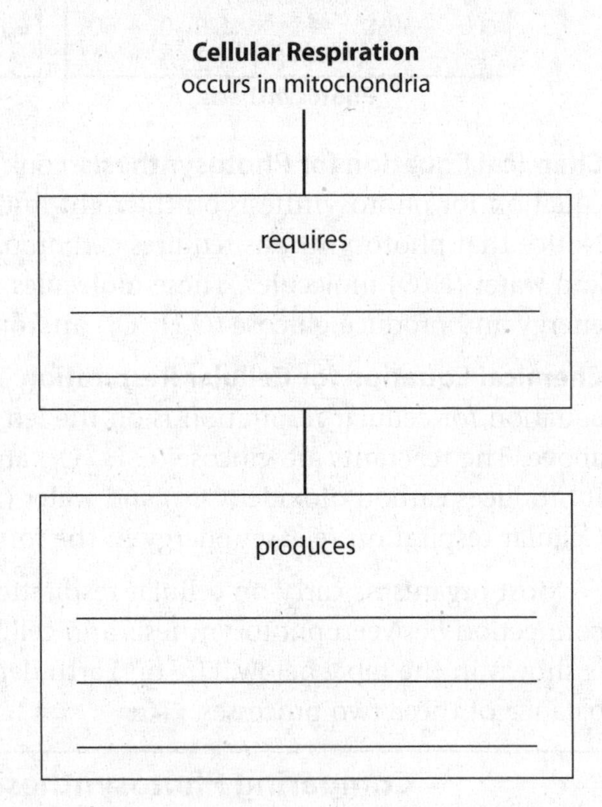

What do you think NOW?

Reread the statements at the beginning of the lesson. Fill in the After column with an A if you agree with the statement or a D if you disagree. Did you change your mind?

Log on to ConnectED.mcgraw-hill.com and access your textbook to find this lesson's resources.

Plant Processes and Reproduction

Plant Responses

Before You Read

What do you think? Read the three statements below and decide whether you agree or disagree with them. Place an A in the Before column if you agree with the statement or a D if you disagree. After you've read this lesson, reread the statements to see if you have changed your mind.

Before	Statement	After
	4. Plants do not produce hormones.	
	5. Plants can respond to their environments.	
	6. All plants flower when nights are 10–12 hours long.	

Key Concepts 

- How do plants respond to environmental stimuli?
- How do plants respond to chemical stimuli?

Read to Learn

Stimuli and Plant Responses

Have you ever been in a dark room when someone suddenly turned on a light? How did you react when the light suddenly came on? You might have shut your eyes or covered them. Organisms can respond to changes in their environments in many different ways. In this lesson, you will learn how plants respond to environmental and chemical stimuli.

Stimuli (STIM yuh li; singular, stimulus) *are any changes in an organism's environment that cause a response.* Many plant responses to stimuli occur slowly. In fact, they are so slow that it is hard to see them happen. The response might occur gradually over a period of hours or days. Light is a stimulus. The stems and leaves of many houseplants grow toward a window. The plants are responding to the light stimulus that comes through the window. This response occurs gradually over many hours. ✓

The response to a stimulus can be quick. For example, a Venus flytrap is a plant with unusual leaves that close when a stimulus, such as a fly, brushes against hairs on the leaf. The trap snaps shut, like jaws, when stimulated by an insect touching the leaf. The insect is trapped inside the plant.

Mark the Text

Identify the Main Ideas Write a phrase beside each paragraph that summarizes the main point of the paragraph. Use the phrases to review the lesson.

✓ Reading Check

1. Determine Why is it sometimes hard to see a plant's response to a stimulus?

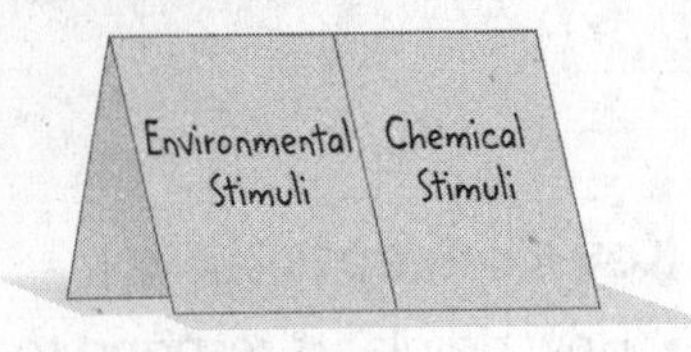

Make a two-tab book to record what you learn about the two types of stimuli that affect plant growth.

Environmental Stimuli

Plants respond to their environments in a variety of ways. In the spring, some trees flower and new, green leaves sprout. In the fall, the same trees drop their leaves. Both are plant responses to environmental stimuli.

Growth Responses

Plants respond to different environmental stimuli. These include light, touch, and gravity. *A* **tropism** (TROH pih zum) *is a response that results in plant growth toward or away from a stimulus.* Positive tropism is growth toward a stimulus. Negative tropism is growth away from a stimulus.

Light Look at the figure below. The plant's growth toward or away from light is a tropism called phototropism. A light-sensing chemical in a plant helps it detect light. Leaves and stems tend to grow in the direction of light.

Recall that photosynthesis occurs in a plant's leaves, and photosynthesis requires light. By growing in the direction of light, the plant's leaves are exposed to more light. Roots generally grow away from light. This usually means that the roots of the plant grow down into the soil and help anchor the plant.

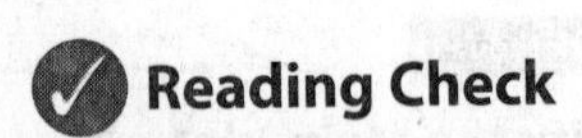

2. Evaluate How is phototropism beneficial to a plant?

______

______

______

Touch Thigmotropism (thihg MAH truh pih zum) is the name given to a plant's response to touch. You might have seen vines growing up the side of a building. Some plants have special structures that respond to touch. These structures are called tendrils. The tendrils wrap around or cling to objects, such as when vine tendrils coil around a blade of grass. This is positive thigmotropism.

Roots display negative thigmotropism. They grow away from objects in soil, enabling them to follow the easiest path through the soil.

Response to Light

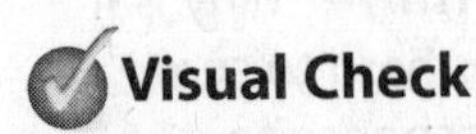

3. Explain Use the figure to explain to a partner what happened to the plant.

Gravity A plant's response to gravity is called gravitropism. Stems grow away from gravity, so they show a negative gravitropism. Roots grow toward gravity, showing a positive gravitropism.

When a seed lands in the soil and starts to grow, its roots will always grow down into the soil. The stem grows up. This will happen even when a seed is grown in a dark chamber. This shows that the response of the root and the stem can occur independently of light.

Flowering Responses

Flowering is a plant response to environmental stimuli. Some plants flower in response to the amount of darkness they are exposed to. **Photoperiodism** *is a plant's response to the number of hours of darkness in its environment.* Scientists once hypothesized that photoperiodism was a response to light. For that reason, the flowering responses are called long-day, short-day, and day-neutral. The names relate to the number of hours of daylight in a plant's environment. Scientists now know that plants respond to the number of hours of darkness.

Long-Day Plants Plants that flower when exposed to less than 10 to 12 hours of darkness are called long-day plants. Long-day plants usually produce flowers in summer. During the summer, the number of hours of daylight is greater than the number of hours of darkness.

Short-Day Plants Short-day plants begin to flower when there are 12 or more hours of darkness. A poinsettia is an example of a short-day plant. Poinsettias tend to flower in the late summer or early fall when the number of hours of darkness is increasing.

Day-Neutral Plants The number of hours of darkness doesn't seem to affect the flowering of some plants. These plants are called day-neutral plants. These plants flower when they reach maturity and the other environmental conditions are right. Roses are day-neutral plants.

Chemical Stimuli

Plants respond to chemical stimuli as well as environmental stimuli. **Plant hormones** *are substances that act as chemical messengers within plants.* These chemicals are produced in tiny amounts. They are called messengers because the chemicals are usually produced in one part of a plant but affect another part of that plant.

Key Concept Check

4. Identify What types of environmental stimuli do plants respond to? Give three examples.

Reading Check

5. State Why are plant responses named according to length of day?

Reading Check

6. Identify How is the flowering of day-neutral plants affected by exposure to hours of darkness?

Auxins

One plant hormone is auxin (AWK sun). There are many different kinds of auxins. Plant cells respond to auxins with increased growth. The growth of leaves toward light is a response to auxin. Auxins concentrate on the dark side of a plant's stem, and these cells grow longer. This causes the stem to grow toward the light. The figure below shows auxin on the left side of the seedling. It causes more growth on the left side, leading the seedling to bend to the right.

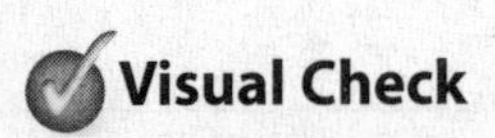

Visual Check

7. Explain Use the figure to explain to a partner how auxins have affected the growth of this seedling.

Response to Auxins

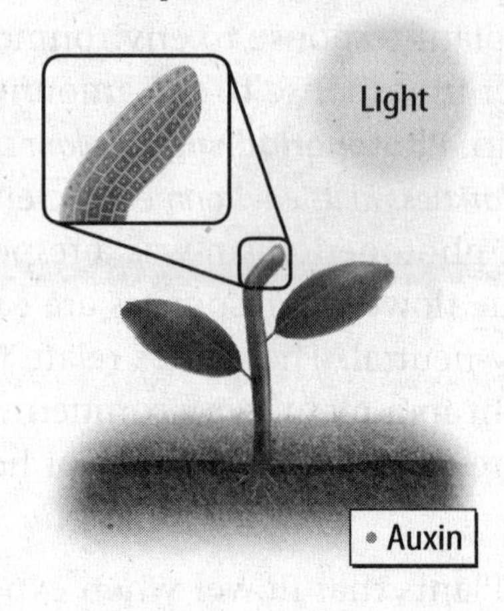

Ethylene

The plant hormone ethylene helps fruit ripen. Ethylene is a gas that can be produced by fruits, seeds, flowers, and leaves. Have you heard the expression "one rotten apple spoils the whole barrel"? This is based on the fact that rotting fruits release ethylene. This can cause other fruits nearby to ripen and possibly rot. Ethylene also can cause plants to drop their leaves.

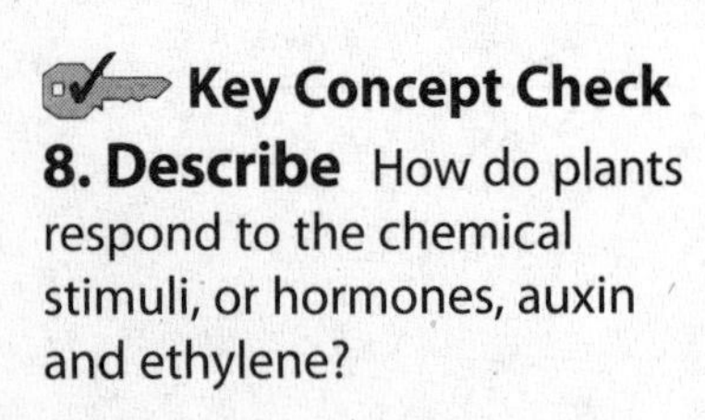

Key Concept Check

8. Describe How do plants respond to the chemical stimuli, or hormones, auxin and ethylene?

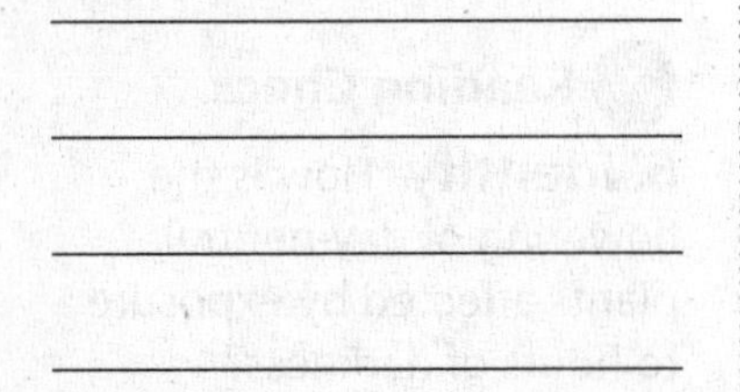

Gibberellins and Cytokinins

Rapidly growing areas of a plant, such as roots and stems, produce gibberellins (jih buh REL unz). These hormones increase the rate of cell division and cell elongation. This results in the increased growth of stems and leaves. Sometimes gibberellins are applied to the outside of plants to encourage plant growth. Fruit-producing plants can be treated with gibberellins to produce more fruit and larger fruit.

Cytokinins (si tuh KI nunz) are another type of hormone. They are produced mostly in root tips. Xylem carries cytokinins to other parts of a plant. Cytokinins increase the rate of cell division. In some plants, cytokinins slow the aging process of flowers and fruits.

Summary of Plant Hormones

Plants produce many different hormones. The hormones discussed in this lesson are groups of similar compounds. Often, two or more hormones interact and produce a plant response. Scientists are still discovering new information about plant hormones.

Humans and Plant Responses

Humans depend on plants for food, fuel, shelter, and clothing. Humans use plant hormones to make plants more productive. Some crops have become easier to grow because humans understand how the plants respond to hormones. For example, bananas and tomatoes can be picked and shipped while they are still green. They can then be treated with ethylene to make them ripen. ✓

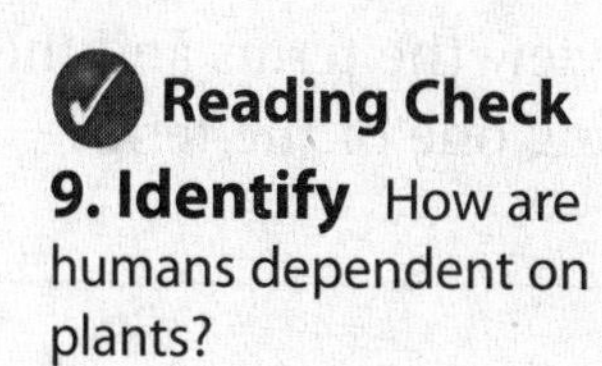

Reading Check

9. Identify How are humans dependent on plants?

____

____

Math Skills

A percentage is a ratio that compares a number to 100. For example, if a plant grows 2 cm per day with no chemical stimulus and 3 cm per day with a chemical stimulus, what is the percentage increase in growth?

Subtract the original value from the final value.

$$3\text{ cm} - 2\text{ cm} = 1\text{ cm}$$

Set up a ratio between the difference and the original value. Find the decimal equivalent.

$$\frac{1\text{ cm}}{2\text{ cm}} = 0.5\text{ cm}$$

Multiply by 100 and add a percent sign.

$$0.5 \times 100 = 50\%$$

10. Use Percentages Without gibberellins, pea seedlings grew to 2 cm in 3 days. With gibberellins, the seedlings grew to 4 cm in 3 days. What was the percentage increase in growth?

____

After You Read

Mini Glossary

photoperiodism: a plant's response to the number of hours of darkness in its environment

plant hormone: a substance that acts as a chemical messenger within plants

stimulus: a change in an organism's environment that causes a response

tropism (TROH pih zum): a response that results in plant growth toward or away from a stimulus

1. Review the terms and their definitions in the Mini Glossary. Write your own sentence using one of the terms.

2. Complete the chart below to summarize what you have learned about environmental stimuli.

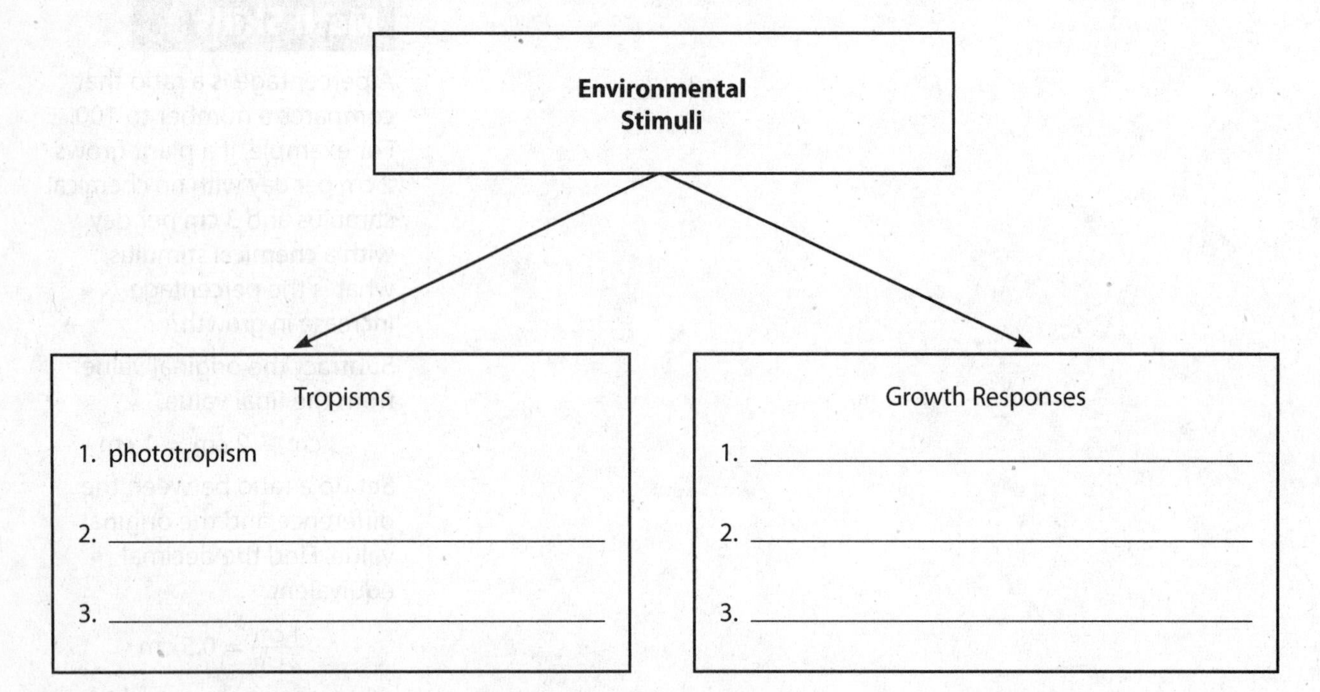

3. How did writing a phrase next to each paragraph help you identify the main ideas?

What do you think NOW?

Reread the statements at the beginning of the lesson. Fill in the After column with an A if you agree with the statement or a D if you disagree. Did you change your mind?

Log on to ConnectED.mcgraw-hill.com and access your textbook to find this lesson's resources.

Plant Processes and Reproduction

Plant Reproduction

Before You Read

What do you think? Read the two statements below and decide whether you agree or disagree with them. Place an A in the Before column if you agree with the statement or a D if you disagree. After you've read this lesson, reread the statements to see if you have changed your mind.

Before	Statement	After
	5. Seeds contain tiny plant embryos.	
	6. Flowers are needed for plant reproduction.	

Read to Learn

Asexual Reproduction Versus Sexual Reproduction

Plants can reproduce either asexually, sexually, or both ways. Asexual reproduction occurs when part of a plant develops into a separate new plant. The new plant is genetically the same as the original, or parent, plant. Irises and daylilies are plants that reproduce asexually through their underground stems. Houseleeks also reproduce asexually. Horizontal stems called stolons grow from the main plant. New plants grow at the ends of the stolons.

Asexual reproduction requires just one parent organism to produce offspring. Sexual reproduction in plants usually requires two parent plants. Sexual reproduction occurs when a plant's sperm combines with a plant's egg. The new plant that results is a genetic combination of its parents.

Alternation of Generations

Most human cells are diploid. Sperm and eggs are the only human haploid cells. The human life cycle includes only a diploid stage. This isn't true for all organisms. Plants, for example, have two life stages called generations. One generation is almost all diploid cells. The other generation has only haploid cells. **Alternation of generations** *occurs when the life cycle of an organism alternates between diploid and haploid generations.*

Key Concepts 
- What is the alternation of generations in plants?
- How do seedless plants reproduce?
- How do seed plants reproduce?

Study Coach

Building Vocabulary Make a vocabulary card for each bold term in this lesson. Write each term on one side of the card. On the other side, write the definition. Use these terms to review the vocabulary for the lesson.

Key Concept Check
1. Define What is alternation of generations in plants?

Visual Check

2. Identify Highlight in one color the diploid generation. Highlight in another color the haploid generation.

Reading Check

3. Explain How do spores grow?

____

____

____

____

REVIEW VOCABULARY

mitosis
the process during which a nucleus and its contents divide

Alternation of Generations

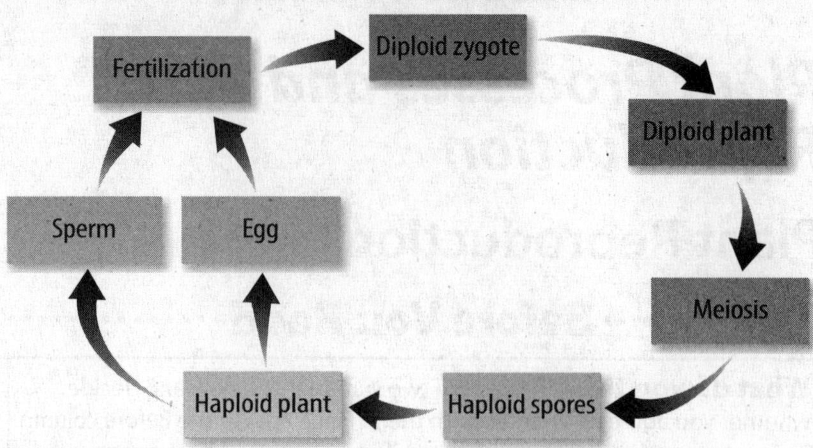

The Diploid Generation

Look at the figure above. A tree or a flower is part of a plant's diploid generation. Meiosis occurs in certain cells in the reproductive structures of a diploid plant. *The daughter cells produced from haploid structures are called* **spores.** Spores grow by mitosis and cell division. They form the haploid generation of a plant.

The Haploid Generation

In most plants, the haploid generation is tiny and lives surrounded by tissues of the diploid plant. In other plants the haploid generation lives on its own. Certain reproductive cells in the haploid generation produce haploid sperm or eggs by mitosis and cell division. Fertilization takes place when a sperm and an egg fuse and form a diploid zygote. The zygote grows into the diploid generation of a plant through mitosis and cell division.

Reproduction in Seedless Plants

Some plants grow from haploid spores, not from seeds. They are known as seedless plants. The first land plants to inhabit Earth were probably seedless plants. Mosses and ferns are examples of seedless plants found on Earth today.

Life Cycle of a Moss

The tiny, green moss plants that grow in moist areas are haploid plants. These plants grow by mitosis and cell division from haploid spores produced by the diploid generation. They have male structures that produce sperm and female structures that produce eggs. After fertilization, the diploid zygote grows by mitosis and cell division into the diploid generation of moss. A diploid moss is tiny and hard to see.

Life Cycle of a Fern

An alternation of generations is also seen in the life cycle of a fern. The leafy, green plants that grow in many forests are the diploid generations. These plants produce haploid spores. The spores grow into tiny haploid plants. The haploid plants produce eggs and sperm that can unite and form the diploid generations.

How do seed plants reproduce?

Most land plants on Earth grow from seeds. There are two groups of seed plants—flowerless seed plants and flowering seed plants.

The haploid generation of all seed plants is within diploid tissue. Separate diploid male reproductive structures produce haploid sperm. Separate diploid female reproductive structures produce haploid eggs. The haploid sperm and the haploid egg join during fertilization.

The Role of Pollen Grains

A **pollen** (PAH lun) **grain** *forms from tissue in a male reproductive structure of a seed plant.* Each pollen grain has a hard outer covering that protects it. All the nutrients the pollen grain needs are contained inside the covering. Pollen grains produce sperm cells. Wind, animals, gravity, or water currents can carry pollen grains to female reproductive structures.

Plants can't move on their own. They do not find mates as most animals do. The male reproductive structures of plants produce a large number of pollen grains. **Pollination** (pah luh NAY shun) *occurs when pollen grains land on a female reproductive structure of a plant that is the same species as the pollen grains.*

The Role of Ovules and Seeds

The female reproductive structure of a seed plant where the haploid egg develops is called the **ovule.** After pollination, sperm enter the ovule and fertilization occurs. A zygote forms and develops into an embryo. *An* **embryo** *is an immature diploid plant that develops from the zygote. An embryo, its food supply, and a protective covering make up a* **seed.** A seed's food supply provides the embryo with the nourishment it needs for its early growth.

Key Concept Check

4. Describe How do seedless plants such as mosses and ferns reproduce?

Reading Check

5. Identify What occurs during fertilization?

Key Concept Check

6. Describe How do seed plants reproduce?

Seed Structures

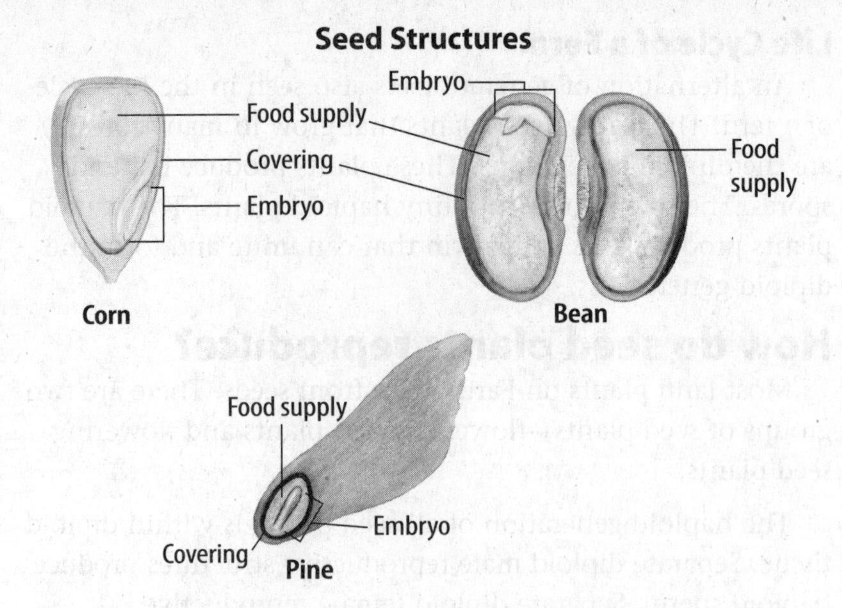

Visual Check

7. Name three things the seeds have in common.

Corn, bean, and pine seeds are shown in the figure above. A seed contains a diploid plant embryo and a food supply protected by a hard outer covering.

FOLDABLES®

Make a two-tab book to record information about reproduction in flowerless and flowering plants.

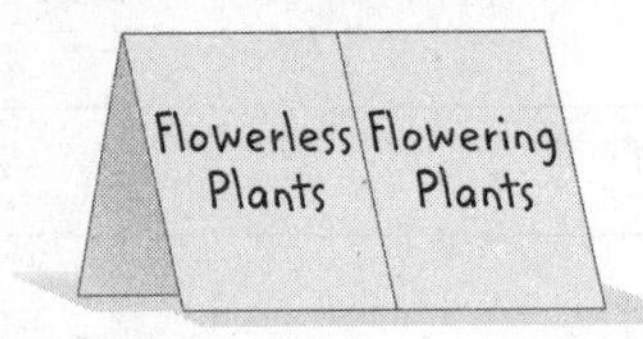

Reproduction in Flowerless Seed Plants

Flowerless seed plants are also known as gymnosperms (JIHM nuh spurmz). The word *gymnosperm* means "naked seed." Gymnosperm seeds are not surrounded by a fruit.

The most common gymnosperms are conifers. Conifers, such as pines, firs, cypresses, redwoods, and yews, are trees and shrubs with needlelike or scalelike leaves. Most conifers are evergreens, meaning they keep their leaves all year.

Life Cycle of a Gymnosperm The life cycle of a gymnosperm includes an alternation of generations. The life cycle is shown in the figure at the top of the next page.

Cones are the male and female reproductive structures of conifers. They contain the haploid generation. Male cones are small structures that produce pollen grains. Female cones can be woody, berrylike, or soft. They produce eggs.

Reading Check

8. Describe Where is the haploid generation of conifers contained?

Male cones release clouds of pollen grains containing the sperm. A zygote forms when a sperm fertilizes an egg. The zygote is the beginning of the diploid generation. Seeds form as part of the female cone.

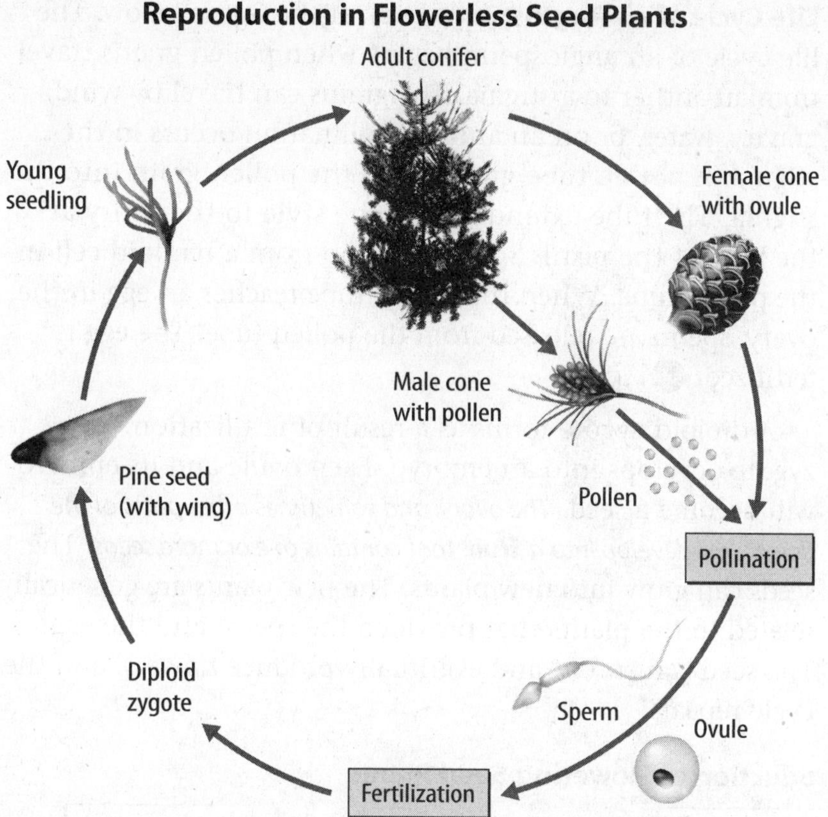

Visual Check

9. Identify Circle each structure of the haploid generation.

Reproduction in Flowering Seed Plants

Flowering seed plants are called angiosperms. Fruits and vegetables come from angiosperms. Many animals depend on angiosperms for food.

Reading Check

10. Name What is another name for flowering seed plants?.

The Flower Reproduction of an angiosperm begins in a flower. Most flowers have male and female reproductive structures. See the figure below.

The male reproductive organ of a flower is the **stamen.** Pollen grains form at the tip of the stamen in the anther. The anther is connected to the base of the flower by the filament. *The female reproductive organ of a flower is the* **pistil.** Pollen can land on the stigma at the tip of the pistil. The stigma is at the top of a long tube called the style. *At the base of the style is the* **ovary,** *which contains one or more ovules.* Each ovule eventually will contain a haploid egg and might become a seed if fertilized.

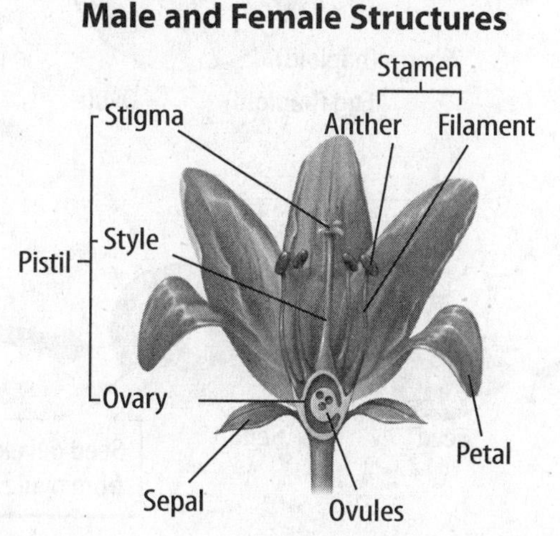

Visual Check

11. Identify In which structure do pollen grains form?

Life Cycle of an Angiosperm Look at the figure below. The life cycle of an angiosperm begins when pollen grains travel from an anther to a stigma. The grains can travel by wind, gravity, water, or on an animal. Pollination occurs in the stigma. A pollen tube grows from the pollen grain into the stigma. The tube extends down the style to the ovary at the base of the pistil. Sperm develop from a haploid cell in the pollen tube. When the pollen tube reaches an egg in the ovary, sperm are released from the pollen tube. The egg is fertilized.

Reading Check

12. Determine Do sperm develop before or after pollination?

____

Visual Check

13. Determine How does sperm in a pollen grain reach an egg in the ovule?

____

____

A diploid zygote forms as a result of fertilization. The zygote develops into an embryo. Each ovule and its embryo will become a seed. *The ovary, and sometimes other parts of the flower, will develop into a* **fruit** *that contains one or more seeds.* The seeds can grow into new plants. The new plants are genetically related to the plants that provided the sperm and the egg. The seed can sprout and eventually produce flowers, and the cycle repeats.

Reproduction in Flowering Seed Plants

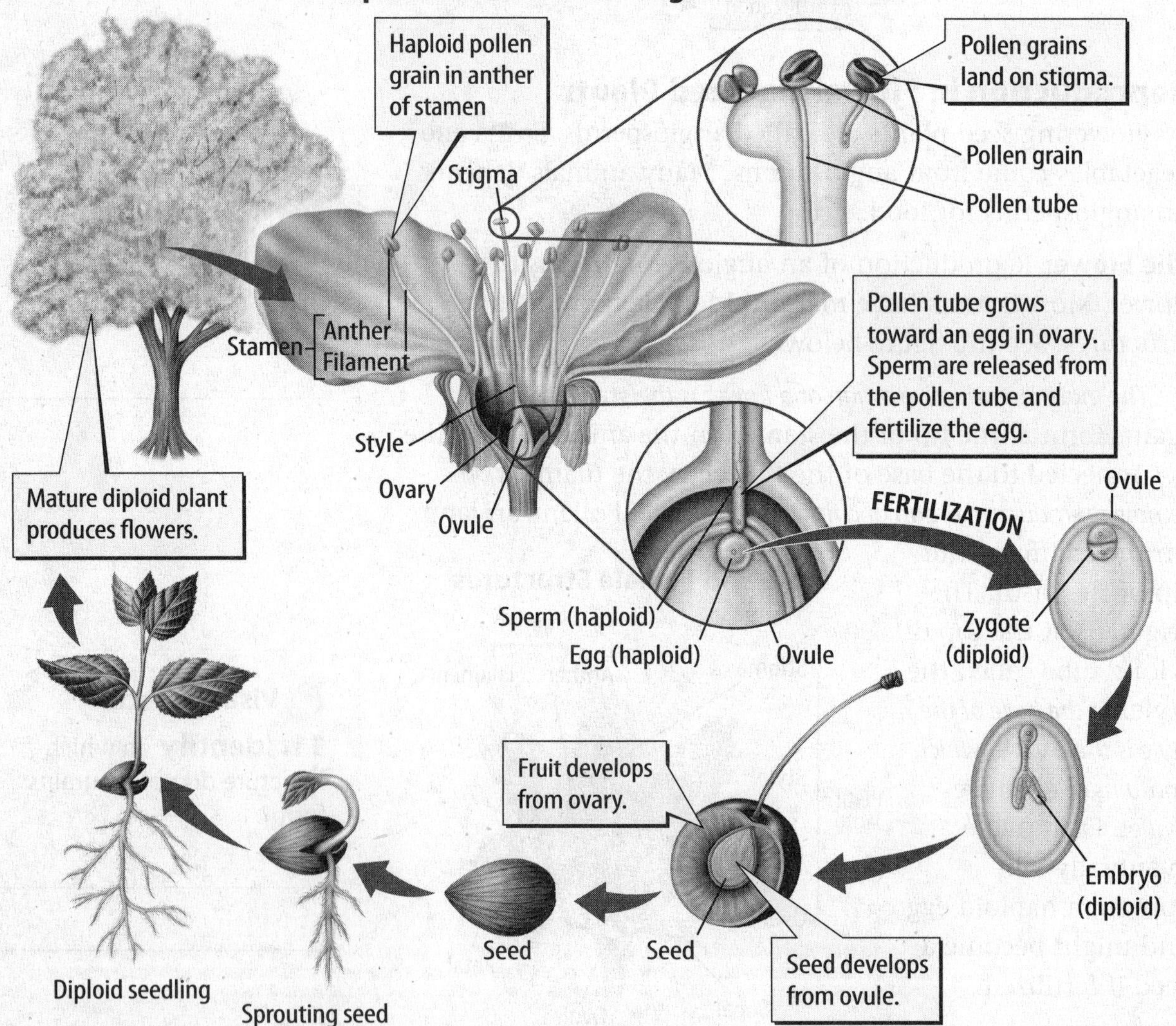

Fruit and Seed Dispersal Fruits and seeds are important sources of food for people and animals. In most cases, seeds of flowering plants are inside fruits. For example, pods are the fruits of pea plants. The peas inside the pods are the seeds of a pea plant. Each ear of corn is made up of many fruits, or kernels. The main part of each kernel is the seed. Strawberries have tiny seeds on the outside of the fruit.

Many fruits are juicy and good to eat, such as an orange or a watermelon. However, some fruits are hard and dry and not good to eat. For example, each parachute-like structure of a dandelion is a dry fruit.

Fruits help protect seeds and help scatter, or disperse, them. The fruits of a dandelion are light and float on air currents which help to scatter the seeds. When an animal eats a fruit, the fruit's seeds can pass through the animal's digestive system with little or no damage to the seed. Imagine what happens when a mouse eats blackberries. The animal digests the fruit but deposits the seeds on the soil with its wastes. The mouse might have traveled some distance before depositing the seeds. This means the mouse helped move the seeds to a location away from the blackberry bush.

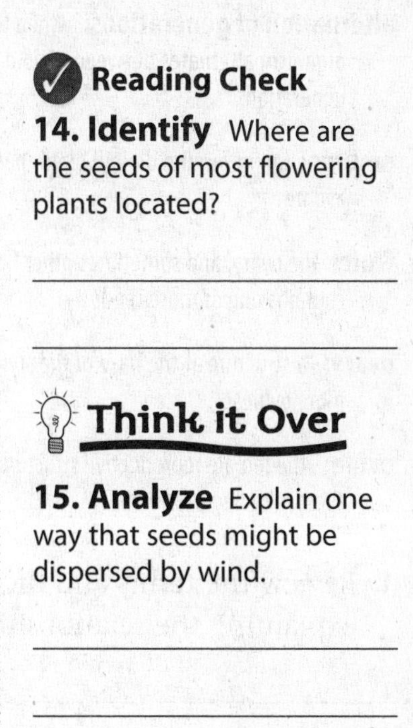

Reading Check

14. Identify Where are the seeds of most flowering plants located?

Think it Over

15. Analyze Explain one way that seeds might be dispersed by wind.

After You Read

Mini Glossary

alternation of generations: when the life cycle of an organism alternates between diploid and haploid generations

embryo: an immature diploid plant that develops from the zygote

fruit: the ovary, and sometimes other flower parts, that contains one or more seeds

ovary: a structure at the base of the style that contains one or more ovules

ovule: the female reproductive structure of a seed plant where the haploid egg develops

pistil: the female reproductive organ of a flower

pollen (PAH lun) grain: forms from tissue in a male reproductive structure of a seed plant

pollination (pah luh NAY shun): occurs when pollen grains land on a female reproductive structure of a plant that is the same species as the pollen grains

seed: an embryo, its food supply, and a protective covering

spore: a daughter cell produced from a haploid structure

stamen: the male reproductive organ of a flower

1. Review the terms and their definitions in the Mini Glossary. Write one sentence explaining the relationship between an embryo and a seed.

____

____

2. Write the letter of each statement below in the correct box to show whether the statement relates to the reproduction of seedless plants, the reproduction of seed plants, or both. The first one has been done for you as an example.

a. Has alternation of generations
b. The haploid egg develops in the ovule.
c. Can be gymnosperms or angiosperms
d. Produces spores.
e. Both pollination and fertilization are part of reproduction.
f. Produces fruit
g. Grows by mitosis and cell division

Seedless Plants
a

Seed Plants
a

What do you think NOW?

Reread the statements at the beginning of the lesson. Fill in the After column with an A if you agree with the statement or a D if you disagree. Did you change your mind?

ConnectED

Log on to ConnectED.mcgraw-hill.com and access your textbook to find this lesson's resources.

Animal Diversity

What defines an animal?

Before You Read

What do you think? Read the two statements below and decide whether you agree or disagree with them. Place an A in the Before column if you agree with the statement or a D if you disagree. After you've read this lesson, reread the statements to see if you have changed your mind.

Before	Statement	After
	1. All animals digest food.	
	2. Corals and jellyfish belong to the same phylum.	

Key Concepts 
- What characteristics do all animals have?
- How are animals classified?

Read to Learn

Animal Characteristics

Although animals have many traits that make them unique, all members of the Kingdom Animalia have the following characteristics:

- Animals are multicellular and eukaryotic.
- Animal cells are specialized for different functions, such as digestion, reproduction, vision, or taste.
- Animals have a protein, called collagen (KAHL uh juhn), that surrounds the cells and helps them keep their shape.
- Animals get energy for life processes by eating other organisms.
- Animals digest their food.

In addition to the characteristics above, most animals reproduce sexually and are capable of movement at some point in their lives.

Mark the Text

Ask Questions As you read, write questions you might have next to each paragraph. Read the lesson a second time and try to answer the questions. When you are done, ask your teacher any questions you still have.

Key Concept Check
1. Identify What characteristics do all animals have?

Animal Classification

Scientists have described and named more than 1.5 million species of animals. Every year thousands more are described and named. Many scientists estimate that Earth is home to millions of animal species that have not yet been discovered. If you discovered a new animal, could you classify it?

Vertebrates and Invertebrates

You could start classifying an animal by finding out if the animal has a backbone. Animals can be grouped into two large categories: vertebrates (VUR tuh brayts) and invertebrates (ihn VUR tuh brayts). *A* **vertebrate** *is an animal with a backbone.* Another name for backbone is spine. Fish, humans, and lizards are examples of vertebrates. *An* **invertebrate** *is an animal that does not have a backbone.* Worms, spiders, snails, and insects are examples of invertebrates. Invertebrates make up most of the animal kingdom—about 95 percent.

Reading Check

2. Differentiate What is the difference between a vertebrate and an invertebrate?

Symmetry

Another step you could take to classify an animal is to determine what kind of symmetry it has. Symmetry describes an organism's body plan. Symmetry can help identify the phylum to which an animal belongs.

An animal with **radial symmetry** *can be divided into two parts that are nearly mirror images of each other anywhere through its central axis.* A radial animal has a top and a bottom but no head or tail. It can be divided along more than one plane and still have two nearly identical halves. Jellyfish, sea stars, and sea anemones have radial symmetry.

An animal with **bilateral symmetry** *can be divided into two parts that are nearly mirror images of each other.* The two sides of a bilateral animal are mirror images of each other. Birds, mammals, reptiles, worms, and insects have bilateral symmetry.

Reading Check

3. Describe What is bilateral symmetry?

An animal with **asymmetry** *cannot be divided into any two parts that are nearly mirror images of each other.* An asymmetrical animal, such as a sponge, does not have a symmetrical body plan.

Molecular Classification

Molecules such as DNA, RNA, and proteins in an animal's cells also can be used for classification. For example, scientists can compare the DNA from two animals to determine if they are related. The more similar the DNA is, the more closely the animals are related.

Key Concept Check

4. Recognize How are animals classified?

Molecular classification has led to discoveries about relationships among species. For example, scientists once classified the grey-faced sengi as a close relative of shrews and voles. Recently, molecular evidence has shown that sengis are more closely related to elephants and aardvarks.

Major Phyla

Scientists classify the members of the animal kingdom into as many as 35 phyla (singular, phylum). The nine major phyla, shown in the figure below, contain 95–99 percent of all animal species. Animals belonging to the same phylum have similar body structures and other characteristics. Only one animal phylum, Chordata (kor DAH tuh), contains vertebrates. The other major phyla contain only invertebrates.

FOLDABLES®

Make a small horizontal four-door book to record your notes about the classification of animals.

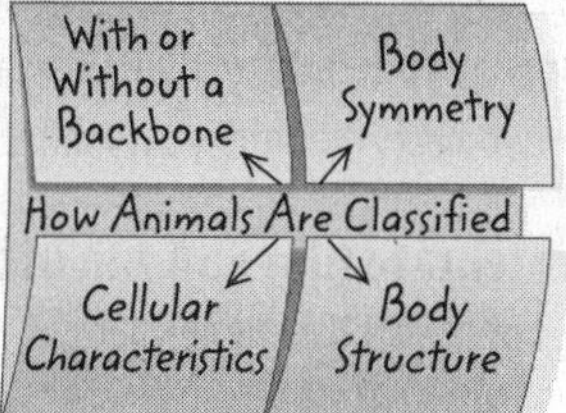

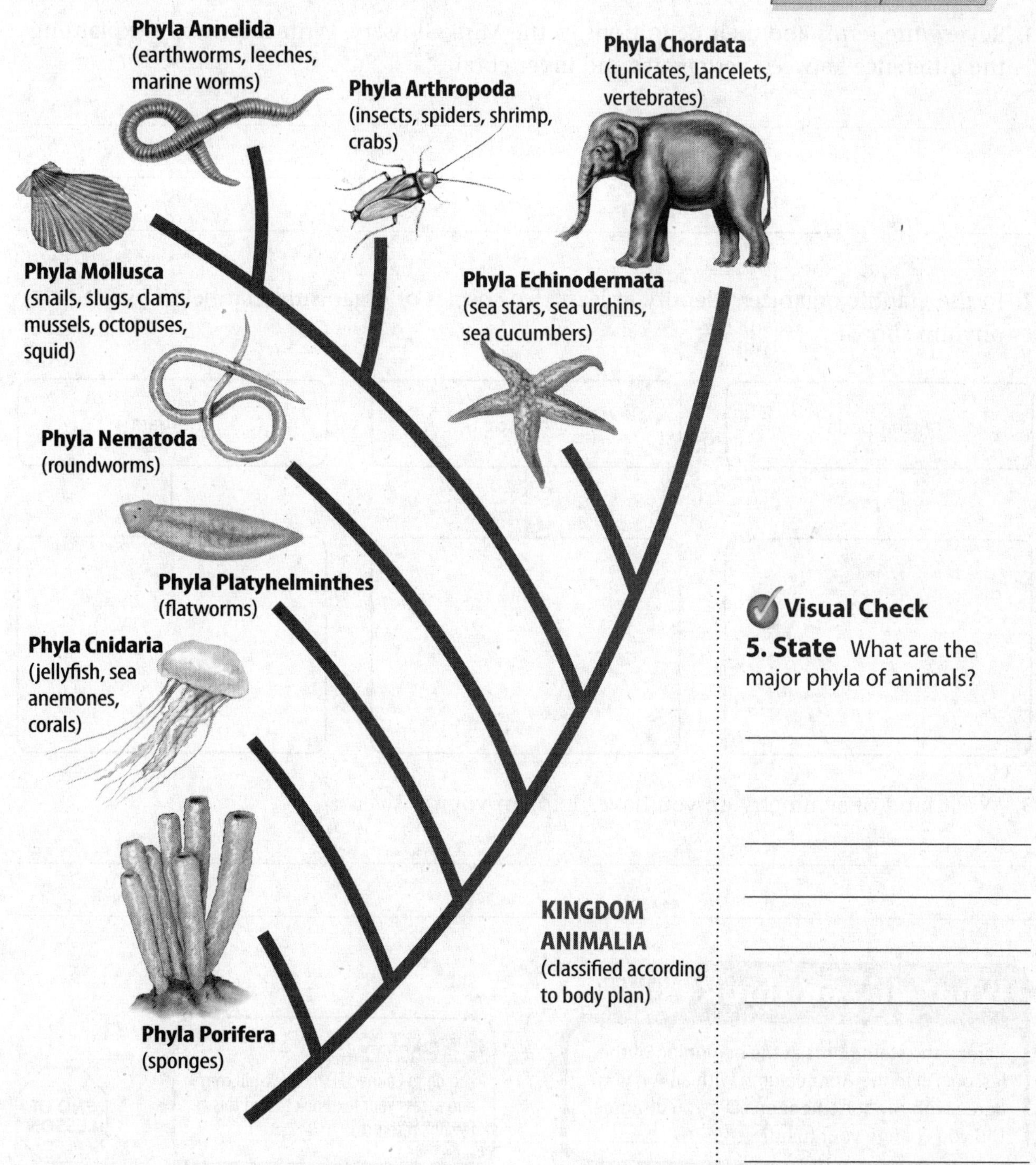

Visual Check

5. State What are the major phyla of animals?

After You Read

Mini Glossary

asymmetry: when an animal cannot be divided into any two parts that are nearly mirror images of each other

bilateral symmetry: when an animal can be divided into two parts that are nearly mirror images of each other

invertebrate (ihn VUR tuh brayt): an animal that does not have a backbone

radial symmetry: when an animal can be divided into two parts that are nearly mirror images of each other anywhere through its central axis

vertebrate (VUR tuh brayt): an animal with a backbone

1. Review the terms and their definitions in the Mini Glossary. Write a sentence explaining the difference between vertebrates and invertebrates.

2. In the graphic organizer, identify at least two species of organisms that belong in each phylum shown.

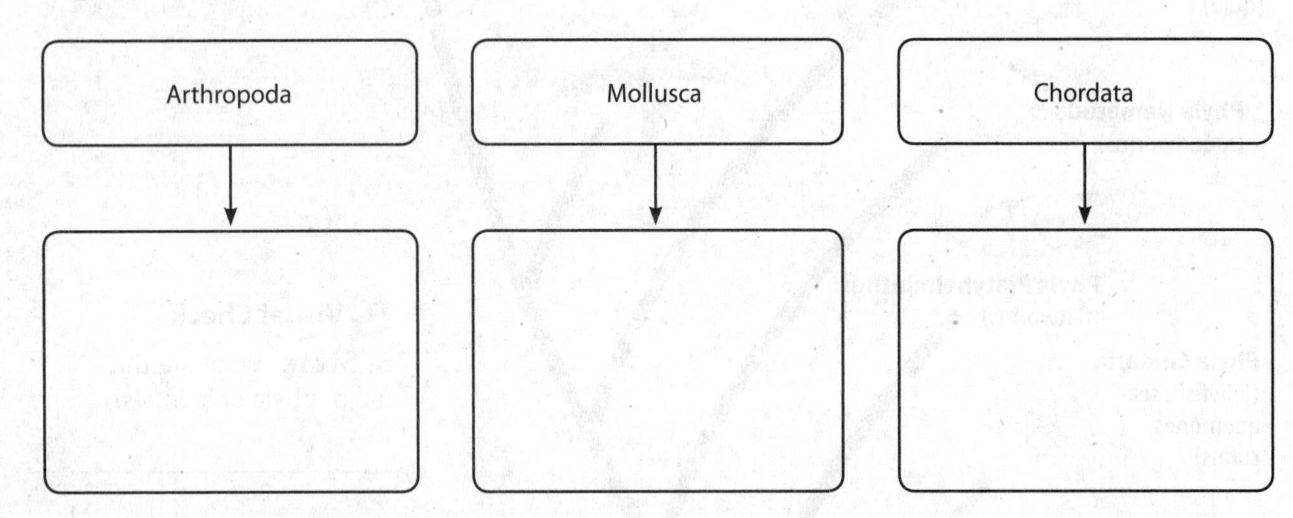

3. What kind of symmetry do you have? Explain your answer.

What do you think NOW?

Reread the statements at the beginning of the lesson. Fill in the After column with an A if you agree with the statement or a D if you disagree. Did you change your mind?

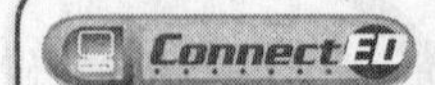

Log on to ConnectED.mcgraw-hill.com and access your textbook to find this lesson's resources.

Animal Diversity

Invertebrate Phyla

Before You Read

What do you think? Read the two statements below and decide whether you agree or disagree with them. Place an A in the Before column if you agree with the statement or a D if you disagree. After you've read this lesson, reread the statements to see if you have changed your mind.

Before	Statement	After
	3. Most animals have backbones.	
	4. All worms belong to the same phylum.	

Key Concepts 

- What are the characteristics of invertebrates?
- How do the invertebrate phyla differ?

Read to Learn

Characteristics of Invertebrates

Invertebrates are animals with no backbone. Most invertebrates have no internal structures to help support their bodies. Invertebrates are also usually smaller and move more slowly than vertebrates. More than 95 percent of all animal species that have been recorded are invertebrates.

Invertebrates are a varied group. Their physical characteristics range from the simple structures of sponges and jellyfish to the more complex bodies of worms, snails, and insects. The animals in each invertebrate phylum have similar body plans and physical characteristics.

Study Coach

Make an outline as you read to summarize the information in the lesson. Use the main headings in the lesson as the main headings in your outline. Use your outline to review the lesson.

ACADEMIC VOCABULARY
internal
(adjective) existing inside something

Key Concept Check
1. Identify What are the characteristics of invertebrates?

Sponges and Cnidarians

The simplest invertebrates are the sponges. Sponges belong to the phylum Porifera. All sponges share several characteristics.

Sponges are asymmetrical and have no tissues, organs, or organ systems. Their cells are specialized for capturing food, digestion, and reproduction. Other cells provide support inside the layers of the sponge. All sponges live in water. Most species live in ocean environments.

The phylum Cnidaria (ni DAR ee uh) includes jellyfish, sea anemones, hydras, and corals. Cnidarians, such as the sea anemone, differ from all other animals based on their unique characteristics.

Cnidarians

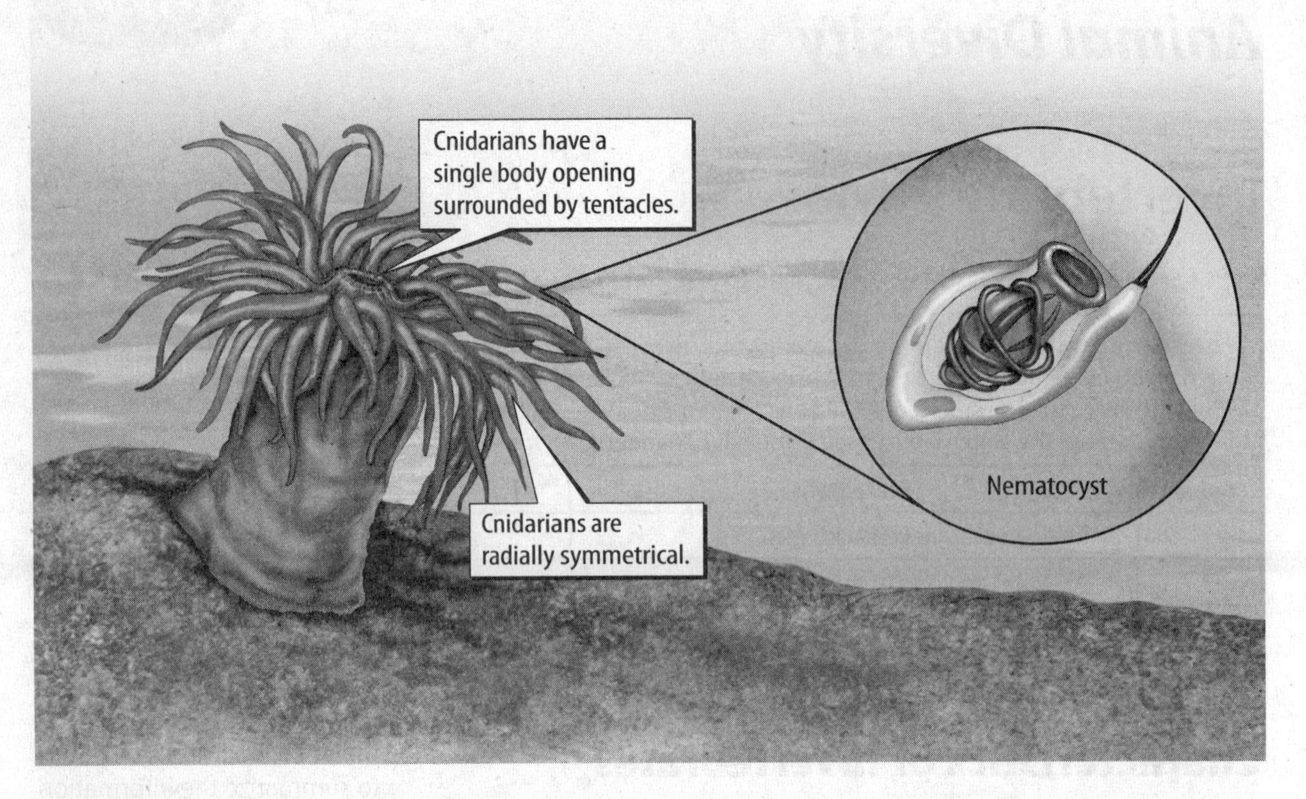

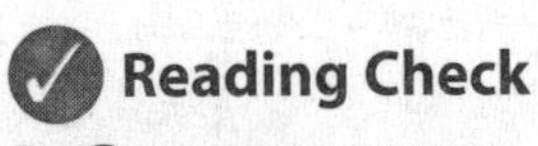

Visual Check
2. Categorize What kind of symmetry do cnidarians have?

____

Reading Check
3. Compare What characteristics do poriferans and cnidarians share?

____

____

____

Cnidarians, such as the sea anemone in the figure above, have no organs or organ systems. Cnidarians have radial symmetry. They have a single body opening surrounded by tentacles. Simple tissues, including muscles, nerves, and digestive tissue, enable cnidarians to survive by moving, reacting to stimuli, and digesting food.

Cnidarians also have specialized cells called nematocysts (NE mah toh sihsts). They use these cells for defense and capturing food. Most species of cnidarians live in ocean environments. All species live in water.

Flatworms and Roundworms

Flatworms are invertebrates that belong to the phylum Platyhelminthes (pla tih hel MIHN theez). All flatworms share several characteristics.

Flatworm Characteristics Flatworms have bilateral symmetry. They have nerve, muscle, and digestive tissues and a simple brain. They have soft and flattened bodies that are usually only a few cells thick. The digestive system of a flatworm has only one opening: a mouth.

Flatworms live in moist environments. Most, like tapeworms, are parasites. They live in or on the bodies of other organisms and rely on them for food. Others are free-living. Many live in oceans or other marine environments.

Roundworm Characteristics Roundworms, also called nematodes, belong to the phylum Nematoda (ne muh TOH duh). Like flatworms, roundworms have bilateral symmetry with nerve, muscle, and digestive tissues and a simple brain. Unlike flatworms, their bodies are round and have a stiff outer covering called a cuticle. A roundworm's digestive system has two openings: a mouth and an anus. Food enters the mouth and is digested as it travels to the anus. Wastes are excreted from the anus. Roundworms live in moist environments. Some species are parasites that live in animals' digestive systems. Free-living roundworms eat material such as fecal matter and dead organisms.

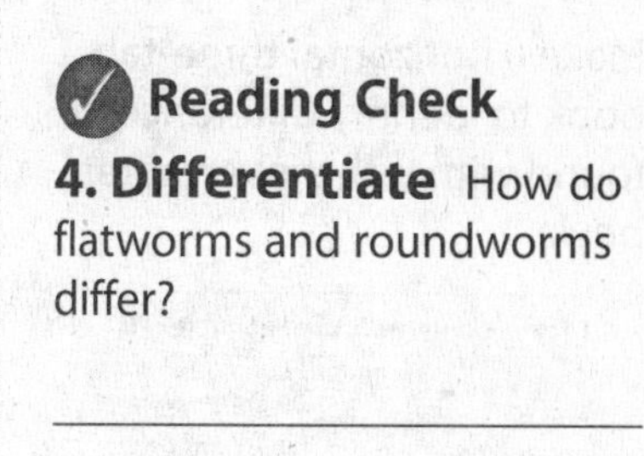

Reading Check

4. Differentiate How do flatworms and roundworms differ?

______

______

______

Mollusks and Annelids

The phylum Mollusca (mah LUS kuh) includes snails, slugs, clams, mussels, octopi, and squid. Mollusks have bilateral symmetry and soft bodies. Some have hard shells that protect their bodies. Mollusks have digestive systems with two openings. The body cavity contains the heart, the stomach, and other organs. The mollusk circulatory system contains blood but no blood vessels. The nervous system includes eyes and other sensory organs as well as simple brains. Mollusks must live in water or other moist environments.

The phylum Annelida includes earthworms, leeches, and marine worms. Annelida worms have bilateral symmetry and soft bodies. Their bodies consist of repeating segments covered with a thin cuticle. Their digestive systems have two openings. Annelids have circulatory systems that are made up of blood vessels that carry blood throughout the body. Their nervous systems include a simple brain. Annelids live in water or moist environments such as soil.

Math Skills

More than 95 percent of animals are invertebrates. This means that 95 out of every 100 animals is some type of invertebrate. Out of every 1,000 animal species, 48 are mollusks. What percentage of animal species are mollusks?

Express the information as a fraction.

$$\frac{48}{1{,}000}$$

Change the fraction to a decimal.

$$\frac{48}{1{,}000} = 0.048$$

Multiply by 100 and add a % sign.

$$0.048 \times 100 = 4.8\%$$

5. Use Percentages Out of 900,000 species of arthropods, 304,200 are beetles. What percentage of arthropods are beetles?

______

Arthropods

The phylum Arthropoda includes insects, spiders, shrimp, crabs, and their relatives. More species belong to this phylum than all the other animal phyla combined. There are more than 1 million identified species of arthropods.

All arthropods have bilateral symmetry and **exoskeletons**—*thick, hard outer coverings that protect and support animals' bodies.* Arthropods have several pairs of jointed appendages. *An* **appendage** *is a structure, such as a leg or an arm, that extends from the central part of the body.* The body parts of arthropods are segmented and specialized for different functions such as flying and eating. Arthropods live in almost every environment on Earth.

Reading Check

6. Summarize What do exoskeletons do?

______

______

______

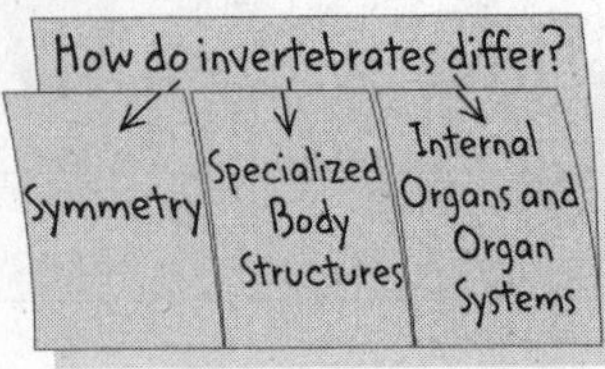

Make a horizontal three-tab book to identify differences found within the invertebrate phyla.

Insects

The largest order of arthropods is the insects. The stag beetle in the figure below is an insect. All insects have three pairs of jointed legs, three body segments, a pair of antennae, and a pair of compound eyes. Many also have one or two pairs of wings. There are 16 major groups of insects, but most insect species belong to one of five groups. About 40 percent of all known species of insects are beetles.

Stag Beetle

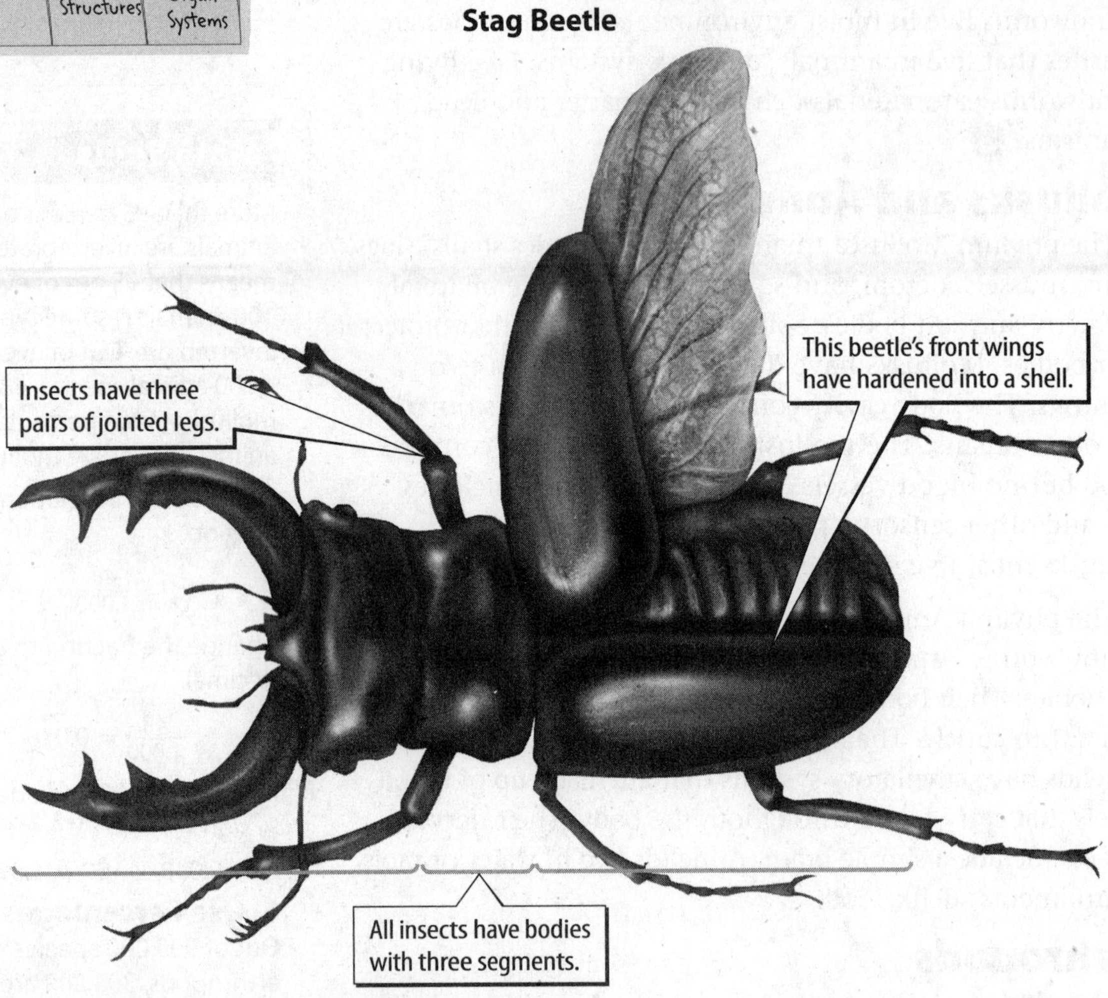

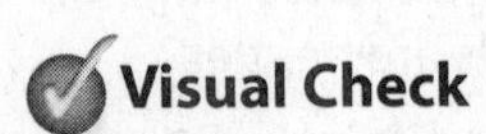

Visual Check

7. Label two additional characteristics of the stag beetle that are common to all insects.

Arachnids

Spiders, ticks, and scorpions are arachnids (uh RAK nudz). All arachnids have four pairs of jointed legs and two body segments. They do not have antennae or wings.

Crustaceans

All crustaceans (krus TAY shunz) have one or two pairs of antennae. They also have jointed appendages in the mouth area that are specialized for biting and crushing food. Crabs, shrimp, lobsters, and their close relatives are crustaceans.

Echinoderms

The phylum Echinodermata (ih kin uh DUR muh tuh) includes sea stars, sea cucumbers, and sea urchins, such as the one in the figure below. Echinoderm (ih KI nuh durm) means "spiny skin." Echinoderms have some features that are not in any of the other invertebrate phyla. They also are more closely related to vertebrates than to any other phyla.

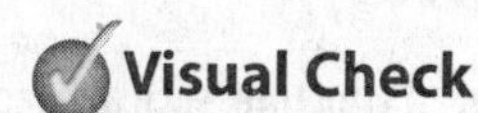

Visual Check

8. Draw a line through the sea urchin to show its radial symmetry.

Sea Urchin

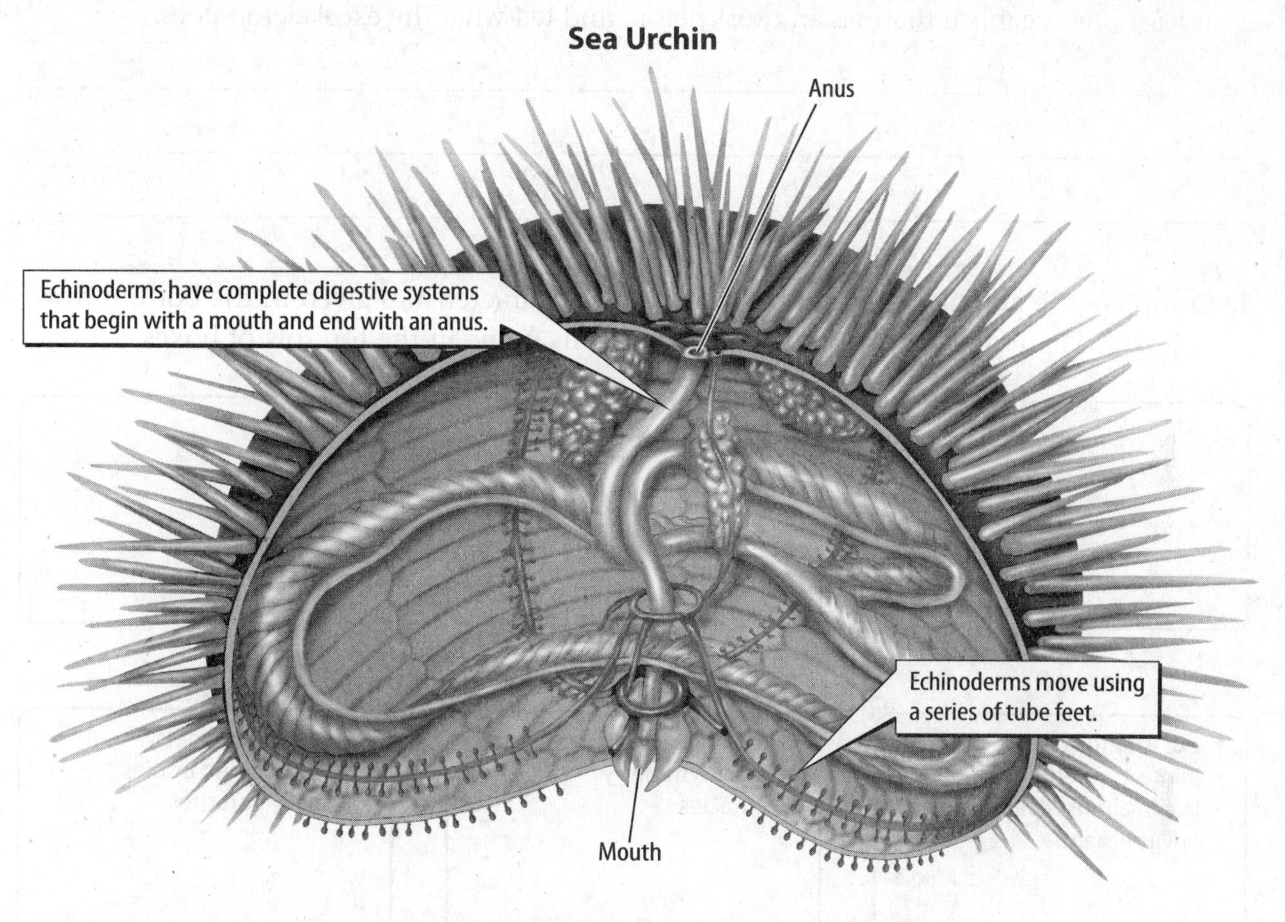

All echinoderms have radial symmetry. Unlike any other phyla, echinoderms have hard plates embedded in the skin. These plates support the body. Thousands of small, muscular, fluid-filled tubes, called tube feet, enable them to move and feed. They also have complete digestive systems including a mouth and an anus. Echinoderms live only in oceans. However, some can survive out of the water for short periods during low tides.

Key Concept Check

9. Distinguish How do the invertebrate phyla differ?

After You Read

Mini Glossary

appendage: a structure, such as a leg or an arm, that extends from the central part of the body

exoskeleton: a thick, hard outer covering that protects and supports an animal's body

1. Review the terms and their definitions in the Mini Glossary. Write a sentence that identifies at least one organism that has an exoskeleton, and tell what the exoskeleton does.

___

___

___

2. Complete the graphic organizer below. Read the characteristics given in the bottom row of boxes. Then identify each phylum and write its name in the top row of boxes.

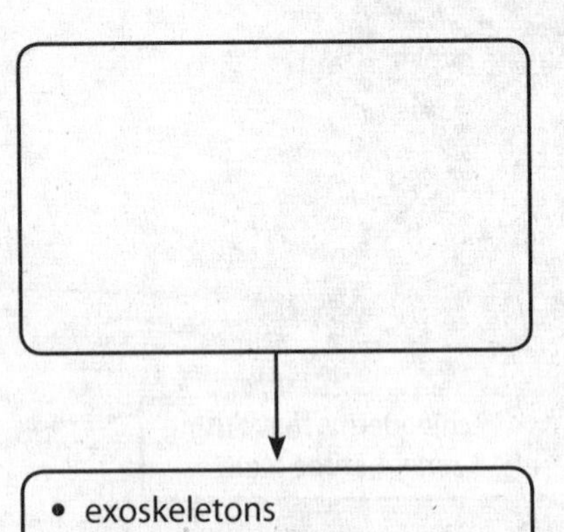

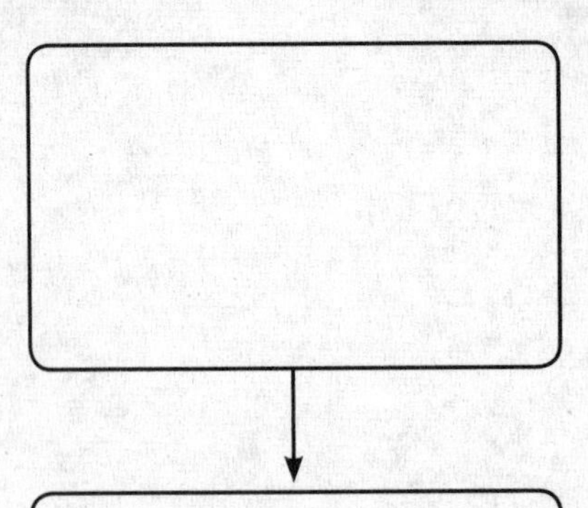

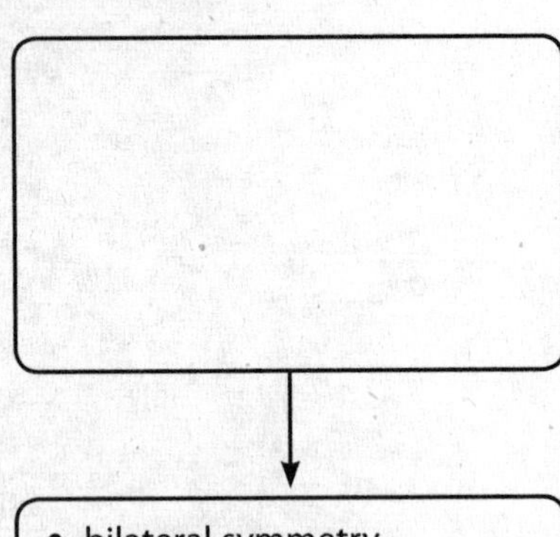

- exoskeletons
- jointed appendages
- live in almost every environment

- no organs or organ systems
- radial symmetry
- tentacles

- bilateral symmetry
- soft flattened bodies
- usually parasites

3. Select a word that appears in the main heading of the outline you made. In the space below, write that word and define it.

___

___

___

What do you think NOW?

Reread the statements at the beginning of the lesson. Fill in the After column with an A if you agree with the statement or a D if you disagree. Did you change your mind?

Log on to ConnectED.mcgraw-hill.com and access your textbook to find this lesson's resources.

Animal Diversity

Phylum Chordata

Before You Read

What do you think? Read the two statements below and decide whether you agree or disagree with them. Place an A in the Before column if you agree with the statement or a D if you disagree. After you've read this lesson, reread the statements to see if you have changed your mind.

Before	Statement	After
	5. All chordates have backbones.	
	6. Reptiles have three-chambered hearts.	

Key Concepts

- What are the characteristics of all chordates?
- What are the characteristics of all vertebrates?
- How do the classes of vertebrates differ?

Read to Learn

Characteristics of Chordates

One way to classify an animal is to check for a backbone. Another way to classify animals is to look for the four characteristics of a chordate (KOR dat). *A* **chordate** *is an animal that has a notochord, a nerve cord, a tail, and structures called pharyngeal* (fer IN jee ul) *pouches at some point in its life.* In vertebrates, these characteristics exist only during embryonic development. *A* **notochord** *is a flexible, rod-shaped structure that supports the body of a developing chordate.* The nerve cord develops into the central nervous system. The pharyngeal pouches are between the mouth and the digestive system.

Most chordates are vertebrates. Two groups of invertebrates, tunicates and lancelets (LAN sluhts), are chordates. A lancelet is shown in the figure to the right. Invertebrate chordates live in salt water. They are usually only a few centimeters long. In vertebrate chordates, such as humans, the notochord develops into a backbone during the growth of an embryo.

Lancelet

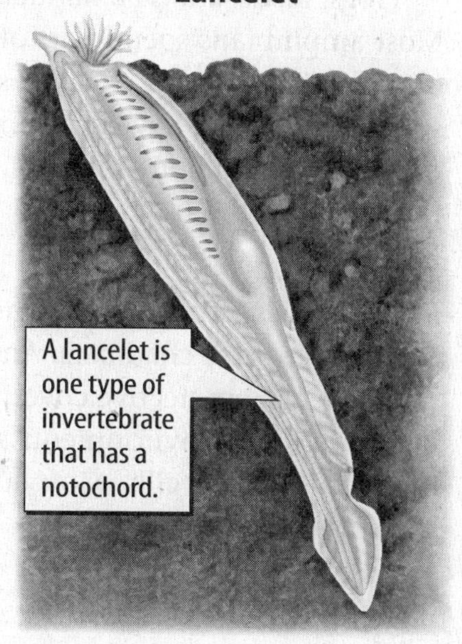

Mark the Text

Sticky Notes As you read, use sticky notes to mark information that you do not understand. Read the text carefully a second time. If you still need help, write a list of questions to ask your teacher.

Visual Check

1. Locate Highlight the lancelet's notochord.

Key Concept Check

2. Identify What are the characteristics of chordates?

FOLDABLES®

Make a vertical five-tab book to identify specific characteristics and examples of vertebrates.

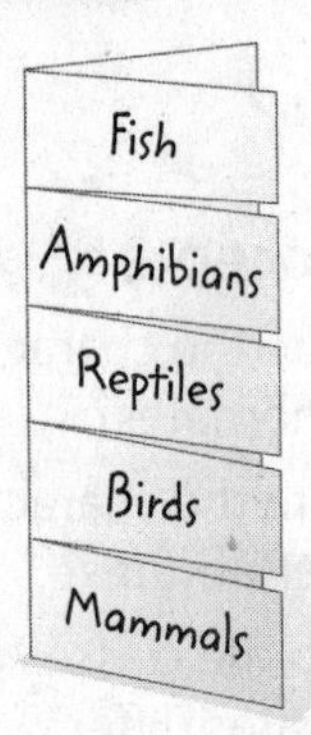

Key Concept Check

3. Recognize What are the characteristics of all vertebrates?

____

____

____

____

____

____

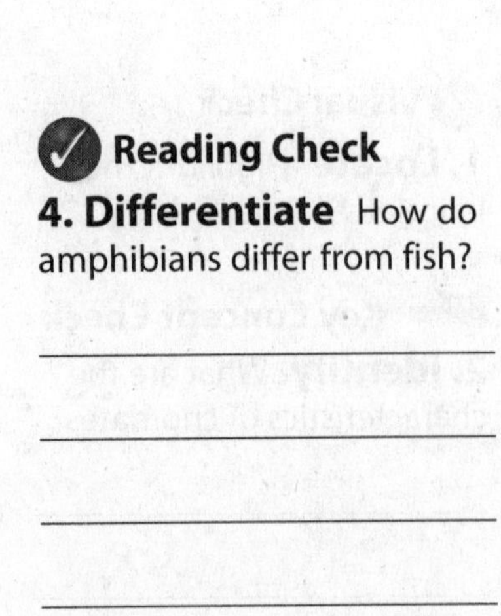

Reading Check

4. Differentiate How do amphibians differ from fish?

____

____

____

____

____

____

Characteristics of Vertebrates

Recall that all vertebrates have a backbone, also called a spinal column or spine. The backbone is a series of structures that surround and protect the nerve cord, or spinal cord. The spinal cord connects all the nerves in the body to the brain. Bones that form a backbone are called vertebrae (VUR tuh bray). If you gently touch the back of your neck, the bones you feel are some of your vertebrae.

Vertebrates have well-developed organ systems. All vertebrates have digestive systems with two openings. They also have circulatory systems that move blood through the body and nervous systems that include brains. The five major groups of vertebrates are fish, amphibians, reptiles, birds, and mammals.

Fish

Most fish spend their entire lives in water. All fish share two important characteristics: gills for absorbing oxygen gas from water and paired fins for swimming. Fish are grouped into one of three classes.

Hagfish and lampreys have no jaws and are in a group called jawless fish. Sharks, skates, and rays are called cartilaginous fish. They have skeletons made of a tough, fibrous tissue called cartilage (KAR tuh lihj). Both jawless and cartilaginous fish have internal structures made of cartilage.

Trout, guppies, perch, tuna, mackerel, and thousands of other species do not have cartilaginous skeletons. Instead, they have bones and are grouped together as bony fish.

Amphibians

Frogs, toads, and salamanders belong to the class Amphibia. Most amphibians spend part of their lives in water and part on land. Their bodies change as they grow older. In many species, the young have different body forms than the adults do. The different body forms of a salamander are shown in the figure on the next page.

Amphibians have skeletons made of bone. They have legs for movement. Their skin is smooth and moist, and their hearts have three chambers. Amphibians lay eggs. The eggs do not have hard protective coverings, or shells. Their eggs must be laid in moist environments such as ponds. The young live in water and have gills. Most adults develop lungs and live on land.

Amphibians

Reptiles

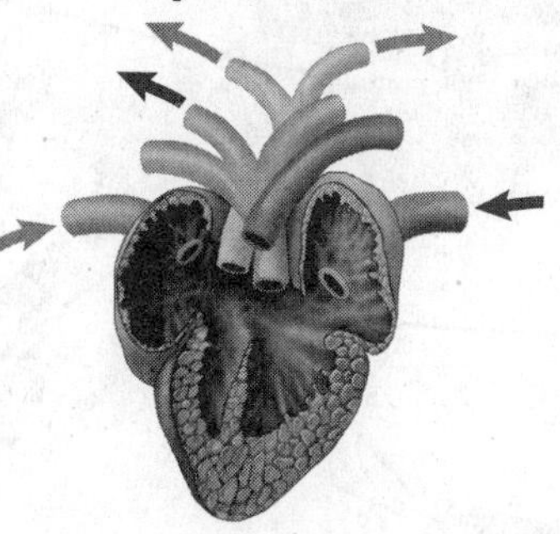

Lizards, snakes, turtles, crocodiles, geckos, and alligators belong to the class Reptilia. All reptiles share several characteristics. The skin of reptiles is waterproof and covered in scales. Like amphibians, most reptiles have three-chambered hearts, as shown in the figure at right. Unlike amphibians, lizards and other reptiles have lungs throughout their lives, not just in adulthood.

Most reptiles lay fluid-filled eggs with leathery shells. Unlike amphibian eggs, reptile eggs are laid on land rather than in water. Young reptiles do not change form as they mature into adult reptiles.

Visual Check

5. Explain How does the body form of this salamander change as it grows?

____

____

____

Visual Check

6. Distinguish How are reptile and amphibian hearts similar?

____

____

Reading Check

7. Contrast How do birds differ from reptiles?

____

____

____

Visual Check

8. Identify How many chambers does a bird's heart have?

____

Birds

All birds, including the owl in the figure below, are in the class Aves. Many birds make nests to hold their eggs, and many have unique calls or songs.

Birds have lightweight bones. Their skin is covered with feathers and scales. Birds also have two legs and two wings. Many birds can fly. They have stiff feathers that enable them to move through the air. Birds that spend a lot of time in the water have oil glands that help water roll off their feathers.

Birds have beaks and do not chew their food. Instead, their digestive systems include gizzards—organs that help grind food into smaller pieces. Their circulatory systems include four-chambered hearts. Birds also lay fluid-filled eggs with hard shells. Birds feed and care for their young.

Birds

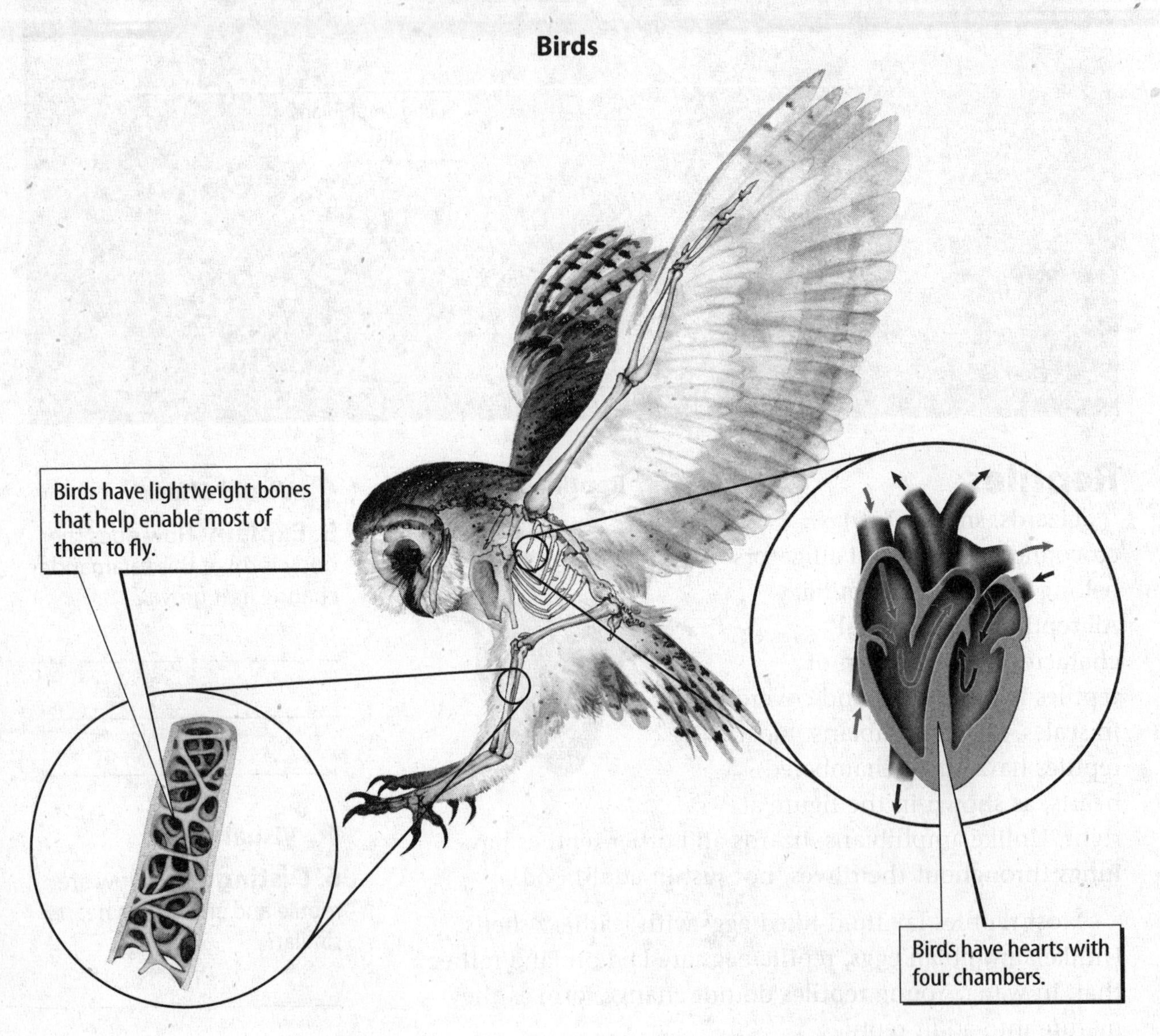

Mammals

Dogs, cats, goats, rats, seals, whales, and humans are among the many vertebrates belonging to the class Mammalia. All mammals have hair or fur covering their bodies. Mammals have teeth to tear and chew their food.

Mammals have complete digestive systems that include a mouth and an anus. Mammals also have a complex nervous system that includes a brain.

The most notable characteristic of mammals, however, is the presence of mammary glands. These glands produce milk that feeds young mammals. Many mammals give birth to live young. A few species of mammals, including the duck-billed platypus, lay eggs.

Think it Over

9. Consider How might fur benefit mammals?

Key Concept Check

10. Classify How do the classes of vertebrates differ?

After You Read

Mini Glossary

chordate (KOR dat): an animal that has a notochord, a nerve cord, a tail, and pharyngeal pouches at some point in its life

notochord: a flexible, rod-shaped structure that supports the body of a developing chordate

1. Review the terms and their definitions in the Mini Glossary. Write a sentence that describes the two groups of chordate invertebrates.

____

____

2. Complete the chart. Provide three characteristics and three examples for each vertebrate group.

Vertebrates	Class Name	Characteristics	Examples of Animals
amphibians			
reptiles			
birds			

What do you think NOW?

Reread the statements at the beginning of the lesson. Fill in the After column with an A if you agree with the statement or a D if you disagree. Did you change your mind?

Log on to ConnectED.mcgraw-hill.com and access your textbook to find this lesson's resources.

Animal Structure and Function

Support, Control, and Movement

Before You Read

What do you think? Read the two statements below and decide whether you agree or disagree with them. Place an A in the Before column if you agree with the statement or a D if you disagree. After you've read this lesson, reread the statements to see if you have changed your mind.

Before	Statement	After
	1. All animals on Earth have internal skeletons.	
	2. All animals that live in the water move using fins.	

Key Concepts 

- How are the types of support alike, and how are they different?
- How do the types of control compare and contrast?
- How do the types of movement compare and contrast?

Study Coach

Identify the Main Ideas Fold a sheet of paper into three columns. Label them *(K)* for what you already know about animal support, control, and movement; *(W)* for what you want to learn; and *(L)* for the facts that you learned. Fill in the third column after you have read this lesson.

Read to Learn

The Importance of Support, Control, and Movement

Think about the different environments in which animals live. Some animals live their entire lives in water. Others live only on land. Regardless of their environments, all animals have the same basic needs: food, water, and oxygen. However, in order to survive in different habitats, animals have different structures that perform the same function.

Fish and birds live in different environments. However, both use structures to obtain oxygen from their environments. In a similar way, animals have different structures for support, control, and movement. Without these, animals could not obtain the things they need to survive. In this lesson, you will read about how animals in different habitats have different structures that provide support and control for their bodies and how they use these structures to move around.

Structures for Support

Organisms have structures to provide support, control, and movement. Most animals are invertebrates, or animals without backbones. Animals with backbones, such as humans, are vertebrates. Vertebrate and invertebrate animals have different types of structures that provide support. 

Reading Check

1. Name What are animals with backbones called?

FOLDABLES®

Make a vertical three-tab book to organize your notes about support structures.

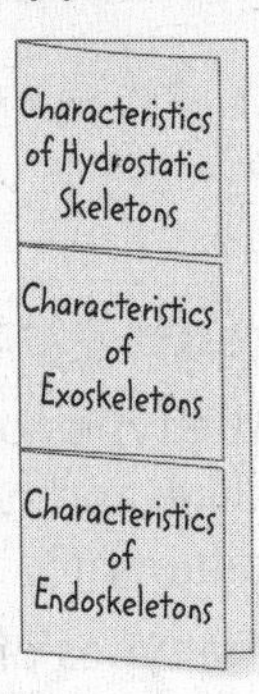

Reading Check

2. State What type of skeleton do jellyfish have?

____

Visual Check

3. Identify What is the fluid-filled cavity of the earthworm called?

____

Hydrostatic Skeletons

Filling a balloon with water gives the balloon shape. This is because the force of the water against the surface of the balloon gives the balloon structure. Just as the water in a balloon provides structure, many organisms have internal fluids that provide support.

A **hydrostatic skeleton** *is a fluid-filled internal cavity surrounded by muscle tissue. The fluid-filled cavity is called the* **coelom** (SEE lum). Muscles that surround the coelom help some organisms move by pushing the fluid in different directions.

Earthworms, like those in the figure below, jellyfish, and sea anemones (uh NE muh neez) are organisms that have hydrostatic skeletons. Organisms with hydrostatic skeletons do not have bones or other hard structures that provide support. ✓

Hydrostatic Skeletons

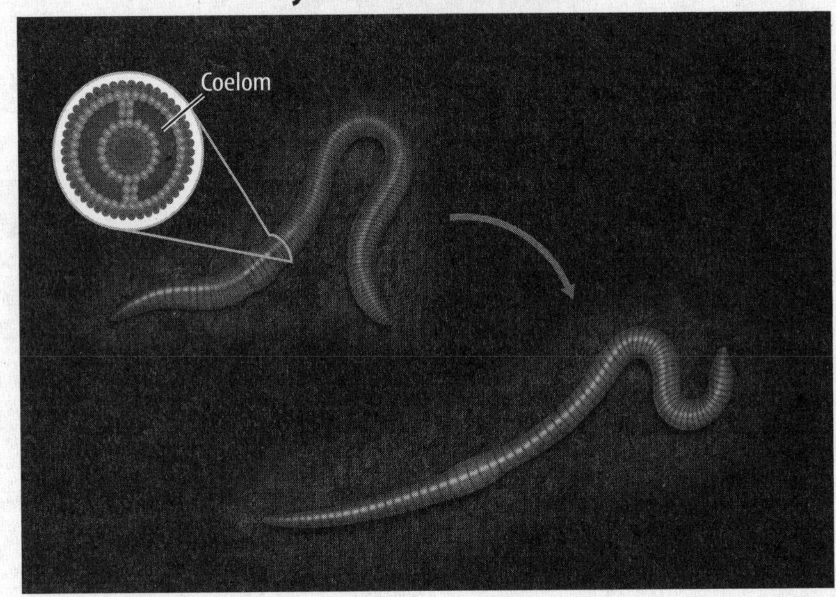

Exoskeletons

Some organisms are supported by structures on the outside of their bodies. Hard outer coverings, called exoskeletons, provide support and protection for many invertebrates. A hard exoskeleton protects internal tissues from predators or damage. The shells of species such as crabs and snails are exoskeletons.

In some species, the exoskeleton does not grow as the animal grows. A lobster must shed its exoskeleton when the exoskeleton becomes too small for the growing lobster. This leaves the animal defenseless until the new exoskeleton forms and hardens.

Endoskeletons

The bodies of many animals have a soft exterior that covers hard internal support structures. Animals such as fish, birds, and mammals—including you—have internal support structures. These structures are called endoskeletons.

The endoskeletons of most animals are made of bone. Some animals, such as sharks, have endoskeletons that are made of cartilage. An endoskeleton protects internal organs and provides the organism with structure and support.

Tortoises and turtles are unique because they have both endoskeletons and exoskeletons. The endoskeleton protects the organs of the animal, and the hard exoskeleton shell protects the animal from predators.

Key Concept Check
4. Describe How are the types of support alike, and how are they different?

Structures for Control

All animals react to changes in their environments. Just as different animals have different structures for support, they also have different control systems.

These control systems are called nervous systems. A nervous system helps protect an animal from harm. It also helps the animal move and find food.

Nerve Nets

Animals with radial symmetry and no brain have nerve nets with a central ring that control their bodies. *A* ***nerve net*** *is a netlike control system that sends signals to and from all parts of the body.*

Reading Check
5. Explain What controls the bodies of animals with radial symmetry and no brain?

Signals sent through the nerve net and ring cause an organism's muscle cells to contract. These contractions help the animal move.

Cnidarians (nih DAYR ee unz) such as jellyfish and sea anemones have nerve nets. The nerve nets sense physical contact and detect food. Nerve nets and rings help these animals move and capture prey.

Nerve Cords

Animals with bilateral symmetry have brains or brainlike structures that detect and respond to their environments. An animal with a brain or a brainlike structure has a nerve cord.

An animal with a nerve cord usually has many neurons that detect changes in its external environment. Signals detected by neurons are sent to the nerve cord, which might initiate a reflex response, and then to the brain for processing.

REVIEW VOCABULARY
neuron
basic functioning unit of the nervous system

Zebra and Planarian Nerve Cords

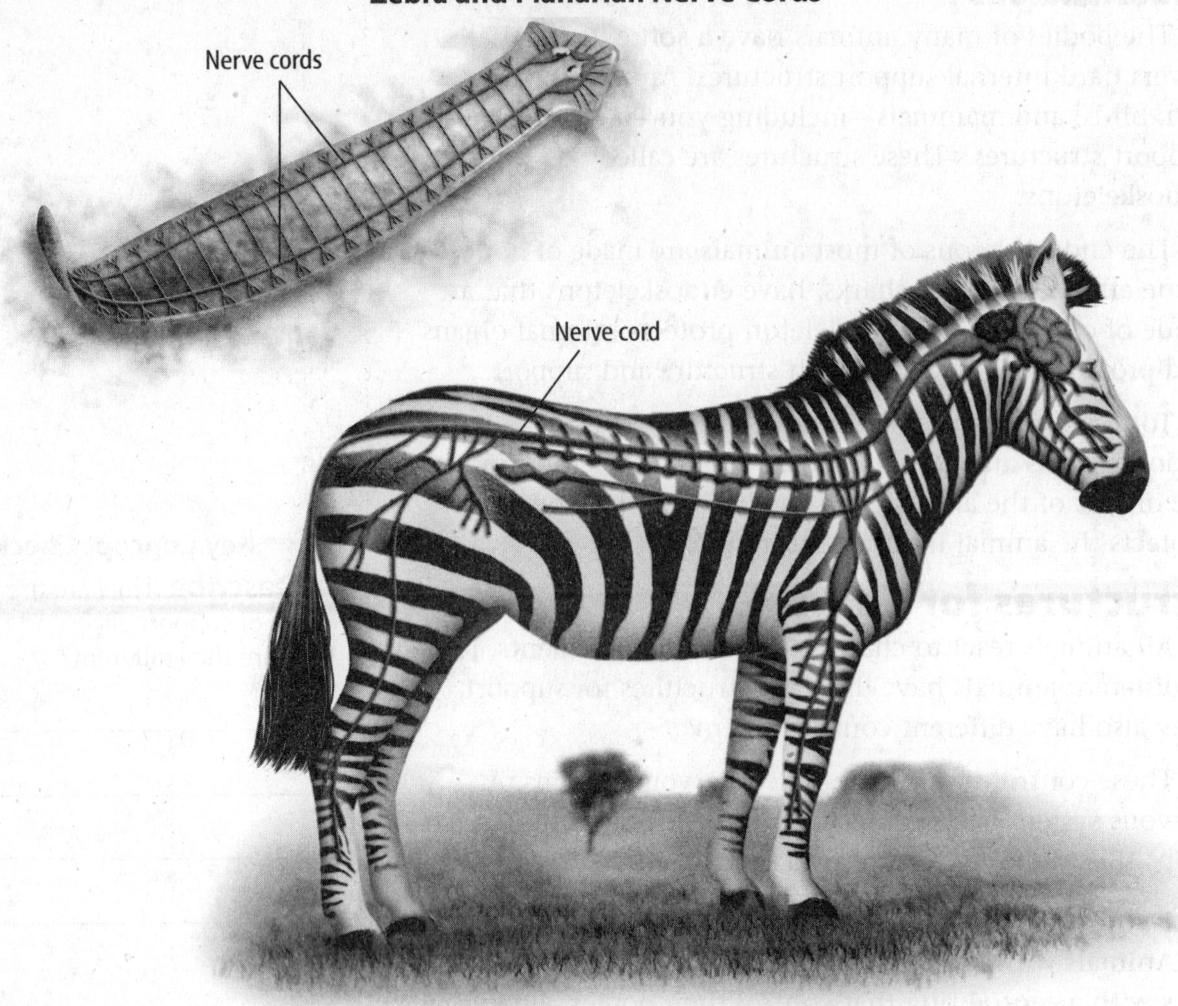

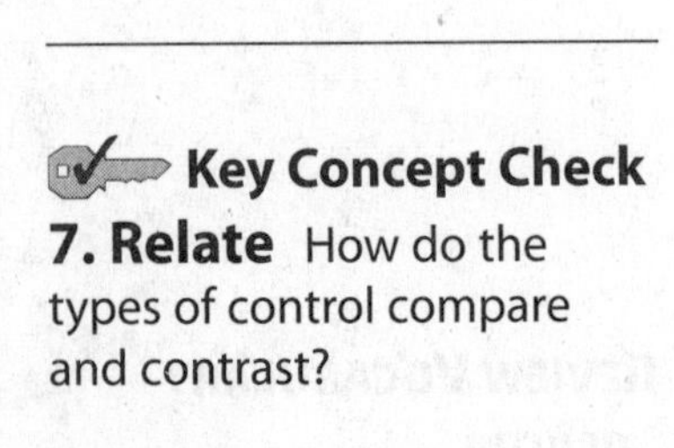

Visual Check

6. Contrast How do the nerve structures of the zebra and the planarian differ?

____

____

____

Key Concept Check

7. Relate How do the types of control compare and contrast?

____

____

____

Just as a telephone wire transmits signals between two buildings, the nerve cord enables signals to move between neurons and the brain. In vertebrates, nerve cords are also called spinal cords. Nerve cords in a zebra and a planarian are shown in the figure above.

Types of Movement

All animals move at some point in their lives. Some animals, such as birds and tigers, move around throughout most of their lives. Other animals, such as sponges, move during only part of their lives.

Movement helps an animal obtain food. Movement also helps an animal escape from danger. Different animals live in many different habitats. These animals have different structures that they use for movement.

Undulate Motion

You know how an animal with legs moves. How does an animal without legs move? *Some animals move in a wavelike motion called* **undulation** (un juh LAY shun).

Snakes, fishes, and eels are some of the animals that move by undulation. These animals have muscles that push their bodies forward. Animals that move by undulating live on land and in water.

Swimming

Many animals that live in water move by swimming. Some animals, such as fish, have fins and tails that move them through water. Other animals, such as octopuses, take in water and then push the water out forcefully to move forward. This process is called jet propulsion. Many organisms, such as humans and dogs, also can swim by moving their arms and legs, even though they do not live in water.

Walking

Most animals that live on land move by walking. The body's weight rests on two, four, six, or eight legs and shifts as the legs move. Some animals, such as rabbits and frogs, also are capable of jumping using their limbs.

Flying

Many animals move through the air by flying. Birds, some insects, and bats use wings to move around. Wings are a type of limb. By moving their wings, animals can lift their bodies and keep them in the air. Animals that have wings also have legs that are used to move around on land.

Wings are not the only structures that enable animals to move through the air. Some animals can glide or move through the air without flapping their limbs.

Large fins enable some species of fish to glide short distances and escape predators. Some squirrels, marsupials, and even some snakes can glide through the air. To glide, they launch themselves from a high point, flatten their bodies or stretch out tissues that form a structure similar to a parachute, and glide down.

Reading Check

8. State How is undulate motion achieved?

Reading Check

9. Describe How do most land animals move?

Key Concept Check

10. Express How do the types of movement compare and contrast?

After You Read

Mini Glossary

coelom (SEE lum): a fluid-filled cavity within a hydrostatic skeleton

hydrostatic skeleton: a fluid-filled internal cavity surrounded by muscle tissue

nerve net: a netlike control system that sends signals to and from all parts of the body

undulation (un juh LAY shun): to move in a wavelike motion

1. Review the terms and their definitions in the Mini Glossary. Write a sentence that describes the characteristics of an animal with a hydrostatic skeleton.

2. Fill in the graphic organizer below to show the types of movement.

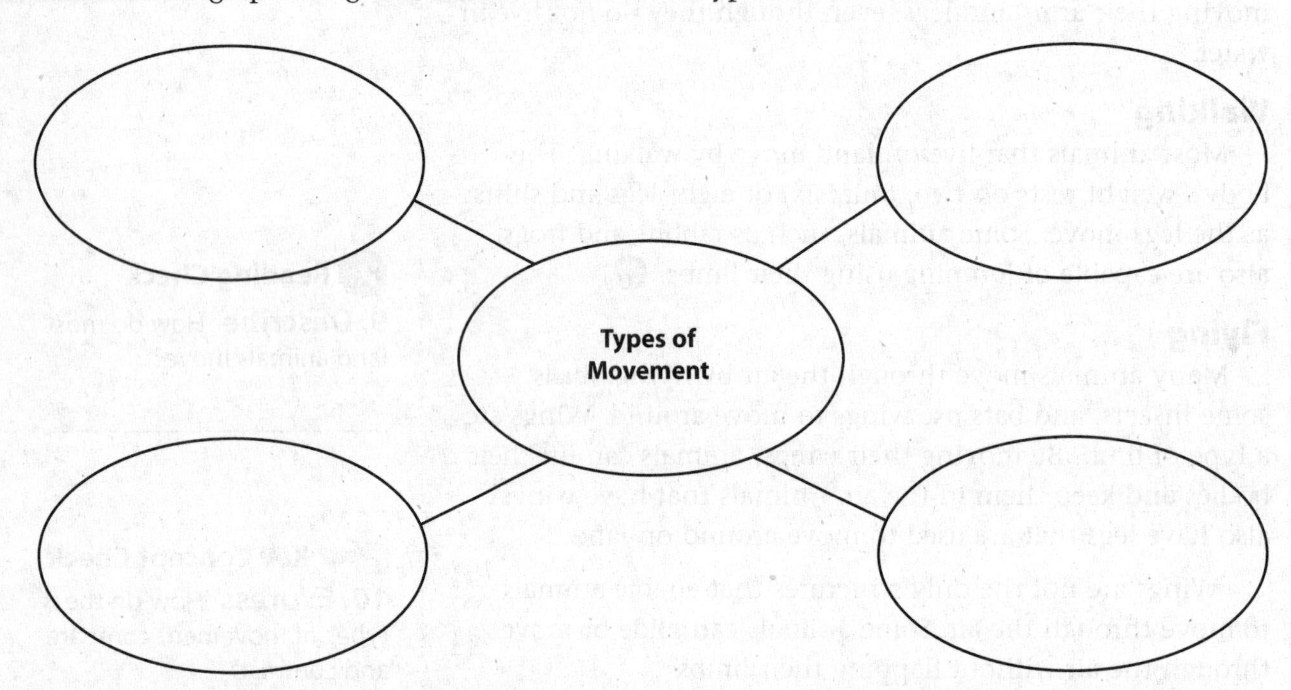

3. Explain how nerve nets and rings help cnidarians move and obtain food.

What do you think NOW?

Reread the statements at the beginning of the lesson. Fill in the After column with an A if you agree with the statement or a D if you disagree. Did you change your mind?

Log on to ConnectED.mcgraw-hill.com and access your textbook to find this lesson's resources.

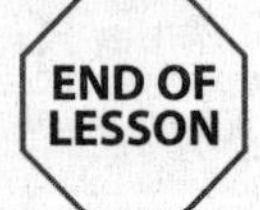

Animal Structure and Function

Circulation and Gas Exchange

Before You Read

What do you think? Read the two statements below and decide whether you agree or disagree with them. Place an A in the Before column if you agree with the statement or a D if you disagree. After you've read this lesson, reread the statements to see if you have changed your mind.

Before	Statement	After
	3. Earthworms have an open circulatory system that transports blood and other substances throughout the body.	
	4. Gills and lungs are different structures that perform the same function.	

Key Concepts 

- How do the types of gas exchange differ?
- What are the differences between open and closed circulatory systems?

Read to Learn

The Importance of Gas Exchange and Circulation

All cells need nutrients and oxygen to survive. Recall that animals obtain nutrients and oxygen from their environment. Organisms must take in these substances and get them to each cell. Structures in animal bodies transport the substances to all cells. They also help remove wastes such as carbon dioxide from the body. Most of an animal cell is made of water. Water also is transported throughout the body, in addition to nutrients, oxygen, and wastes.

As with support, control, and movement, different animals have different structures that exchange gases and move substances throughout the body. An animal's system depends on the animal's habitat. In this lesson, you will read about the different structures that animals have that help cells exchange gases and get nutrients and oxygen.

Mark the Text

Ask Questions As you read, write questions you might have next to each paragraph. Read the lesson a second time and try to answer the questions. When you are done, ask your teacher any questions you still have.

Gas Exchange

All animals must take in oxygen and eliminate carbon dioxide to survive. Oxygen must enter the body so cells and tissues can use it for life processes. However, different animals have various structures for gas exchange. ✓

Reading Check

1. State Why is gas exchange important to all animals?

FOLDABLES®

Make a vertical book to organize your notes about gas exchange and circulatory systems.

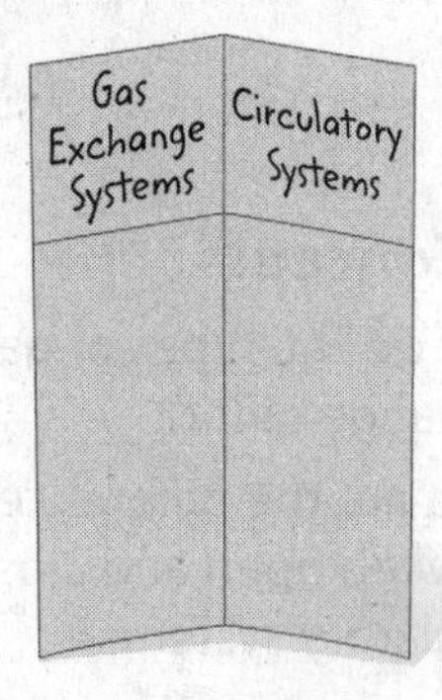

Diffusion

The basic process of gas exchange requires no structures. This process is called diffusion. **Diffusion** *is the movement of substances from an area of higher concentration to an area of lower concentration.*

Simple animals such as sponges, whose bodies are only a few cell layers thick, have no special gas exchange structures. Diffusion occurs through all parts of the body. Oxygen passes directly into cells from the environment.

In a similar manner, waste gases leave cells and enter the environment. Other animals exchange gas by diffusion and also have specialized structures for gas exchange.

Spiracles

Some organisms exchange gases through the sides of their bodies. **Spiracles** *are tiny holes on the surface of an organism where oxygen enters the body and carbon dioxide leaves the body.* Insects such as beetles and arachnids such as spiders have spiracles.

Although beetles and spiders have spiracles, they have different tissues that transport oxygen throughout the body. Beetles have structures called tracheal (TRAY kee ul) tubes, and spiders have folded structures called book lungs to take in oxygen.

Tracheal tubes, such as the ones shown in the figure at the top of the next page, are hoselike structures that branch into smaller tubes. Much as a river branches off into smaller streams, smaller branches of tracheal tubes help oxygen get to more places in the body. In contrast, book lungs are stacks of folded wall-like structures. Although tracheal tubes and book lungs look different, both are used for gas exchange.

Gills

Most animals that live in the water have gills for gas exchange. **Gills** *are organs that enable oxygen to diffuse into an animal's body and carbon dioxide to diffuse out.*

See the figure at the top of the next page. In aquatic animals such as fish, water enters the mouth and passes through the gills. Gill filaments take in the oxygen from the water. The oxygen is then transported to the rest of the body. Gills also remove carbon dioxide from the body. Like other organs in the body, gills are surrounded by capillaries that help transport oxygen and carbon dioxide to and from cells.

Math Skills

A proportion is an equation with two ratios that are equivalent. Use proportions to solve problems such as the following: Veins hold about 55 percent of the body's blood. What is an organism's blood volume if the veins hold 2.6 L?

Set up the proportion.

$$\frac{55\%}{2.6\text{ L}} = \frac{100\%}{x\text{L}}$$

Cross-multiply.

$$55x = 260$$

Divide both sides by 55.

$$\frac{55x}{55} = \frac{260}{55} = 4.7\text{ L}$$

$$x = 4.7\text{ L}$$

2. Use Proportions If a normal, complete heart cycle takes 0.8 s, how many cycles would the heart make in one day?

____

Structures for Gas Exchange

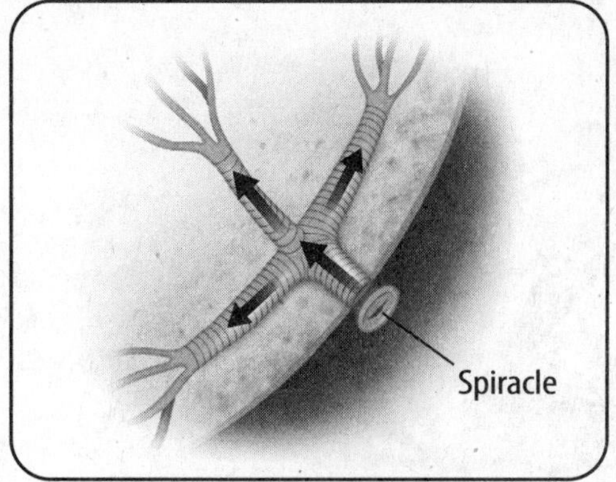

Tracheal tubes

Gills

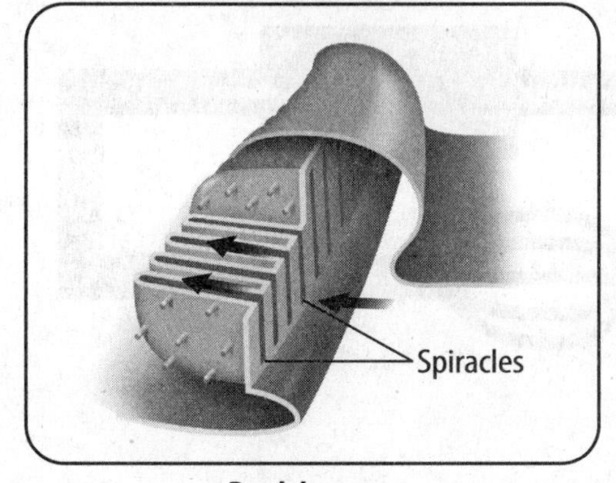

Book lungs

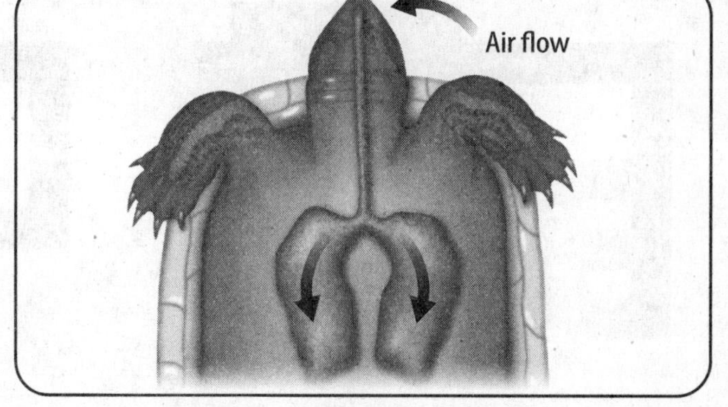

Lungs

Lungs

Many animals that live on land, including the turtle shown in the figure above, and some types of fish and snails have lungs for gas exchange. Lungs are baglike organs that can be filled with air.

Once the lungs fill with air, oxygen diffuses into the capillaries within the lungs' tissues and carbon dioxide diffuses out of the animal's body. Recall that capillaries transport oxygen to other cells in the body through the circulatory system.

Visual Check

3. Recognize What structures for gas exchange involve spiracles?

Key Concept Check

4. Contrast How do the organs used for gas exchange differ?

Circulation

You have just read about how gases are exchanged between animals and the environment. After an animal takes in oxygen, the oxygen has to travel to all parts of the body.

Much like pipes in a house help transport water to the kitchen and bathrooms, an animal's circulatory system helps materials move through the body. Different animals have different circulatory systems. The type of circulatory system an animal has often determines how quickly blood moves through the animal.

Open and Closed Circulatory Systems

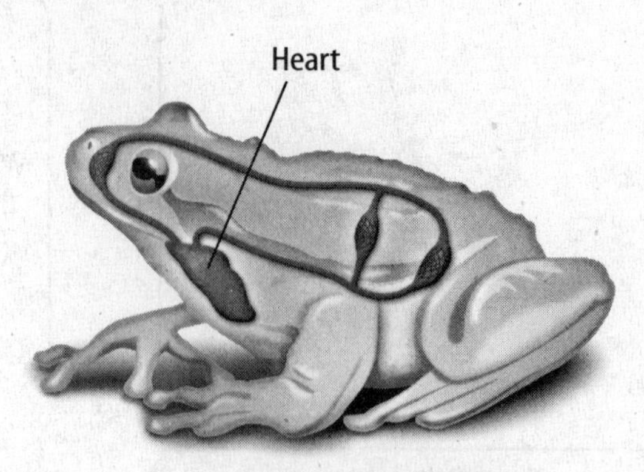

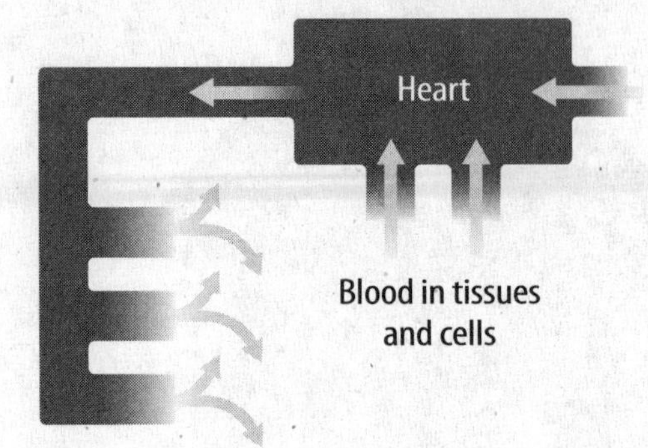

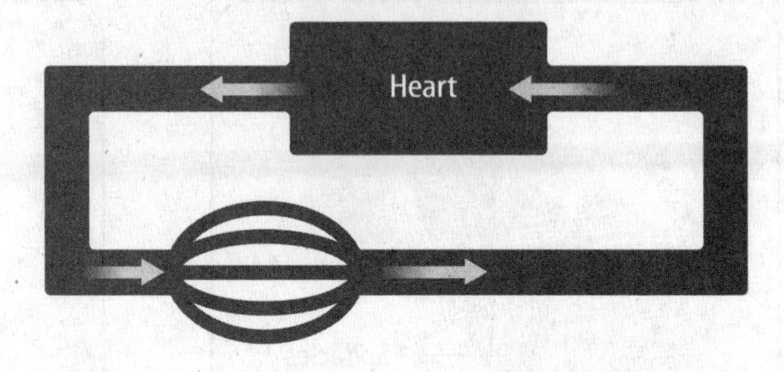

Visual Check

5. Name Which of the animals has an open circulatory system?

____

Reading Check

6. Identify What helps blood move through the body in an open circulatory system?

____

Open Circulatory Systems

Snails, insects, and many other invertebrates have open circulatory systems. *An* **open circulatory system** *is a system that transports blood and other fluids into open spaces that surround organs in the body.*

In an open circulatory system, such as the circulatory system of a bee shown in the figure above, oxygen and nutrients in blood can enter all tissues and cells directly. Carbon dioxide and other wastes are taken up by blood surrounding the organs and removed from the body.

Muscles help move blood through the body in an open circulatory system. It can take a long time for blood to move through an open circulatory system.

Closed Circulatory Systems

Some animals transport materials through another type of system. The tree frog shown in the figure at the top of this page has a closed circulatory system. *A* **closed circulatory system** *is a system that transports materials through blood using vessels.*

Vessels and Muscles Vessels help animals with closed circulatory systems move blood and other substances through the body faster than an open circulatory system. As in an open circulatory system, muscles help blood move in a closed circulatory system. However, in a closed circulatory system, the muscles surround the blood vessels. These muscles contract and push blood through the vessels. They also can change the amount of blood flow.

Red Blood Cells A closed circulatory system keeps plasma and red blood cells that carry oxygen separated from other fluids and structures in the body. Small blood vessels called capillaries surround organs and help oxygen and nutrients move from the circulatory system to cells in organs.

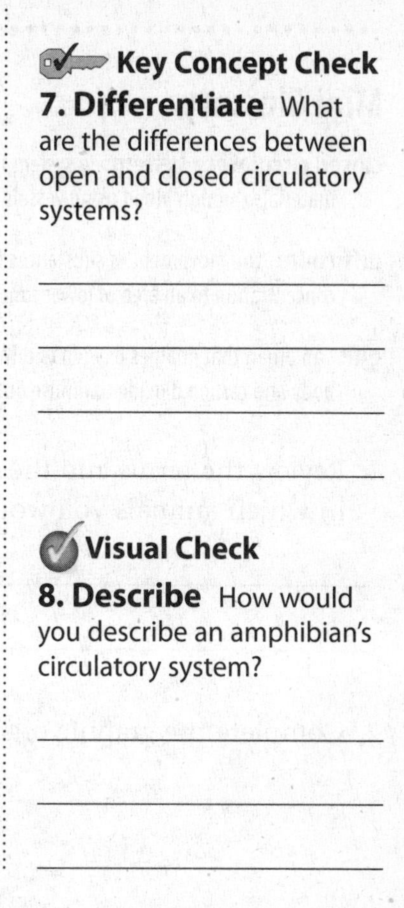

Key Concept Check
7. Differentiate What are the differences between open and closed circulatory systems?

Visual Check
8. Describe How would you describe an amphibian's circulatory system?

Chambered Hearts

Different animals have hearts with different numbers of compartments called chambers. The figure below shows hearts with different numbers of chambers. Fish have hearts with two chambers. Amphibians and most reptiles have hearts with three chambers. Birds and mammals such as cats, dogs, and humans have hearts with four chambers. Almost all animals with three- or four-chambered hearts have lungs.

Two-, Three-, and Four-Chambered Hearts

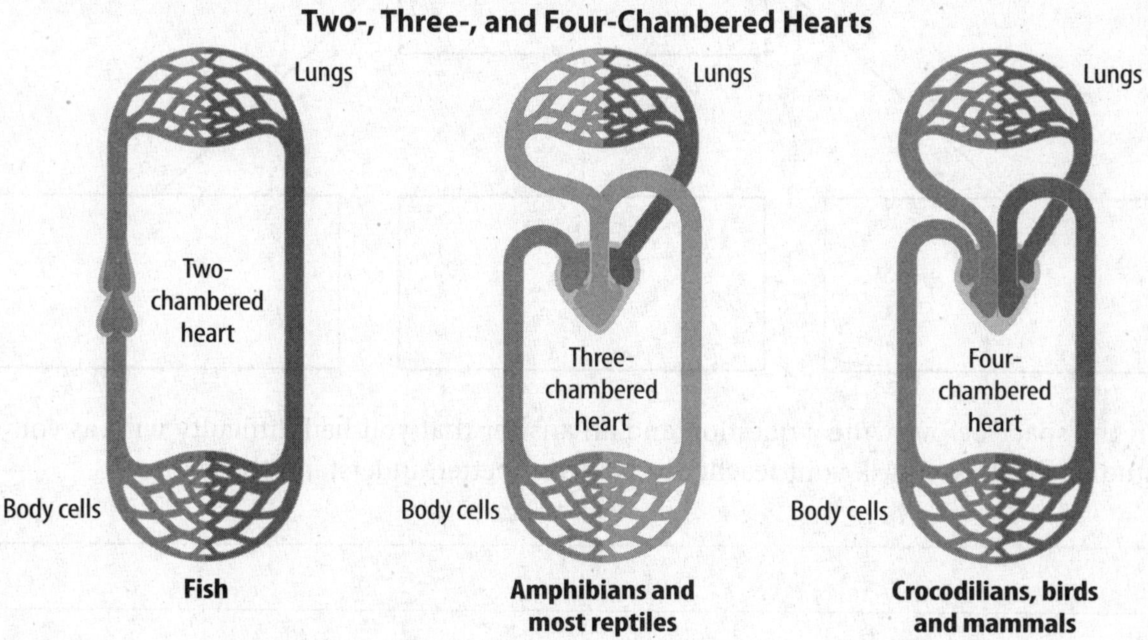

After You Read

Mini Glossary

closed circulatory system: a system that transports materials through blood using vessels

diffusion: the movement of substances from an area of higher concentration to an area of lower concentration

gill: an organ that enables oxygen to diffuse into an animal's body and carbon dioxide to diffuse out

open circulatory system: a system that transports blood and other fluids into open spaces that surround organs in the body

spiracle: a tiny hole on the surface of an organism where oxygen enters the body and carbon dioxide leaves the body

1. Review the terms and their definitions in the Mini Glossary. Write a sentence explaining in which animals you would find spiracles.

2. Complete the graphic organizer to identify the structures used in gas exchange.

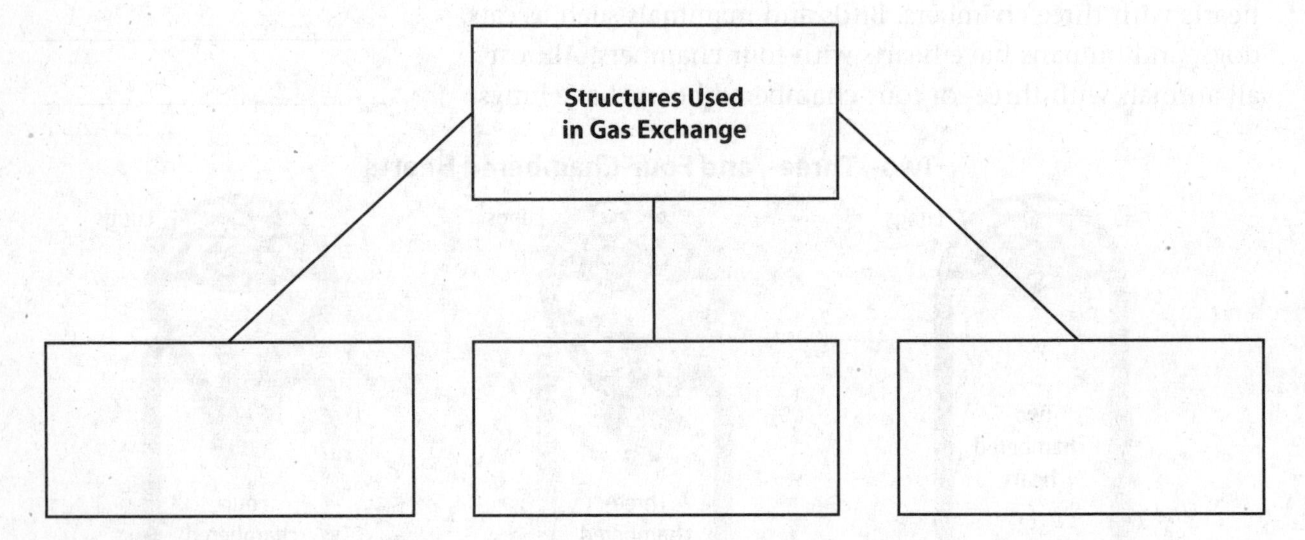

3. In the space below, write a question and an answer that you had difficulty with as you studied the lesson. Ask your teacher to help you better understand those ideas.

What do you think NOW?

Reread the statements at the beginning of the lesson. Fill in the After column with an A if you agree with the statement or a D if you disagree. Did you change your mind?

Log on to ConnectED.mcgraw-hill.com and access your textbook to find this lesson's resources.

Animal Structure and Function

Digestion and Excretion

Before You Read

What do you think? Read the two statements below and decide whether you agree or disagree with them. Place an A in the Before column if you agree with the statement or a D if you disagree. After you've read this lesson, reread the statements to see if you have changed your mind.

Before	Statement	After
	5. The shape of an animal's teeth depends on its diet.	
	6. Excretion is used only by animals that live on land.	

Key Concepts 

- How are an animal's structures for feeding and digestion related to its diet?
- How do the excretory structures of aquatic and terrestrial animals differ?

Read to Learn

The Importance of Digestion and Excretion

Animals need nutrients to survive. They obtain nutrients from food through digestion. Digestion is the process of breaking down food into molecules that cells can absorb and use. After all nutrients are taken in, the body removes its waste products by excretion. Excretion is important for survival because it removes harmful substances from the body. Different animals have different structures to obtain and process nutrients and to remove wastes.

Mark the Text

Building Vocabulary Skim this lesson and circle any words you do not know. If you still do not understand a word after reading the lesson, look it up in the dictionary. Keep a list of these words and definitions to refer to when you study other chapters.

Digestion

Animals have different structures for digestion, depending on what type of food they eat. For example, an animal that eats only seeds has a different set of structures for digestion than an animal that eats only meat.

The first step of digestion usually happens when an animal chews its food. The animal's body further breaks down the food in the digestive system in various ways, depending on the animal's diet. For example, a cow has structures that break down grass and other plant matter. As you will read in this lesson, different animals have different structures to obtain and break down food.

Reading Check

1. Explain How do animals obtain nutrients?

Structures for Feeding

For many animals, the first step in feeding is obtaining food. As with other functions of the body, animals have different structures to find and chew their food. You often can tell what type of diet an animal eats by looking at the structures that it has for feeding.

Teeth Many animals have teeth—one type of structure used for feeding. Animals have different types of teeth to process different diets.

See the figure below. Animals that eat plants often have wide teeth for chewing grass and other plants. Some have a few sharp teeth that can be used to cut through twigs. Animals that eat insects have teeth with sharp points for chewing. ✓

✓ Reading Check
2. Relate Animals that eat plants have what kind of teeth?

Animals that eat only meat have several types of teeth. Teeth in the front of the mouth bite and hold food. Pointed teeth in the rear of the mouth cut up food. Animals that eat both plants and meat have sharp teeth that they use for cutting up food and wide, flat teeth that they use for grinding up food.

✓ Visual Check
3. Describe the teeth of an animal that eats both meat and plants.

Animal Teeth

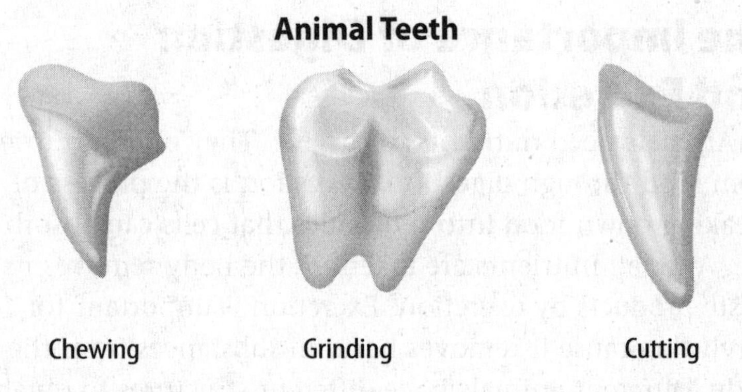

Filter Feeding Some animals take in food suspended in water. These animals have structures for filter feeding. They take in the water with the food, push the water out through a filtering structure, and then eat the organisms that remain.

Some animals, such as certain whales, take a mouthful of water and push it out through baleen (bay LEEN). Baleen is a material, similar to the bristles of a broom, that filters out tiny food organisms in the water. ✓

✓ Reading Check
4. Explain What is filter feeding?

Other Types of Filter Feeding Certain types of sharks and fish filter food through their gills. Some animals, such as clams, filter feed without moving. They filter food from the water that moves around them.

However, many filter feeders move around to find food. When flamingoes filter feed, they eat shrimp that are filtered through their beaks from the water they take in.

Mouthparts Some animals, particularly insects, have specialized mouthparts for eating. Butterflies and moths have a long, tubelike mouthpart to get nectar from flowers. Ants and certain beetles have crushing jaws for ripping plant and animal matter.

Structures for Digestion

After food is broken down into smaller parts by chewing, it is broken down into even smaller parts during digestion. Most animals have organs that form a specialized system for digestion.

For example, many animals have stomachs and intestines for the digestion of food. The structures of the stomach and the intestine vary, depending on the animal's diet.

Animals such as cows and sheep that eat a lot of plant material have stomachs with several chambers. In each of these chambers, the tough plant material is processed so the animal can digest it.

Crops Some animals store their food in a crop before digesting it. *A* ***crop*** *is a specialized structure in the digestive system where ingested material is stored.*

Many birds and insects have crops. Leeches, snails, and earthworms also have crops where they store undigested food. The crop in a leech can store blood and expands up to five times its body size.

Gizzards Animals without teeth that eat hard foods such as seeds sometimes have gizzards. *A* ***gizzard*** *is a muscular pouch similar to a stomach that is used to grind food.*

Some animals with gizzards, including certain birds, swallow rocks with their food. The rocks help with the grinding action that breaks up the food.

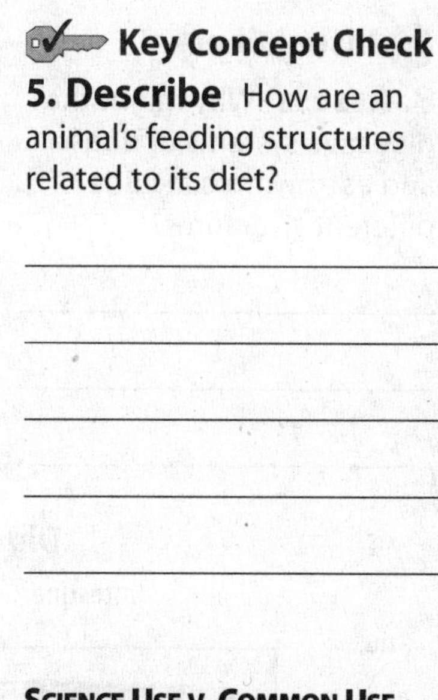

Key Concept Check

5. Describe How are an animal's feeding structures related to its diet?

SCIENCE USE V. COMMON USE

crop

Science Use a digestive system structure where material is stored

Common Use a plant or animal product that can be grown or harvested

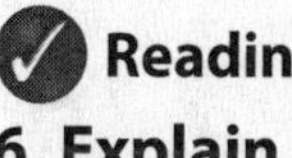

Reading Check

6. Explain Why do some animals swallow rocks with their food?

 **Key Concept Check**
7. Describe How are an animal's structures for digestion related to its diet?

____

____

____

Visual Check
8. Assess Why might the digestive systems of dogs and earthworms include different structures?

____

____

____

Absorption

Whether an animal stores food before digestion or not, it must take the nutrients into its body in order to use them. **Absorption** *is the process in which nutrients from digested food are taken into the body*. Absorption happens as food moves through the digestive system. The digestive systems of a dog and an earthworm are shown in the figure below.

Many animals have digestive systems that contain enzymes. Enzymes are chemicals that help break food into small parts so that cells can absorb the nutrients.

In addition to enzymes, many organisms have structures that enable absorption. Many animals absorb nutrients in the intestine. The structures that move the nutrients throughout an animal's body after they are absorbed also differ.

In animals with a closed circulatory system, the capillaries that surround the intestine transport nutrients throughout the body. Recall that blood surrounds the organs of animals with open circulatory systems. In this case, nutrients enter the blood directly after absorption.

Digestive Systems of an Earthworm and Dog

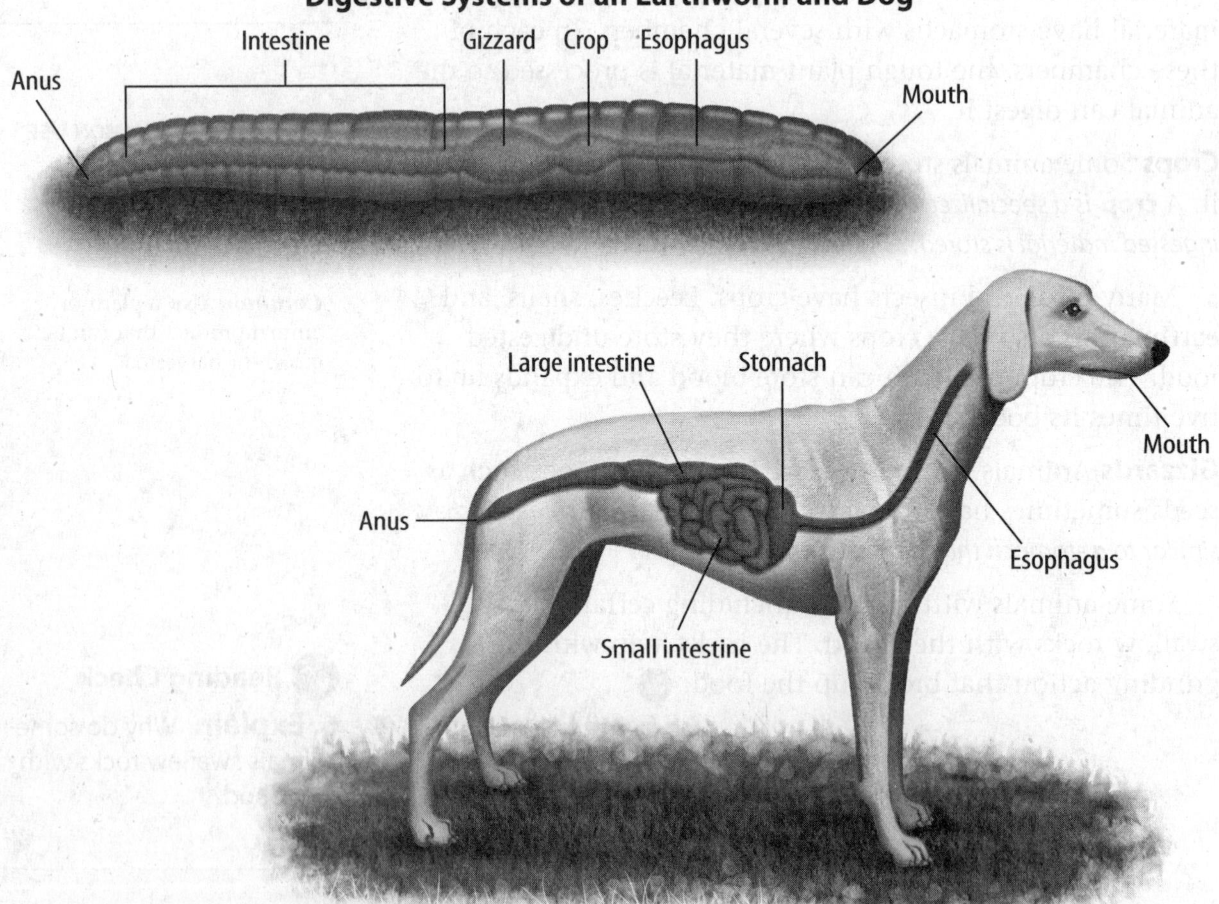

Excretion

The process of excretion removes waste materials from the body. Different animals excrete different types of wastes. The types of wastes animals excrete depend on the environments where they live.

Diffusion

Recall that gas exchange occurs due to diffusion. While diffusion can bring in oxygen, it also can release carbon dioxide.

Some organisms, such as sponges, have no filtering mechanisms in their bodies. Rather, these animals excrete waste materials as water moves in and out of their pores.

Excretion in Aquatic Animals

Fish, along with many other animals that live in aquatic environments, have kidneys that remove liquid wastes. Most of the waste the kidneys remove is water.

The kidneys of fish also excrete other wastes, such as ammonia. The gills of fish remove carbon dioxide. Solid waste leaves the body of a fish in the form of feces.

Excretion in Terrestrial Animals

Like aquatic animals, terrestrial animals have kidneys. However, terrestrial animals excrete less water than aquatic animals when removing wastes. Instead of excreting ammonia, most animals that live on land excrete urea as a waste product.

Birds also excrete wastes, but they conserve water by excreting uric acid instead of ammonia or urea. Land animals excrete carbon dioxide through the lungs. They also excrete solid waste as feces.

FOLDABLES

Make a vertical tri-fold Venn book to compare and contrast the excretion process in animals from different habitats.

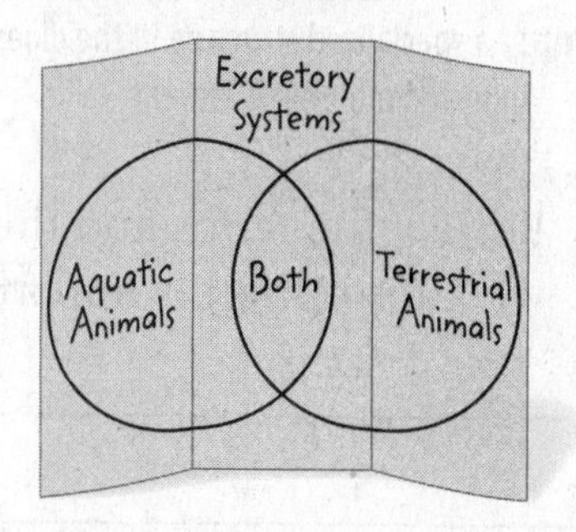

Reading Check

9. State How do organisms without filtering mechanisms excrete waste materials?

Key Concept Check

10. Contrast How do the excretory structures of aquatic and terrestrial animals differ?

After You Read

Mini Glossary

absorption: the process in which nutrients from digested food are taken into the body

crop: a specialized structure in the digestive system where ingested material is stored

gizzard: a muscular pouch similar to a stomach that is used to grind food

1. Review the terms and their definitions in the Mini Glossary. Write a sentence explaining why absorption is important in the body.

____

____

2. Complete the diagram to identify what is excreted by the given structures in fish.

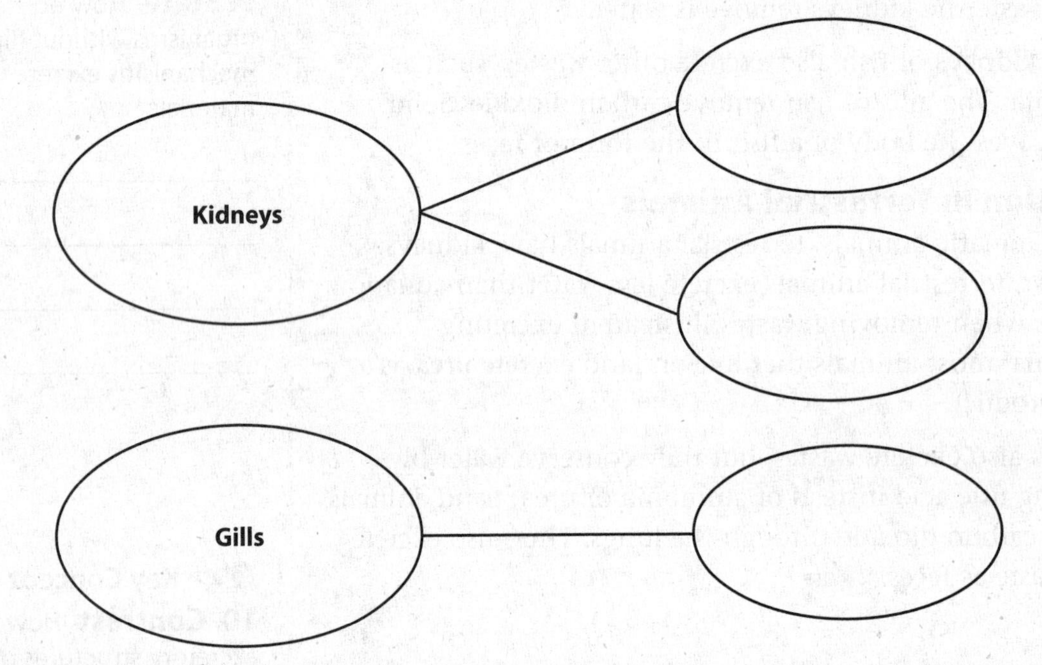

3. Write a sentence explaining what determines the types of wastes that animals excrete.

____

____

What do you think NOW?

Reread the statements at the beginning of the lesson. Fill in the After column with an A if you agree with the statement or a D if you disagree. Did you change your mind?

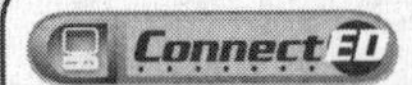

Log on to ConnectED.mcgraw-hill.com and access your textbook to find this lesson's resources.

Animal Behavior and Reproduction

Types of Behavior

Before You Read

What do you think? Read the two statements below and decide whether you agree or disagree with them. Place an A in the Before column if you agree with the statement or a D if you disagree. After you've read this lesson, reread the statements to see if you have changed your mind.

Before	Statement	After
	1. Animals react to their environments.	
	2. All animal behavior is instinctive.	

Key Concepts 
- How do behaviors help animals maintain homeostasis?
- How are animal behaviors classified?

Read to Learn

What is a behavior?

Dogs receive information about their surroundings by sniffing. Dogs have a much more developed sense of smell than humans have. A dog's nose has about 220 million scent receptors. A human's nose has only about 5 million.

The act of sniffing is a common dog behavior. *A* **behavior** *is the way an organism reacts to other organisms or to its environment.* Behaviors might be carried out by individual animals, such as a dog sniffing. Behaviors also might be carried out by groups of animals of the same species. When a flock of birds flies together, it is a group behavior. Recall that organisms' bodies work to maintain a steady internal state called homeostasis. Behaviors are a way to maintain homeostasis when the environment changes.

Study Coach

Create a Quiz Write five questions about animal behavior. Exchange quizzes with a partner. After taking the quizzes, discuss your answers. Read more about the topics you don't understand.

Key Concept Check
1. Explain How do behaviors help animals maintain homeostasis?

______

______

______

Stimuli and Responses

When an animal carries out a behavior, it is reacting to a stimulus (STIHM yuh lus; plural, stimuli), or change. A stimulus can be external or internal. The weather is an external stimulus. Hunger is an internal stimulus. Scents coming from the pavement or a tree are external stimuli for a dog. A dog's response to the stimuli is sniffing.

REVIEW VOCABULARY
stimulus
a change in an organism's environment that causes a response

Stimuli

Stimuli can come in many forms and result in different behaviors. Changes in the external environment, such as a temperature change or a rainstorm, can affect an animal's behavior. Hunger, thirst, illness, and other changes in an animal's internal environment are stimuli, too.

Responses to Change

Animals respond to changes and maintain homeostasis in different ways. For example, when the weather gets cooler, an organism might respond with a specific behavior. Birds must keep their bodies at the same temperature year-round. During warm weather, a bird's feathers are close to its body. When the weather gets cooler, a bird fluffs its feathers. This traps a layer of air around the bird's body. The air helps keep the bird warm. The cooler weather is the stimulus. The bird's feather fluffing is a response.

Animals also respond to internal stimuli, such as illnesses. If an animal is sick, its body might respond with a fever. The fever increases the animal's body temperature and might help the animal fight a disease. Vomiting is another response to an internal stimulus. If a dog eats something from the garbage, it might vomit to get the material out of its body. This behavior helps the dog maintain homeostasis by removing something that could cause an illness.

Stress

Have you ever seen an animal run away when a human got too close? The human caused the animal to become stressed. The animal reacted by running away. Some animals, such as antelopes, will almost always run away when they feel threatened. When an animal senses a danger, its body prepares to either fight or run away from the threat. This behavior is called the fight-or-flight response.

Not all animals run away from danger. Some animals react in a different way. A wild male horse might attack another male horse in the same area to protect its herd. Some animals, like rats, will run from danger but will fight if cornered.

Think it Over

2. Apply Circle the internal stimulus.

a. fewer hours of daylight
b. a headache
c. broken pavement on the bike trail

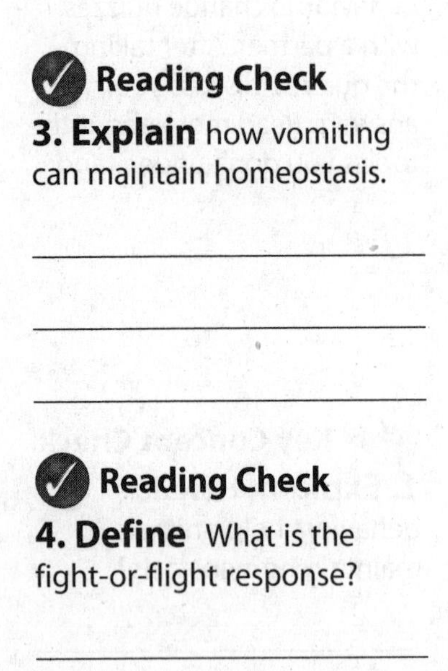

Reading Check

3. Explain how vomiting can maintain homeostasis.

____

____

____

Reading Check

4. Define What is the fight-or-flight response?

____

____

____

Innate Behaviors

As you have read, behaviors are responses to some type of stimulus. An animal's behaviors are a combination of those behaviors that are learned and those that are inherited. Inherited behaviors are not linked to past experiences. *A behavior that is inherited rather than learned is called an* **innate behavior.**

An innate behavior happens automatically the first time an animal responds to a certain stimulus. For example, when tadpoles hatch, they already know how to swim. They do not have to watch other tadpoles swim. Tadpoles can swim away from danger and find food as soon as they hatch.

Animals with short life spans have mostly innate behaviors. Insects are able to find food and mates and avoid danger early in their lives. A cricket's ability to chirp and a moth's attraction to light are innate behaviors. These types of behaviors make it possible for animals to survive without learning from another animal.

Reflexes

What happens to the pupils in your eyes when you go into a dimly lit room? Soon your pupils get larger. This happens automatically. You don't have to think about it. This is the simplest type of innate behavior. It is called a reflex. A reflex is an automatic response that does not involve a message from the brain.

Animals have reflexes, too. For example, an armadillo will jump straight upward about 1 m when startled. The sudden movement of the jumping armadillo often startles predators, and the armadillo is able to escape.

Instincts

Reflexes happen quickly and involve one behavior. Some innate behaviors involve a number of steps performed in a specific order. *A complex pattern of innate behaviors is called an* **instinct** (IHN stingt). Finding food, running away from danger, and grooming are some behaviors that are instincts in many animals.

FOLDABLES®

Make a four-door shutterfold book to compare and contrast animal behaviors.

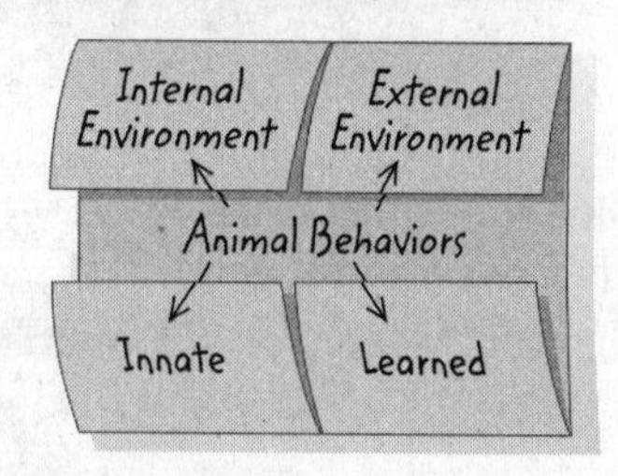

Think it Over

5. Infer Why are innate behaviors important for animals with short life spans?

____

____

____

____

Instinctive Feeding Pattern of an Egg-Eating Snake

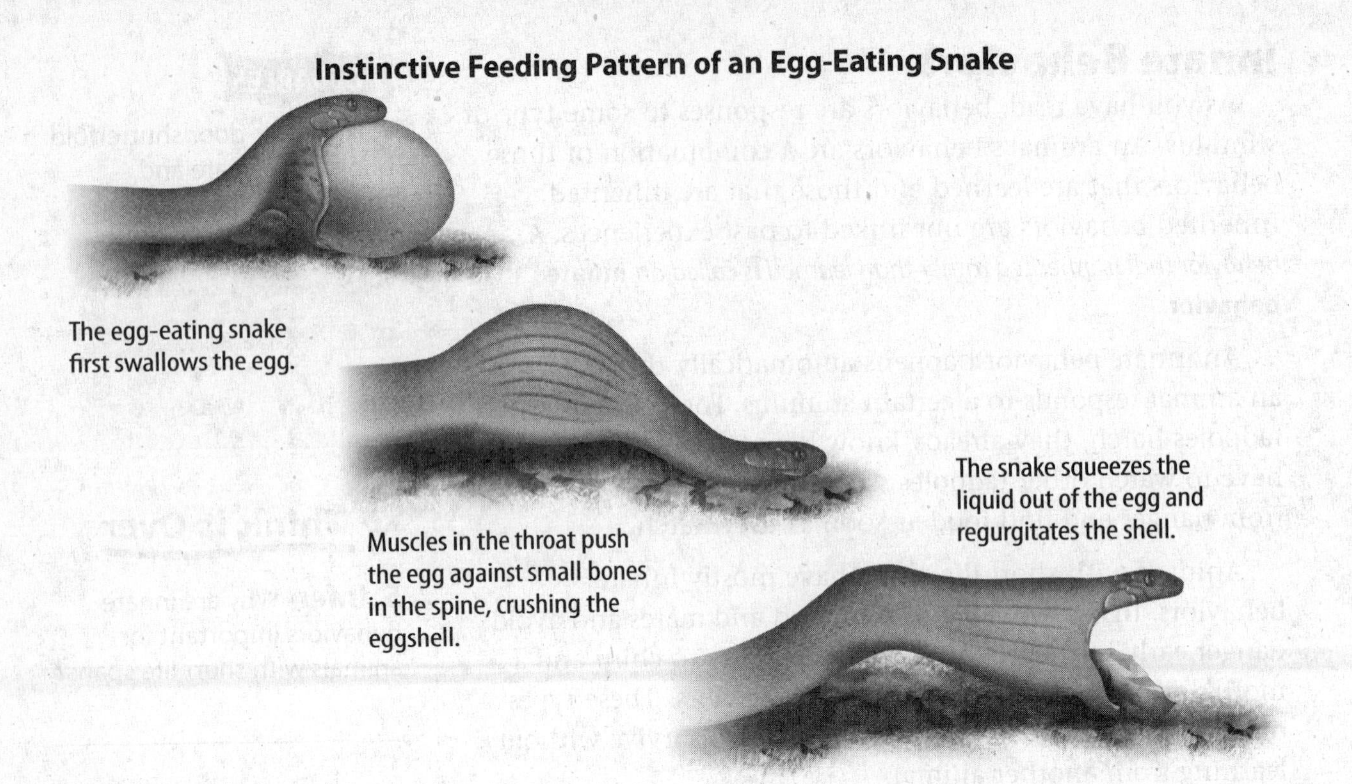

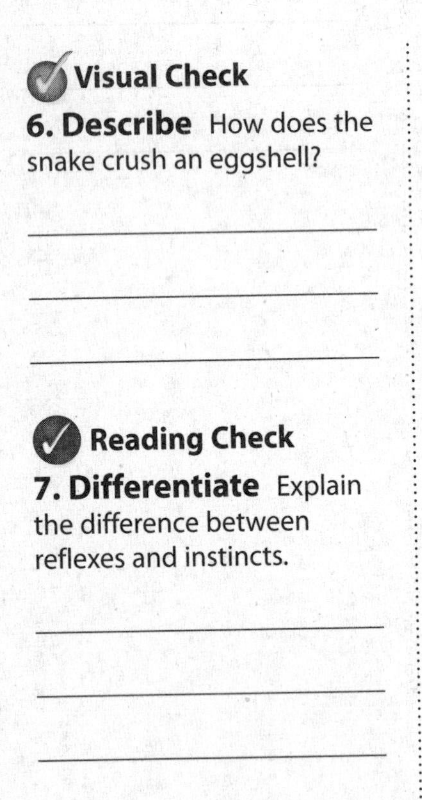

Visual Check

6. Describe How does the snake crush an eggshell?

____

____

____

Reading Check

7. Differentiate Explain the difference between reflexes and instincts.

____

____

____

____

Instincts, such as web spinning in spiders, may take hours or days to complete and are usually made up of many behaviors. The figure above shows the feeding behavior of an egg-eating snake. The snake swallows the egg, crushes the shell, and regurgitates the shell pieces. Together, all of these behaviors that make up the snake's eating pattern are an instinct.

Behavior Patterns

Many animal behaviors change in response to the change of seasons. In warm weather, there is plenty of food and water, and animals have no difficulty keeping warm. As the weather becomes cooler, food and water supplies might decrease. Animals might have difficulty surviving.

Migration Some animals move to other locations when seasons change. *This instinctive, seasonal movement of animals from one place to another is called* **migration.** Animals migrate to find food and water when the seasons change or to return to specific breeding locations. Many birds migrate long distances. The map on the next page shows where ruby-throated hummingbirds live and one migratory path they use. They fly about 805 km nonstop to reach their summer or winter territory.

Hibernation Other animals do not leave an area when temperatures get colder. Snowy owls and snowshoe hares are able to search for food in winter because of their body coverings. Their feathers and fur provide protection from the cold. Other animals respond to cold temperatures and limited food supplies by hibernating. **Hibernation** *is a response in which an animal's body temperature, activity, heart rate, and breathing rate decrease during periods of cold weather.*

Migration of Ruby-Throated Hummingbirds

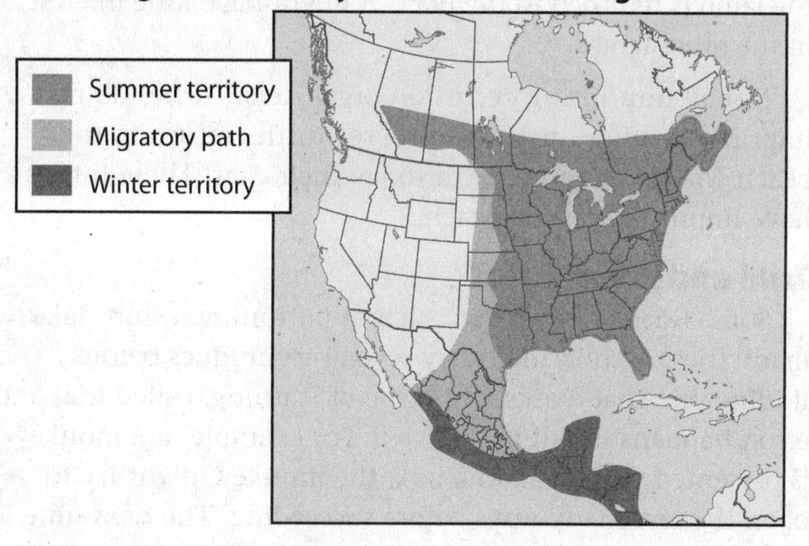

Chipmunks, some bat species, and prairie dogs are a few types of animals that hibernate. Hibernating animals live on the fat that was stored in their bodies before hibernation. In some hibernating rodents, up to 50 percent of their body weight is fat.

The internal temperatures of reptiles and other animals change with the environment. These animals do not hibernate. Rather, they enter a hibernationlike state. In areas such as deserts, many animals become less active when temperatures become high. This period of inactivity is called estivation (es tuh VAY shun).

Learned Behaviors

Service dogs help humans by opening doors or turning on light switches. How are these dogs able to do such amazing things? Dogs and all other mammals, birds, reptiles, amphibians, and fish learn. This means that these animals develop new behaviors through experience or practice. Invertebrates, such as mollusks, insects, and arthropods, also can learn. However, most of their behaviors are innate.

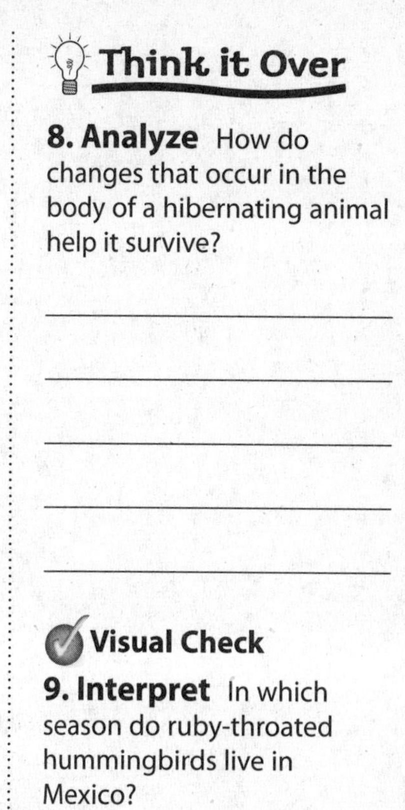

Think it Over

8. Analyze How do changes that occur in the body of a hibernating animal help it survive?

Visual Check

9. Interpret In which season do ruby-throated hummingbirds live in Mexico?

Reading Check

10. Contrast How are learned behaviors different from innate behaviors?

Imprinting

Young birds and mammals usually follow their mothers around. This helps protect them from danger and helps them find food. How do they learn to follow their mothers? **Imprinting** *occurs when an animal forms an attachment to an organism or place within a specific time period after birth or hatching.* Once a young animal has imprinted itself on an organism, it will usually not attach itself to another. For example, a lamb might become imprinted on a human who fed it from a bottle. Once the lamb is returned to the flock, it might have little interest in the other lambs.

Not all imprinting occurs on organisms. Turtles do not imprint on other turtles. Female sea turtles return to the beach where they were born to lay their eggs. These turtles have imprinted on the beach.

Trial and Error

Some learned behaviors, such as buttoning a shirt, take many tries. A child might try several techniques before finding one that works. This type of learning, called trial and error, happens in animals as well. For example, if a monkey is presented with food in a box, the monkey might try to open the box many ways before succeeding. The next time it encounters a similar box, it will remember how to open the box. It will not retry the techniques that did not work.

11. Apply A wasp stings a bird. The next time the bird sees a wasp, the bird avoids it. What stimulus does the bird associate with the wasp?

Conditioning

Another way that animals might learn new behaviors is through conditioning. *In* **conditioning,** *behavior is modified so that a response to one stimulus becomes associated with a different stimulus.* Some fish learn to come to the surface of the water when a hand is held over the water. They have learned that the hand often holds food. Some birds learn to avoid stinging wasps through conditioning.

Cognitive Behavior

Thinking, reasoning, and solving problems are cognitive behaviors. Humans use cognitive behavior to solve problems and plan for the future. Experiments with primates, dolphins, elephants, and ravens suggest that these animals also might use cognitive behaviors. Studies done with ravens showed the birds could figure out how to get meat by pulling a string attached to the food. Other animals appear to show cognitive behaviors such as using tools to get food. For example, sea otters use rocks to crack the shells of clams and mussels.

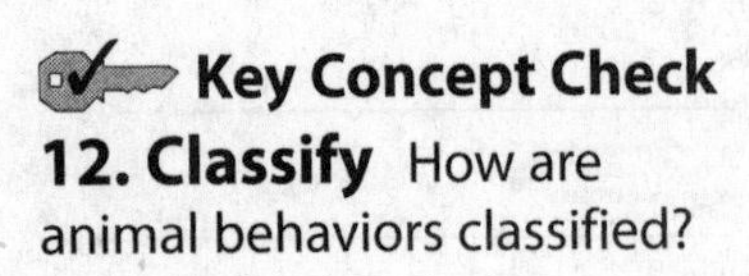

12. Classify How are animal behaviors classified?

After You Read

Mini Glossary

behavior: the way an organism reacts to other organisms or to its environment

conditioning: modifying behavior so that a response to one stimulus becomes associated with a different stimulus

hibernation: a response in which an animal's body temperature, activity, heart rate, and breathing rate decrease during periods of cold weather

imprinting: the process in which an animal forms an attachment to an organism or place within a specific time period after birth or hatching

innate behavior: behavior that is inherited rather than learned

instinct (IHN stingt): a complex pattern of innate behaviors

migration: an instinctive, seasonal movement of animals from one place to another

1. Review the terms and their definitions in the Mini Glossary. Write a sentence that describes one behavior that conditioning produced in you.

2. Complete each stimulus-response pair in the diagram below with the terms provided.

seasonal change **look for food** **fever** **danger**

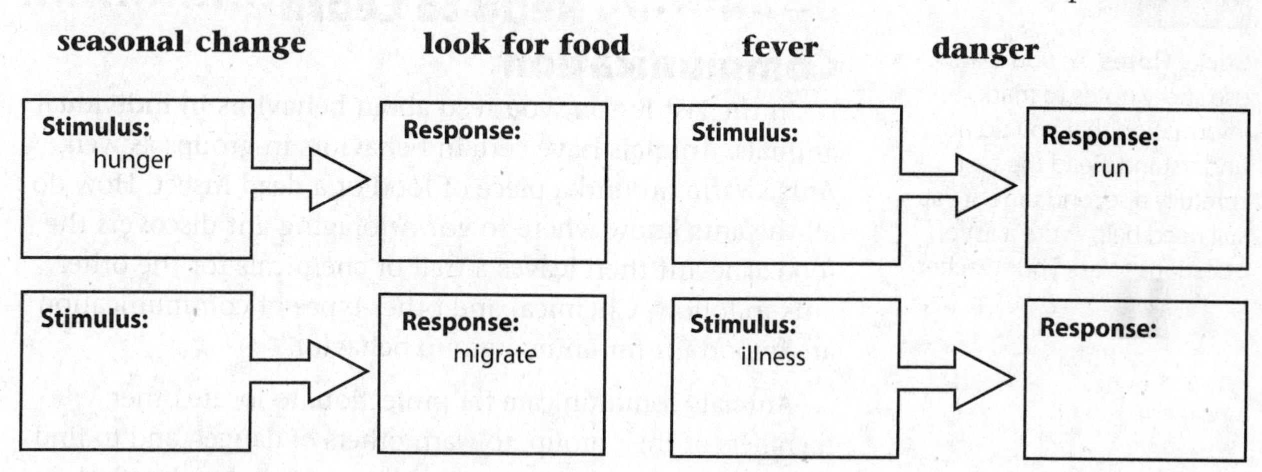

3. Write one question from your partner's quiz that you answered incorrectly. Then write the correct answer.

What do you think NOW?

Reread the statements at the beginning of the lesson. Fill in the After column with an A if you agree with the statement or a D if you disagree. Did you change your mind?

Log on to ConnectED.mcgraw-hill.com and access your textbook to find this lesson's resources.

CHAPTER 13

LESSON 2

Animal Behavior and Reproduction

Interacting with Others

Key Concepts
- How do animals communicate?
- How do animals interact in societies?

Before You Read

What do you think? Read the two statements below and decide whether you agree or disagree with them. Place an A in the Before column if you agree with the statement or a D if you disagree. After you've read this lesson, reread the statements to see if you have changed your mind.

Before	Statement	After
	3. Some animals give off light to communicate with each other.	
	4. Animals always fight to protect their territories.	

Mark the Text

Sticky Notes As you read, use sticky notes to mark information that you do not understand. Read the text carefully a second time. If you still need help, write a list of questions to ask your teacher.

Read to Learn

Communication

In the last lesson, you read about behaviors in individual animals. Animals have certain behaviors in groups as well. Ants swarm around a piece of food or a dead insect. How do all the ants know where to go? A foraging ant discovers the food. The ant then leaves a trail of chemicals for the other ants to follow. Chemical and other types of communication are important for animal group behavior.

Animals communicate for protection, to locate other members of their group, to warn others of danger, and to find mates. Animals communicate using sound, light, chemicals, and body language. An animal might communicate with other animals of the same species, or it might communicate with different species in the same area.

Key Concept Check
1. Describe How do animals communicate?

____

____

____

____

Sound

Birds, amphibians, reptiles, and mammals are some of the animals that communicate with sound. Dolphins make a wide variety of sounds, including whistles and grunts. Each sound has a different meaning to the other dolphins, such as excitement, play, or danger. Although many animals make calls, some animals produce sound in other ways. The male ruffed grouse uses its wings to beat the air. This creates a drumming noise to attract a mate. Many insects, such as cicadas and crickets, also produce sounds to attract mates.

Light

Some animals use a tool called bioluminescence (BI oh lew muh NE sunts) to communicate in the dark. **Bioluminescence** *is the ability of certain living organisms to give off light.* Chemical reactions in the animal's body produce the light. You might have seen fireflies as they blink out a code to attract females in the area.

Most animals that use bioluminescence live in the ocean. In the dimly lit zone of the ocean, up to 90 percent of fish and crustaceans use bioluminescence. Some fish use bioluminescence to lure prey into their mouths. Others have pockets of bioluminescent bacteria in their cheeks. The light from these bacteria help the fish attract mates.

Chemicals

Many animals produce chemicals, called pheromones (FER uh mohnz), to communicate. *A* **pheromone** *is a chemical that is produced by one animal and influences the behavior of another animal of the same species.* When pheromones are released to the environment, they can signal the presence of danger, food, or mates. They can even communicate the borders of a territory.

Some moths release pheromones into the air that attract mates. Male dogs mark surfaces with pheromones that identify their territory to other dogs. Recall the ants that you read about in the beginning of this lesson. Ants leave a trail of one type of pheromone that leads other ants to food. They produce different pheromones that warn other ants of danger.

Body Language

You can often tell a person's mood by looking at his or her face or body position. The person is using body language to communicate his or her mood. Animals also communicate with body language. Some parrots bob their heads when they are content. They crouch with their heads down when they are sick or stressed. Wolves communicate excitement, aggression, and other moods through facial expressions. This body language helps an animal communicate with other members of its species. The figure at the top of the next page illustrates some wolf expressions.

FOLDABLES®

Make a two-tab book to record information about animal communication and animal societies.

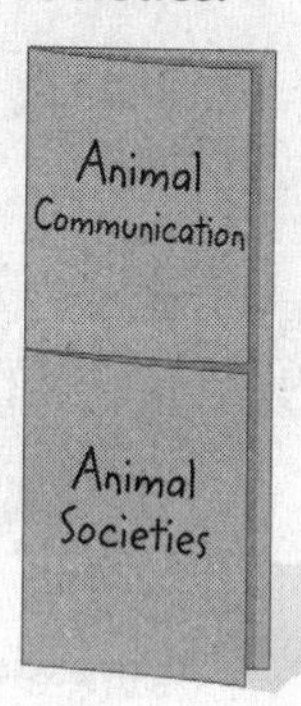

Reading Check

2. Summarize how animals use pheromones.

____

____

____

____

Wolf Body Language

Aggression:
- ears forward
- narrowed or staring eyes
- body tense and upright

Playfulness:
- ears relaxed
- wide open eyes
- relaxed body

Fear:
- ears laid back
- narrowed eyes
- body crouched low

Visual Check

3. Interpret If a wolf has narrowed eyes and ears laid back, what mood is it communicating?

____

Societies and Behaviors

Have you ever seen a flock of birds flying together? Animals live in groups for many reasons, such as for protection and obtaining food. *A* ***society*** *is a group of animals of the same species living and working together in an organized way.*

The societies of some animals have a strict structure. Members have specific roles. Spotted hyenas live in large groups of up to 90 members. The members work together to hunt and defend their kill. Other animal societies are less organized. Each member might serve different roles. Some species of animals stay close together only at certain times of the year, such as for breeding or migration.

Dominance and Submission

Dominance determines the organization of spotted hyena societies. This means that the members are organized according to their social status compared to the status of other members. The animal with the highest social status, the dominant animal, has power over the animals below it. Animals with a lower status than a dominant animal are submissive to that animal. In a spotted hyena society, females are most dominant, then cubs, and then males. Dominance also is important in groups of other animals, such as wolves, chickens, and some primates.

SCIENCE USE V. COMMON USE

submission

Science Use the condition of being humble or compliant

Common Use something presented to another for review

Dominant Behaviors Dominance also might help reduce fighting among animals living in a society. For example, hyenas rarely hurt each other while fighting with other members of their society. Less dominant members usually submit to, or stop fighting, more dominant ones.

Submissive Behaviors A submissive animal might mimic the behavior of a young animal to show that it is not a threat. For example, submissive wolves roll over or crouch. Less dominant hens move out of the way of the dominant hen.

Territorial Behaviors

Animals might set up and defend an area for feeding, mating, and raising young called a **territory.** Some insects and most vertebrates have a territory. Some animals identify their territories by making noises. Others establish their claim to a territory by making physical changes, such as by scraping bark off trees or marking their territory with pheromones, urine, or feces.

Animals defend the borders of their territory from other members of their species. If the borders are crossed, the animal first might attempt to scare or intimidate the invading animal. For example, cats puff up their fur and appear more threatening to intruders. If the animal does not leave, the defender might use aggression. **Aggression** *is a forceful behavior used to dominate or control another animal.*

When animals fight another member of the same species, they usually do not try to cause serious harm to the other animal. For example, giraffes have the ability to kick fiercely. They use this ability to defend against predators such as lions. These attacks can be deadly. However, when two male giraffes show aggression toward each other, they push at each other with their necks. This behavior is common and rarely fatal.

Courtship

Animals have specialized behaviors that help them find and attract a mate. They often compete with others of the same species for a mate. Some animals, such as female gypsy moths, release pheromones that attract males. Other animals, such as frogs and birds, use mating songs that gain the attention of mates. Some male birds bring the female a gift of food. A male tern brings a fish to a female tern. Male fiddler crabs wave their enlarged claws and skitter across the ocean floor to get the attention of a female fiddler crab. Male bowerbirds build elaborate nests during courtship. They add brightly colored objects to the nests to attract a mate.

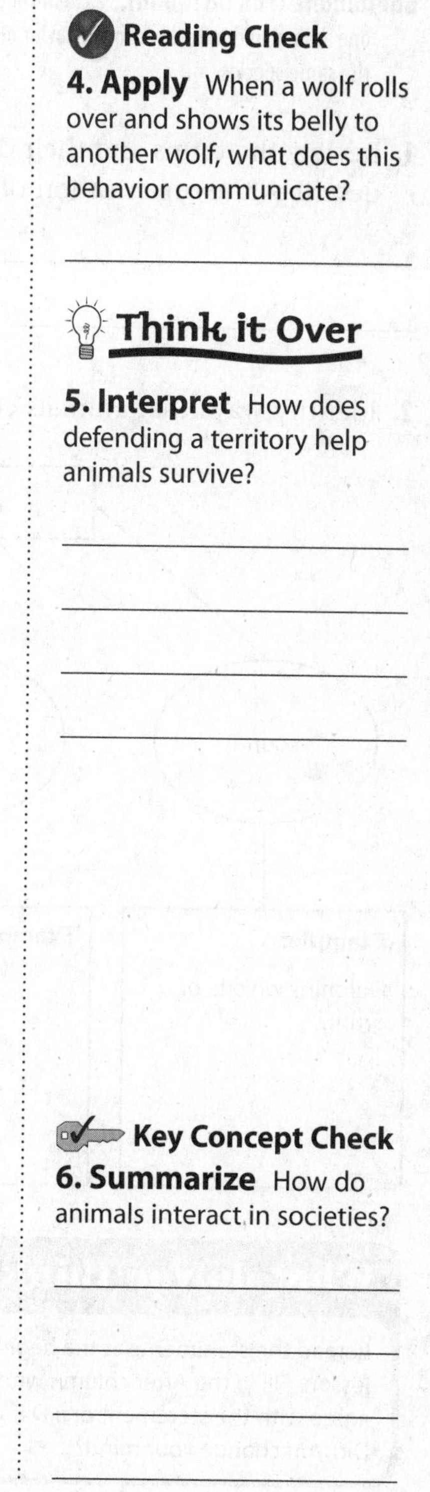

After You Read

Mini Glossary

aggression: a forceful behavior used to dominate or control another animal

bioluminescence (BI oh lew muh NE sunts): the ability of certain living organisms to give off light

pheromone (FER uh mohn): a chemical that is produced by one animal and influences the behavior of another animal of the same species

society: a group of animals of the same species living and working together in an organized way

territory: an area for feeding, mating, and raising young that animals set up and defend

1. Review the terms and their definitions in the Mini Glossary. Write a sentence that describes the organization of a spotted hyena society.

2. Identify ways that animals communicate and give an example of each method.

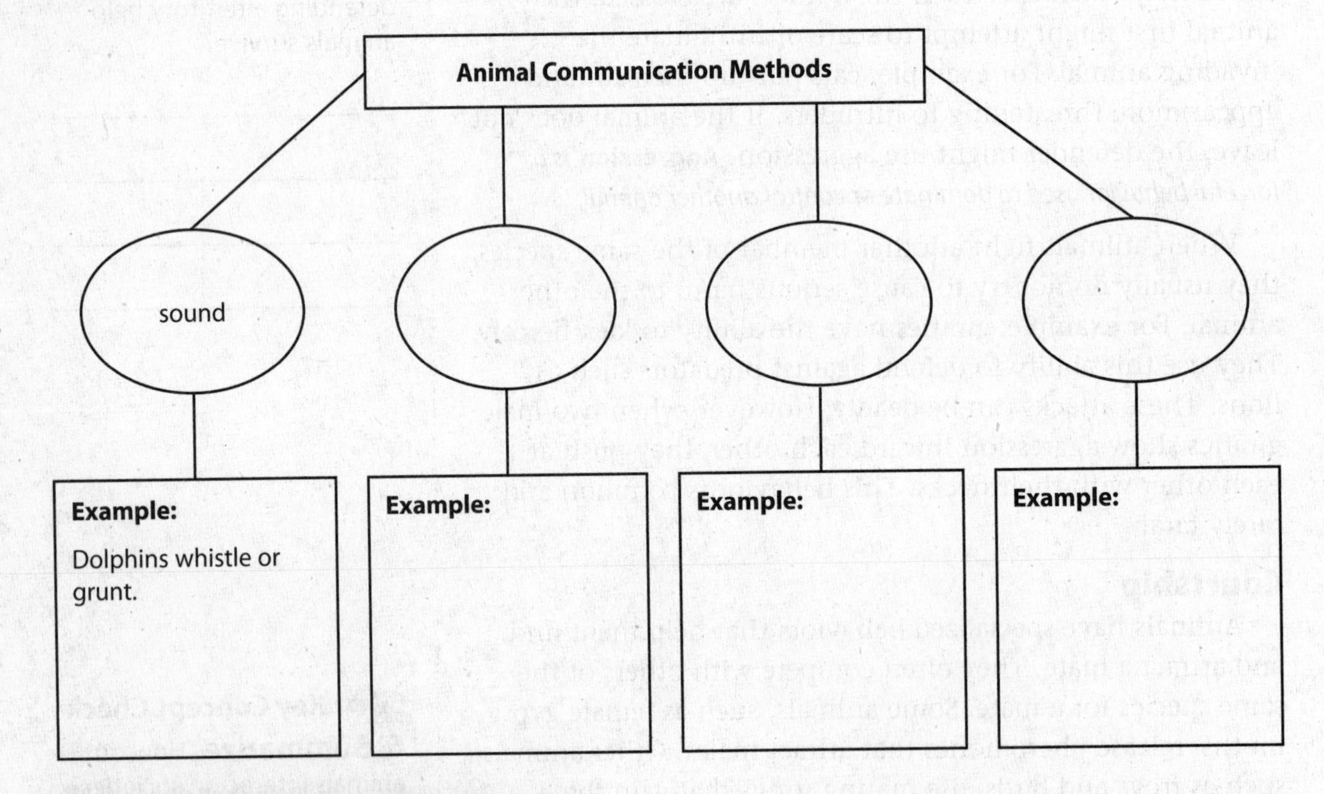

What do you think NOW?

Reread the statements at the beginning of the lesson. Fill in the After column with an A if you agree with the statement or a D if you disagree. Did you change your mind?

Log on to ConnectED.mcgraw-hill.com and access your textbook to find this lesson's resources.

Animal Behavior and Reproduction

Animal Reproduction and Development

Before You Read

What do you think? Read the two statements below and decide whether you agree or disagree with them. Place an A in the Before column if you agree with the statement or a D if you disagree. After you've read this lesson, reread the statements to see if you have changed your mind.

Before	Statement	After
	5. During sexual reproduction, a sperm cell and an egg cell join.	
	6. Some animals develop inside the mother.	

Key Concepts 

- What are the roles of male and female reproductive organs?
- How do the two types of fertilization differ?
- What are the different types of animal development?

Read to Learn

Sexual Reproduction

Have you ever found tiny, beadlike structures on the underside of a leaf? They might be eggs laid by a butterfly or some other insect. Eggs are an important part of the life cycle of many animals. *In* **sexual reproduction,** *the genetic material from two different cells—a sperm and an egg—combine, producing an offspring.* Most animals reproduce sexually. Some can reproduce asexually, without a sperm and an egg joining.

Male and female animals of the same species often look different from each other. In mammals and birds, males are often larger or more colorful than females. For example, a male lion has a ruff of fur around his neck and is larger than a lioness.

Mark the Text

Identify Main Ideas Write a phrase beside each paragraph that summarizes the main point of the paragraph. Use the phrases to review the lesson.

ACADEMIC VOCABULARY

cycle
(noun) a series of events that regularly recur and lead back to the starting point

Male Reproductive Organs

The reproductive system of an animal includes specialized reproductive organs that produce sperm or eggs. Male animals have **testes** (TES teez; singular, testis), *the male reproductive organs that produce sperm.* Sperm are reproductive cells with tails. The tails make it possible for sperm to swim through fluid to reach an egg cell. Most male animals have two testes located inside the body.

1. Identify Circle the structure that produces egg cells.

Key Concept Check

2. Specify What are the functions of male and female reproductive organs?

FOLDABLES

Make a vertical three-tab book with a Venn diagram on the front to compare internal and external fertilization.

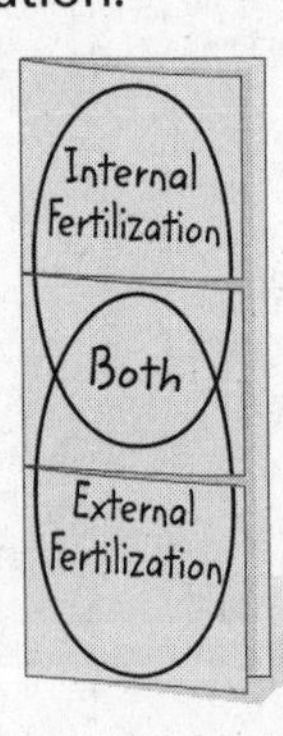

Reproductive Organs of Chickens

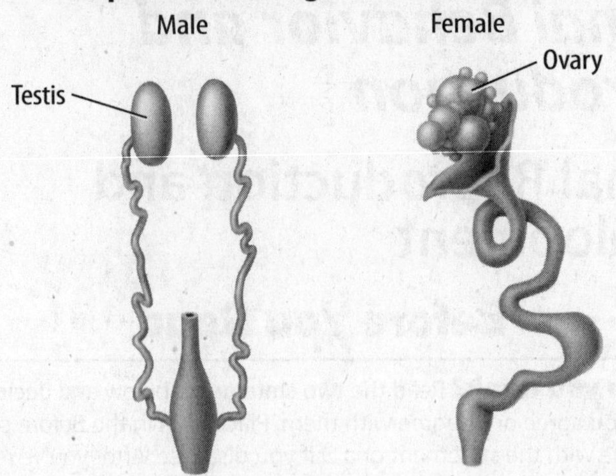

Female Reproductive Organs

Female animals have **ovaries** (OH va reez), *the female reproductive organs that produce egg cells.* Most female animals have two ovaries. Female birds have only one ovary, as shown above. Egg cells are larger than sperm and cannot move on their own. Many female mammals are born with all the eggs they will ever have.

Fertilization

Sexual reproduction requires **fertilization**—*the joining of an egg cell with a sperm cell.* Half of the genetic material is in a sperm cell, and half is in an egg cell. *When a sperm cell fertilizes an egg cell, the new cell that forms is called a* **zygote** (ZI goht). The zygote develops into a new organism. It has the genetic material from the sperm cell and the egg cell.

Not all animals fertilize their eggs in the same way. The eggs of some animals are fertilized inside the mother's body. Some are fertilized outside the body.

Internal Fertilization When fertilization occurs inside the body of an animal, it is called internal fertilization. For many animals, the male has a specialized structure that can deposit sperm in or near the female's reproductive system. The sperm swim to the egg cells. Earthworms, spiders, insects, reptiles, birds, and mammals have internal fertilization.

An embryo develops from a fertilized egg, or zygote. With internal fertilization, the embryo is protected and nourished inside the female's body. This increases the chance that an embryo will survive, develop into an adult, and reproduce.

External Fertilization A female frog deposits unfertilized eggs under water. A male frog releases its sperm above the eggs as the female lays them. Fertilization that occurs outside the body of an animal is called external fertilization. When a sperm reaches an egg, fertilization takes place. Animals that reproduce using external fertilization include jellyfish, clams, sea urchins, sea stars, many species of fish, and amphibians.

Most animals that reproduce using external fertilization do not care for the fertilized eggs or for the newly hatched young. As a result, eggs and young are exposed to predators and other dangers in the environment. This reduces their chances of surviving. For successful reproduction, animals with external fertilization produce a large number of eggs. This ensures that at least a few offspring will survive to become adults.

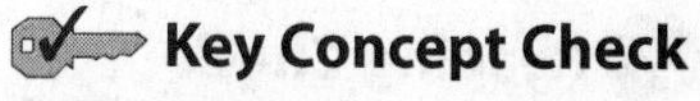

Key Concept Check

3. Contrast How do the two types of fertilization differ?

Development

The zygote produced by fertilization is only the beginning of an animal's development. The zygote grows by mitosis and cell divisions and becomes an embryo, which is the next stage in an animal's development.

A growing embryo needs nourishment and protection from predators and other dangers in the environment. Different animals have different ways of supplying the needs of an embryo. In some animals, the embryo develops outside the mother's body. In others, the embryo develops inside the mother's body.

External Development

Animals that develop outside the mother usually are protected inside an egg. In most instances, one embryo develops inside each egg. Most eggs contain a yolk that provides food for the developing embryo. A covering surrounds the egg. The covering protects the embryo, helps keep it moist, and discourages predators.

Eggs laid by lizards, snakes, and other reptiles have a tough, leathery covering. A tough, jellylike substance usually surrounds eggs laid under water, such as those laid by frogs. Bird eggs have a hard covering called a shell.

A grass snake's embryos develop outside the mother. The grass snake protects its eggs while the embryos develop.

Math Skills

A ratio can be used to compare data about sperm and eggs. For example, the head of a human sperm cell averages 5 μm. (A μm, or micron, is one-millionth of a meter.) If the tail of the sperm cell measures 50 μm, what is the ratio of tail to head?

Set up the two numbers as a ratio by writing them in any of the following forms:

$$50 \text{ to } 5;\ 50{:}5;\ \frac{50}{5}$$

Reduce the numbers to their lowest form.

Divide each side by 5. The ratio is 10 to 1 or 10:1 or $\frac{10}{1}$.

4. Use Ratios A fruit fly sperm cell is about 1.8 mm in length. What is the ratio of a human sperm cell to a fly sperm cell?

Reading Check
5. Specify Where does a snake embryo get nourishment if it develops inside its mother?

Reading Check
6. Define What is gestation?

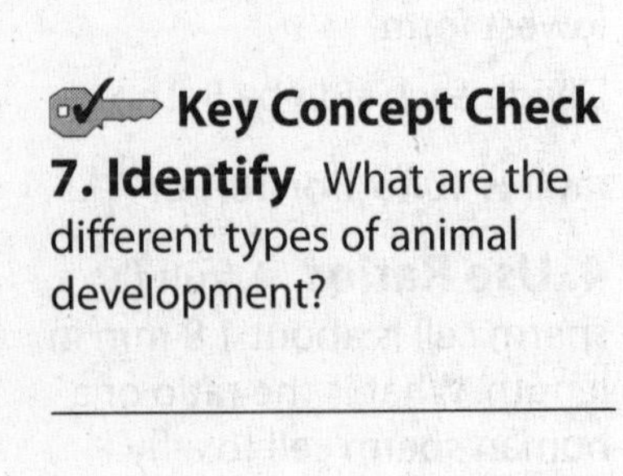

Key Concept Check
7. Identify What are the different types of animal development?

Internal Development

The embryos of some animals, including most mammals, develop inside the mother. These embryos get nourishment from the mother. An organ or a tissue transfers nourishment from the mother to the embryo.

Other embryos, such as those of some snakes, insects, and fish, develop in an egg with a yolk while it is inside the mother. For these animals, the yolk, not the mother, provides nourishment for the developing young. The young hatch from the eggs while they are inside the mother and then leave the mother's body.

Gestation

The length of time between fertilization and birth of an animal is called gestation (jeh STAY shun). Gestation varies from species to species and usually relates to the size of the animal—the smaller the animal, the shorter its gestation period. For example, gestation for a mouse is about 21 days, for a human it is about 266 days, for an elephant it is about 600 days and for a kangaroo is 35 days.

A kangaroo measures only about 2.5 cm long at birth. A newborn kangaroo crawls into a pouch on the mother's body. Most of its development occurs inside this pouch. The young kangaroo feeds and grows inside the pouch until it is large enough to live on its own.

Metamorphosis

Some animals, including amphibians and many animals without backbones, go through more than one phase of development. **Metamorphosis** (me tuh MOR fuh sihs) *is a developmental process in which the form of the body changes as an animal grows from an egg to an adult.*

The metamorphosis of a ladybug and the metamorphosis of a frog are shown in the figure on the next page. During its development, the ladybug goes from egg to larva to pupa to adult.

The tadpole is the larval stage of a frog. A tadpole hatches from an egg. It grows legs and loses its tail as it develops into an adult frog.

Larva and adult forms often have different lifestyles. The larva of the frog lives only in water. The adult frog can live on land or in water.

Metamorphosis

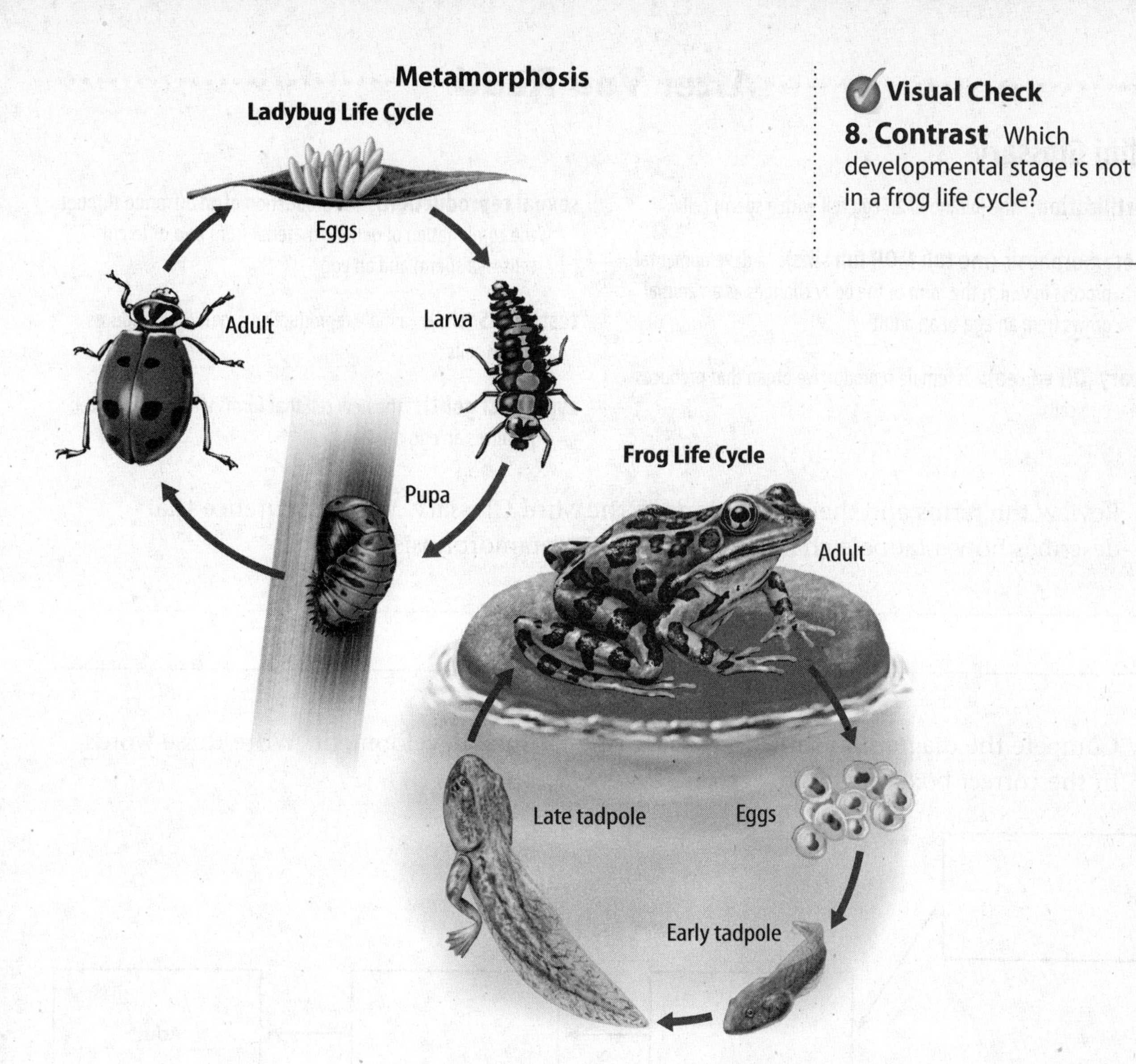

Visual Check

8. Contrast Which developmental stage is not in a frog life cycle?

After You Read

Mini Glossary

fertilization: the joining of an egg cell with a sperm cell

metamorphosis (me tuh MOR fuh sihs): a developmental process in which the form of the body changes as an animal grows from an egg to an adult

ovary (OH va ree): a female reproductive organ that produces egg cells

sexual reproduction: the production of an offspring through the combination of genetic material from two different cells—a sperm and an egg

testis (TES tihs): a male reproductive organ that produces sperm cells

zygote (ZI goht): the new cell that forms when a sperm cell fertilizes an egg cell

1. Review the terms and their definitions in the Mini Glossary. Write a sentence that describes how a tadpole changes form during metamorphosis.

2. Compete the diagram to show the sequence of animal development. Write these words in the correct boxes: embryo, egg cell, zygote, sperm cell.

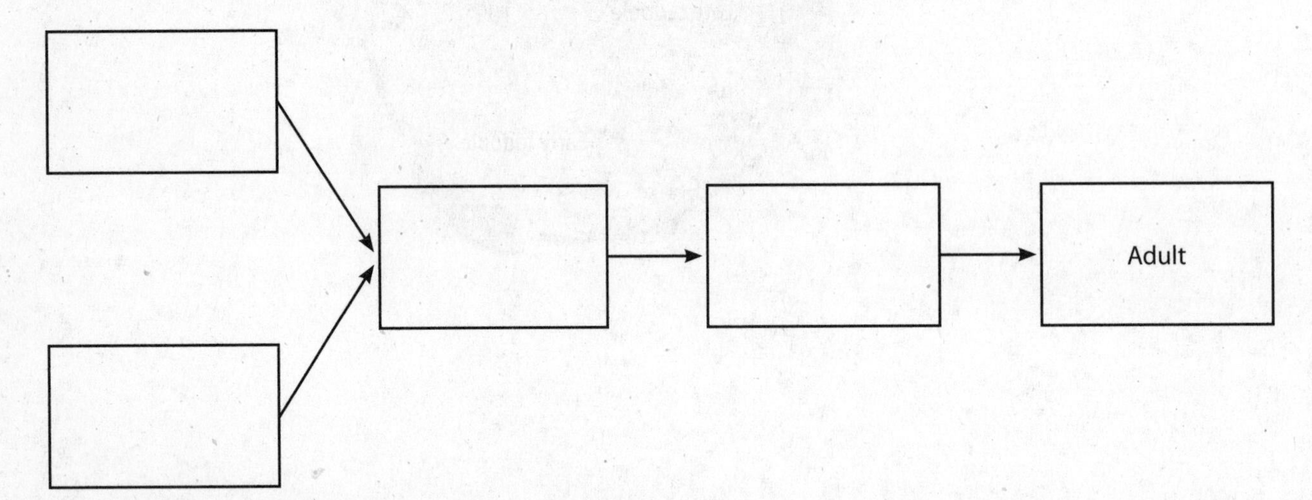

3. In external development, how does the egg benefit the offspring developing inside it?

What do you think NOW?

Reread the statements at the beginning of the lesson. Fill in the After column with an A if you agree with the statement or a D if you disagree. Did you change your mind?

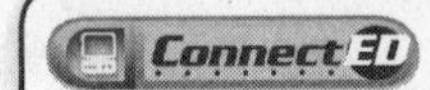

Log on to ConnectED.mcgraw-hill.com and access your textbook to find this lesson's resources.

Structure and Movement

The Skeletal System

Before You Read

What do you think? Read the two statements below and decide whether you agree or disagree with them. Place an A in the Before column if you agree with the statement or a D if you disagree. After you've read this lesson, reread the statements to see if you have changed your mind.

Before	Statement	After
	1. Bones protect internal organs.	
	2. Bones do not change during a person's lifetime.	

Key Concepts
- What does the skeletal system do?
- How do the parts of the skeletal system work together?
- How does the skeletal system interact with other body systems?

Read to Learn

Functions of the Skeletal System

Squeeze your hands and arms. The hard parts that you feel are parts of your skeleton. When you think of your skeleton, you probably think of bones. Your skeleton is part of your skeletal system and is made up of more than 200 bones. *The* **skeletal system** *contains bones as well as other structures that connect and protect the bones and that support other functions in the body.*

Your skeletal system performs several important functions. It supports your body and helps you move. It protects the organs in your body, such as your lungs and heart. The skeletal system also makes and stores important materials needed by your body.

Mark the Text

Identify Main Ideas As you read, highlight the functions of the skeletal system. Use another color to highlight the different parts of a bone.

Support

Imagine trying to stack cubes of gelatin. The gelatin cubes would be hard to stack because they do not have any support structures inside of them. Without bones, your body would be like the gelatin cubes. Bones support your body. They help you sit up and stand. They make it possible for you to lift your legs to walk up stairs.

Reading Check

1. Explain How do bones act as a support system?

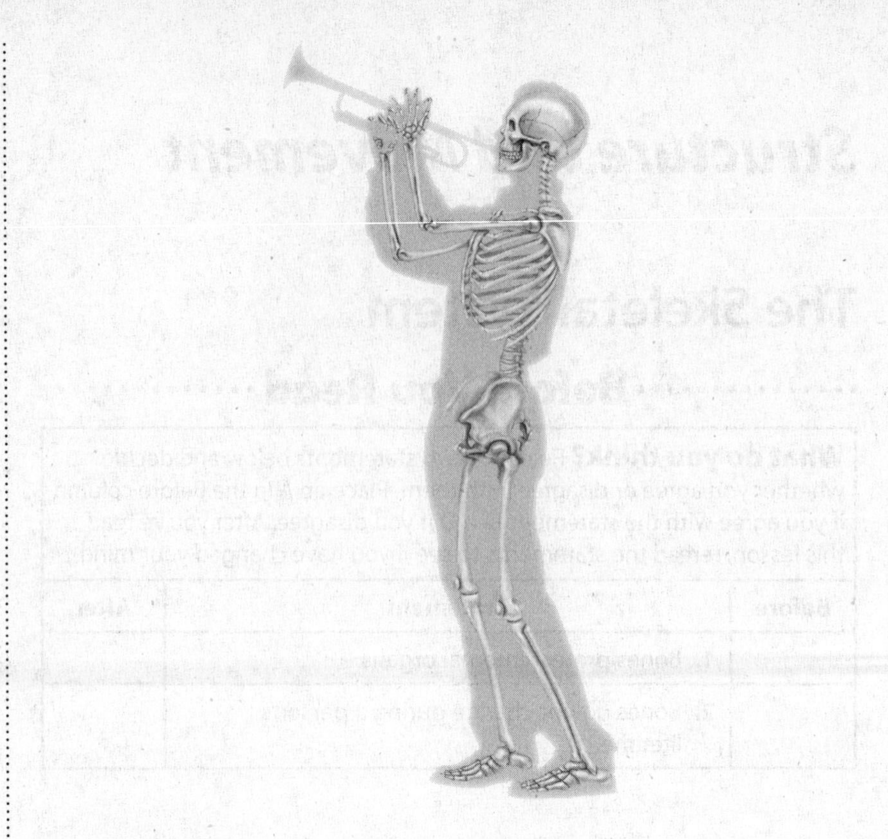

Visual Check

2. Explain How does the skeletal system help a person play a musical instrument?

Movement

Different parts of your body can move in different ways because of your skeletal system, as shown in the figure above. Your knee bends when you kick a soccer ball. However, your shoulders move in a different way when you raise your arms to catch the same ball. Bones can move because they are attached to muscles. Your skeletal system and your muscular system work together and move your body.

Protection

Feel the top of your head. The hard structure you feel is your skull. The skull protects the soft tissue of your brain from damage. Other bones help protect your spinal cord, heart, lungs, and other organs in your body.

SCIENCE USE V. COMMON USE
tissue
Science Use a group of the same type of cells that perform a specific function within an organism

Common Use a soft, thin paper

Production and Storage

Another function of your skeletal system is to make and store materials needed by your body. Red blood cells are produced inside some of your bones. Bones also store fat and calcium. Calcium is a mineral needed for strong bones. It is also used in many other cellular processes. When the body needs calcium, it is released from the bones into the blood.

Key Concept Check

3. Name the major functions of the skeletal system.

Structure of Bones

A bone is an organ made of living tissue. There are two main types of bone tissue: compact bone and spongy bone.

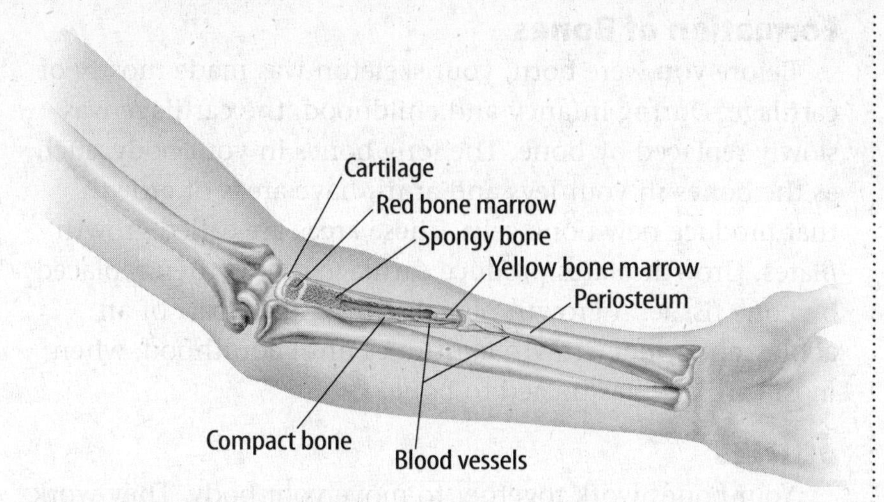

Compact Bone Tissue

Find the compact bone tissue in the figure above. The hard, outer area of long bones is made up mostly of compact bone tissue. This tissue is a thick, dense web of fibers.

Spongy Bone Tissue

Spongy bone tissue is located near the ends of long bones, such as the arm bone in the figure above. A short bone, such as the one in your wrist, is mostly spongy bone tissue. The small holes in spongy bone tissue make it look like a sponge. Because of these holes, spongy bone tissue is not as dense as compact bone tissue.

Bone Marrow

The inside of most bones is made up of a soft tissue called bone marrow (MER oh). There are two types of bone marrow. Red bone marrow is the tissue where red blood cells are made. It is found in the spongy ends of long bones and in some flat bones, such as the ribs. Yellow bone marrow stores fat. Yellow bone marrow is found in the longest part of long bones. 

Cartilage

The strong, flexible tissue that covers the ends of bones is called **cartilage** (KAR tuh lihj). Cartilage is shown in the figure above. Cartilage keeps the surfaces of bones from rubbing against each other. It protects bones and reduces friction in joints.

Periosteum

The parts of a bone that are not covered by cartilage are covered by periosteum (per ee AHS tee um). *The* **periosteum** *is a membrane that surrounds bone.* This thin tissue has blood vessels, nerves, and cells that make new bone tissue. It helps bones function and grow properly. It also helps a bone heal after an injury.

 **Visual Check**

4. Identify Circle the red bone marrow in the bone shown in the figure.

Reading Check

5. Contrast What is the difference between red bone marrow and yellow bone marrow?

____

____

FOLDABLES

Make a half-book, then draw and describe what a bone is made of.

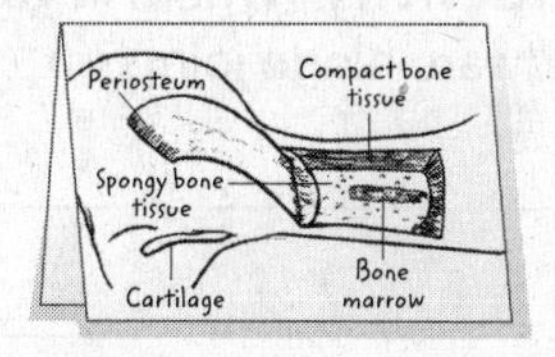

Formation of Bones

Before you were born, your skeleton was made mostly of cartilage. During infancy and childhood, the cartilage was slowly replaced by bone. The long bones in your body, such as the bones in your legs and arms, have areas of growth that produce new bone cells. These areas are called growth plates. Growth plates produce cartilage that is then replaced by bone tissue. A growth plate is the weakest part of an adolescent bone. Growth continues until adulthood, when most cartilage has turned to bone.

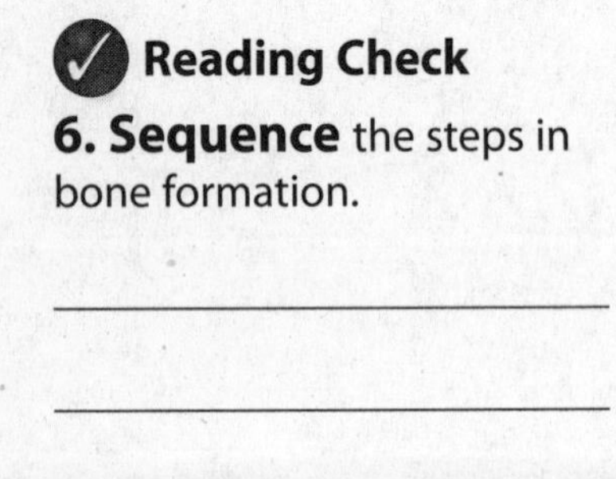

Reading Check
6. Sequence the steps in bone formation.

Joints

Your bones work together to move your body. They work together at places called joints. *A* **joint** *is where two or more bones meet.* Joints provide flexibility and enable the skeleton to move. **Ligaments** (LIH guh munts) *are tissues that connect bones to other bones.* When the bones in joints move, ligaments stretch and keep the bones in place. Ligaments connect bones at joints, but they do not protect bones. Cartilage protects the ends of bones. Your skeletal system has two types of joints—immovable joints and movable joints.

Immovable Joints

Some parts of your skeleton are made of bones that are connected but do not move. These are called immovable joints. Your skull has several immovable joints.

Movable Joints

You are able to move your hands and bend your body because of movable joints. You can move in many ways because of your body's movable joints. The three main types of movable joints and the ligaments that hold them together are shown in the table below.

Key Concept Check
7. Explain How do ligaments and cartilage help your skeletal system function?

Types of Movable Joints

Joint	Description	Examples
Ball and socket	Bones can move and rotate in nearly all directions.	hips and shoulders
Hinge	Bones can move back and forth in a single direction.	fingers, elbows, knees
Pivot	Bones can rotate.	neck, lower arm below the elbow

Visual Check
8. Identify Where in your body are hinge joints?

Bone Injuries and Diseases

Because bones are made of living tissue, they can be injured. Bones can break. They can also develop disease.

Broken Bones

A break in a bone is called a fracture (FRAK chur). Broken bones can repair themselves, but it takes a long time. A broken bone must be held together while it heals. Sometimes a person wears a cast to hold a bone in place while it heals. Sometimes metal plates and screws hold a bone together while it heals.

9. Analyze What are some ways to help a fractured bone heal?

Arthritis

Arthritis (ar THRI tus) *is a disease in which joints become irritated or inflamed, such as when cartilage in joints is damaged or wears away.* When the joints become irritated, it can be painful to move. Arthritis is most common in adults. It can also affect children.

Osteoporosis

Another common bone disease is osteoporosis (ahs tee oh puh ROH sus). **Osteoporosis** *is a disease that causes bones to weaken and become brittle, or easily broken.* Osteoporosis can change a person's skeleton and cause fractures. Osteoporosis is most common in women over the age of 50.

Healthy Bones

One of the best ways to keep bones healthy is to exercise. Certain types of exercise, such as running, walking, and lifting weights, place weight on your bones. These types of exercises help make bones strong and build new bone tissue.

A balanced diet also helps keep bones healthy. Bones need calcium and vitamin D most of all. Calcium makes bones strong. It is also important for cell processes. If you do not have enough calcium in your diet, your body will use the calcium stored in your bones. This can make your bones weak. Vitamin D helps the body use calcium.

The Skeletal System and Homeostasis

Homeostasis is an organism's ability to keep its internal conditions stable. Homeostasis requires that all body systems work together properly. Your skeletal system helps your body maintain homeostasis by supplying calcium to your nerves, heart, and muscles so they can function properly. Bones also help you respond to unpleasant stimuli, such as a buzzing mosquito. Working together with muscles, bones enable you to move away from or even swat a mosquito.

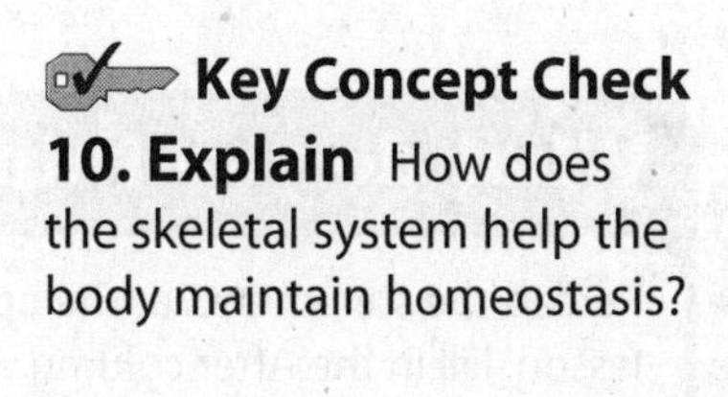

10. Explain How does the skeletal system help the body maintain homeostasis?

After You Read

Mini Glossary

arthritis (ar THRI tus): a disease in which joints become irritated or inflamed, such as when cartilage in joints is damaged or wears away

cartilage (KAR tuh lihj): a strong, flexible tissue that covers the ends of bones

joint: where two or more bones meet

ligament (LIH guh munt): tissue that connects bones to other bones

osteoporosis (ahs tee oh puh ROH sus): a disease that causes bones to weaken and become brittle

periosteum (per ee AHS tee um): a membrane that surrounds bone

skeletal (SKE luh tul) system: contains bones as well as other structures that connect and protect the bones and that support other functions in the body

1. Review the terms and their definitions in the Mini Glossary. Write a sentence or two that describes the relationship between the skeletal system and joints.

____

____

____

2. On the blanks in the diagram below, write the parts of the bone.

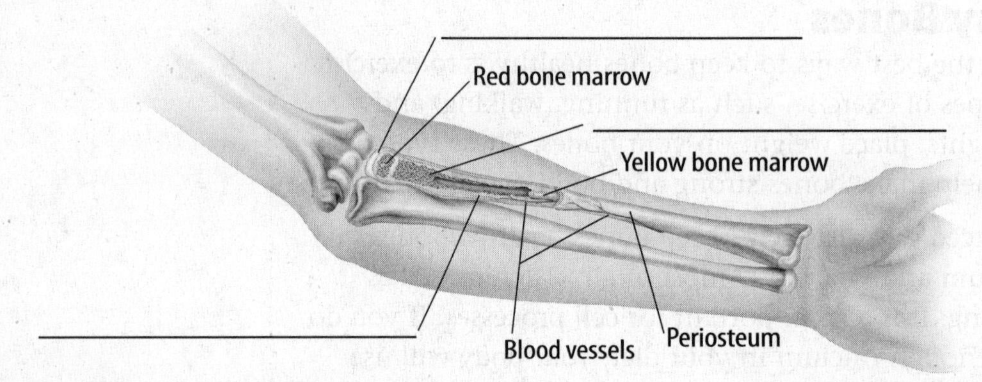

3. How did highlighting information about the skeletal system help you understand what you read?

____

____

What do you think NOW?

Reread the statements at the beginning of the lesson. Fill in the After column with an A if you agree with the statement or a D if you disagree. Did you change your mind?

Log on to ConnectED.mcgraw-hill.com and access your textbook to find this lesson's resources.

END OF LESSON

Structure and Movement

The Muscular System

Before You Read

What do you think? Read the two statements below and decide whether you agree or disagree with them. Place an A in the Before column if you agree with the statement or a D if you disagree. After you've read this lesson, reread the statements to see if you have changed your mind.

Before	Statement	After
	3. The same type of muscle that moves bones also pumps blood through the heart.	
	4. Muscles cannot push bones.	

Key Concepts 

- What does the muscular system do?
- How do types of muscle differ?
- How does the muscular system interact with other body systems?

Read to Learn

Functions of the Muscular System

A **muscle** *is made of strong tissue that can contract in an orderly way.* Your muscular system is made of different types of muscles and has several functions. Muscles help you move, help protect your body, help keep your body stable, and help maintain your body temperature.

All muscle tissues are made of cells that contract. When the cells of a muscle contract, the muscle tissues become shorter. The muscle tissues return to their original length when the cells relax.

Study Coach

Use an Outline As you read, make an outline to summarize the information in the lesson. Use the main headings in the lesson as the main headings in the outline. Complete the outline with the information under each heading.

Movement

Many of your muscles help you move. Most of these muscles attach to bones. These muscles make your skeleton move. When muscles contract, they move bones. This movement can be fast, such as when you run. The movement can also be slow, such as when you stretch.

You have many muscles in your body that are not attached to bones. Contractions in these muscles cause blood and food to move through your body. Contractions also make your heart beat and cause the hair on your arms to stand on end when you get goose bumps.

Key Concept Check

1. Identify What is one major function of the muscular system?

______

______

Stability

What happens when you start to lose your balance? Your muscles pull in different directions to help you get your balance back. Muscles that are attached to your bones support your body and help you keep your balance.

Tendons attach muscles to bones. Look at the back of your hand and move your fingers. Do you see the cordlike structures moving under your skin? These are tendons. Tendons work with muscles and keep your joints in place when your body moves. Tendons also help hold your body in a correct posture, or shape. ✓

Reading Check

2. Explain What is the function of tendons?

Protection

Muscles protect your body. They cover most of your skeleton. Muscles also cover most of the organs inside your body. Muscles are like a layer of padding. They surround your abdomen, chest, and back, and protect your internal organs.

Temperature Regulation

Your muscular system helps your body keep your internal temperature within a certain range. Have you ever felt cold and then started shivering? Shivering is muscles rapidly contracting. This changes chemical energy to thermal energy. The released thermal energy helps maintain your body's temperature. This is important because a human's body temperature must stay around 37°C in order for the body to function properly. Muscles also change chemical energy to thermal energy during exercise. This is why you feel warm after physical activity.

FOLDABLES®

Make the following tri-fold book to organize your notes on the three types of muscles.

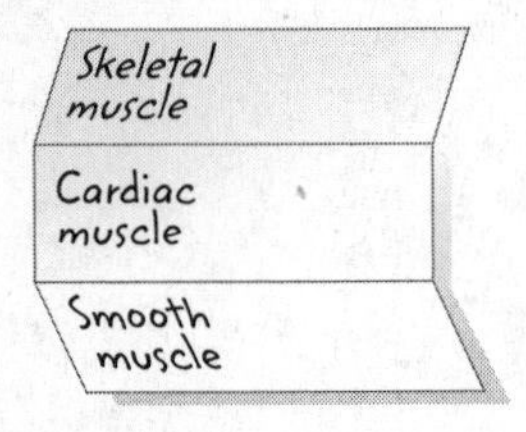

Types of Muscles

Your body has three different types of muscles—skeletal, smooth, and cardiac. They are shown in the figures below. Each type of muscle is specialized for a different function.

Skeletal	Smooth	Cardiac

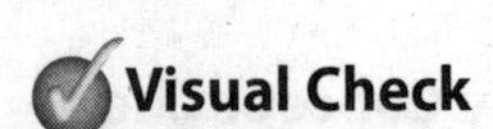

Visual Check

3. Locate Where is cardiac tissue? (Circle the correct answer.)

a. in the stomach

b. in the arm

c. in the heart

Skeletal Muscle

Muscle that attaches to bones is **skeletal muscle.** Skeletal muscles are also called voluntary muscles. **Voluntary muscles** *are muscles that you can consciously control.* For example, you can control whether or not you lift your leg. The contractions of skeletal muscles can be quick and powerful, such as when you run. However, contracting these muscles for a long time can tire them or make them cramp.

How Skeletal Muscles Work Skeletal muscles work by pulling on bones. Muscles cannot push on bones. Instead, muscles work in pairs and move the body. The figure below shows how an arm's biceps (BI seps) and triceps (TRI seps) muscles work as a pair. The arm on the left shows that as the biceps muscle contracts, the muscle shortens and pulls the lower arm up. The arm on the right shows that when biceps muscle relaxes, the triceps muscle contracts and the lower arm is pulled down. ✓

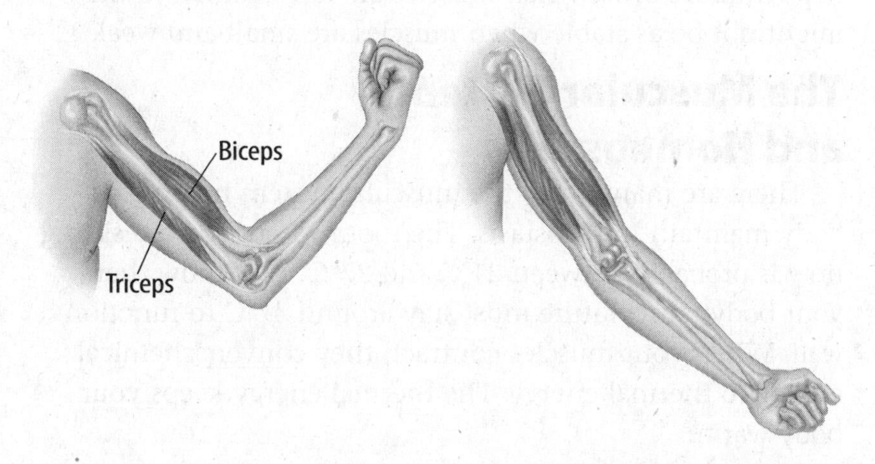

Changes in Skeletal Muscles Your skeletal muscles can change throughout your lifetime. If you exercise, your muscle cells get larger. Then, your entire muscle becomes larger and stronger.

Cardiac Muscle

Your heart is made of **cardiac** (KAR dee ak) **muscles,** *which are found only in the heart.* A cardiac muscle is a type of **involuntary muscle,** *which is muscle you cannot consciously control.* As cardiac muscles contract and relax, they pump blood through your heart and through vessels throughout your body. Cardiac muscle cells have branches with discs at their ends. These discs send signals to other cardiac muscle cells. The signals cause all the cardiac muscle cells to contract at almost the same time.

✓ **Reading Check**

4. Describe Why must skeletal muscles work in pairs?

____

____

____

✓ **Visual Check**

5. Explain What happens when the bicep muscle contracts?

____

____

____

Smooth Muscle

Smooth muscles line your blood vessels and many of your organs. **Smooth muscles** *are involuntary muscles named for their smooth appearance.* Contraction of smooth muscles helps move material through your body, such as food in your stomach. Smooth muscles also control the movement of blood through your vessels.

Key Concept Check
6. Name the three types of muscles.

Healthy Muscles

Recall that a good diet helps keep your bones healthy. Your muscles need a healthful diet, too. All of your muscles use energy when they contract. This energy comes from the food you eat. Eating a diet full of nutrients such as protein, fiber, and potassium can help keep muscles strong.

Exercise also helps keep your muscles healthy. Muscle cells get smaller and weaker without exercise. Weak muscles can increase the risk of heart disease. Bone injuries can happen more often when muscles are not healthy. Joints might not be as stable when muscles are small and weak.

Think it Over
7. Evaluate Why is it important to keep your muscles healthy?

The Muscular System and Homeostasis

There are many ways the muscular system helps your body maintain homeostasis. The room where you are sitting now is probably between 21°C and 27°C. You know that your body temperature must stay around 37°C to function well. When your muscles contract, they convert chemical energy to thermal energy. The thermal energy keeps your body warm.

ACADEMIC VOCABULARY
convert
(verb) to change something into a different form

When you exercise, your cells use more oxygen and release more waste, such as carbon dioxide. The cardiac muscles in your heart help maintain homeostasis by contracting more often. When your heart contracts faster, it pumps more blood and more oxygen is carried to the cells.

Key Concept Check
8. Describe How do muscles help maintain homeostasis in the body?

After You Read

Mini Glossary

cardiac (KAR dee ak) muscle: muscle that is only in the heart

involuntary muscle: muscle that you cannot consciously control

muscle: strong tissue that can contract in an orderly way

skeletal muscle: the type of muscle that attaches to bones

smooth muscle: involuntary muscle named for its smooth appearance

voluntary muscle: muscle that you can consciously control

1. Review the terms and their definitions in the Mini Glossary. Write a sentence that explains the difference between voluntary muscle and involuntary muscle.

2. Complete the table about skeletal, smooth, and cardiac muscles.

Muscle Type	Description	Function	Location in the Body
Skeletal	Cells look like they have stripes.	help the body move	
Smooth		help move things through the body	organs
Cardiac	smaller than skeletal muscles		heart

3. Describe how muscles protect your body.

What do you think NOW?

Reread the statements at the beginning of the lesson. Fill in the After column with an A if you agree with the statement or a D if you disagree. Did you change your mind?

Log on to ConnectED.mcgraw-hill.com and access your textbook to find this lesson's resources.

CHAPTER 14
LESSON 3

Structure and Movement

The Skin

Key Concepts 

- What does the skin do?
- How do the three layers of skin differ?
- How does the skin interact with other body systems?

Mark the Text

Main Idea and Details As you read, circle each heading that contains the word *skin.* Then, underline one sentence from each paragraph that follows that heading and its subheadings and teaches you something about skin.

Reading Check

1. Explain What would happen to your body if you had no skin?

______

______

______

Before You Read

What do you think? Read the two statements below and decide whether you agree or disagree with them. Place an A in the Before column if you agree with the statement or a D if you disagree. After you've read this lesson, reread the statements to see if you have changed your mind.

Before	Statement	After
	5. Skin helps regulate body temperature.	
	6. Skin is made of two layers of tissue.	

Read to Learn

Functions of the Skin

When you touch your face or arm, you are touching the outer layer of your skin. Skin is the largest organ of the body. It is part of the integumentary (ihn teh gyuh MEN tuh ree) system. *The* ***integumentary system*** *is made up of all of the external coverings of the body, including the skin, nails, and hair.* Like your bones and muscles, skin has many different functions in your body.

Protection

Skin covers your bones and muscles. Skin protects them from the outside environment. It keeps your body from drying out in sunlight and wind. Skin also protects the cells and tissues under the skin from damage. Skin keeps dirt, bacteria, viruses, and other substances from entering your body.

Sensory Response

Imagine you closed your eyes and felt two objects—a brick and a piece of paper. You would be able to feel the difference. The brick would feel rough, and the paper would feel smooth. Your skin has sensory receptors that detect texture. Sensory receptors in the skin also detect temperature and sense pain. The more sensory receptors there are in an area of skin, the more sensitive the skin is.

Temperature Regulation

Skin helps control body temperature. Skin has tiny holes, or pores, on its outer surface. When you exercise, sweat comes from these pores. Sweating is one way that skin maintains normal body temperature. As sweat evaporates, excess thermal energy leaves the body and the skin cools.

Another way the skin maintains body temperature is by releasing thermal energy from blood vessels. When your body temperature begins to increase, such as when you are exercising, blood vessels near the skin's surface dilate, or enlarge. This increases the surface area of the blood vessels and releases more thermal energy.

Production of Vitamin D

If your skin is exposed to sunlight, it can make vitamin D. Your body needs vitamin D to help it absorb calcium and phosphorus. Vitamin D also helps bones grow.

Elimination

Normal cellular processes produce waste products. The skin helps eliminate, or get rid of, some of these wastes. Water, salts, and other waste products are removed through the pores of the skin. Elimination occurs all the time, but you probably only notice it when you sweat.

Structures of the Skin

The skin that you see and feel on your body is the outermost layer of your skin. Below it are two other layers of skin. Each layer, as shown in the figure below, has a different structure and function.

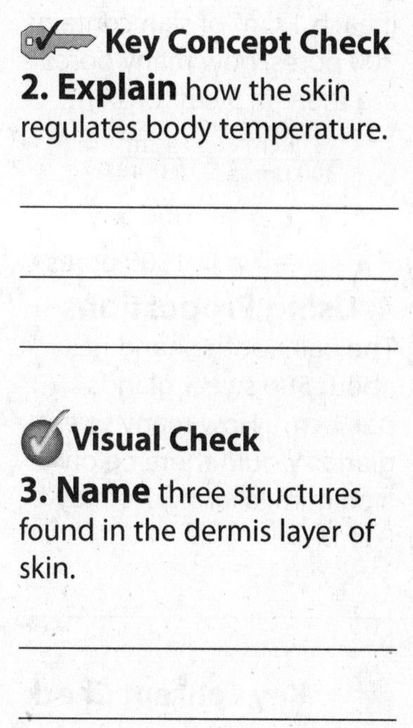

Key Concept Check

2. Explain how the skin regulates body temperature.

____

____

____

Visual Check

3. Name three structures found in the dermis layer of skin.

____

____

____

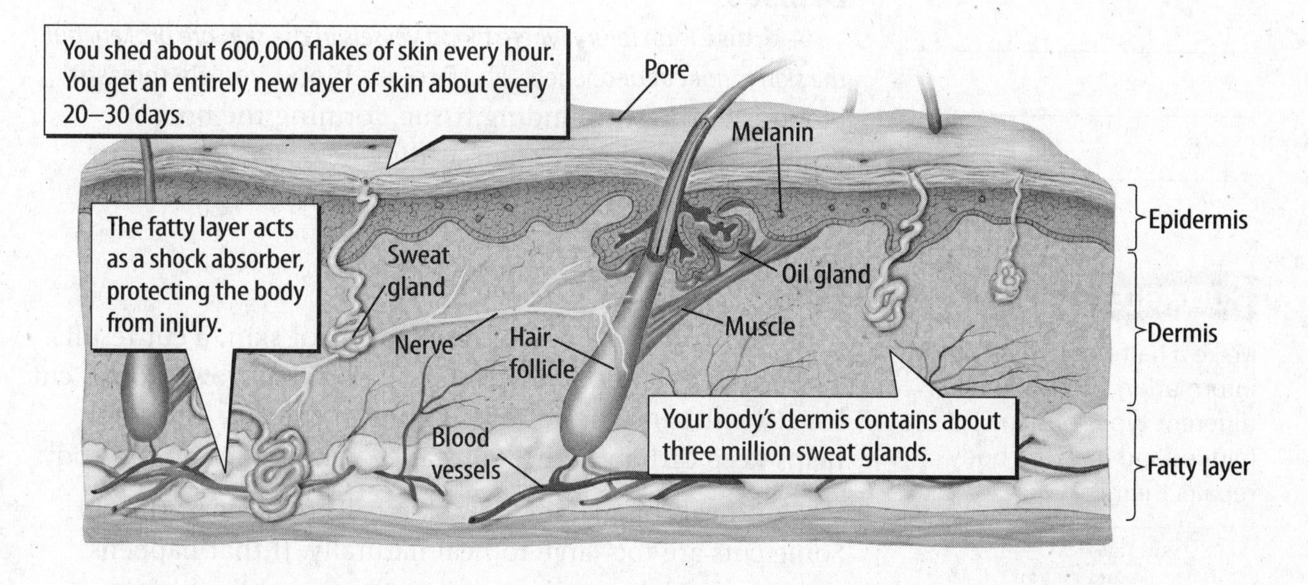

Epidermis

The **epidermis** (eh puh DUR mus) *is the outermost layer of skin and the only layer in direct contact with the outside environment.* The epidermis is thin but tough. The epidermis on your eyelids is thinner than a sheet of paper. Cells of the epidermis are constantly shed, or gotten rid of, and replaced by new cells.

One important function of the epidermis is the production of melanin (MEH luh nun). **Melanin** *is a pigment that protects the body by absorbing some of the Sun's damaging ultraviolet rays.*

Dermis

Below the epidermis is the dermis. *The* **dermis** *is a thick layer of tissue that gives skin strength, nourishment, and flexibility.* The dermis contains sweat glands, blood vessels, nerves, hair follicles, and muscles. When the muscles in the dermis contract, you get goose bumps.

Fatty Layer

The innermost layer of skin is sometimes called the fatty layer. It insulates the body, keeping it warm. It also acts as a protective padding and stores energy. This layer can be thin or very thick, depending on where it is on the body.

Skin Injuries and Repair

Skin is often injured because it is exposed to the outside environment. You might have injured your skin by falling down or bruising it. Your body has different ways to repair skin. The type of repair depends on the type of injury and how serious it is.

Bruises

A **bruise** *is an injury where blood vessels in the skin are broken, but the skin is not cut or opened.* The broken blood vessels release blood into the surrounding tissue, forming the bruise. Bruises usually change color as they heal. This change in color is due to chemical changes in the blood under the skin's surface.

Cuts

When you break one or more layers of skin, a cut results. Cuts often cut blood vessels, too. Blood that flows from a cut usually thickens and forms a scab over the cut. The scab helps keep dirt and other substances from entering the body.

Skin heals by making new skin cells that repair the cut. Some cuts are too large to heal naturally. If that happens, stitches might be needed to close the cut while it heals.

Math Skills

The ratios $\frac{5}{1}$ and $\frac{25}{5}$ are equivalent, so they can be written as the proportion $\frac{5}{1} = \frac{25}{5}$. When ratios form a proportion, the cross products are equal. In the above proportion, $5 \times 5 = 25 \times 1$. You can use cross products to find a missing term. For example, if each 1 cm^2 of skin contains 300 pores, how many pores are there in 5 cm^2 of skin?

$$\frac{1\ cm^2}{300\ pores} = \frac{5\ cm^2}{n\ pores}$$

$$1 \times n = 300 \times 5$$

$$n = 1500\ pores$$

4. Using Proportions The palm of the hand has about 500 sweat glands per 1 cm^2. How many sweat glands would there be on a palm measuring 7 cm by 8 cm?

____

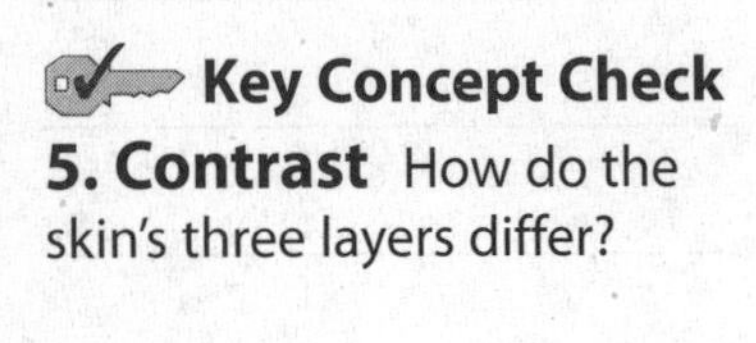

Key Concept Check
5. Contrast How do the skin's three layers differ?

____

____

____

FOLDABLES®

Make a half-book to record information about the different types of skin injuries and how the body repairs them.

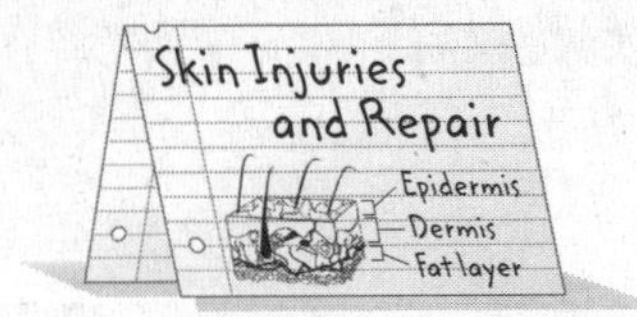

Burns

A burn is an injury to your skin or tissues that can be caused by touching hot objects. Touching extremely cold objects, chemicals, electricity, radiation (such as sunlight), or friction (rubbing) can also cause burns. The three degrees, or levels, of burns are described in the table below. ✓

Burn Type	Description	Symptoms	Healing Time
First-degree burn	damages top layer of skin	pain, redness, swelling	5–7 days without scarring
Second-degree burn	damages top and lower layers of skin	pain, redness, swelling, blistering	2–6 weeks with some scarring
Third-degree burn	damages all three layers of skin and sometimes the tissue below skin	black or white charred skin, might be numb as a result of damaged nerves	several months with scarring, might need surgery

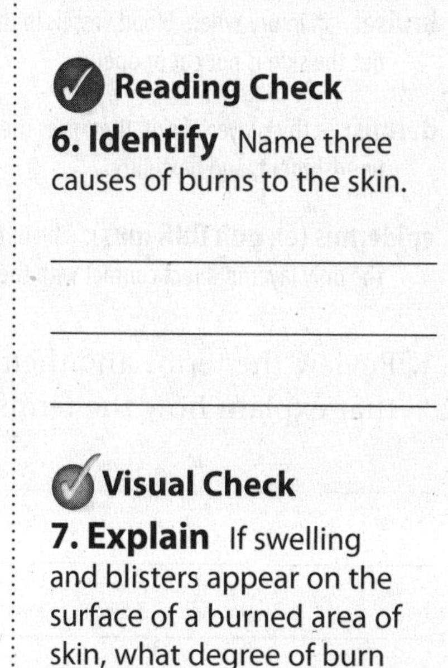

Reading Check

6. Identify Name three causes of burns to the skin.

Visual Check

7. Explain If swelling and blisters appear on the surface of a burned area of skin, what degree of burn would it be?

Healthy Skin

One way to keep your skin healthy is to protect it from sunlight. The ultraviolet (UV) rays in sunlight can cause permanent damage to the skin. Damage to the skin can include dry skin, wrinkles, and skin cancer. You can protect your skin from the UV rays in sunlight by using sunscreen. You can also wear clothing, such as a hat or long-sleeved shirt, to protect your skin.

Another way to keep your skin healthy is to eat a balanced diet. You can also use gentle soaps to clean your skin and lotion to keep your skin moist.

The Skin and Homeostasis

You have read that the skin can make vitamin D and that it protects the body from outside substances. Both of these functions help regulate the body's internal environment.

The skin also works with other body systems to maintain homeostasis. The skin and circulatory system help cool the body when it becomes overheated. The skin also works with the nervous system and muscular system to help the body react to stimuli. For example, if you touch a hot pan, receptors in your skin sense pain. This triggers nerve cells to send a message to your brain. Your brain then sends a message to your muscles to move your hand away.

Key Concept Check

8. State Give two examples of how the skin interacts with other body systems to help maintain homeostasis.

After You Read

Mini Glossary

bruise: an injury where blood vessels in the skin are broken, but the skin is not cut or opened

dermis: a thick layer of skin that gives the skin strength, nourishment, and flexibility

epidermis (eh puh DUR mus): the outer layer of skin and the only layer in direct contact with the outside environment

integumentary (ihn teh gyuh MEN tuh ree) system: made up of all of the external coverings of the body, including the skin, nails, and hair

melanin (MEH luh nun): a pigment that protects the body by absorbing some of the Sun's damaging ultraviolet rays

1. Review the terms and their definitions in the Mini Glossary. Write two or three sentences that explain how the terms *integumentary system, epidermis, dermis,* and *bruise* are related.

2. Rewrite each phrase shown below in the correct part of the Venn diagram to compare and contrast the epidermis and the dermis.

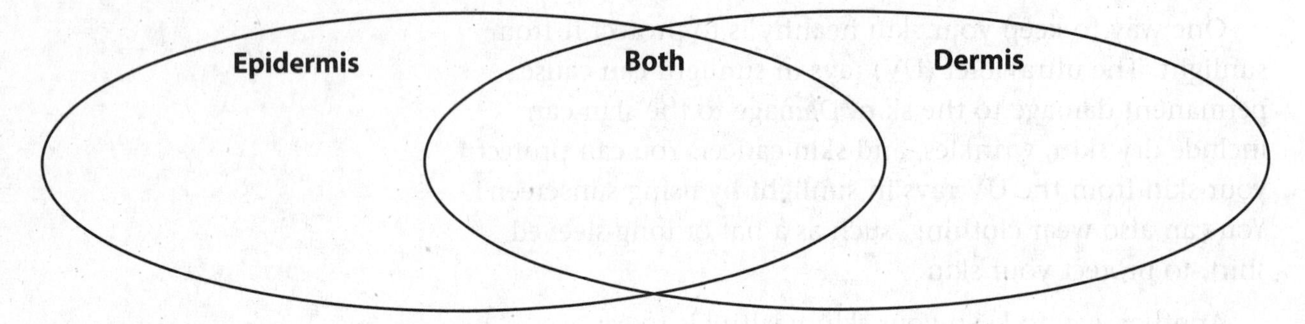

- can be burned by extreme heat
- thick, middle layer of skin
- thin, top layer of skin

3. How does the number of sensory receptors affect an area of skin?

What do you think NOW?

Reread the statements at the beginning of the lesson. Fill in the After column with an A if you agree with the statement or a D if you disagree. Did you change your mind?

Log on to ConnectED.mcgraw-hill.com and access your textbook to find this lesson's resources.

Digestion and Excretion

Nutrition

Before You Read

What do you think? Read the two statements below and decide whether you agree or disagree with them. Place an A in the Before column if you agree with the statement or a D if you disagree. After you've read this lesson, reread the statements and see if you have changed your mind.

Before	Statement	After
	1. An activity such as sleeping does not require energy.	
	2. All fats in food should be avoided.	

Key Concepts 

- Why do you eat?
- Why does your body need each of the six groups of nutrients?
- Why is eating a balanced diet important?

Read to Learn

Why do you eat?

How do you decide what to eat and when to eat it? You could survive for weeks without food. However, you might feel hungry just a few hours after eating a meal.

Hunger is your body's way of telling you that it needs food. Why does your body need food? Food contains energy. Food gives your body the energy and nutrients it needs to survive.

Study Coach

Identify Main Ideas Fold a sheet of paper into three columns. Label the columns *K, W,* and *L*. Write what you already know about nutrition under the *K*. Write what you want to learn under the *W*. After you read this lesson, write what you learned about nutrition under the *L*.

Energy

Every activity you do, such as riding a bike and even sleeping, uses energy. Your digestive system processes the food you eat. It releases the energy from this processed food. The energy is used for cellular processes and all activities that you do.

Measuring Energy The amount of energy in food is measured in Calories. *A* **Calorie** *(Cal) is the amount of energy it takes to raise the temperature of 1 kg of water by 1°C.*

How much energy do foods contain? Each food contains a different amount of energy. One grape contains 2 Cal. One slice of cheese pizza might have 220 Cal. All foods give your body energy to use.

Think it Over

1. Determine Which food would provide your body with more energy—a grape or a slice of cheese pizza? Why?

Your Energy Needs The amount of energy a person needs depends on several factors. Some of those factors are weight, age, activity level, and gender. For example, a person with a mass of 68 kg usually burns more Calories than a person with a mass of 45 kg. Playing soccer requires more energy than playing a video game. How does the food you eat supply you with energy? The energy comes from nutrients.

Nutrients

Food is made of nutrients. Nutrients are substances that provide energy and materials for cell development, growth, and repair. People need different types and amounts of nutrients.

A person's nutrient requirements depend on the person's age, gender, and activity level. Toddlers need more fat in their diets than older children do. Women need more calcium and iron than men do. Older men and women need more calcium than younger adults. Active people need more protein than people who are less active. There are six groups of nutrients. Each group plays a role in maintaining your health.

Key Concept Check
2. Explain Why do you eat?

______

______

______

Groups of Nutrients

The six groups of nutrients are proteins, carbohydrates, fats, vitamins, minerals, and water. Each nutrient has a different function in the body. To be healthy, you need to eat nutrients from each group every day. Different foods contain different combinations of these nutrients. A healthful diet includes a variety of foods.

Think it Over
3. Describe What would you include in a lunch that you wanted to be high in protein?

______

______

Proteins

Most of the tissues in your body are made of proteins. *A **protein** is a large molecule that is made of amino acids and contains carbon, hydrogen, oxygen, nitrogen, and sometimes sulfur.*

Proteins have many functions. Proteins relay signals between cells and protect against disease. They also provide support to cells and speed up chemical reactions. All of these functions are needed to maintain homeostasis. Homeostasis keeps an organism's internal condition consistent even when there are changes to the organism's environment.

Reading Check
4. Identify How does your body obtain amino acids that cannot be made in cells?

______

______

______

Combinations of 20 different amino acids make up the proteins in your body. Your cells can make more than half of these amino acids. The rest of the amino acids must come from the foods that you eat. Some foods that are good sources of protein are red meat, eggs, beans, and peanut butter.

Carbohydrates

Pasta, bread, and potatoes all have high levels of carbohydrates (kar boh HI drayts). **Carbohydrates** *are molecules made of carbon, hydrogen, and oxygen atoms and are usually the body's main source of energy.* Carbohydrates are usually found in one of three forms—starches, sugars, or fibers. All carbohydrates are made of sugar molecules. These molecules are linked together like a chain. It is best to eat foods that have carbohydrates from whole grains. Whole grains are easy to digest. Other foods that are high in carbohydrates include red beans, vegetables, and fruit.

Fats

You need a certain amount of fat in your diet and on your body to stay healthy. **Fats,** *also called lipids, provide energy and help the body absorb vitamins.* They are a major part of cell membranes. Body fat helps to insulate against cold temperatures. Most people get plenty of fat in their diets. Too much fat in your diet can lead to health problems. Only about 25–35 percent of the Calories you consume should be fats. ✓

Fats are classified as either saturated or unsaturated. A diet high in saturated fats can increase levels of cholesterol. High levels of cholesterol can increase the risk of heart disease. Most of the fat in your diet should come from unsaturated fats. Foods such as fish, nuts, and liquid vegetable oils are sources of unsaturated fats.

Vitamins

Vitamins *are nutrients that are needed in small amounts for growth, regulating body functions, and preventing some diseases.* You can get most of the vitamins that your body needs by eating a well-balanced diet. If you do not get enough of one or more vitamins in your diet, then you might develop symptoms of vitamin deficiency. The symptoms depend on which vitamin you are not getting enough of. The table below lists some of the vitamins people need in their diets. ✓

Vitamin	Good Sources	Health Benefit
Vitamin B_2	milk, meats, vegetables	helps release energy from nutrients
Vitamin C	oranges, broccoli, tomatoes	tissue growth and repair
Vitamin A	carrots, milk, sweet potatoes, broccoli	enhances night vision, helps maintain skin and bones

Reading Check

5. Describe Why do you need a certain amount of fat in your diet?

Reading Check

6. Explain Why do you need vitamins in your diet?

Visual Check

7. Identify What foods are good sources of vitamin A?

Minerals

In addition to vitamins, you need other nutrients called minerals. **Minerals** *are inorganic nutrients—nutrients that do not contain carbon—that help the body regulate many chemical reactions.*

In a way, minerals are similar to vitamins. If you do not consume enough of certain minerals, you might develop a mineral deficiency. Minerals are essential for maintaining a healthy body. Some of the minerals that you need in your diet are listed in the table below.

Mineral	Good Sources	Health Benefit
Calcium	milk, spinach, green beans	builds strong bones and teeth
Iron	meat, eggs, green beans	helps carry oxygen throughout the body
Zinc	meat, fish, wheat/grains	aids in the formation of protein

Visual Check
8. Identify Which mineral does milk contain?

Water

You learned that your body is mostly water. Water is necessary for the chemical reactions that occur in your body. Your body takes in water when you eat and drink.

However, you lose water when you sweat, urinate, and breathe. To stay healthy, it is important to replace the water that your body loses. Your body loses more water if you live in a warm climate, if you exercise, or if you become sick.

When the water that you lose is not replaced, you might become dehydrated. Dehydration can be serious. Symptoms of dehydration include thirst, headache, weakness, dizziness, and little or no urination.

Key Concept Check
9. Discuss Why does your body need nutrients?

Healthful Eating

Imagine walking through a grocery store. Each aisle in the store contains hundreds of different foods. Some foods provide more health benefits than others. A large number of choices can make it difficult to choose foods that are part of a healthful diet.

To be a healthful eater, you need to be a smart shopper. Smart shoppers make grocery lists before shopping. They buy products that are high in nutrients. Foods that are high in nutrients come from the major food groups. The major food groups include grains, vegetables, fruits, oils, milk products, and meat and beans.

FOLDABLES®
Make a chart to organize information about the major food groups and to list examples of each.

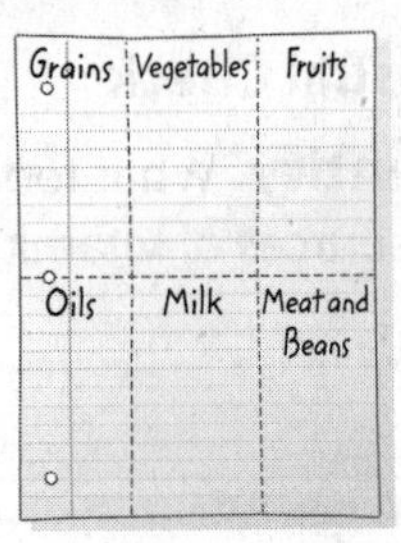

A Balanced Diet

A healthful diet includes carbohydrates, proteins, fats, vitamins, minerals, and water. But how do you know how much of each food group you need to eat? The table below lists the daily recommended amounts of each food group for 9- to 13-year-olds.

The nutrient-rich foods that you choose might be different from the nutrient-rich foods eaten by people in China, Kenya, or Mexico. People usually eat foods that are grown and produced in their region. Regardless of where you live, eating a balanced diet ensures that your body has the nutrients it needs to function.

Food Group	Daily Amount, males, 9–13 years old	Daily Amount, females, 9–13 years old	Examples of Foods
Grains	6-ounce equivalent	5-ounce equivalent	whole-wheat flour, rye bread, brown rice
Vegetables	$2\frac{1}{2}$ cups	2 cups	broccoli, spinach, carrots
Fruits	$1\frac{1}{2}$ cups	$1\frac{1}{2}$ cups	apples, strawberries, oranges
Oils	5 teaspoons or less	5 teaspoons or less	canola oil, olive oil, avocados
Milk	3 cups	3 cups	milk, cheese, yogurt
Meat and Beans	5 ounces or less	5 ounces or less	fish, beans, lean beef, lean chicken

Food Labels

What foods would you buy to follow the recommended guidelines in the table above? Most grocery stores sell many varieties of bread, milk, meat, and other types of food. How do you know what nutrients these foods contain? You can look at their food labels. Food labels help you determine the amount of protein, carbohydrates, oils, and other substances in food. Be careful when reading a food label. The label lists a food's nutrients per serving, not per container. You should also look at the number of servings per container. The number of servings is also included on the food label.

Key Concept Check

10. Explain Why is eating a balanced diet important?

Visual Check

11. Identify How many cups of vegetables should a 9- to 13-year-old female have in her daily diet? (Circle the correct answer.)

a. $1\frac{1}{2}$
b. 2
c. $2\frac{1}{2}$

Think it Over

12. Discuss why it is important to read food labels.

After You Read

Mini Glossary

Calorie (Cal): the amount of energy it takes to raise the temperature of 1 kg of water by 1°C

carbohydrate (kar boh HI drayt): a molecule that is made of carbon, hydrogen, and oxygen atoms and is usually the body's main source of energy

fat: also called a lipid, it provides energy and helps the body absorb vitamins

mineral: an inorganic nutrient—a nutrient that does not contain carbon—that helps the body regulate many chemical reactions

protein: a large molecule that is made of amino acids and that contains carbon, hydrogen, oxygen, nitrogen, and sometimes sulfur

vitamin: a nutrient that is needed in small amounts for growth, regulating body functions, and preventing some diseases

1. Review the terms and their definitions in the Mini Glossary. Write a sentence describing why it is important to include carbohydrates in your diet.

____

____

____

2. Complete the table below by listing the nutrients that you might get from the breakfast foods. Keep in mind that there might be more than one nutrient in each food.

Food	scrambled eggs	one slice of cheese	two slices of whole-wheat toast	small glass of orange juice	two slices of bacon
Nutrients		protein		vitamins	

3. Plan a meal that contains a food from each of the six food groups. Write your menu on the lines below.

____

____

What do you think NOW?

Reread the statements at the beginning of the lesson. Fill in the After column with an A if you agree with the statement or a D if you disagree. Did you change your mind?

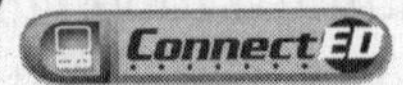

Log on to ConnectED.mcgraw-hill.com and access your textbook to find this lesson's resources.

Digestion and Excretion

The Digestive System

Before You Read

What do you think? Read the two statements below and decide whether you agree or disagree with them. Place an A in the Before column if you agree with the statement or a D if you disagree. After you've read this lesson, reread the statements and see if you have changed your mind.

Before	Statement	After
	3. Digestion begins in the mouth.	
	4. Energy from food stays in the digestive system.	

Key Concepts 
- What does the digestive system do?
- How do the parts of the digestive system work together?
- How does the digestive system interact with other systems?

Read to Learn

Functions of the Digestive System

What did you eat for breakfast today? Do you ever think about what happens to the food you eat?

Someone might have told you to take small bites and chew your food thoroughly. The size of the chewed food particles can affect how quickly your food is digested.

As soon as the food enters your mouth, it begins its journey through your digestive system. All the food you eat goes through a process with four steps—ingestion, digestion, absorption, and elimination. All four steps happen in the organs and tissues of the digestive system in the following order.

- Food is ingested. Ingestion is the act of eating, or putting food in your mouth.
- Food is digested. **Digestion** *is the mechanical and chemical breakdown of food into small particles and molecules that your body can absorb and use.*
- Nutrients and water in the food are absorbed, or taken in, by cells. Absorption occurs when the cells of the digestive system take in small molecules of digested food.
- Undigested food is eliminated. Elimination is the removal of undigested food and other wastes from your body. 

Mark the Text

Building Vocabulary As you read, circle all the words you do not understand. Highlight the part of the text that helps you define these words.

Key Concept Check
1. Describe What does the digestive system do?

Types of Digestion

Before your body can absorb nutrients from food, the food must be broken into small molecules by digestion. There are two types of digestion—mechanical and chemical. *In* **mechanical digestion,** *food is physically broken into smaller pieces.* Mechanical digestion happens when you chew, mash, and grind food with your teeth and tongue. Smaller pieces of food are easier to swallow and have more surface area than larger pieces. This helps with chemical digestion. *In* **chemical digestion,** *chemical reactions break down pieces of food into small molecules.*

Enzymes

Chemical digestion cannot occur without substances called enzymes (EN zimez). **Enzymes** *are proteins that help break down larger molecules into smaller molecules. Enzymes also speed up, or catalyze, the rate of chemical reactions.* Without enzymes, some chemical reactions would be too slow or would not occur at all. There are many kinds of enzymes. Each one is specialized to help break down a specific molecule at a specific location.

REVIEW VOCABULARY

chemical reaction
a process in which a compound is formed or broken down

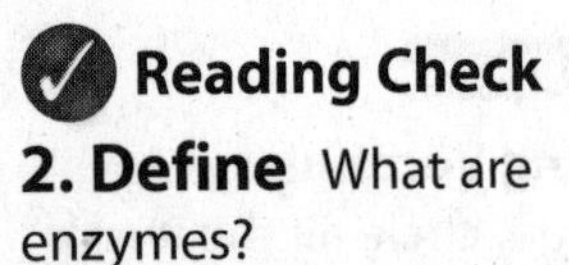

Reading Check

2. Define What are enzymes?

The Role of Enzymes in Digestion

Nutrients in food are made of different molecules, such as carbohydrates, proteins, and fats. Many of these molecules are too large for your body to use. But, because these large molecules are made of long chains of smaller molecules joined together, they can be broken down into smaller pieces.

The digestive system produces enzymes that are specialized to help break down each type of food molecule. For example, the enzyme amylase helps break down carbohydrates. The enzymes pepsin and papain help break down proteins. Fats are broken down with the help of the enzyme lipase. The figure below shows how an enzyme helps break down food molecules into smaller pieces.

Notice in the figure that the food molecule breaks apart, but the enzyme does not change. Therefore, the enzyme can immediately be used to break down another food molecule.

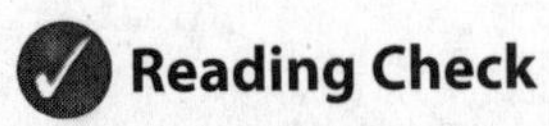

Reading Check

3. Explain What happens to an enzyme after it breaks down a food molecule?

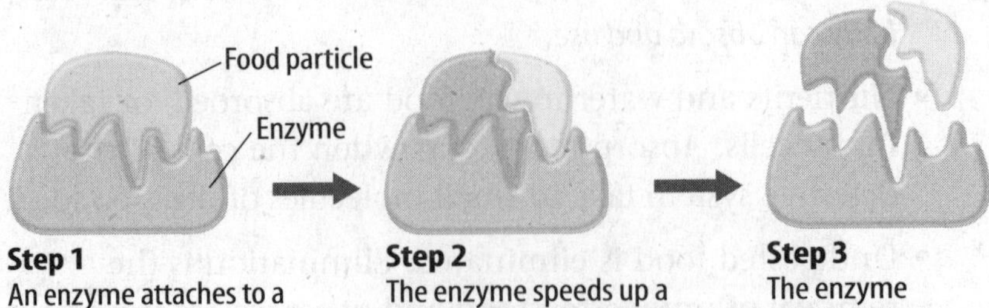

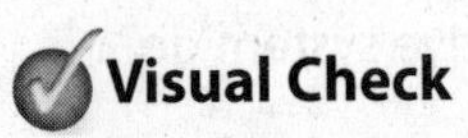

Visual Check

4. Identify Circle the food molecule that is in the process of being broken down.

Organs of the Digestive System

For your body to use the nutrients in the foods you eat, the nutrients must pass through your digestive system. There are two parts to your digestive system: the digestive tract and the other organs that help the body break down and absorb food. These organs, shown at the bottom of this page, include the tongue, salivary (SA luh ver ee) glands, liver, gallbladder, and pancreas.

The digestive track extends from the mouth to the anus. It has different organs that are connected by tubelike structures. Each of these organs is specialized for a certain function.

The Mouth

Mechanical digestion of food begins in your mouth. Your teeth and tongue mechanically digest food as you chew. But even before chewing begins, your salivary (SA luh ver ee) glands produce saliva (suh LI vuh) at the thought of food. They produce more than 1 L of saliva every day. Saliva contains an enzyme that helps break down carbohydrates. It also contains substances that neutralize acidic foods and a slippery substance that makes food easier to swallow.

The Esophagus

When you swallow a bite of food, it enters your esophagus (ih SAH fuh gus). *The* ***esophagus*** *is a muscular tube that connects the mouth to the stomach. Food moves through the esophagus and the rest of the digestive tract by waves of muscle contractions called* **peristalsis** (per uh STAHL sus). Peristalsis is similar to squeezing the bottom of a toothpaste tube. This forces toothpaste to the top of the tube. Muscles in the esophagus contract and relax. This action pushes partially digested food down the esophagus and into the stomach. ✓

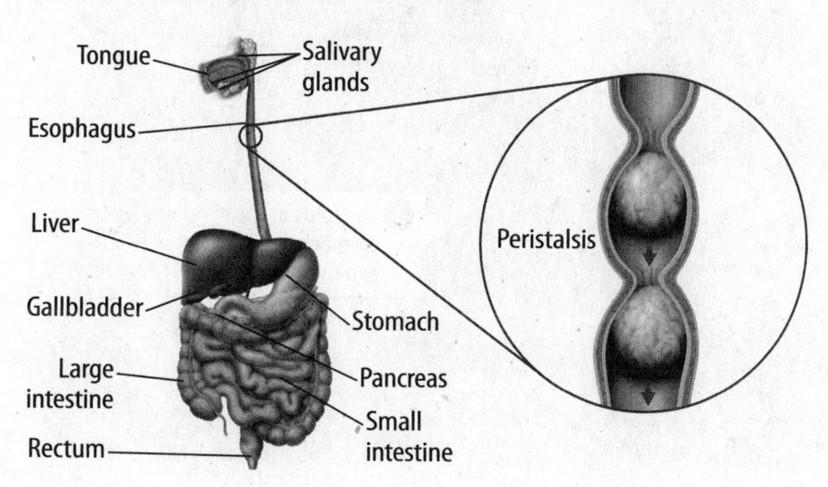

FOLDABLES®

Create a shutterfold book to illustrate the organs of the digestive system and their functions.

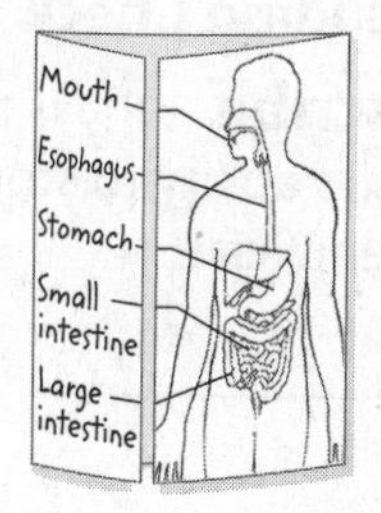

Reading Check

5. Define What is the esophagus?

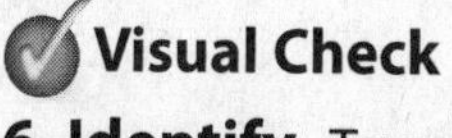

Visual Check

6. Identify Trace with a highlighter the organ that connects the mouth to the stomach.

The Stomach

Partially digested food leaves the esophagus and enters the stomach. The stomach, as shown below, is a large, hollow organ. One function of the stomach is to store food. This allows you to go many hours between meals. The stomach is like a balloon. It can stretch when filled. An adult stomach can hold about 2 L of food and liquid.

Reading Check

7. Describe Why is the stomach's ability to store food beneficial?

The stomach also helps with chemical digestion. The walls of the stomach are extremely folded. These folds enable the stomach to expand and hold large amounts of food. The cells in these folds produce chemicals that help break down proteins.

The stomach contains an acidic fluid called gastric juice. Gastric juice makes the stomach acidic. Acid helps break down some of the structures that hold plant and animal cells together, such as the cells in meat, lettuce, and tomatoes. Gastric juice also contains pepsin. Pepsin is an enzyme that helps break down the proteins in foods into amino acids. Food and gastric juices mix as muscles in the stomach contract through peristalsis. As food mixes with gastric juice in the stomach, it forms *a thin, watery liquid called* **chyme** (KIME).

Visual Check

8. Identify Where does food go after it leaves the stomach?

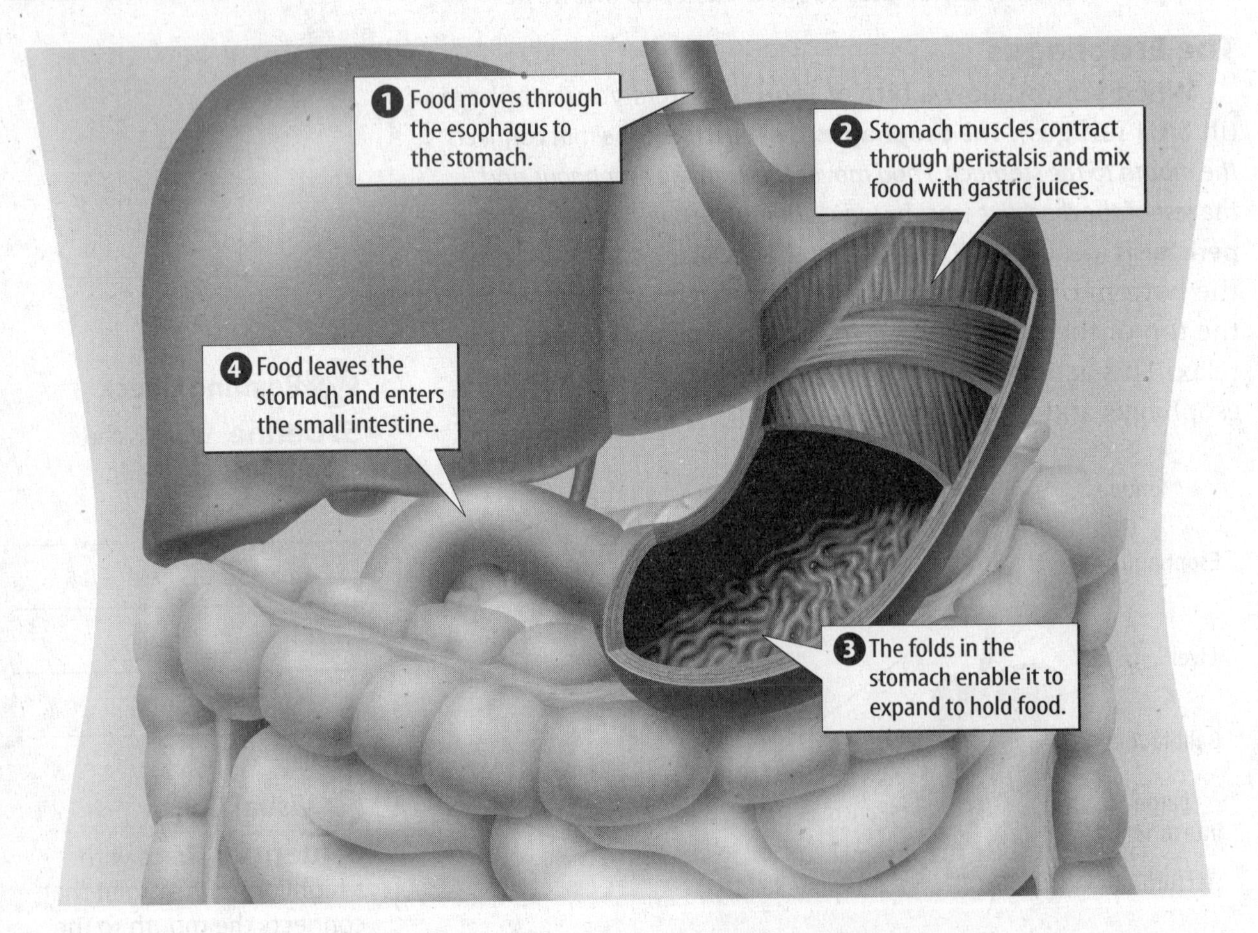

The Small Intestine

Chemical digestion begins in the mouth and the stomach. But most chemical digestion occurs in the small intestine. The small intestine is a long tube that is connected to the stomach. Chemical digestion and nutrient absorption take place in the small intestine. The small intestine is named for its small diameter—about 2.5 cm. The small intestine is about 7 m long.

Chemical digestion of proteins, carbohydrates, nucleic acids, and fats takes place in the first part of the small intestine, called the duodenum (doo uh DEE num). The remainder of the small intestine absorbs nutrients from food. Like the stomach, the wall of the intestine has many folds. *The folds of the small intestine are covered with fingerlike projections called* **villi** (VIH li) (singular, villus). Each villus contains small blood vessels. Nutrients in the small intestine diffuse into the blood through these blood vessels. You can see the small vessels in a villus in the figure below.

Locate the pancreas and liver in the figure below. These organs produce substances that enter the small intestine and help with chemical digestion. The pancreas produces the enzyme amylase and a substance that neutralizes stomach acid. The enzyme amylase helps break down carbohydrates. The liver produces bile. Bile makes it easier to digest fats. The gallbladder stores bile until it is needed in the small intestine.

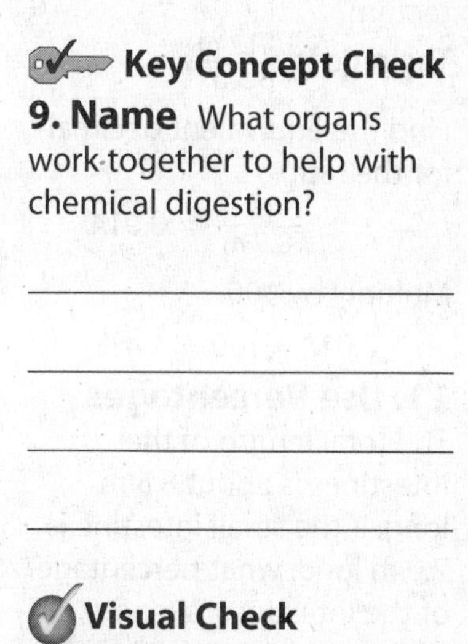

Key Concept Check

9. Name What organs work together to help with chemical digestion?

Visual Check

10. Identify the structures in the folds of the small intestine.

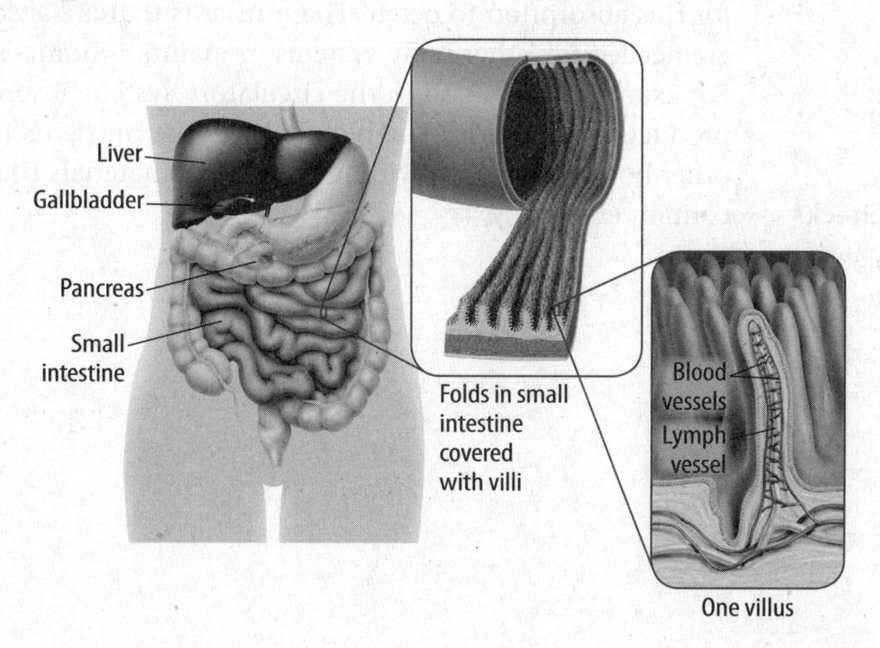

Math Skills

A percentage is a ratio that compares a number to 100. For example, the total length of the intestines is 8.5 m. That value represents 100%. If the rectum is 0.12 m long, what percentage of the intestines is made up of the rectum?

The ratio is $\frac{0.12 \text{ m}}{8.5 \text{ m}}$.

Find the equivalent decimal for the ratio.

$$\frac{0.12 \text{ m}}{8.5 \text{ m}} = 0.014$$

Multiply by 100.

$$0.014 \times 100 = 1.4\%$$

11. Use Percentages The total length of the intestines is about 8.5 m long. If the small intestine is 7.0 m long, what percentage of the intestines is made up of the small intestine?

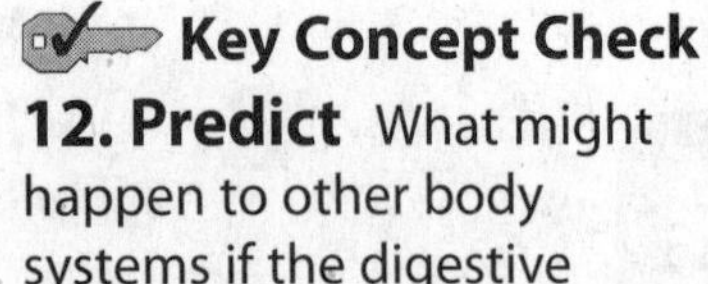

Key Concept Check

12. Predict What might happen to other body systems if the digestive system did not function properly?

The Large Intestine

Any food that is not absorbed in the small intestine moves by peristalsis into the large intestine. The large intestine is also called the colon. The large intestine has a larger diameter (about 5 cm) than the small intestine. It is much shorter than the small intestine, however. It is only about 1.5 m long.

Most of the water in ingested foods and liquids is absorbed in the small intestine. As food travels through the large intestine, even more water is absorbed. Materials that pass through the large intestine are the waste products of digestion. The waste products become more solid as even more water is absorbed. Peristalsis continues to force the semisolid waste into the rectum, the last section of the large intestine. Muscles in the rectum and anus control the release of this semisolid waste, called feces (FEE seez).

Bacteria and Digestion

You might think that all bacteria are harmful. However, some bacteria have an important role in the digestive system. Bacteria in the intestines digest food and produce important vitamins and amino acids. Bacteria in the intestines are necessary for proper digestion. Without these bacteria, food would not be digested well.

The Digestive System and Homeostasis

Recall that nutrients from food are absorbed in the small intestine. The digestive system must be functioning properly for this absorption to occur. The nutrients that are absorbed are needed for other body systems to maintain homeostasis. For example, the blood in the circulatory system absorbs the products of digestion. The blood carries the nutrients to all other body systems, providing them with materials that contain energy.

After You Read

Mini Glossary

chemical digestion: the breakdown of food into small molecules by chemical reactions

chyme (KIME): a thin, watery liquid; formed when food mixes with gastric juices in the stomach

digestion: the mechanical and chemical breakdown of food into small particles and molecules that your body can absorb and use

enzyme (EN zime): a protein that helps break down larger molecules into smaller molecules, and also helps speed up, or catalyze, the rate of chemical reactions

esophagus (ih SAH fuh gus): a muscular tube that connects the mouth to the stomach

mechanical digestion: the physical breakdown of food into smaller pieces

peristalsis (per uh STAHL sus): waves of muscle contractions that move food through the esophagus and the rest of the digestive tract

villi (VIH li): fingerlike projections that cover the folds of the small intestine

1. Review the terms and their definitions in the Mini Glossary. Write a sentence describing the difference between mechanical and chemical digestion.

___

___

2. Fill in the table below, highlighting the main structures and function of the organs of the digestive system.

Organ	Function	Special Features
Mouth		teeth, tongue, saliva
Esophagus	moves food from mouth to stomach	
Stomach		makes chyme with gastric juice
Small intestine		villi
Large intestine		highly folded

What do you think NOW?

Reread the statements at the beginning of the lesson. Fill in the After column with an A if you agree with the statement or a D if you disagree. Did you change your mind?

Log on to ConnectED.mcgraw-hill.com and access your textbook to find this lesson's resources.

Digestion and Excretion

The Excretory System

Key Concepts
- What does the excretory system do?
- How do the parts of the excretory system work together?
- How does the excretory system interact with other body systems?

Before You Read

What do you think? Read the two statements below and decide whether you agree or disagree with them. Place an A in the Before column if you agree with the statement or a D if you disagree. After you've read this lesson, reread the statements and see if you have changed your mind.

Before	Statement	After
	5. Several human body systems work together to eliminate wastes.	
	6. Blood contains waste products that must be removed from the body.	

Study Coach

Identify the Main Ideas As you read each paragraph, write the main idea on a sheet of paper or in your notebook to study later.

Read to Learn

Functions of the Excretory System

You have read about nutrients in food. These nutrients are necessary to maintain health. You also have read about how the digestive system processes the food that you eat.

Your body does not use all the food that you take in. The unused food parts are waste products. These wastes are processed by the excretory system. *The* ***excretory system*** *collects and eliminates wastes from the body and regulates the level of fluid in the body.*

Key Concept Check
1. Describe What does the excretory system do?

____

____

____

____

Collection and Elimination

In your home, you probably collect waste in several places. You might have a trash can in the kitchen and another one in the bathroom. The furnace might have a filter that collects dust from the air. Your body also collects wastes. The digestive system collects waste products in the intestines. The circulatory system collects waste products in the blood.

When the trash cans in your home are full of waste, you must take the trash outside. The waste in your body also must be removed. Waste that is not removed, or eliminated, from your body can become toxic, or poisonous, and damage your organs.

Regulation of Liquids

The excretory system also regulates the level of fluids in the body. Recall that water is an essential nutrient for your body. Some of the water in your body is lost when waste is eliminated. The excretory system controls how much water leaves the body through elimination. This ensures that neither too much nor too little water is lost.

Types of Excretion

Your body excretes, or eliminates, different substances from different body systems. The excretory system is made of four body systems.

- The digestive system collects and removes undigested solids from the foods you eat.
- The urinary system processes, transports, collects, and removes liquid wastes from the body.
- The respiratory system removes carbon dioxide and water vapor from the body.
- The integumentary system, which includes the skin, secretes excess salt and water from the body through sweat glands. ✓

Organs of the Urinary System

The urinary system processes, stores, and removes liquid wastes from the body. It helps maintain homeostasis. The organs of the urinary system include two kidneys, two ureters, the bladder, and the urethra. These organs work together to process, transport, collect, and excrete liquid wastes. Most functions of the urinary system occur in the kidneys. ✓

The Kidneys

The bean-shaped organ that filters, or removes, wastes from blood is the **kidney.** You have two kidneys, one on each side of your body. They are near the back wall of your abdomen, above your waist, and below your rib cage.

Each kidney is about the size of your fist. Kidneys are dark red in color because of the large amount of blood that passes through them.

Kidney Functions The kidneys have several functions. This lesson will discuss the role of the kidneys in the urinary system. But the kidneys have other important functions. They produce hormones that stimulate the production of red blood cells. They also control blood pressure and help control calcium levels in the body.

FOLDABLES®

Use a four-door book to summarize information about the functions of the body systems that make up the excretory system.

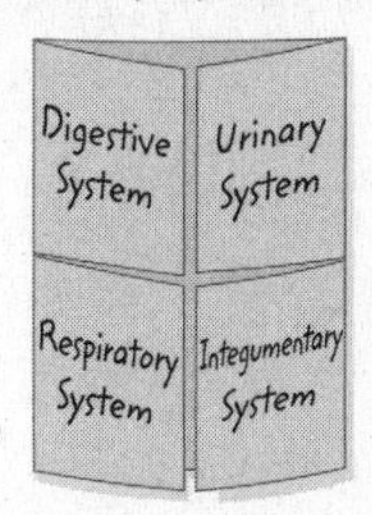

✓ Reading Check

2. Name What body systems make up the excretory system?

______

______

______

______

✓ Reading Check

3. Describe What is the function of the urinary system?

______

______

______

WORD ORIGIN

nephron
from Greek *nephros*, means "kidney"

The Kidneys' Role in the Urinary System The kidneys contain blood vessels and nephrons (NEH frahnz). **Nephrons** *are networks of capillaries and small tubes, or tubules, where filtration of blood occurs.* Each kidney contains about one million nephrons.

Blood contains waste products, salts, and sometimes toxins from cells that need to be removed from the body. As blood passes through the kidneys, they filter these products from the blood. *When blood is filtered, a fluid called* **urine** *is produced.* The kidneys filter the blood and produce urine in two stages.

1. **First Filtration** Blood is constantly circulating and filtering through the kidneys. In one day, the kidneys filter about 180 L of blood plasma, or the liquid part of blood. That is enough liquid to fill ninety 2-L bottles. You have about 3 L of blood plasma in your body. This means that your kidneys filter your entire blood supply about 60 times each day. The first filtration occurs in the nephrons. There, groups of capillaries filter water, sugar, salts, and wastes out of the blood.
2. **Second Filtration** What would happen if all of the liquid from the first filtration were excreted? Your body would quickly dehydrate, and important nutrients would be lost. To regain some of this water, the kidneys filter the liquid collected in the first filtration again. The second filtration occurs in small tubes in the nephrons. During the second filtration, up to 99 percent of the water and nutrients from the first filtration are separated out and reabsorbed into the blood. The remaining liquid and waste products form urine. On average, an adult excretes about 1.5 L of urine per day.

Think it Over

4. Evaluate Why are two filtrations necessary?

____

____

____

____

The Ureters, Bladder, and Urethra

When garbage piles up in a trash can, the can must be emptied. In a similar way, the urine produced by your body cannot stay in your kidneys. *Urine leaves each kidney through a tube called the* **ureter** (YOO ruh tur).

Each of your kidneys has a ureter. Both ureters drain into your bladder. *The* **bladder** *is a muscular sac that holds urine until the urine is excreted.* Your bladder expands and contracts like a balloon when it fills or empties. An adult bladder can hold about 0.5 L of urine.

Urine leaves the bladder through a tube called the **urethra** (you REE thruh). The urethra contains circular muscles called sphincters (SFINGK turz) that control the release of urine.

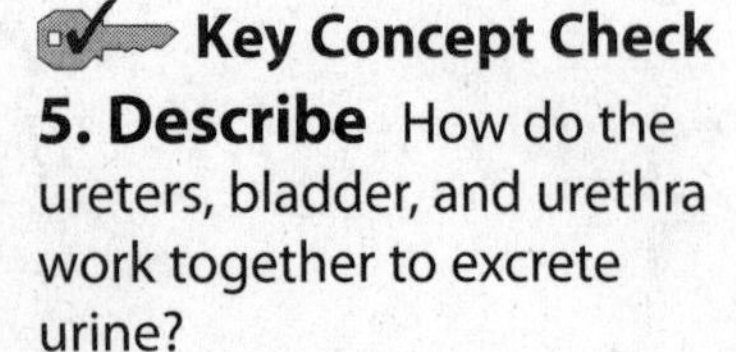

Key Concept Check

5. Describe How do the ureters, bladder, and urethra work together to excrete urine?

____

____

Urinary Disorder	Description	Possible Causes
Kidney disease	The nephrons are damaged and the ability of the kidneys to filter blood is reduced. In the beginning stages, there might not be symptoms.	diabetes, high blood pressure, poisons, trauma
Urinary tract infection	Infections usually occur in the bladder or urethra but can be in the kidneys and ureters. Symptoms might include burning during urination, small and frequent urination, and blood in urine.	bacteria in the urinary system
Kidney stones	Kidney stones are solid substances that form in the kidneys. The most common type is made of calcium. Stones that pass through the urinary system can be very painful.	calcium buildup in the kidneys
Bladder control problems	Urine is released from the bladder involuntarily. These problems occur in women more often than in men.	urinary tract infections, muscle weakness, prostate enlargement

Urinary Disorders

A urinary disorder is an illness that affects one or more organs of the urinary system. Some urinary disorders are described above. Several of these disorders are common. Urinary tract infections are a leading cause of doctor visits.

The Excretory System and Homeostasis

You have already read about some of the ways that the excretory system helps to maintain homeostasis. For example, the excretory system filters wastes from the blood. The blood is part of the circulatory system. A buildup of wastes in the circulatory system would be toxic to your body.

Homeostasis is also maintained by the removal of wastes from the digestive system. Wastes would damage your body if the excretory system did not remove them from the digestive system.

The excretory system also interacts with the nervous system. The hypothalamus is an area of the brain that helps to maintain homeostasis. One function of the hypothalamus is to control the secretion of some hormones. One hormone causes the tubules of the kidneys to absorb more water from the blood. This helps the body control fluid levels. Water is kept in the blood instead of being excreted in the urine.

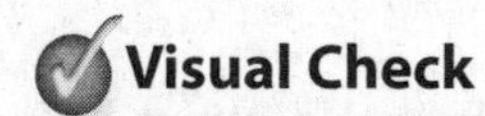

Visual Check

6. Identify Which urinary disorder results from bacteria in the urinary system? (Circle the correct answer.)

a. kidney disorder

b. urinary tract infection

c. kidney stones

ACADEMIC VOCABULARY

area
(noun) a part of something that has a particular function

Key Concept Check

7. Analyze How does the excretory system interact with the nervous system?

After You Read

Mini Glossary

bladder: a muscular sac that holds urine until the urine is excreted

excretory system: collects and eliminates wastes from the body and regulates the level of fluid in the body

kidney: a bean-shaped organ that filters, or removes, wastes from blood

nephrons (NEH frahnz): networks of capillaries and small tubes, or tubules, where filtration of blood occurs

ureter (YOO ruh tur): a tube through which urine leaves each kidney

urethra (yoo REE thruh): a tube through which urine leaves the bladder

urine: a fluid that is produced when blood is filtered

1. Review the terms and their definitions in the Mini Glossary. Write a sentence that describes how the excretory system helps maintain homeostasis.

2. Complete the chart below by naming the body systems that form the excretory system.

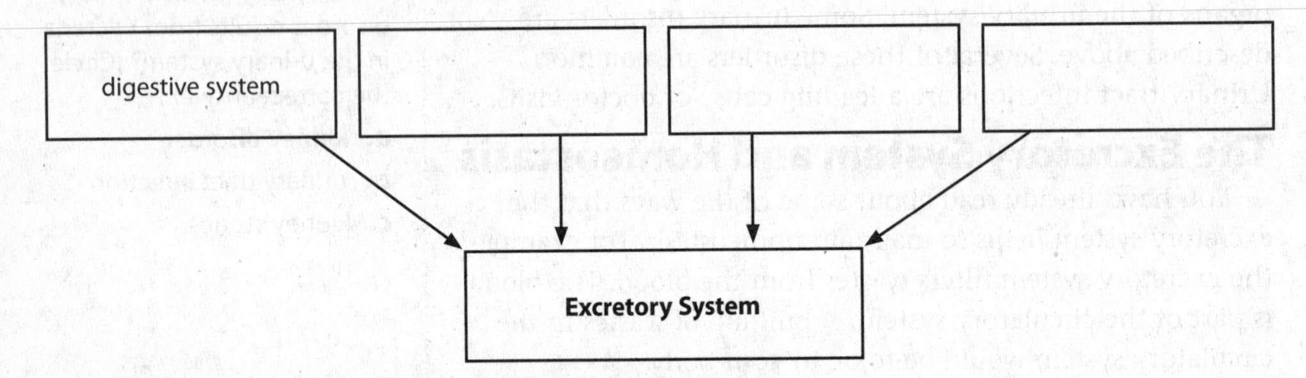

3. Review the main ideas that you wrote about the excretory system. Write a sentence about the most interesting idea to you.

What do you think NOW?

Reread the statements at the beginning of the lesson. Fill in the After column with an A if you agree with the statement or a D if you disagree. Did you change your mind?

Log on to ConnectED.mcgraw-hill.com and access your textbook to find this lesson's resources.

Respiration and Circulation

The Respiratory System

Before You Read

What do you think? Read the two statements below and decide whether you agree or disagree with them. Place an A in the Before column if you agree with the statement or a D if you disagree. After you've read this lesson, reread the statements to see if you have changed your mind.

Before	Statement	After
	1. Breathing and respiration are the same.	
	2. Lungs are the only parts of the body that use oxygen.	

Key Concepts 

- What does the respiratory system do?
- How do the parts of the respiratory system work together?
- How does the respiratory system interact with other body systems?

Read to Learn

Functions of the Respiratory System

If you have ever held your breath, you probably took deep breaths afterward. Those breaths were your body's way of getting the oxygen it needs. **Breathing** *is the movement of air into and out of the lungs*. Breathing is how your respiratory system takes in oxygen and gets rid of carbon dioxide.

Mark the Text

Identify the Main Ideas As you read, highlight the main idea of each section. Then use a different color to highlight a detail or example to help you understand the main idea. Use your highlighting to review this lesson.

Taking in Oxygen

Think about the plumbing pipes that bring water into a house. Your respiratory system is similar. It is a system of organs that bring oxygen into your body. Oxygen is important for life. In fact, your brain will tell your body to breathe even if you try not to breathe. Your brain responds when your body needs oxygen.

Why is oxygen so important? Every cell in your body needs oxygen for cellular respiration. Cellular respiration is a series of chemical reactions. During cellular respiration, oxygen and sugars react. This reaction releases energy that a cell can use.

REVIEW VOCABULARY

cellular respiration
a series of chemical reactions that transforms the energy in food molecules to usable energy

Eliminating Carbon Dioxide

A house's plumbing pipes also remove wastewater. In a similar way, your respiratory system removes waste gases from your body. Carbon dioxide is one of the waste gases that your respiratory system removes. If waste gases are not removed, cells cannot function.

Key Concept Check

1. Explain What does the respiratory system do?

Make an eight-tab vocabulary book. Use it to organize your notes on the organs of the respiratory system and their functions.

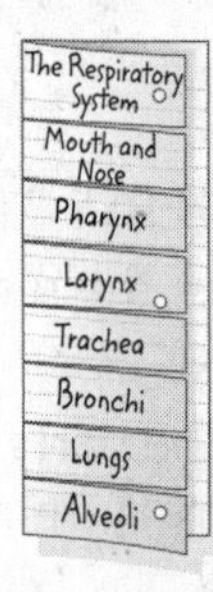

Key Concept Check

2. Explain What function do cilia have in the respiratory system?

Visual Check

3. Identify Which part of the respiratory system contains bronchi?

4. Apply What takes place in your respiratory system when you make sounds?

Organs of the Respiratory System

The figure below shows air moving into and out of the respiratory system. Air enters your respiratory system through your mouth and nose. Your nose warms and moistens the air. Hairs and sticky mucus in your nose help trap dust and dirt from the air.

Cilia are hairlike structures that line your nose and most other airways in your respiratory system. Wavelike motions of the cilia carry trapped particles away from your lungs. The cilia help keep harmful particles from getting very far into your respiratory system.

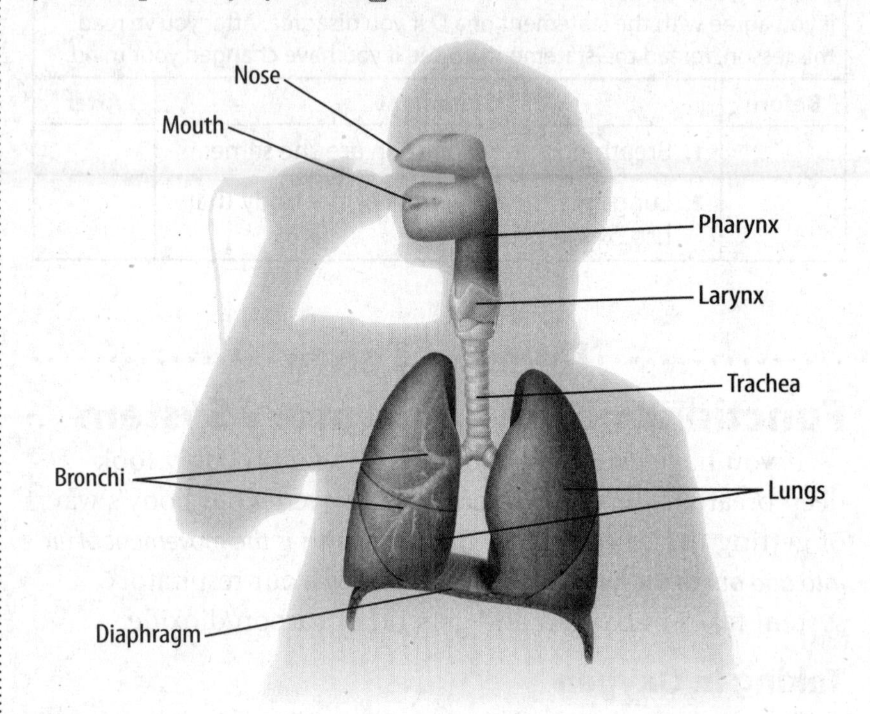

Pharynx

After air enters your mouth and nose, it passes into your throat. *The* **pharynx** (FER ingks) *is a tubelike passageway at the top of your throat that receives air, food, and liquids from your mouth or nose.* The epiglottis (eh puh GLAH tus) is a flap of tissue at the bottom of the pharynx. It keeps food and liquids out of the rest of your respiratory system.

Larynx and Trachea

Air passes from your pharynx into a triangle-shaped area called the voice box, or **larynx** (LER ingks). Your vocal cords are in your larynx. They consist of two thick folds of tissue. Your vocal cords vibrate and make sounds as air passes over them.

Air moves from your larynx into your trachea (TRAY kee uh). *Your* **trachea** *is a tube that is held open by C-shaped rings of cartilage.*

Bronchi and Lungs

Your trachea branches into two narrower tubes called **bronchi** (BRAHN ki) (singular, bronchus) *that lead into your lungs.* **Lungs** *are the main organs of the respiratory system.* Inside your lungs, your bronchi branch into smaller and narrower tubes called bronchioles.

Alveoli

In your lungs, your bronchioles end in *microscopic sacs, or pouches, called* **alveoli** (al VEE uh li) (singular, alveolus), *where gas exchange occurs.* During gas exchange, oxygen from the air you breathe moves into your blood. Carbon dioxide from your blood moves into your alveoli.

Look at the alveoli in the figure below. Alveoli look like bunches of grapes at the ends of the bronchioles. Like tiny balloons, your alveoli fill with air when you breathe in. They contract and release air when you breathe out. Notice in the figure below that blood vessels surround each alveolus. The walls of alveoli are only one cell thick. Red blood cells drop off carbon dioxide and pick up oxygen as they move through these blood vessels.

The thin walls of alveoli and the large surface areas make it possible for a high rate of gas exchange. If you could spread out all the alveoli in your lungs onto a flat surface, they would cover an area bigger than your classroom. Every time you breathe, alveoli help your body take in billions of oxygen molecules. They also help your body get rid of billions of carbon dioxide molecules.

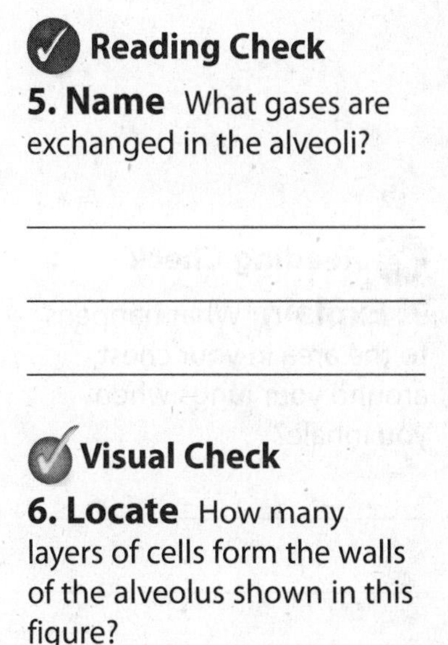

Reading Check

5. Name What gases are exchanged in the alveoli?

Visual Check

6. Locate How many layers of cells form the walls of the alveolus shown in this figure?

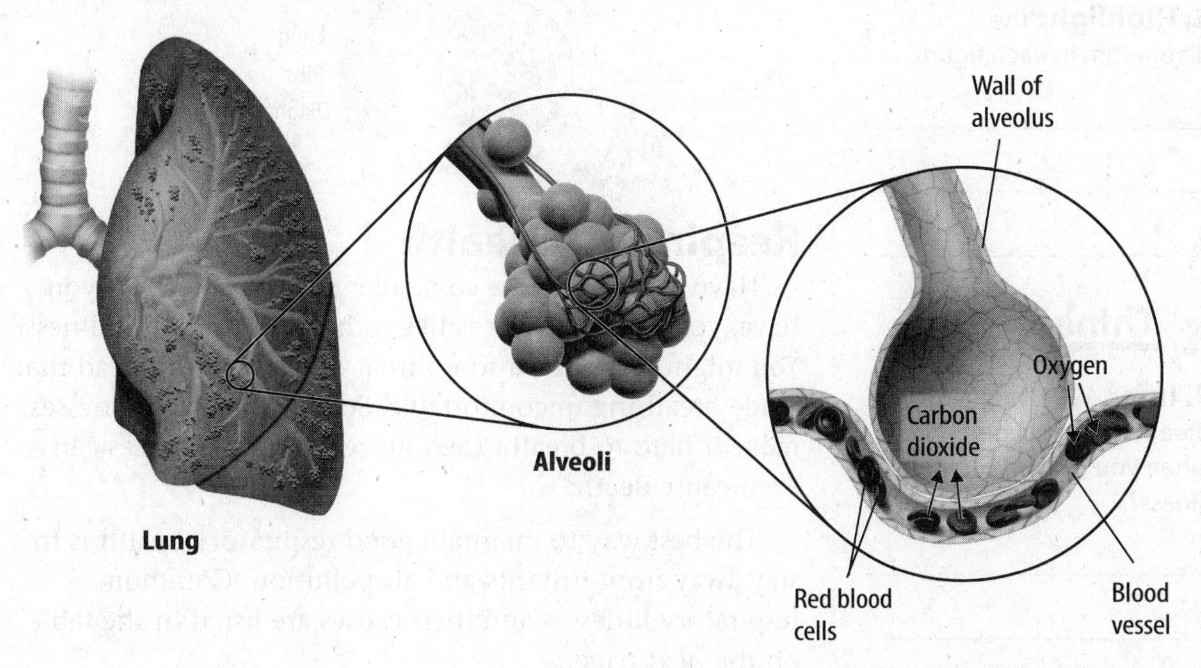

Gas exchange in alveoli

Breathing and Air Pressure

How does your body know when to breathe? When high levels of carbon dioxide build up in your blood, your nervous system tells your body to breathe out, or exhale. After you exhale, you breathe in, or inhale. How does this happen?

Below your lungs is a large muscle called a **diaphragm** (DI uh fram) *that contracts and relaxes and moves air in and out of the lungs*. Air moves into and out of your lungs with the help of your diaphragm. When your diaphragm moves, it changes the air pressure inside your chest. Breathing occurs because of these changes in air pressure.

When you inhale, your diaphragm contracts and moves down, as shown below. The space around your lungs gets bigger, or expands. The increased space reduces the air pressure in your chest. Air rushes into your lungs until the pressure inside your chest equals the air pressure outside it. ✓

When you exhale, your diaphragm relaxes and moves up. The space around your lungs gets smaller, or reduces. Air pressure in your chest increases. Waste gases rush out of your lungs.

Reading Check

7. Explain What happens to the area in your chest around your lungs when you inhale?

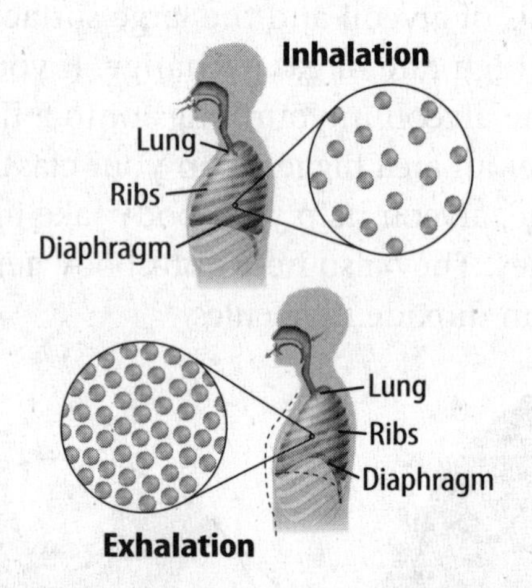

Visual Check

8. Highlight the diaphragm in each figure.

Respiratory Health

Have you ever had a cold, allergies, or asthma? If you have, you know what it is like to have a respiratory illness. You might have had a sore throat or a stuffed-up head that made breathing uncomfortable. Some respiratory illnesses make it hard to breathe. Serious respiratory illnesses can even cause death.

The best way to maintain good respiratory health is to stay away from irritants and air pollution. Common respiratory illnesses and their causes are listed in the table on the next page.

Think it Over

9. Infer Why might breathing be uncomfortable when you have a respiratory illness?

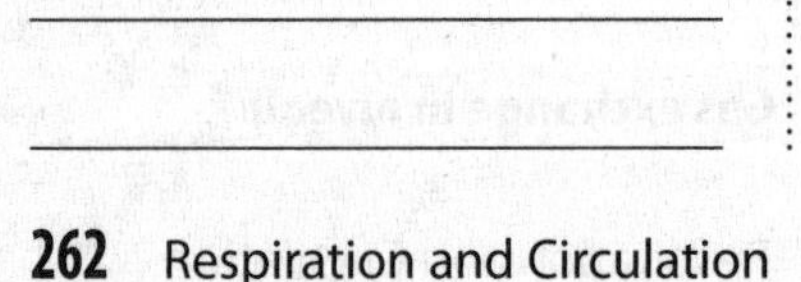

Respiratory Illnesses		
Illness	**Causes**	**Symptoms**
Colds, flu	viruses	congestion, runny nose, watery eyes, coughing, sneezing
Bronchitis (brahn KI tus)	viruses, bacteria	coughing and tiredness resulting from mucus blocking the bronchi and bronchioles, slowing air movement
Pneumonia (noo MOH nyuh)	viruses, bacteria	trouble breathing as a result of fluid in the alveoli that slows gas exchange
Asthma (AZ muh)	dust, smoke, pollen, pollution	trouble breathing as a result of swollen airways and increased mucus
Emphysema (em fuh SEE muh)	smoking	coughing, tiredness, loss of appetite, and weight loss resulting from destruction of alveoli
Lung cancer	smoking	coughing, trouble breathing, chest pain

The Respiratory System and Homeostasis

You have read in this lesson that the muscular system works with your respiratory system so you can breathe. The muscles and respiratory system working together bring oxygen into your lungs and remove carbon dioxide from your lungs.

In the next lesson, you will learn how the circulatory system and the respiratory system work together to bring oxygen to body cells. These systems also work to remove carbon dioxide from cells. The muscular system, the respiratory system, and the circulatory system help maintain homeostasis in your body.

Visual Check

10. Analyze According to the table, which respiratory illness involves the destruction of alveoli?

Key Concept Check

11. Describe How do the respiratory and muscular systems work together to maintain your body's homeostasis?

After You Read

Mini Glossary

alveoli (al VEE uh li): the microscopic sacs, or pouches, at the ends of bronchioles in the lungs, where gas exchange occurs (singular, alveolus)

breathing: the movement of air into and out of the lungs

bronchi (BRAHN ki): the two narrower tubes branching from the trachea that lead into the lungs (singular, bronchus)

diaphragm (DI uh fram): a large muscle below the lungs that contracts and relaxes and moves air in and out of the lungs

larynx (LER ingks): a triangle-shaped area where air passes from the pharynx; also called the voice box

lungs: the main organs of the respiratory system

pharynx (FER ingks): a tubelike passageway at the top of the throat that receives air, food, and liquids from the mouth or nose

trachea (TRAY kee uh): a tube past the larynx that is held open by C-shaped rings of cartilage

1. Review the terms and their definitions in the Mini Glossary. Write a sentence that explains the relationship between the alveoli and the bronchi.

____

____

2. Fill in the blanks in the diagram below to create a "map" of your respiratory system.

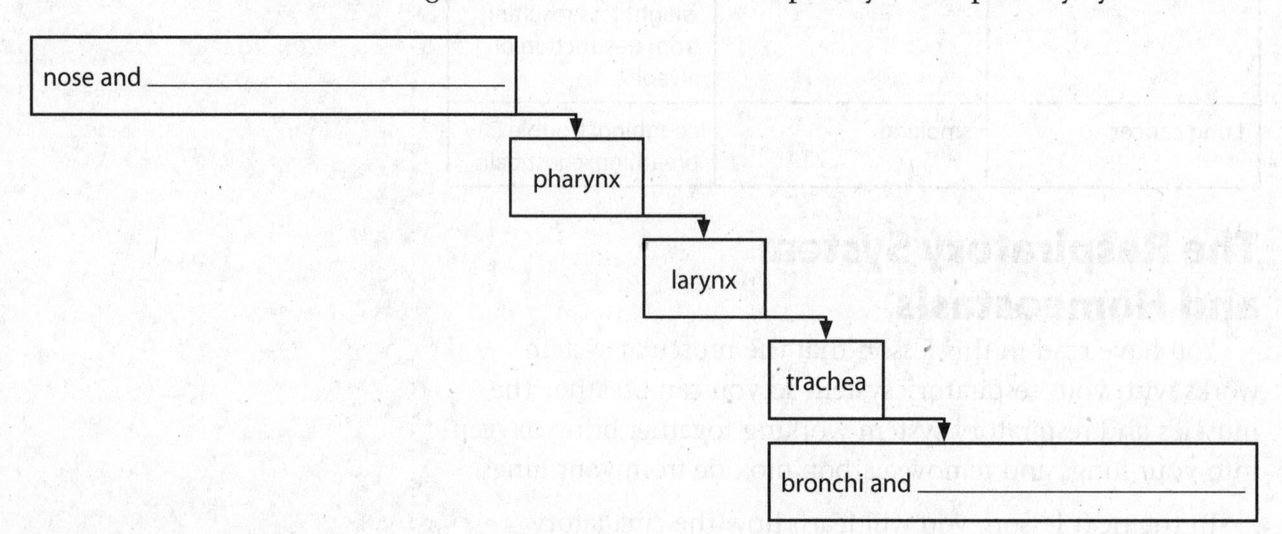

3. Compare breathing and cellular respiration.

____

____

What do you think NOW?

Reread the statements at the beginning of the lesson. Fill in the After column with an A if you agree with the statement or a D if you disagree. Did you change your mind?

Log on to ConnectED.mcgraw-hill.com and access your textbook to find this lesson's resources.

Respiration and Circulation

The Circulatory System

Before You Read

What do you think? Read the two statements below and decide whether you agree or disagree with them. Place an A in the Before column if you agree with the statement or a D if you disagree. After you've read this lesson, reread the statements to see if you have changed your mind.

Before	Statement	After
	3. There are four chambers in a human heart.	
	4. Blood travels in both directions in veins.	

Key Concepts 

- What does the circulatory system do?
- How do parts of the circulatory system work together?
- How does the circulatory system interact with other body systems?

Read to Learn

Functions of the Circulatory System

Have you ever looked at a road map of the United States? A complex network of roads and highways crisscrosses the country. This network of roads is important for transporting people and materials from one place to another. The circulatory (SUR kyuh luh tor ee) system is similar to this road network. Your circulatory system is important for transporting materials from one part of your body to another.

Study Coach

Make Flash Cards Write each boldfaced word on one side of a flash card. Write the definition on the other side. Use the cards to quiz yourself on the words and their meanings.

Transportation

Trucks move food, fuel, and other products from factories and farms to markets and businesses around the country. Your circulatory system is like the network of roads that the trucks travel on. Your blood cells are like the vehicles that travel on these roads. In your circulatory system, blood carries food, water, oxygen, and other materials to your body's cells and tissues.

Elimination

Blood carries away waste materials, just as garbage trucks haul away trash. As blood travels through your circulatory system, it picks up carbon dioxide. Remember that carbon dioxide is produced during cellular respiration. Blood also picks up wastes produced by all the other chemical reactions that take place inside your cells.

Key Concept Check

1. Describe What does the circulatory system do?

____

____

____

____

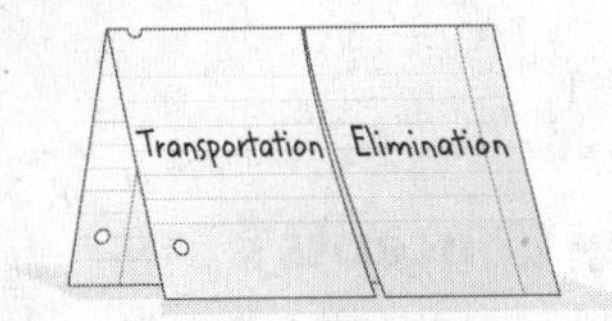

FOLDABLES

Make a two-tab book to use for taking notes about the circulatory system and its organs.

Visual Check

2. Identify the chamber of the heart in the bottom left of the figure. What is it called?

____

____

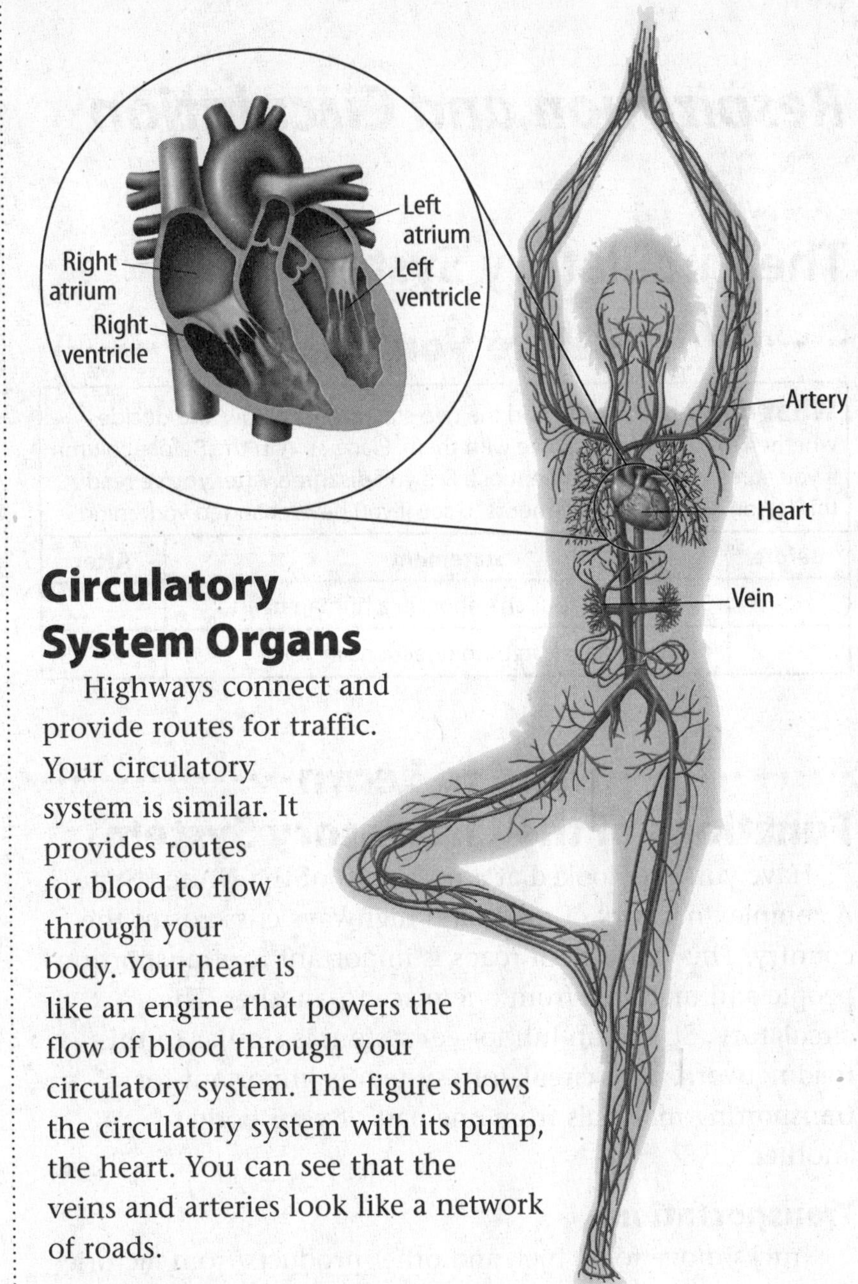

Circulatory System Organs

Highways connect and provide routes for traffic. Your circulatory system is similar. It provides routes for blood to flow through your body. Your heart is like an engine that powers the flow of blood through your circulatory system. The figure shows the circulatory system with its pump, the heart. You can see that the veins and arteries look like a network of roads.

The Heart

Can you feel your heart beating? Even when you do not notice it, your heart is at work. The heart is a muscle that pushes blood through the circulatory system. A human heart beats an average of 70 to 75 times per minute, every minute of life. It slows when you sleep. It speeds up when you exercise or are scared.

Reading Check

3. Discuss What does the heart do?

____

____

____

Your heart has four main chambers—two upper chambers and two lower chambers. Look at the figure above to see the chambers. *Blood enters the upper two chambers of the heart, called the* **atria** (AY tree uh) (singular, atrium). *Blood leaves the heart through the lower two chambers of the heart, called the* **ventricles** (VEN trih kulz).

Blood Vessels

Blood travels through your blood vessels and reaches every cell in your body. There are three main types of blood vessels. All three are shown in the figure below.

Arteries *A vessel that takes blood away from the heart is an* **artery.** Look at the artery in the figure below. Blood pressure in the arteries is high because arteries are near the pumping action of the heart. Artery walls are thick and can stand up to the high pressure of the flowing blood. The aorta is the largest artery. It carries a large volume of blood. Arteries branch into smaller vessels called arterioles.

Capillaries Arterioles branch into tiny capillaries, as shown below. **Capillaries** *are tiny blood vessels that deliver supplies to individual cells and take away waste materials.* Capillaries are the smallest blood vessels in the circulatory system. Many capillary walls are only one cell thick. Thin walls make it possible for molecules of oxygen, food, water, and waste products to move between blood and body cells.

Veins *A vessel that brings blood toward the heart is a* **vein.** The pressure in veins is lower than in arteries. This is because capillaries separate veins from the pumping action of the heart. Because there is less pressure in the veins, there is a greater chance that blood could flow backward. Veins have one-way valves that prevent blood from moving backward and keep it moving toward the heart.

Capillaries join and form larger vessels called venules. Venules join and form veins. The inferior vena cava is the largest vein. It carries blood from the lower half of your body to your heart.

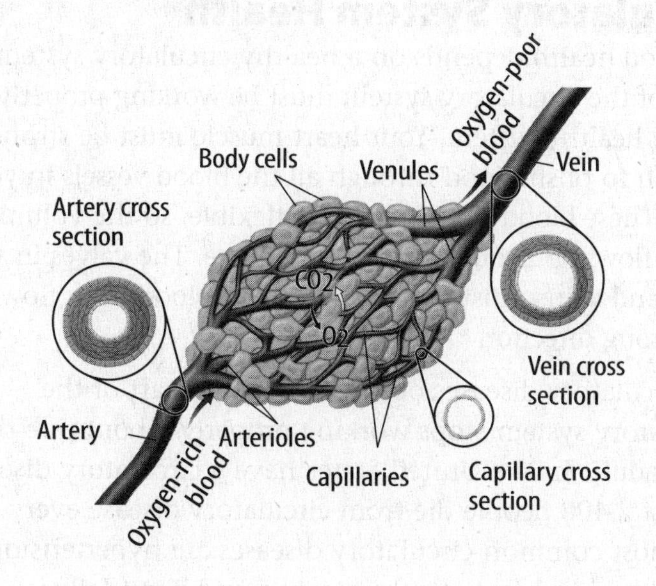

Key Concept Check

4. Explain How do the heart and the blood vessels work together?

Visual Check

5. Compare Which is thicker—the artery walls or the vein walls?

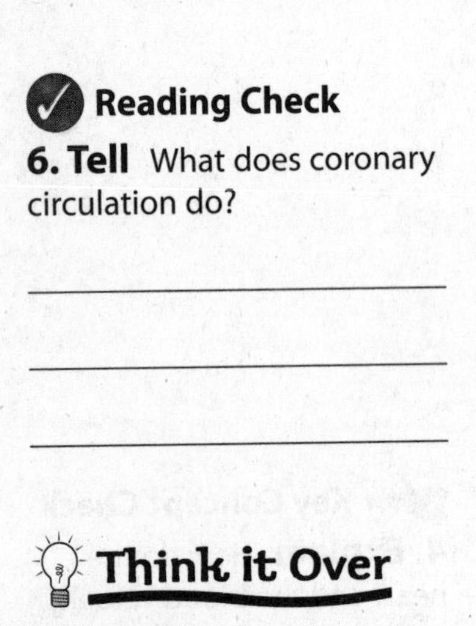

Reading Check

6. Tell What does coronary circulation do?

Think it Over

7. Contrast How do pulmonary and systemic circulation differ?

Types of Circulation

Your circulatory system moves blood throughout your entire body. There are three different types of circulation. The first type takes blood to the body. The second type takes blood to the heart. The third type takes blood to and from the lungs.

Systemic Circulation

Blood leaves your heart and travels to your body. **Systemic circulation** *is the network of vessels that carry blood from the heart to the body and from the body back to the heart.*

Coronary Circulation

The heart is a thick organ made of many layers of cells. Most heart cells do not come into contact with the blood inside the heart. *A network of arteries and veins called* **coronary circulation** *supplies blood to all the cells of the heart.* Coronary circulation provides oxygen and nutrients to the cells of the heart. It also removes carbon dioxide from the blood. Some of these vessels are inside the heart, while others are on the outside of the heart.

Pulmonary Circulation

Blood moves back and forth between the heart and the lungs. *The network of vessels that carries blood to and from the lungs is called* **pulmonary circulation.** Pulmonary circulation carries oxygen-poor blood, or blood low in oxygen, from the heart to the lungs. It also carries oxygen-rich blood, or blood high in oxygen, from the lungs back to the heart. Blood that enters the heart from the lungs is then pushed to the rest of the body.

Circulatory System Health

Good health depends on a healthy circulatory system. All parts of the circulatory system must be working properly to have a healthy system. Your heart muscle must be strong enough to push blood through all the blood vessels in your body. These blood vessels must be flexible, so the volume of blood flowing through them can change. The valves in your heart and veins must work well to keep blood from flowing in the wrong direction.

Circulatory diseases occur when some part of the circulatory system stops working properly. About one-third of all adults in the United States have a circulatory disease. Almost 2,400 people die from circulatory disease every day. The most common circulatory diseases are hypertension, athersclerosis, heart attacks, strokes, and heart failure.

Hypertension

When the ventricles of the heart contract, they push blood into the arteries. The arteries bulge a little because blood presses against their sides. This bulging of an artery is what you feel when you check your pulse. This pressure is called blood pressure.

Normal blood pressure is considered to be 120 mm Hg (millimeters of mercury) or less during the contraction of the ventricles. It is 80 mm Hg or less after the contraction. Normal blood pressure is written as 120/80 mm Hg. Blood pressure higher than 140/90 mm Hg is called hypertension, or high blood pressure. Hypertension can weaken the artery walls and make them less flexible.

Atherosclerosis

Atherosclerosis (a thuh roh skluh ROH sus) *is the buildup of fatty material within the walls of arteries*. Fat deposits can keep blood from flowing well in the arteries. The deposits can also break loose, flow to a narrower artery, and block it. The figure below shows the results. A blockage in the heart can cause a heart attack. A blockage in a blood vessel in the brain can cause a stroke. 

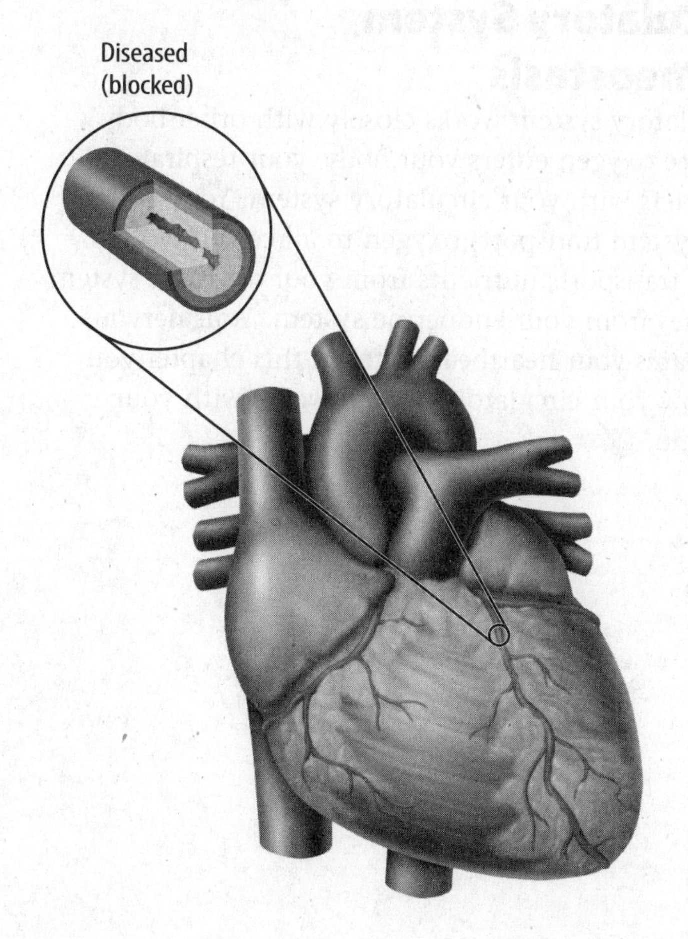

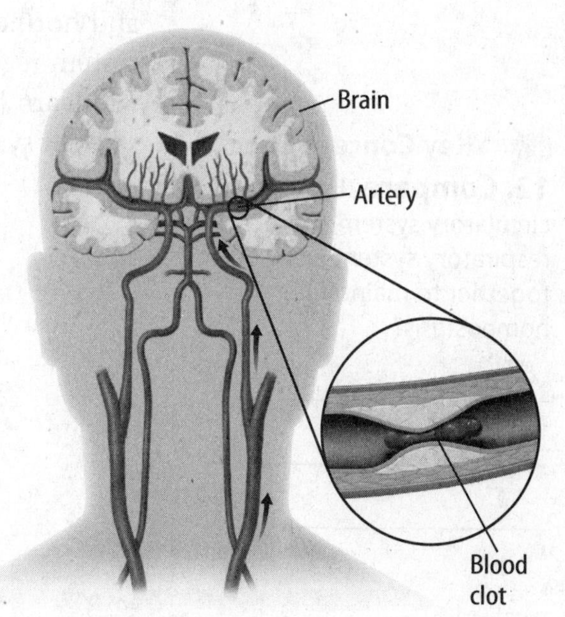

Think it Over

8. Apply A neighbor says that she has high blood pressure. What is another name for that condition?

____

____

Reading Check

9. Identify What happens when an artery in the brain is blocked?

____

____

Visual Check

10. State What circulatory diseases are shown in the figures?

____

____

Heart Attacks, Strokes, and Heart Failure

A heart attack happens when part of the heart muscle dies or is damaged. Heart attacks occur when not enough oxygen reaches cells in the heart. Most heart attacks occur when a blood vessel in the heart is blocked.

11. Cause and Effect What causes most heart attacks? (Circle the correct answer.)

a. lack of oxygen

b. blocked blood vessel

c. damage to the heart muscle

A stroke happens when part of the brain dies or is damaged. Most strokes are caused when not enough oxygen reaches cells in the brain. A stroke might occur if a blood clot blocks a blood vessel in the brain.

Heart failure occurs when the heart is not working as well as it should. It can happen because of a heart attack, a problem with heart valves, or diseases that damage the heart.

Preventing Circulatory System Disorders

Some risk factors for circulatory system diseases cannot be avoided. For example, if one of your parents has a circulatory disease, you might have a slightly higher risk of getting a similar disease. However, most risk factors can be controlled by making good life choices. You can eat a healthful diet, control your weight, exercise, and not smoke.

ACADEMIC VOCABULARY
factor
(noun) something that helps produce a result

The Circulatory System and Homeostasis

The circulatory system works closely with other body systems. Once oxygen enters your body, your respiratory system interacts with your circulatory system. Your circulatory system transports oxygen to all cells in your body. It also transports nutrients from your digestive system and hormones from your endocrine system. Your nervous system regulates your heartbeat. Later in this chapter you will learn how your circulatory system works with your skeletal system.

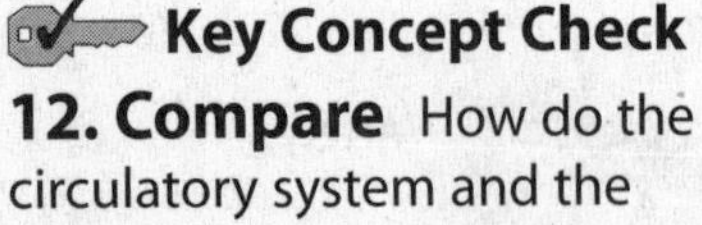

12. Compare How do the circulatory system and the respiratory system work together to maintain homeostasis?

After You Read

Mini Glossary

artery: a vessel that takes blood away from the heart

atherosclerosis (a thuh roh skluh ROH sus): the buildup of fatty material within the walls of arteries

atria (AY tree uh): the upper two chambers of the heart through which blood enters the heart (singular, atrium)

capillary: a tiny blood vessel that delivers supplies to individual cells and takes away their waste materials

coronary circulation: the network of arteries and veins that supplies blood to all the cells of the heart

pulmonary circulation: the network of vessels that carries blood to and from the lungs

systemic circulation: the network of vessels that carry blood from the heart to the body and from the body back to the heart

vein: a vessel that brings blood toward the heart

ventricles (VEN trih kulz): the lower two chambers of the heart through which blood leaves the heart

1. Review the terms and their definitions in the Mini Glossary. Write a sentence that compares arteries and veins.

2. Fill in the diagram below to show the functions of the circulatory system.

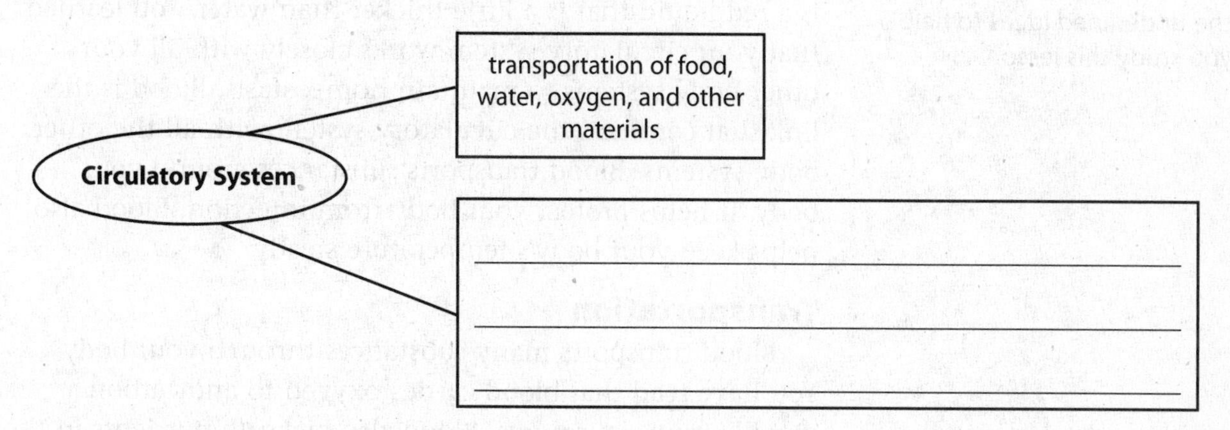

3. Describe how the three types of circulation carry blood throughout the body.

What do you think NOW?

Reread the statements at the beginning of the lesson. Fill in the After column with an A if you agree with the statement or a D if you disagree. Did you change your mind?

Log on to ConnectED.mcgraw-hill.com and access your textbook to find this lesson's resources.

Respiration and Circulation

Blood

Key Concepts

- What does the blood do?
- How do the parts of the blood differ?

Before You Read

What do you think? Read the two statements below and decide whether you agree or disagree with them. Place an A in the Before column if you agree with the statement or a D if you disagree. After you've read this lesson, reread the statements to see if you have changed your mind.

Before	Statement	After
	5. All blood cells are red.	
	6. Blood plasma is just water.	

Mark the Text

Find the Main Idea Find and underline the main idea in each paragraph. Review the underlined ideas to help you study this lesson.

Read to Learn

Functions of Blood

Have your ever had an injury that caused bleeding? Blood is a red liquid that is a little thicker than water. You learned that your circulatory system works closely with all your other body systems to maintain homeostasis. Blood is the link that connects the circulatory system with all the other body systems. Blood transports substances around your body. It helps protect your body from infection. Blood also helps keep your body's temperature steady.

Transportation

Blood transports many substances through your body. You have read that blood carries oxygen to and carbon dioxide from your lungs. Blood also picks up nutrients in the small intestine and carries them to all body cells. It transports hormones that are produced by the endocrine system. Blood carries waste products to the excretory system. Most of the substances are dissolved in the liquid part of blood.

Protection

Some blood cells fight infection. They help protect you from harmful organisms, such as bacteria, viruses, fungi, and parasites. Blood also contains materials that help repair torn blood vessels and heal wounds. When you get a cut or a scrape, materials in your blood help protect your body from losing too much blood.

Reading Check

1. Explain In what ways does your blood protect you?

____

____

____

Temperature Regulation

Blood helps your body stay at a temperature of about 37°C. When your body temperature is too high, blood vessels near the surface of your skin widen. This increases blood flow to your skin's surface and releases more thermal energy into the air. Your body cools down. When your body temperature lowers, blood vessels at your skin's surface get narrower. This decreases blood flow to your skin's surface and reduces the amount of thermal energy that is lost to the air. Your body warms up.

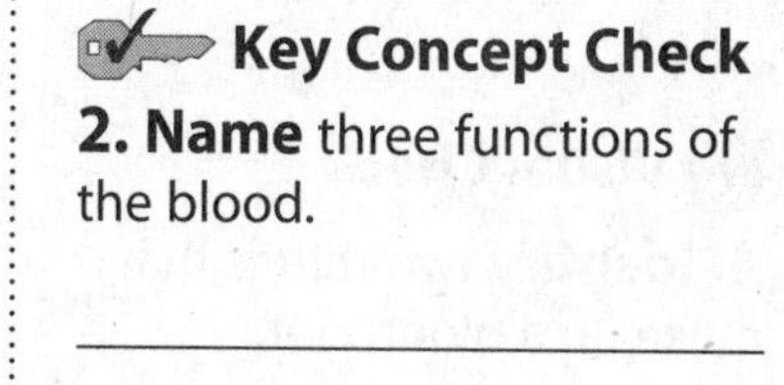

Key Concept Check

2. Name three functions of the blood.

____

____

____

Parts of Blood

Blood is a tissue because it is made up of different kinds of cells that work together. The figure shows blood's four main parts: red blood cells, white blood cells, platelets, and plasma. Most adults have about 70 mL of blood per kilogram of body weight. An average adult has about five to six liters of blood.

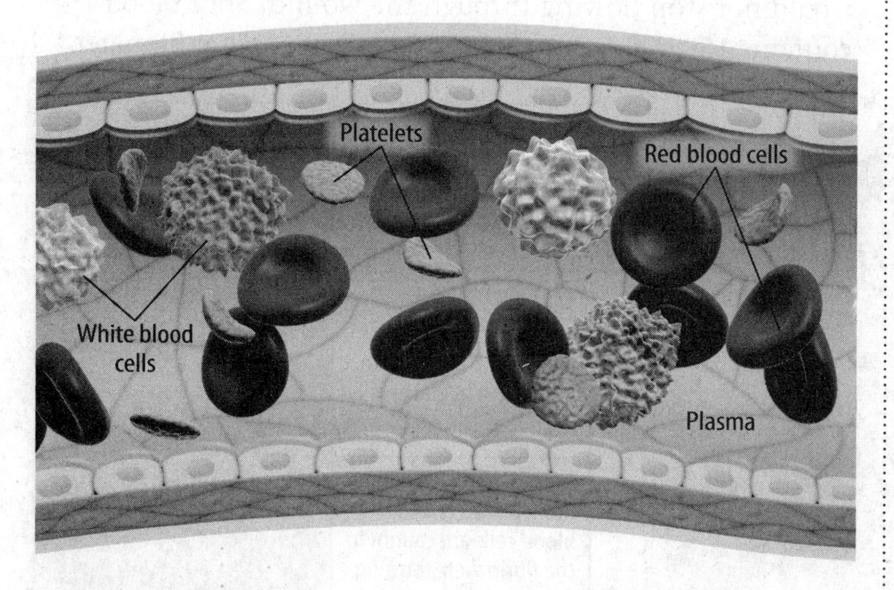

Visual Check

3. Locate the four main parts of blood. Circle an example of each of them.

Red Blood Cells

Every cubic milliliter of your blood contains four to six million red blood cells, or erythrocytes (ih RITH ruh sites). Red blood cells are made mostly of iron-rich protein molecules called hemoglobin (HEE muh gloh bun). In the alveoli of the lungs, oxygen attaches to the hemoglobin. The hemoglobin releases the oxygen when red blood cells enter the capillaries and get close to body cells.

Look at the red blood cells in the figure above. How would you describe their shape? You might say that they look like doughnuts without holes. This flattened disk shape gives red blood cells more surface area. They can carry more oxygen than they could if they were round like a ball. Red blood cells wear out after a few months, so your body produces new red blood cells all the time.

FOLDABLES®

Make a four-door book to organize information about the parts of blood and their functions.

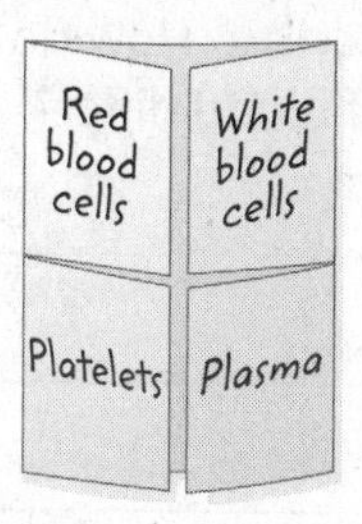

White Blood Cells

Your blood contains several kinds of white blood cells, or leukocytes (LEW kuh sites). White blood cells protect your body from illness and infection. Some attack viruses, bacteria, fungi, and parasites that might invade your body.

Most white blood cells last only a few days. Your body is always replacing them. You have fewer white blood cells—5,000 to 10,000 per cubic millimeter—than red blood cells.

Visual Check

4. Identify two things that make up a blood clot.

Platelets

What happens if you get a cut? The cut, or wound, bleeds for a short time. Then the blood clots, as shown below.

Platelets *are small, irregularly shaped pieces of cells in the blood that plug wounds and stop bleeding.* Platelets produce proteins that help make the plug stronger. Without platelets, blood would not stop flowing through the wound. Your blood contains 150,000 to 440,000 platelets per cubic millimeter.

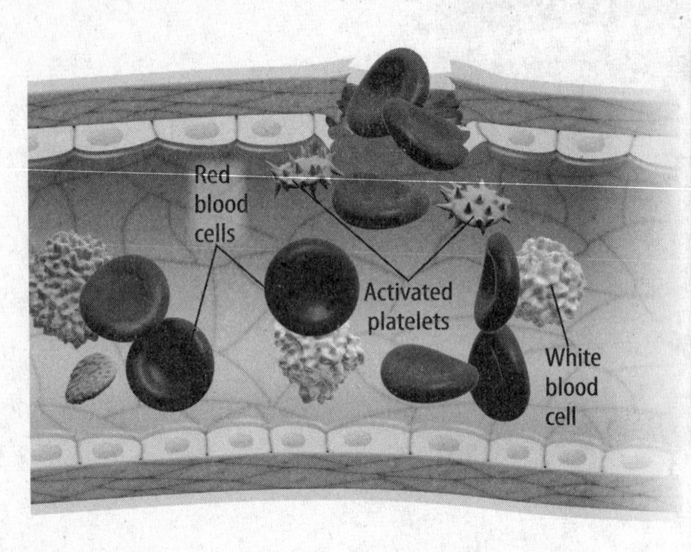

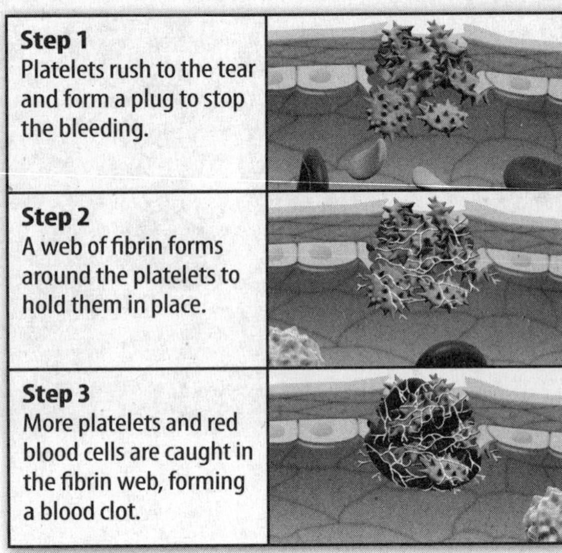

Plasma

The yellowish, liquid part of blood, called **plasma,** *transports blood cells.* Plasma is 90 percent water. It helps thin the blood. Blood has to be thin to move through small blood vessels. Plasma contains many dissolved molecules that travel along in the blood. They include salts, vitamins, sugars, minerals, proteins, and cellular wastes.

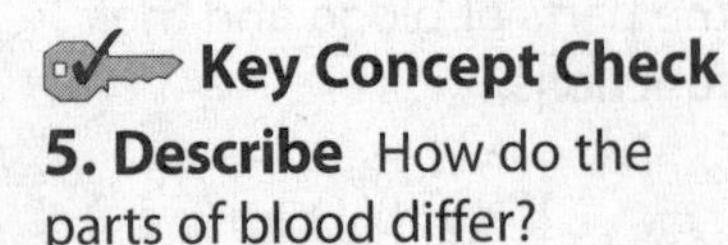

Key Concept Check

5. Describe How do the parts of blood differ?

Plasma also helps control the activities of cells in your body. Plasma carries chemical messengers that control the amounts of salts and glucose that enter cells.

Blood Types

Do you know someone who has donated blood? Doctors use donated blood to help people who have lost too much blood from an injury or surgery. A blood transfusion is the transfer of one person's blood to another person. All human blood has the same four parts—red blood cells, white blood cells, platelets, and plasma. But you cannot receive a blood transfusion from just anyone. Different people have different blood types.

The ABO System

You inherited your blood type from your parents. Blood type refers to the type of proteins, or antigens, on red blood cells. The table below shows the four human blood types: A, B, AB, and O. As you can see, type A blood cells have the A antigen. Type B blood cells have the B antigen. Type AB blood cells have both A and B antigens. Type O blood cells have no antigens.

Visual Check

6. Solve Which blood type has more types of clumping proteins than the others?

Blood Type	Type A	Type B	Type AB	Type O
Antigens on red blood cells	A A A A A A	B B B B B B	B A A B B A	
Percentage of US population with this blood type	42	10	4	44
Clumping proteins in plasma	Anti-B	Anti-A	None	Anti-A and anti-B
Blood type(s) that can be RECEIVED in a transfusion	A or O	B or O	A or B or AB or O	O only
This blood type can DONATE TO these blood types	A or AB	B or AB	AB only	A or B or AB or O

If different antigens are introduced through a blood transfusion, the red blood cells will clump together and no longer function. Clumps form because of clumping proteins in blood plasma, shown in the table above. The type of clumping proteins in your blood determines what blood type you could safely get in a transfusion.

A, B, and O blood types have clumping proteins in their plasma. A person with type A blood has anti-B clumping proteins that attack type B antigens and cause type B red blood cells to clump together. Type AB blood has no clumping proteins. People with type AB blood can receive any blood type because it has no clumping proteins. Type O blood has anti-A and anti-B proteins. People with type O blood can donate blood to anyone.

Reading Check

7. State What usually happens when two different types of blood mix?

The Rh Factor

Another protein found on red blood cells is a chemical marker called the **Rh factor.** Some people have this protein on their red blood cells. People who have this protein are Rh positive. People without this protein are Rh negative. If Rh positive blood mixes with Rh negative blood, clumping can result. Blood types usually have a plus (+) or a negative (–) sign to show whether the person is Rh positive or negative. For example, a person with an A+ blood type has red blood cells with A antigens and the Rh factor. Someone with O– blood has no antigens and no Rh factor.

Reading Check

8. Determine What kinds of antigens are found in AB+ blood?

____

____

____

Blood Disorders

Sometimes a person's blood does not function as it should. People with hemophilia do not have a protein needed to clot blood. They bleed at the same rate as other people. However, their bleeding does not stop as quickly as it does for other people.

People with anemia have low numbers of red blood cells or have red blood cells that do not contain enough hemoglobin. Their blood might not carry as much oxygen as their bodies need.

Bone marrow is the soft tissue in the center of bones. It produces red blood cells. Cancer of the bone marrow is called leukemia. Leukemia can slow or prevent blood cell formation. Leukemia can lead to anemia and a damaged immune system.

People who inherit sickle-cell disease have red blood cells shaped like crescents, or sickles (old-fashioned farm tools with curved blades). As shown in the figure below, sickle-shaped cells do not move through blood vessels as easily as normal, disk-shaped cells do. They form clumps that can block blood vessels. Sickle cells can keep oxygen from reaching tissues and cause sickle-cell anemia.

Math Skills

If percentages refer to the same factor, they can be added or subtracted. For example, you could add the percentages of people with each of the four blood types:

42% + 10% + 4% + 44% = 100%

You could also subtract to find the percentage of people who do not have type O blood:

100% – 44% = 56%

9. Use Percentages Forty-four percent of people have type O blood. If 7 percent of people have type O blood and are Rh negative, what percent has type O Rh positive blood?

____

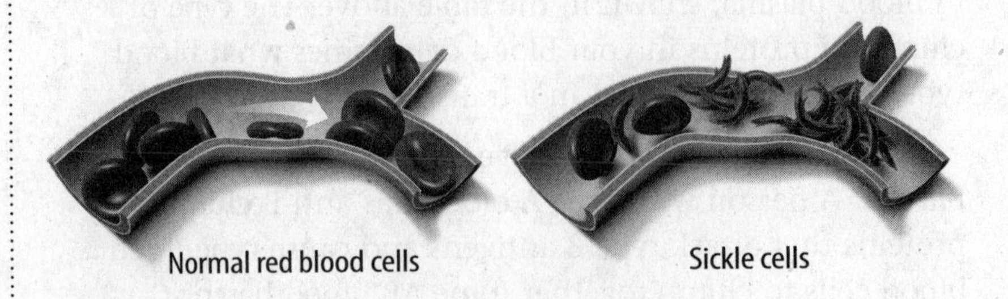

Visual Check

10. Explain Why doesn't blood flow smoothly in a person with sickle-cell disease?

____

____

After You Read

Mini Glossary

plasma: the yellowish, liquid part of blood that transports blood cells

platelets: small, irregularly shaped pieces of cells in the blood that plug wounds and stop bleeding

Rh factor: a protein found on red blood cells that is a chemical marker

1. Review the terms and their definitions in the Mini Glossary. Write a sentence using your own words to explain the role of platelets in the formation of blood clots.

2. Fill in the table below to identify the parts of blood.

Parts of Blood	What They Do
Red blood cells	
	protect from illness and infection
Platelets	
	thin the blood; carry dissolved molecules around the body

3. Imagine that you have one of the blood disorders discussed in this lesson. Write a paragraph to explain your disorder and tell how it affects your blood.

What do you think NOW?

Reread the statements at the beginning of the lesson. Fill in the After column with an A if you agree with the statement or a D if you disagree. Did you change your mind?

Log on to ConnectED.mcgraw-hill.com and access your textbook to find this lesson's resources.

Respiration and Circulation

The Lymphatic System

Key Concepts
- What does the lymphatic system do?
- How do the parts of the lymphatic system work together?
- How does the lymphatic system interact with other body systems?

Before You Read

What do you think? Read the two statements below and decide whether you agree or disagree with them. Place an A in the Before column if you agree with the statement or a D if you disagree. After you've read this lesson, reread the statements to see if you have changed your mind.

Before	Statement	After
	7. Lymph nodes are only in the neck.	
	8. The lymphatic system helps fight infections to maintain a healthy body.	

Study Coach

Make an Outline
Summarize the information in this lesson. The headings should be the main points in the outline. Under each heading, list the details or examples that help explain the main idea.

Read to Learn

Functions of the Lymphatic System

When you are sick, you might notice small, swollen structures under your jaw. These structures in your neck can become swollen as they work to fight off an infection in your body.

The **lymphatic system** *is part of the immune system and helps destroy microorganisms that enter the body.* Your lymphatic system works closely with your circulatory system. Both systems move liquids through the body, and both contain white blood cells. However, the lymphatic system and the circulatory system do different tasks. The lymphatic system has four main functions:

- It absorbs some of the tissue fluid that collects around cells.
- It absorbs fats from the digestive system and transports them to the circulatory system.
- It filters dead cells, viruses, bacteria, and other unneeded particles from tissue fluid. After filtering, it returns the tissue fluid to the circulatory system.
- It helps fight off illnesses and infections. The lymphatic system includes structures in which your body's white blood cells develop.

Key Concept Check
1. Explain What does the lymphatic system do?

____

____

____

____

____

Parts of the Lymphatic System

The lymphatic system is a network of vessels and organs that runs throughout your body. The lymphatic system includes lymph vessels. The lymph vessels carry fluid. The fluid is also part of the system. When the fluid reaches the area beneath the collarbone, it re-enters the circulatory system. The lymphatic system includes several other structures, as shown below. Refer to the figure as you read about the parts of the lymphatic system.

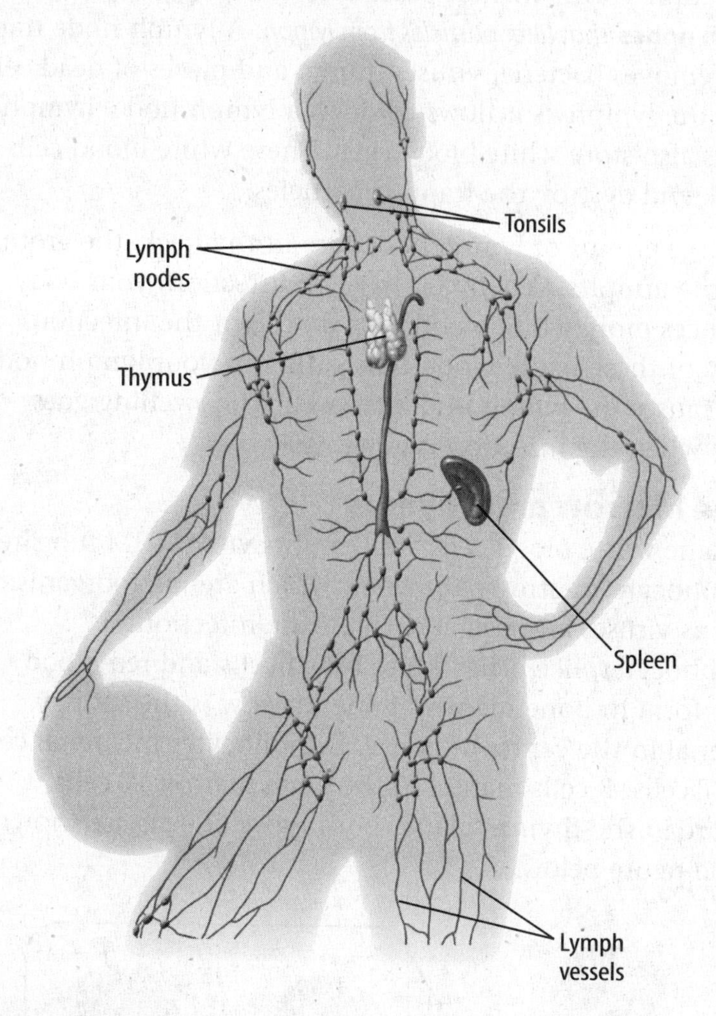

Lymph

Water, white blood cells, and dissolved materials such as salts and glucose leak out of capillary walls. They move into the space that surrounds tissue cells. This fluid is called tissue fluid.

The cells absorb the materials they need from tissue fluid and release wastes into it. About 90 percent of the tissue fluid is reabsorbed by the capillaries. *About 10 percent of the tissue fluid is absorbed by the lymph vessels and is called* ***lymph.***

SCIENCE USE V. COMMON USE

vessel

Science Use a tube through which a body fluid travels

Common Use a container for holding something

FOLDABLES®

Fold a sheet of paper into a small book and use it to organize your notes about the parts of the lymphatic system and their functions.

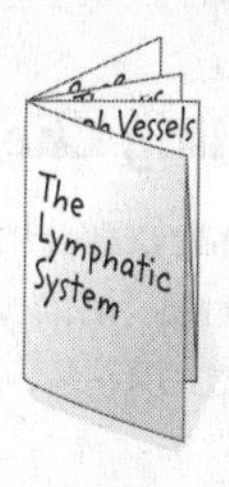

Visual Check

2. Identify What organs of the lymphatic system are in the throat?

Reading Check

3. Define What is lymph?

Lymph Vessels

The lymphatic system forms a network of lymph vessels. The network of lymph vessels looks like the network of blood vessels in the circulatory system. Lymph vessels absorb and transport lymph. Lymph is pushed through the lymph vessels by contractions of the muscles in your body. The heart does not pump lymph through the body.

Lymph Nodes

Lymph vessels include *clusters of small, spongy structures called* **lymph nodes** *that filter particles from lymph*. A lymph node traps and removes bacteria, viruses, fungi, and pieces of dead cells from the lymph as it flows through a lymph node. Lymph nodes also store white blood cells. These white blood cells attack and destroy the trapped particles.

Large groups of lymph nodes are in the neck, the groin, and the armpits. When you have an infection, your body produces more white blood cells that fight the infection. Many of these white blood cells gather in your lymph nodes. This causes the lymph nodes to swell. The swelling goes away when the infection is gone. ✓

✓ Reading Check

4. Summarize How do the lymph nodes help your body fight infection?

____

____

____

Bone Marrow and Thymus

Some white blood cells are lymphocytes (LIHM fuh sites). Lymphocytes destroy pathogens, which are microorganisms such as viruses and bacteria that cause infection. Lymphocytes, like other white blood cells and red blood cells, form in bone marrow. Bone marrow is the spongy material in the center of bones. Lymphocytes include B cells and T cells. B cells mature in the bone marrow. T cells mature in the thymus gland. Both types of cells are shown in the figure below.

✓ Visual Check

5. Predict If a person has a disease that affects his or her bone marrow, what kinds of cells might be affected?

____

____

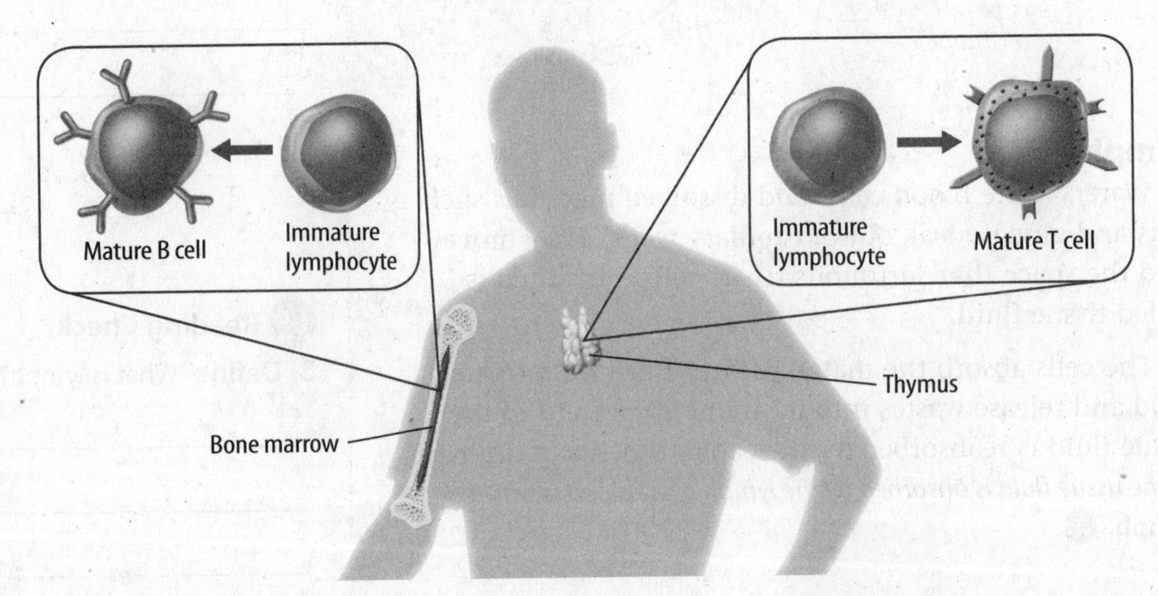

Thymus *The* **thymus** *is the organ of the lymphatic system in which T cells complete their development.* Immature T cells move from the bone marrow to the thymus.

B Cells and T Cells Fight Infection T cells in the thymus develop the ability to recognize and destroy body cells that have been infected by microorganisms. Mature B cells and T cells move into the lymph and blood where they help fight infection.

Spleen

You learned that red blood cells live for only a few months. *The* **spleen** *is an organ of the lymphatic system that recycles worn-out red blood cells and produces and stores lymphocytes.* The spleen also stores blood and platelets. If a person is injured and loses a lot of blood, the spleen can release stored blood and platelets into the circulatory system.

Tonsils

Tonsils are clusters of lymph tissue on the sides of your throat. They help protect your body from infection. They trap and destroy bacteria and other pathogens that enter your nose and mouth. However, your body can function without tonsils.

Lymph Diseases and Disorders

Damage to the lymphatic system from injury or surgery can keep tissue fluid from draining into lymph vessels. Tissue fluid can build up around cells and cause swelling. Lack of activity can also cause swelling. When the muscles do not push lymph through the lymph vessels, lymph can build up.

If the cells of your tonsils become infected, your tonsils will swell. This condition is tonsillitis.

The uncontrolled production of white blood cells is a type of cancer called lymphoma. Cancer of the lymph nodes is a related disease called Hodgkin's lymphoma.

The Lymphatic System and Homeostasis

The lymphatic system helps your body maintain homeostasis by keeping tissue fluids from building up around cells. It supports your circulatory system by cleaning fluids and replacing them in the bloodstream. It also helps you stay healthy by fighting infection.

Key Concept Check
6. Discuss How do bone marrow and the thymus work together?

Reading Check
7. Name a disease that involves the lymphatic system.

Key Concept Check
8. Identify How does the lymphatic system support the circulatory and immune systems?

After You Read

Mini Glossary

lymph: about 10 percent of the tissue fluid that is absorbed by the lymph vessels

lymph nodes: clusters of small, spongy structures in lymph vessels that filter particles from lymph

lymphatic system: part of the immune system that helps destroy microorganisms that enter the body

spleen: an organ of the lymphatic system that recycles worn-out red blood cells and produces and stores lymphocytes

thymus: the organ of the lymphatic system in which T cells complete their development

1. Review the terms and their definitions in the Mini Glossary. Write a sentence that summarizes the role of the spleen in the lymphatic system.

____

____

2. Complete the table below by describing the parts of the lymphatic system and what they do.

Parts of the Lymphatic System	
Lymph nodes	
Bone marrow and thymus	
Lymph	

3. How did making a detailed outline of the lesson help you keep track of the functions of the lymphatic system?

____

____

What do you think NOW?

Reread the statements at the beginning of the lesson. Fill in the After column with an A if you agree with the statement or a D if you disagree. Did you change your mind?

Log on to ConnectED.mcgraw-hill.com and access your textbook to find this lesson's resources.

Immunity and Disease

Diseases

Before You Read

What do you think? Read the two statements below and decide whether you agree or disagree with them. Place an A in the Before column if you agree with the statement or a D if you disagree. After you've read this lesson, reread the statements to see if you have changed your mind.

Before	Statement	After
	1. Some diseases are infectious, and others are noninfectious.	
	2. Cancer is an infectious disease.	

Key Concepts 

- Why do we get diseases?
- How do the two types of diseases differ?

Read to Learn

Disease Through History

Thousands of years ago, a doctor might have treated a disease by scraping a person's skull with a rock until a hole formed. The doctor made the hole in the skull to let the illness escape. Archaeologists know this because they have discovered skulls with smooth holes in them. Bone growth around the holes shows that people lived after these holes were made. This treatment may sound strange today. However, it was an accepted treatment for disease thousands of years ago.

Study Coach

Discuss Main Ideas Read the first paragraph about disease. Then take turns with a partner saying something about what you have learned. Repeat this process with the other paragraphs in this lesson.

Today we know that many diseases are caused by bacteria and viruses. *Disease-causing agents, such as bacteria and viruses, are called* **pathogens.** Pathogens have always caused illnesses. However, it has been only during the last few hundred years that scientists have understood the relationship between pathogens and diseases. Before then, little was known about disease and immunity, and superstitions were common. ✓

Reading Check

1. Identify List two types of pathogens.

____

____

Early Research on Diseases

During the eighteenth and nineteenth centuries, doctors learned a lot about the causes and treatment of diseases. The research and experiments performed by scientists such as Edward Jenner, John Snow, and Anton van Leeuwenhoek saved many lives.

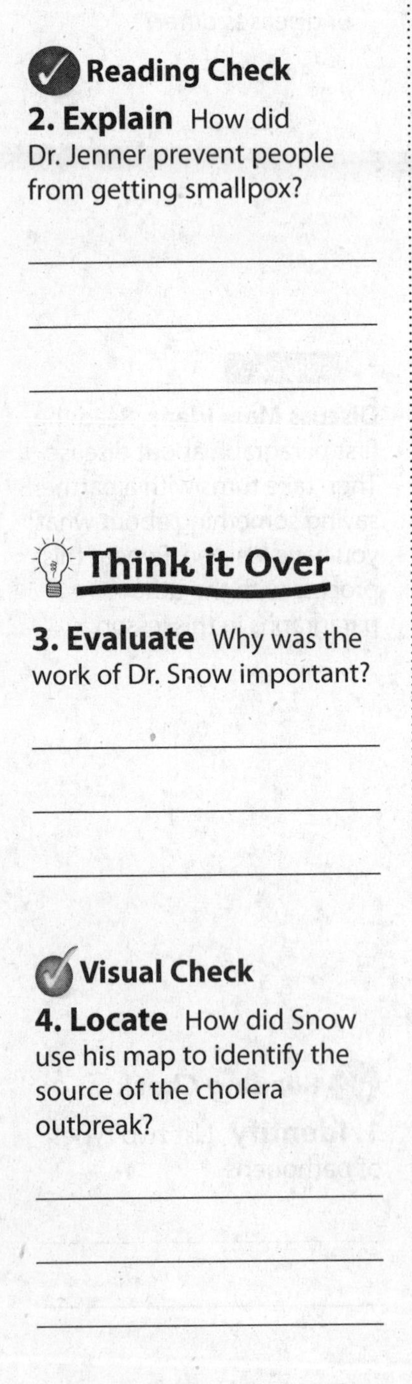

Reading Check

2. Explain How did Dr. Jenner prevent people from getting smallpox?

Think it Over

3. Evaluate Why was the work of Dr. Snow important?

Visual Check

4. Locate How did Snow use his map to identify the source of the cholera outbreak?

First Vaccination

In 1796, Edward Jenner developed the first vaccination. A vaccination is a procedure that helps the body defend itself against disease. Jenner knew that women who milked cows often got a mild disease called cowpox.

However, these women did not get smallpox, a deadly disease related to cowpox. Jenner, who was a doctor, cut the arm of a young boy and placed pus from a cowpox sore in the cut. Two weeks later, Jenner infected the boy with smallpox. Like the women who milked the cows, the boy never got sick from smallpox. Although the smallpox vaccine saved many lives, people did not understand why or how it worked.

Connecting Disease with a Source

In the mid-1800s, people realized that there was a connection between pathogens and disease. Cholera is a disease of the intestinal tract caused by a bacterium. Dr. John Snow connected the disease to a bacterium he found in water. He did this by tracking the origin of a cholera outbreak on a map similar to the one below. Using the map, Snow tracked the outbreak to a single water pump in the city of London. He had the pump closed. Almost immediately, the number of new cholera cases decreased. When Snow looked at the water, he saw a microscopic organism that he thought caused the disease. Not everyone agreed with Snow, but people were beginning to think that pathogens did exist.

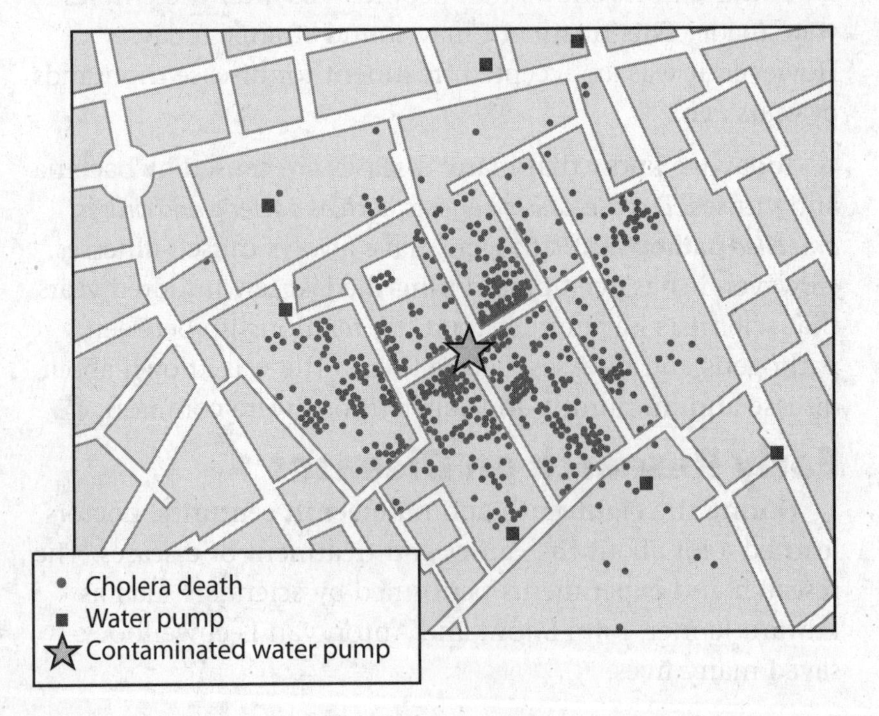

The Development of Microscopes

One of the reasons people were slow to accept the idea of pathogens is because people could not see the pathogens. However, the development of microscopes changed that. In the late 1600s, Anton van Leeuwenhoek (LAY vuh hook) made one of the first microscopes. Using the microscope, he observed pond water and saw moving organisms. Van Leeuwenhoek had discovered bacteria. Because he did not share how he made the lenses, bacteria were not observed again until the nineteenth century.

Connecting Bacteria to Infections

As first, scientists thought that wounds caused bacteria to appear. Louis Pasteur experimented with bacteria in the mid-1800s. He realized that this idea was backwards. Instead, bacteria caused the tissue in the wound to decay. He found that he could kill bacteria in boiling liquids. **Pasteurization** *is the process in which a food is heated to a temperature that kills most harmful bacteria*. The pasteurization process used today is based on the work of Pasteur.

Joseph Lister used Pasteur's discoveries to make surgery safer for patients. He found that a substance called carbolic acid killed bacteria. He developed a system of spraying carbolic acid throughout an operating room during surgery. Infection and death from surgery decreased greatly. In the late 1800s, doctors improved on Lister's idea. They used carbolic acid to sterilize tools before surgery. They also used steam to sterilize the clothes and linens used during surgery. Sterilization completely kills the bacteria.

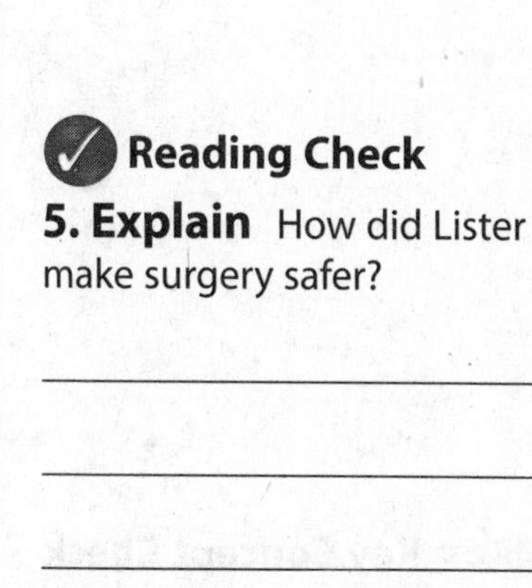

Reading Check

5. Explain How did Lister make surgery safer?

Discovering Disease Organisms

In 1867, Robert Koch developed a set of rules that could be used to find out if certain bacteria caused an illness. The research based on these rules convinced most scientists that bacteria were disease-causing pathogens. Koch's rules are described in the following figure.

Visual Check

6. Describe According to Koch's rules, what must happen when a pathogen is placed in a healthy animal?

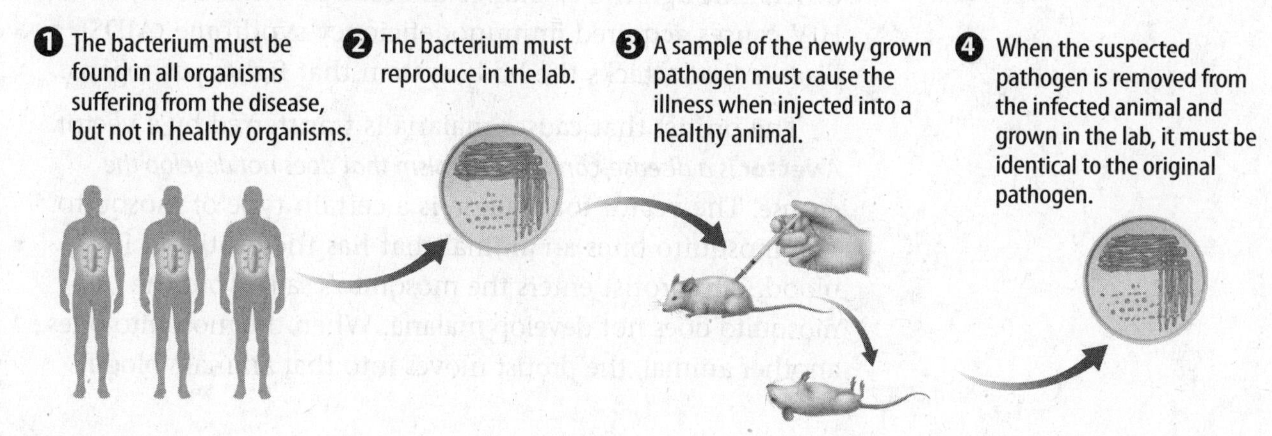

Human Diseases and the Pathogens That Cause Them	
Pathogens	**Diseases Caused**
Viruses	flu, colds, chickenpox, and AIDS
Bacteria	ear infections, strep throat, pneumonia, meningitis, whooping cough, and syphilis, a sexually transmitted disease
Fungi	athlete's foot, ringworm, and yeast infections
Protists	malaria, African sleeping sickness, and dysentery

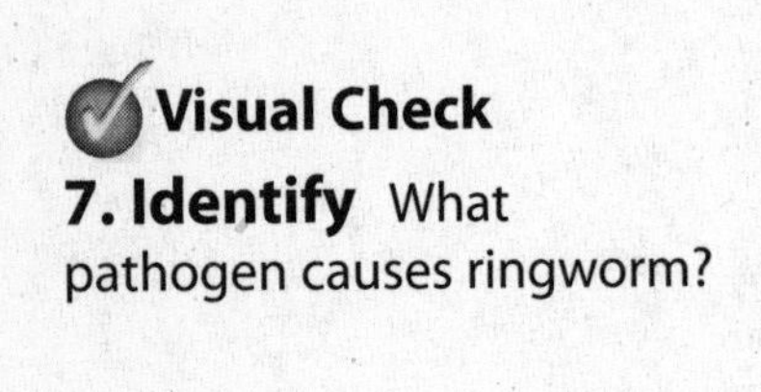

Visual Check
7. Identify What pathogen causes ringworm?

Bacteria aren't the only pathogens that cause disease. Viruses, fungi, and protists can also cause disease.

The table above shows diseases caused by different pathogens. Pathogens can be passed through water and food and carried by insects. They can also be passed directly among people by physical contact, sneezing, coughing, or exchange of bodily fluids. Some pathogens require a host to reproduce.

Types of Diseases

Have you ever heard someone say, "I caught a cold"? The common cold is contagious. This means that the pathogens can be passed from person to person. Not all diseases are caused by pathogens. Some diseases are the result of inherited traits. Other diseases can be caused by factors such as your environment and the choices you make about diet, exercise, and sleep.

Key Concept Check
8. Explain Why do we get diseases?

Infectious Diseases

Diseases caused by pathogens that can be transmitted from one person to another are **infectious diseases.** For example, flu and cold viruses can be passed through direct contact, such as by shaking hands.

The human immunodeficiency virus (HIV) can pass to others through the exchange of blood or other bodily fluids. HIV causes acquired immunodeficiency syndrome (AIDS), a disease that attacks the body system that fights pathogens.

The protist that causes malaria is transferred by a vector. *A* **vector** *is a disease-carrying organism that does not develop the disease.* The vector for malaria is a certain type of mosquito. The mosquito bites an animal that has the protist in its blood. The protist enters the mosquito's saliva, but the mosquito does not develop malaria. When the mosquito bites another animal, the protist moves into that animal's blood.

Noninfectious Diseases

A disease that cannot pass from person to person is a **noninfectious disease.** For example, you cannot catch lung cancer from a person who has lung cancer. Pathogens do not directly cause noninfectious diseases. There are two common causes of noninfectious diseases:

- genetics, or traits inherited in your DNA from your biological parents, and
- environmental conditions, including lifestyle choices.

A person who has a genetic trait might develop a noninfectious disease because of environmental conditions. It is the combination of genetics and environment that causes the disease to develop.

Childhood Diseases Noninfectious diseases that affect children are mainly a result of genetics. Cystic fibrosis is a genetic disease. It causes the body to produce mucus that is thicker than normal. This affects breathing and other body functions. The gene that causes cystic fibrosis is recessive. This means a child must inherit this recessive gene from each parent to be affected by cystic fibrosis. The parents, called carriers, do not have to have the disease.

For many children with genetic disorders, environmental conditions can make the disease worse. A poor diet, air pollution, and lack of exercise make the symptoms of cystic fibrosis worse.

Other Diseases Many noninfectious diseases that affect adults are mainly a result of environmental causes and life choices. For example, an unhealthful diet, obesity, a lack of regular exercise, and smoking cause most cases of heart disease. Osteoporosis is a disease in which bones become weak and less dense. People inherit a tendency to develop osteoporosis. However, years of poor lifestyle choices, such as an unhealthful diet, lack of calcium and vitamin D, smoking, and a lack of exercise can lead to weakened bones.

Cancer Tumors form when cells reproduce uncontrollably. **Cancer** *is a disease in which cells reproduce uncontrollably without the usual signals to stop.* For example, lung-cancer tumors form in the lungs. The tumors decrease normal lung function. People can inherit forms of genes that make them more likely to develop lung cancer. But if they are not exposed to harmful environmental conditions such as poor air quality and smoking, they might not develop lung cancer.

FOLDABLES®

Make the following two-tab book, then use it to compare infectious diseases and noninfectious diseases.

Infectious Diseases | Noninfectious Diseases

Reading Check

9. Identify List some causes of noninfectious diseases.

Key Concept Check

10. Explain How do infectious diseases and noninfectious diseases differ?

After You Read

Mini Glossary

cancer: a disease in which cells reproduce uncontrollably without the usual signals to stop

infectious disease: a disease caused by pathogens that can be transmitted from one person to another

noninfectious disease: a disease that cannot pass from person to person

pasteurization: the process in which a food is heated to a temperature that kills most harmful bacteria

pathogen: a disease-causing agent, such as a bacterium or a virus

vector: a disease-carrying organism that does not develop the disease

1. Review the terms and their definitions in the Mini Glossary. Write a sentence that uses the following terms: *cancer, noninfectious disease,* and *pathogen.*

2. Complete the chart below by filling in one discovery for each century that helped scientists understand more about pathogens and disease. The last frame has been done for you.

Discoveries in the 1600s	Discoveries in the 1700s	Discoveries in the 1800s
		Dr. John Snow made a connection between cholera and the bacterium that causes it.

3. Reread the paragraphs about infectious and noninfectious diseases. Write one fact about each type of disease that you learned from the paragraphs.

What do you think NOW?

Reread the statements at the beginning of the lesson. Fill in the After column with an A if you agree with the statement or a D if you disagree. Did you change your mind?

Log on to ConnectED.mcgraw-hill.com and access your textbook to find this lesson's resources.

Immunity and Disease

The Immune System

Before You Read

What do you think? Read the two statements below and decide whether you agree or disagree with them. Place an A in the Before column if you agree with the statement or a D if you disagree. After you've read this lesson, reread the statements to see if you have changed your mind.

Before	Statement	After
	3. The immune system helps keep the body healthy.	
	4. All immune responses are specific to the invading germs.	

Key Concepts 
- What does the immune system do?
- How do the parts of the immune system work together?
- How does the immune system interact with other body systems?

Read to Learn

Functions of the Immune System

Your body is always exposed to pathogens that cause diseases. In Lesson 1, you read that disease-causing agents are pathogens. Pathogens include bacteria, viruses, fungi, and protists. They are in the air, on objects, and in water. Your immune system works to protect your body from pathogens. There are many barriers to keep pathogens from entering your body.

Sometimes pathogens get past your body's first barriers. When this happens, your immune system has defenses to stop the pathogens. For example, there are cells in your body that can destroy pathogens. The immune system works with other body systems and helps keep you healthy, even as the environment outside your body changes.

You can help your immune system fight pathogens and environmental influences. By making healthful choices, you support your immune system. Wise, healthful choices include eating healthful food, getting enough sleep, exercising regularly, and using sunscreen. Remember that the choices you make every day affect how well your immune system functions.

Mark the Text

Building Vocabulary Read all of the headings in this lesson and circle any words you don't understand. As you read, highlight the sections of text that help you understand what the words mean. Use the circled words and highlighted text to review the lesson.

Key Concept Check
1. Describe What does the immune system do?

______

______

______

FOLDABLES®

Make the following layered book and use it to organize information about the immune system's lines of defense.

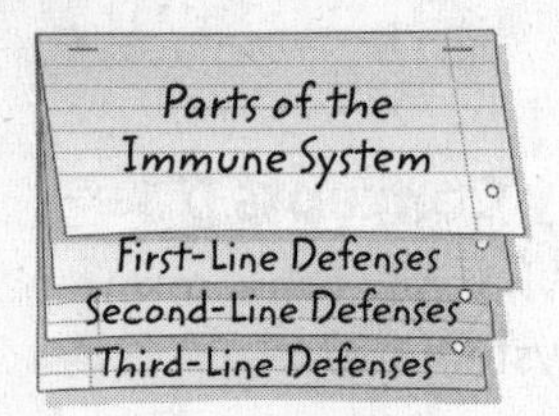

Parts of the Immune System

Different parts of your body work together to keep pathogens from making you sick. The integumentary system (skin), the respiratory system, the circulatory system, the digestive system, and the nervous system all work with the immune system and protect you against disease.

First-Line Defenses

First-line defenses keep germs from reaching the parts of your body where they can make you sick. Skin, hair, mucus, and acids are first-line defenses. They are effective against many types of pathogens. An immune defense that protects against more than one type of pathogen is a nonspecific defense. Many of the first-line defenses are nonspecific defenses.

Skin Your skin is often the first nonspecific defense that protects you from pathogens. Your skin keeps dirt and germs from entering your body. Sweat and acids from skin cells kill some bacteria. Natural oils make skin waterproof, so you can easily wash it.

You encounter pathogens every day. Your skin stops most of them from entering your body. Pathogens, such as cold and flu viruses, can survive for short periods on objects such as telephones and doorknobs. The pathogens can transfer to your hands when you touch these objects. Once pathogens are on your hands, they can enter your body through your mouth, nose, eyes, or a cut. Washing your hands often with soap and water easily removes most pathogens from your hands.

Your skin protects you from other dangers. It forms a chemical called melanin that protects you from the Sun's ultraviolet (UV) rays. Nerve endings in your skin can help protect you from injury. They help you sense the warmth of a stove or the sharpness of a pin. ✓

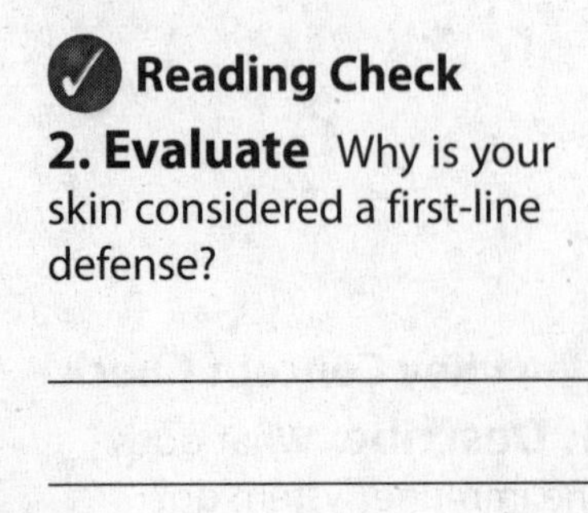

Reading Check

2. Evaluate Why is your skin considered a first-line defense?

____

____

____

Respiratory System You can inhale pathogens from the air through your nose or mouth. The hairs in your nose help protect you by trapping dirt and pathogens. This keeps pathogens from reaching the rest of your respiratory system.

Small, hairlike structures called cilia also trap pathogens. Cilia move pathogens up and out of the upper respiratory system. When pathogens get past the cilia, mucus is the next line of defense. Mucus traps pathogens. The pathogens are removed from the respiratory system when you cough, sneeze, or swallow.

Digestive System Pathogens can enter your digestive system on or in the food you eat. The digestive system is effective at stopping bacteria from making you sick. The stomach has strong acids inside it. Stomach acids destroy many pathogens. The digestive system contains mucus, much like the respiratory system. Mucus in the digestive system traps disease-causing bacteria and viruses.

Sometimes pathogens are not destroyed by stomach acids, and they can cause you to feel sick. In these cases, your digestive system can reverse the usual direction of muscle contractions, causing you to vomit and remove pathogens. Other times, muscle contractions speed up, and pathogens are removed through diarrhea. ✓

Reading Check

3. Name the ways the digestive system helps defend the body against pathogens.

____

____

____

Circulatory System and Nervous System Your circulatory system also protects you from pathogens. Pathogens can be moved through the circulatory system to organs that fight infection.

The nervous system and the circulatory system work together and increase your body's temperature to fight pathogens more effectively. Some foreign substances trigger the brain to increase body temperature. When this occurs, blood vessels narrow and a fever develops. Many pathogens cannot survive at this higher temperature. For those that do survive, the fever brings another line of defense—white blood cells. White blood cells are part of the second-line defenses against pathogens.

ACADEMIC VOCABULARY

occur
(verb) to come into existence

Second-Line Defenses

Sometimes pathogens get past the first-line defenses. When they do, the next line of defense goes into action. Like the first-line defense, second-line defenses are nonspecific. They fight against any type of pathogen.

White Blood Cells The spongy tissue in the center of your bones is called bone marrow. White blood cells form in the bone marrow. When pathogens get past first-line defenses, white blood cells attack them.

White blood cells flow through the circulatory system. However, they do most of their work attacking pathogens in the fluid outside blood vessels. They fight infection in several different ways. Some white blood cells surround and destroy bacteria directly. Others release chemicals that make it easier to kill pathogens. Another type of white blood cell produces proteins. The proteins destroy viruses and other foreign substances that get past the first-line defenses. ✓

Reading Check

4. Identify three ways in which white blood cells fight infection.

____

____

____

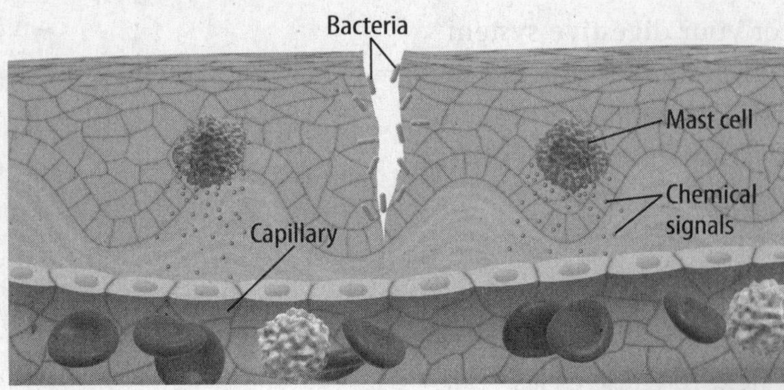

1 When skin is torn or cut, the damaged tissue tells mast cells to release chemical signals into the surrounding tissue.

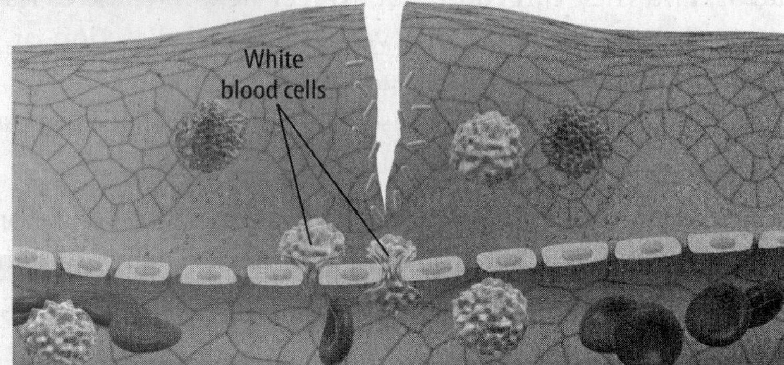

2 Chemical signals attract white blood cells to the area and cause the capillary to dilate. White blood cells and plasma leak into the tissue, causing swelling.

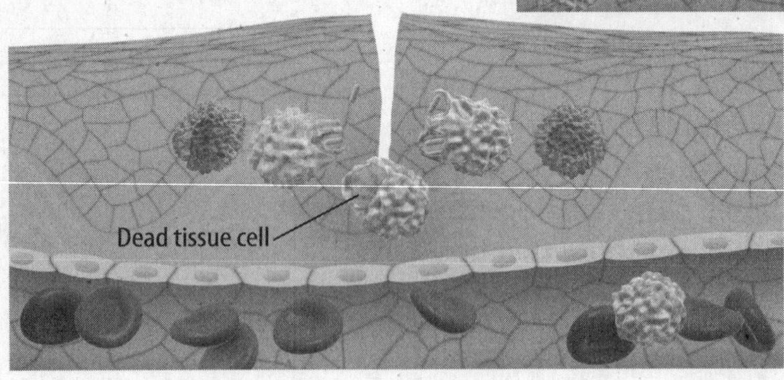

3 The white blood cells surround and take in the bacteria and any dead cells. The tissue heals. The white blood cells and the plasma flow back into the capillary and swelling decreases.

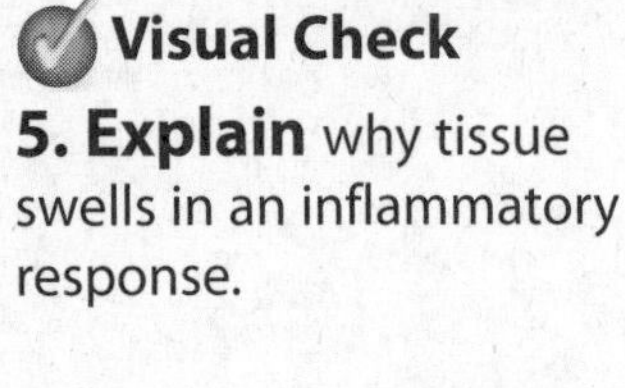

Visual Check

5. Explain why tissue swells in an inflammatory response.

____

____

____

____

____

Inflammatory Response When you are injured, your body produces an inflammatory response, causing inflammation. **Inflammation** *is a process that causes an area to become red and swollen.* If the injury is to the surface of the skin, there is an inflammatory response, as shown above. First, damaged cells release a protein that signals the capillaries to widen. Blood flows to the injury site. The site becomes red and warm. Second, swelling occurs as plasma and white blood cells leak into the area. Third, the white blood cells break down damaged cells and destroy any bacteria that might have entered the wound. The inflammatory response cleans the area of the injury. Because of the inflammatory response, the damaged tissue is able to heal.

Third-Line Defenses

If the first- and second-line defenses do not destroy all invading pathogens, another type of immune response occurs. Third-line defenses are specific to foreign substances. Often, the three lines of defense work together.

Antigens and Antibodies *An* **antigen** *is a substance that causes an immune response.* An antigen can be on the surface of a pathogen. *Proteins called* **antibodies** *can attach to the antigen and make it useless.*

Certain white blood cells, called B cells and T cells, form antibodies. **B cells** *form and mature in the bone marrow and secrete antibodies into the blood.* **T cells** *form in the bone marrow and mature in the thymus gland. They produce a protein antibody that becomes part of a cell membrane.*

Antibodies match with specific antigens, as shown below. Once your body has developed antibodies to an antigen, the antibodies will respond rapidly when the same pathogen invades your body again. This information is stored in antibodies on white blood cells called memory B cells.

Key Concept Check

6. Explain How do the parts of the immune system work together?

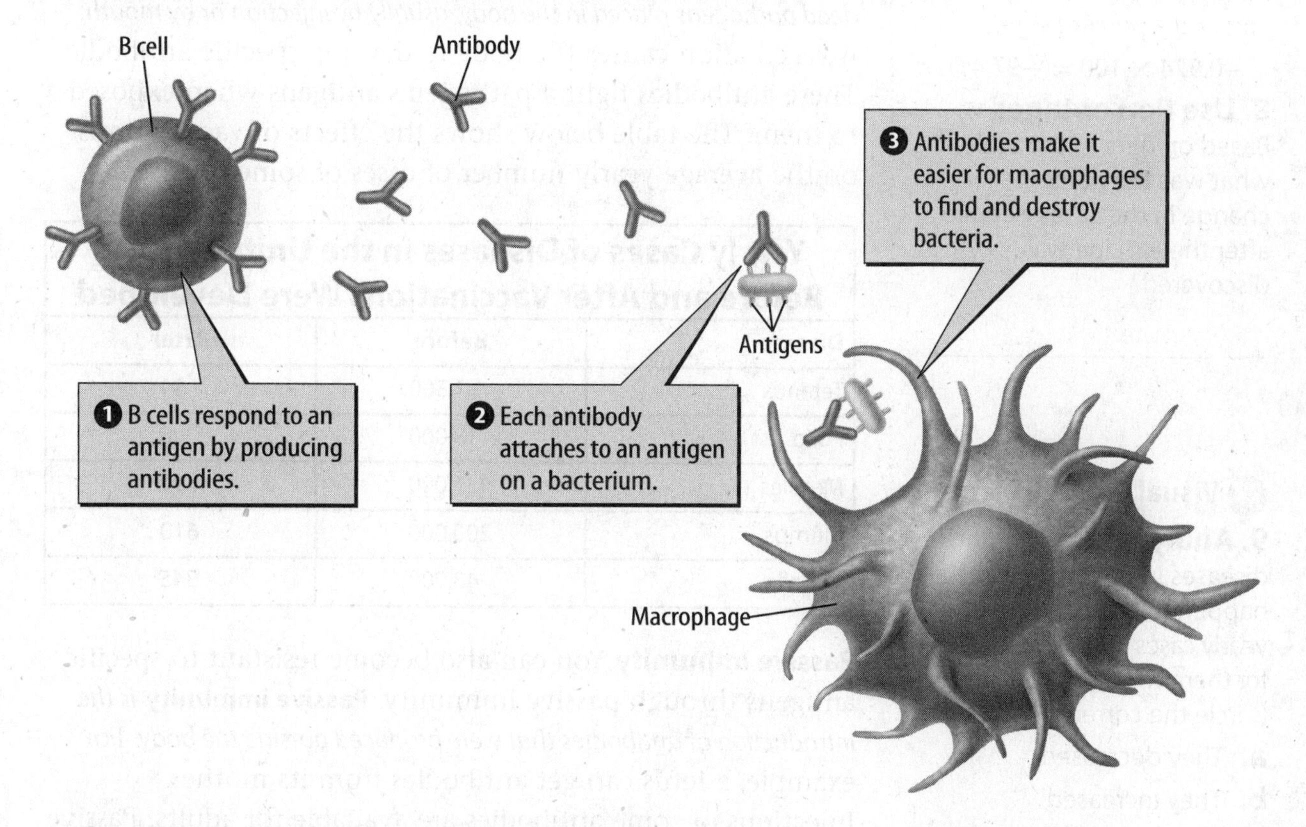

An **allergy** *is an overly sensitive immune response to common antigens.* Most people do not produce antibodies to the proteins in dog saliva. However, the antigens in dog saliva do cause some people to have an immune response. These people have an allergy. Their bodies treat the dog saliva as if it were a pathogen. People with allergies experience inflammation and increased mucus production.

Visual Check

7. Describe What do B cells do?

Math Skills

If tetanus cases fell from 1,300 cases before the vaccine was developed to 34 cases after the vaccine was developed, what percent change does this represent?

Subtract the starting value from the final value.

$$34 - 1{,}300 = -1{,}266$$

Divide the difference by the starting value.

$$\frac{-1{,}266}{1{,}300} = -0.974$$

Multiply the answer by 100 and add a percent sign.

$$-0.974 \times 100 = -97.4\%$$

8. Use Percentages Based on data in the table, what was the percent change in the cases of polio after the vaccine was discovered?

____

9. Analyze For all of the diseases in the table, what happened to the number of yearly cases after vaccinations for them were developed? (Circle the correct answer.)

a. They decreased.
b. They increased.
c. They stayed the same.

Key Concept Check

10. Explain How does the immune system interact with other body systems?

____

____

____

Immunity

The resistance to specific pathogens is **immunity.** There are two types of immunity—active immunity and passive immunity.

Active Immunity *Your body produces antibodies in response to an antigen in* **active immunity.** Your body recognizes the antigen, and the matching antibodies quickly respond. You can develop active immunity through illness or infection. Even after an infection or illness, antibodies remain in your body. This is why you usually get diseases such as chicken pox only once. You can get other diseases, such as a cold, over and over. There are many different cold viruses that cause similar symptoms.

You can also develop antibodies if you are exposed to an antigen through a vaccination. *A* **vaccination** *is weakened or dead pathogens placed in the body, usually by injection or by mouth.* A vaccination causes the body to develop specific antibodies. These antibodies fight a pathogen's antigens when exposed to them. The table below shows the effects of vaccinations on the average yearly number of cases of some diseases.

Yearly Cases of Diseases in the United States Before and After Vaccinations Were Developed

Disease	Before	After
Tetanus	1,300	34
Polio	18,000	8
Measles	425,000	90
Mumps	200,000	610
Rubella	48,000	345

Passive Immunity You can also become resistant to specific antigens through passive immunity. **Passive immunity** *is the introduction of antibodies that were produced outside the body.* For example, a fetus can get antibodies from its mother. Injections of some antibodies are available for adults. Passive immunity is temporary. The body does not continue to make these antibodies.

The Immune System and Homeostasis

You are exposed to many different pathogens every day. Your immune system works to maintain your body's homeostasis. Body systems, including the circulatory system and the respiratory system, work together to protect against pathogens.

After You Read

Mini Glossary

active immunity: when the body produces antibodies in response to an antigen

allergy: an overly sensitive immune response to common antigens

antibody: a protein that attaches to an antigen and makes it useless

antigen: a substance that causes an immune response

B cell: a cell that forms and matures in the bone marrow and secretes antibodies into the blood

immunity: the resistance to specific pathogens

inflammation: a process that causes an area to become red and swollen

passive immunity: the introduction of antibodies that were produced outside the body

T cell: a cell that forms in the bone marrow and matures in the thymus gland; it produces a protein antibody that becomes part of a cell membrane

vaccination: weakened or dead pathogens placed in the body, usually by injection or by mouth

1. Review the terms and their definitions in the Mini Glossary. Compare and contrast active immunity and passive immunity.

2. Complete the concept web to identify second-line defenses that you have in your body.

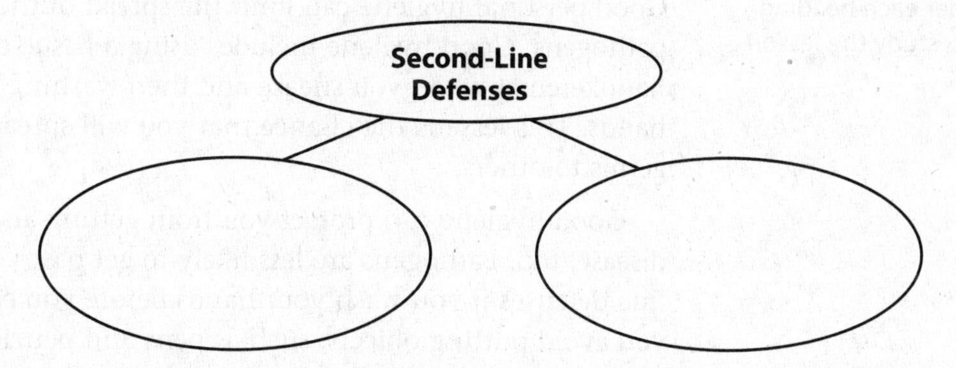

3. How does the respiratory system fight pathogens?

What do you think NOW?

Reread the statements at the beginning of the lesson. Fill in the After column with an A if you agree with the statement or a D if you disagree. Did you change your mind?

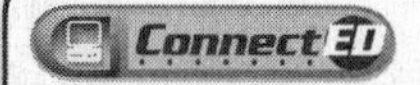

Log on to ConnectED.mcgraw-hill.com and access your textbook to find this lesson's resources.

Immunity and Disease

Staying Healthy

Key Concepts
- How can healthful habits and healthful choices affect diseases?
- How do sanitation practices affect human health?
- How can chemicals affect the human body?

Before You Read

What do you think? Read the two statements below and decide whether you agree or disagree with them. Place an A in the Before column if you agree with the statement or a D if you disagree. After you've read this lesson, reread the statements to see if you have changed your mind.

Before	Statement	After
	5. Exercise and sleep can help keep you healthy.	
	6. Chemicals make you sick and should not be used.	

Study Coach

Make an Outline As you read, make an outline using the heads in the text. Fill in your outline with a few details under each heading to help you study the lesson.

Read to Learn

Healthful Habits

Infectious diseases, such as colds and flu, are common because the pathogens are passed from person to person. Good personal hygiene can limit the spread of these pathogens. Good hygiene includes using a tissue or handkerchief when you sneeze and then washing your hands. This lessens the chance that you will spread your germs to others.

Good hygiene can protect you from getting an infectious disease, too. Pathogens are less likely to get past your first-line defenses if you wash your hands before you eat and if you avoid putting objects, such as pens and pencils, in your mouth.

FOLDABLES®

Make a half-book and use it to record information about habits and choices that can help you stay healthy.

Healthful Choices

You can make other choices that can help keep you healthy. Healthful choices about the environment can protect you from many infectious and noninfectious diseases.

Diet If you eat healthful foods, your immune system can work more efficiently against pathogens. A healthful diet can even protect you against noninfectious diseases such as osteoporosis and heart disease. A healthful diet, a healthful weight, and regular exercise have been linked with disease prevention.

Sun Protection Skin cancer is a noninfectious disease. The ultraviolet (UV) rays from the Sun damage skin cells and can cause them to reproduce uncontrollably. Wearing protective clothing and using sunscreen block UV rays and limit the damage from sunlight.

Alcohol and Tobacco Many types of cancer are related to excessive drinking of alcohol and to tobacco use. Lung cancer is linked to smoking or poor air quality. Healthful choices include not smoking or chewing tobacco and limiting or avoiding alcoholic beverages.

Health and Sanitation

Improved cleanliness in public areas has increased overall health. Connecting cleanliness with health was the key.

Food Preparation

Improved sanitation in food preparation has led to better health. Employees must wash their hands regularly and keep equipment clean. Inspections help catch problems early and protect consumers from most pathogens.

Waste Management

In the past, cities had no plumbing or sewer systems. People often dumped their waste and garbage in the streets. Today, landfills and sewer systems keep our streets and households cleaner. This cleanliness slows the spread of infectious diseases.

Health and Chemicals

Some chemicals, such as those in sunscreen, protect us from noninfectious diseases. Other chemicals are used in medicines. ***Antibiotics** are medicines that stop the growth and reproduction of bacteria.* Chemicals are also used to destroy cancer cells. *These medicines, used in a type of treatment called **chemotherapy,** kill the cells that are reproducing uncontrollably.*

Chemicals in paints and pesticides can be harmful to our health. They should be used and disposed of properly. More than 50 chemicals in cigarettes have been linked to cancer.

Health and the Environment

Some chemicals that are harmful to our health are in our environment. Before 1978, lead was used in many paints. If the dried paint flaked, it released lead into the air. Inhaling the lead caused noninfectious kidney and nervous system diseases. Asbestos fibers are safe when contained in products such as ceiling and floor tiles. But if the tiles are broken and the fibers are released, they can cause cancer.

Key Concept Check
1. Evaluate How can healthful habits and healthful choices affect diseases?

Key Concept Check
2. Explain how sanitation practices affect human health.

Key Concept Check
3. Describe How can chemicals affect the human body?

After You Read

Mini Glossary

antibiotic: a medicine that stops the growth and reproduction of bacteria

chemotherapy: a treatment that uses chemicals to kill cancer cells that are reproducing uncontrollably

1. Review the terms and their definitions in the Mini Glossary. Use each term in a sentence to explain how chemicals can affect human health.

2. For each area listed below, give an example of a healthful choice you can make to help keep your immune system healthy.

Personal Hygiene: ______

Diet: ______

Sun Protection: ______

Alcohol and Tobacco Use: ______

3. List three examples of sanitation practices that led to improved human health.

What do you think NOW?

Reread the statements at the beginning of the lesson. Fill in the After column with an A if you agree with the statement or a D if you disagree. Did you change your mind?

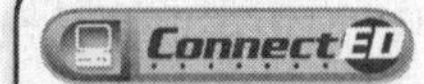

Log on to ConnectED.mcgraw-hill.com and access your textbook to find this lesson's resources.

Control and Coordination

The Nervous System

Before You Read

What do you think? Read the two statements below and decide whether you agree or disagree with them. Place an A in the Before column if you agree with the statement or a D if you disagree. After you've read this lesson, reread the statements to see if you have changed your mind.

Before	Statement	After
	1. The nervous system contains two parts—the central nervous system and the peripheral nervous system.	
	2. The autonomic nervous system controls voluntary functions.	

Key Concepts

- What does the nervous system do?
- How do the parts of the nervous system work together?
- How does the nervous system interact with other body systems?

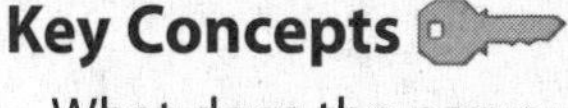

K-W-L Fold a sheet of paper into three columns. Label them *(K)* what you know about the nervous system, *(W)* what you want to learn, and *(L)* the facts that you learned. Fill in the third column after you have read this lesson.

Read to Learn

Functions of the Nervous System

Have you ever had goose bumps on your arms when you were cold? These bumps form when muscle cells in your skin respond to cold temperatures. These muscle cells contract, or shorten. Then bumps form, and the hairs on your arms rise up. The raised hairs trap air. The trapped air insulates your skin and helps you feel warmer.

How do your muscle cells know to contract? When you first feel the cold, a message is sent to your brain. Your brain then sends a message to your skin's muscle cells, and the goose bumps form.

The part of an organism that gathers, processes, and responds to information is called the **nervous system.** Your nervous system gets information from your five senses. These senses are vision, hearing, smell, taste, and touch. You will read more about the senses in Lesson 2. ✓

Reading Check

1. Identify What does your nervous system use to gather information?

The nervous system reacts quickly. It can receive information, process it, and respond in less than one second. Signals received by the nervous system can travel as fast as some airplanes. This is around 400 km/h.

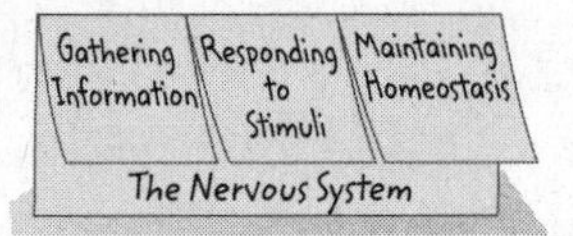

Make a three-tab book as shown to organize your notes about the functions of the nervous system.

Think it Over

2. Apply Imagine that you duck when you hear a loud noise overhead. What function of the nervous system does ducking show?

___

___

___

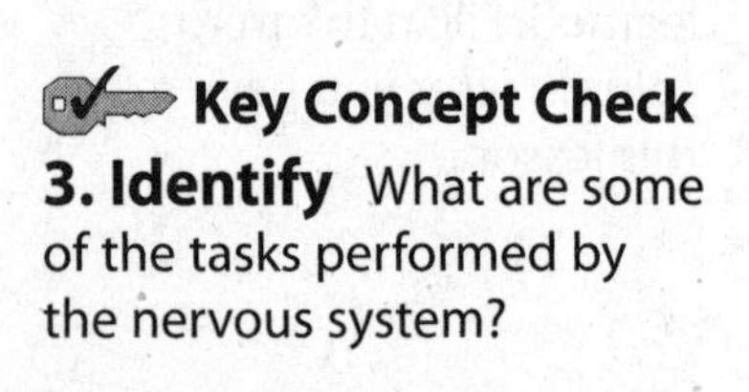

Key Concept Check

3. Identify What are some of the tasks performed by the nervous system?

___

___

___

Gathering Information

Imagine a driver stopping a car quickly so it doesn't hit a ball that has rolled into the street. The sight of the ball is a stimulus (STIHM yuh lus) (plural, stimuli). *A* ***stimulus*** *is a change in an organism's environment that causes a response.* The driver's nervous system gathered and interpreted the sight of the ball and caused her body to react by braking.

Responding to Stimuli

How would you react if you were riding your bike and a squirrel ran in front of you? You might quickly put on the brakes. Or you might change direction to avoid the squirrel. These reactions are ways that your nervous system helps you respond to a stimulus from the environment.

Your nervous system receives many stimuli at the same time. The type of response you make often depends on how your nervous system processes the information.

Maintaining Homeostasis

Imagine again the driver who stops her car quickly so she doesn't hit a ball in the street. The driver responds to the stimulus of the moving ball. Her nervous system probably also causes her heart to beat faster and her breathing to speed up. These changes help the driver react quickly.

People are always reacting to changes in their environments. Therefore, their nervous systems help maintain homeostasis, or the regulation of their internal environments. For example, the driver's nervous system must signal her heart to beat slower and her breathing to slow down to restore homeostasis after safely reacting to the ball.

Neurons

The basic functioning units of the nervous system are called nerve cells, or **neurons** (NOO rahnz). Neurons help different parts of your body communicate with each other. Without looking down, how do you know whether you are walking on sand or on a sidewalk? Neurons in your feet are connected to other neurons. These neurons send information to your brain about the surface you are walking on.

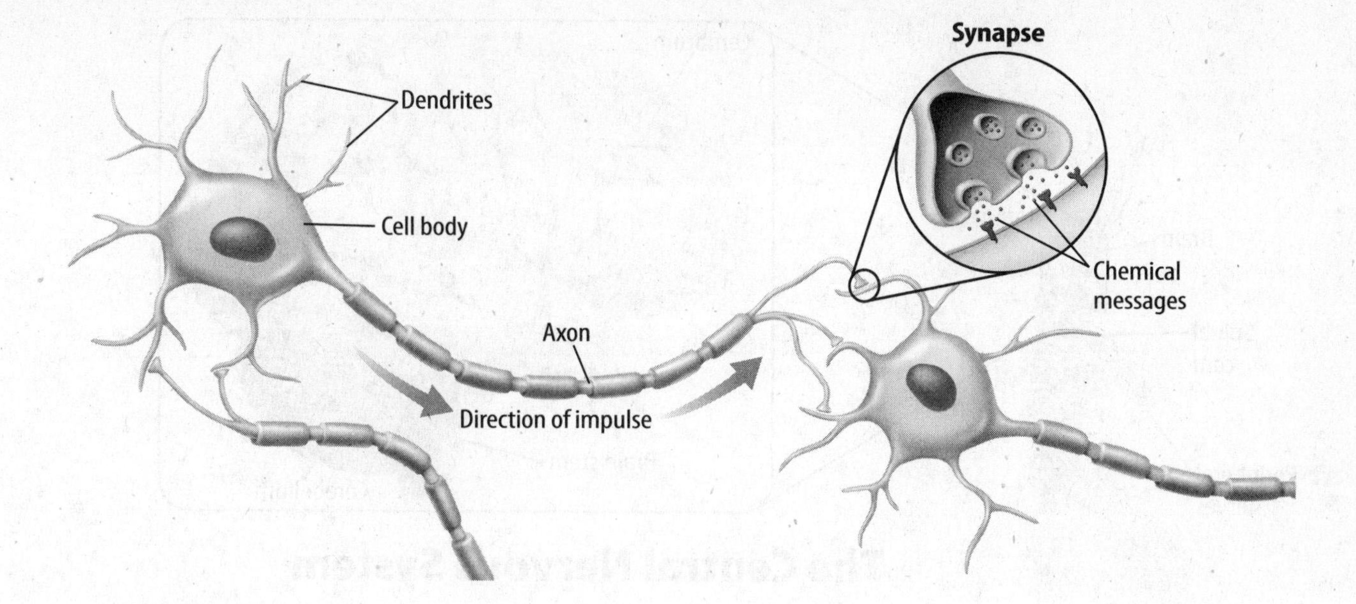

The figure above shows a neuron. A neuron has three parts: dendrites (DEN drites), a cell body, and an axon (AK sahn). A dendrite receives information from another neuron or from another cell in your body. A cell body processes information. An axon sends the information to another neuron or cell in your body.

Visual Check

4. Label the second neuron in the figure with the following terms: *axon, cell body,* and *dendrite.*

Types of Neurons

There are three types of neurons that work together. They send and receive information throughout your body. You have sensory neurons, motor neurons, and interneurons. Sensory neurons send information about your environment to your brain or spinal cord. Motor neurons send information from your brain or spinal cord to your tissues and organs. Interneurons make a connection, or bridge, between sensory and motor neurons.

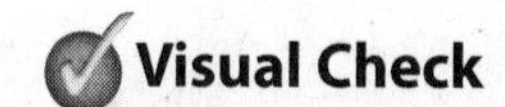

5. Apply Which kind of neuron sends information to your brain after you burn your finger?

____

____

Synapses

The gap between two neurons is called a **synapse** (SIH naps). Look again at the figure above. In the close-up of the synapse, you can see that neurons communicate across synapses by releasing chemicals. These chemicals carry information from the axon of one neuron to a dendrite of another neuron. Most synapses are between an axon of one neuron and a dendrite of another neuron. Information usually goes in just one direction. 

Reading Check

6. Recall In a neuron, the message usually goes from which of the following? (Circle the correct answer.)

a. from an axon to a dendrite

b. from a dendrite to an axon

c. from an axon to an axon

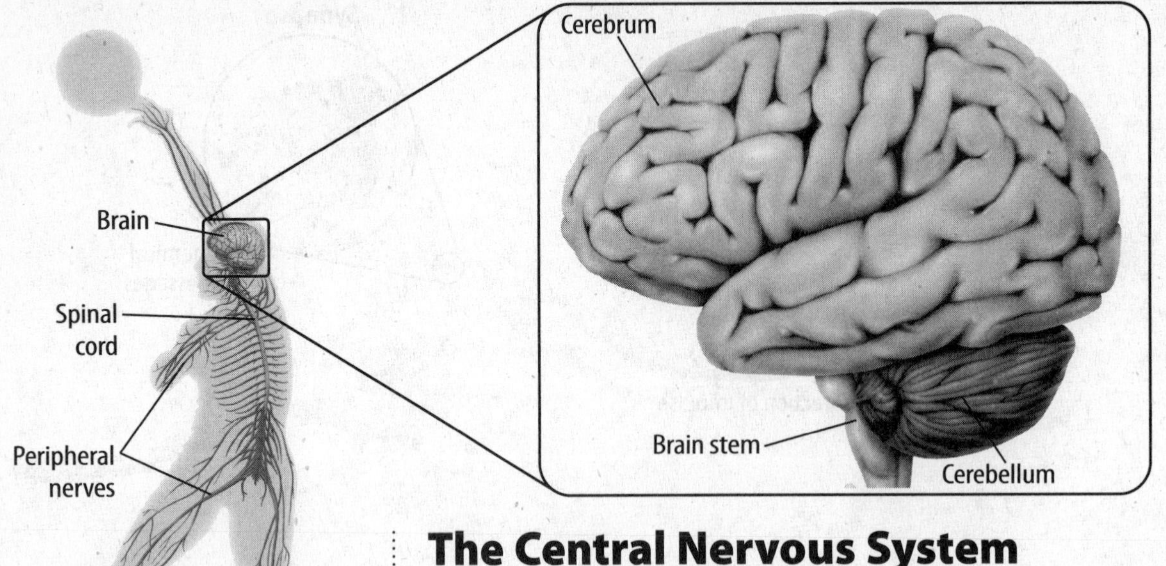

The Central Nervous System

Your nervous system has two parts: the central nervous system and the peripheral (puh RIH frul) nervous system. *The* ***central nervous system (CNS)*** *is made up of the brain and the spinal cord.* The CNS receives, processes, stores, and transfers information.

The Brain

The brain is the control center of your body. Your brain receives information, processes the information, and sends out a response. The brain also stores some information as memories.

The Cerebrum *The part of the brain that controls memory, language, and thought is the* **cerebrum** (suh REE brum). The cerebrum also processes information from the senses of touch and sight. The cerebrum is the largest and most complex part of the brain.

Look at the cerebrum in the figure above. The surface has many folds. The folds make it possible for many neurons to fit into a small space. If you could unfold the cerebrum, you would find that it has as much surface area as a large pillowcase.

Visual Check

7. Identify In the enlarged picture of the brain, highlight in the area that controls memories.

Think it Over

8. Generalize If someone had a disease that affected the cerebrum, which of the following would be affected? (Circle the correct answer.)

a. breathing

b. talking and writing

c. walking

The Cerebellum *The part of the brain that coordinates voluntary muscle movement and regulates balance and posture is the* **cerebellum** (ser uh BEH lum). Voluntary muscle movements are the ones that you control. For example, you tie your shoe or pedal your bike because you make your muscles do what is needed. Locate the cerebellum in the figure on the previous page and in the figure below.

The cerebellum also stores information about the movements you do often, such as tying your shoe or pedaling a bike. Because your brain stores this information, you can repeat the movements faster and more accurately. For example, the more you tie your shoe or pedal a bike, the faster and better you can do these things.

The Brain Stem Some of your body's functions, such as digestion and the beating of your heart, are involuntary. These functions are called involuntary because they happen without your controlling them.

The area of the brain that controls involuntary functions is the **brain stem.** Your brain stem also controls sneezing, coughing, and swallowing. The brain stem connects the brain to the spinal cord. Both are shown in the figure below.

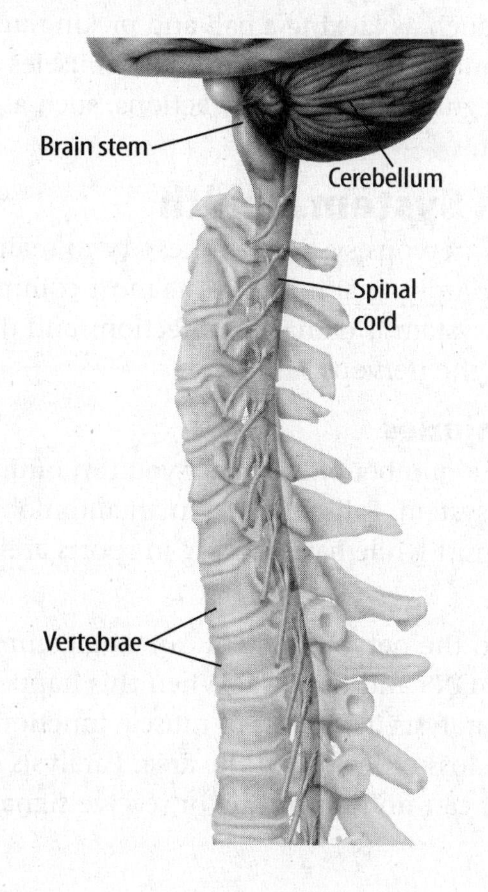

Math Skills

A nerve impulse from your hand travels at about 119 m/s. How long does it take the signal from your hand to reach your spinal cord if the distance is 0.40 m? You can use proportions to solve the problem.

Set up a proportion.

$$\frac{119 \text{ m}}{1 \text{ s}} = \frac{0.40 \text{ m}}{y \text{ s}}$$

Cross multiply.

$$119 \text{ m} \times y \text{ s} = 0.40 \text{ m} \times 1 \text{ s}$$

Solve for y by dividing both sides by 119 m.

$$y = \frac{0.40 \text{ m} \cdot \text{s}}{119 \text{ m}}$$

$$y = 0.003 \text{ s}$$

9. Use Proportions One giraffe neuron has an axon 4.6 m long that extends from its toe to the base of its neck. How long will it take a nerve impulse to travel this distance at a speed of 75 m/s?

____

Visual Check

10. Recall Circle the two major parts of the central nervous system.

Think it Over

11. Predict Name something else your body does that is involuntary and therefore might be controlled by your brain stem.

____

____

____

The Spinal Cord

*The **spinal cord** is a tubelike structure of neurons.* You read earlier that a neuron is a nerve cell that sends and receives information in your body. Because the neurons extend to other areas of the body, the brain can send out and get information. The spinal cord can be thought of as an information highway between the brain and the rest of the body. Cars use a highway to move quickly from one city to another. Neurons in the spinal cord send information quickly back and forth between the brain and other body parts. Bones called vertebrae protect the spinal cord.

The Peripheral Nervous System

Recall that the nervous system is made up of both the central nervous system (CNS) and the peripheral nervous system. *The **peripheral nervous system (PNS)** has sensory neurons and motor neurons that transmit information between the CNS and the rest of the body.*

The PNS has two parts: the somatic system and the autonomic system. The somatic system controls skeletal muscles. Neurons of the somatic system communicate between the CNS and skeletal muscles. They cause voluntary movements such as kicking a ball and picking up a book. The autonomic system controls smooth muscles and cardiac muscles. It regulates involuntary actions, such as the beating of your heart.

Key Concept Check
12. Discuss How does the PNS work with the CNS?

Nervous System Health

A healthy nervous system is necessary to maintain homeostasis. A physical injury is the most common way that the nervous system is damaged. Infections and diseases can also damage the nervous system.

ACADEMIC VOCABULARY
physical
(adjective) relating to material things

Physical Injuries

There are a number of ways that you can injure and harm the nervous system. Falling, being in an automobile accident, and getting hurt while participating in sports are just a few of them.

Injuries to the nervous system can stop communication between the CNS and the PNS. When this happens, paralysis can occur. Paralysis is the loss of muscle function and sometimes a loss of feeling in the area. Paralysis occurs in the area that can no longer send or receive signals.

Reading Check
13. Define What word do we use to refer to the loss of muscle function and sometimes feeling?

Preventing Injuries

Imagine that you are walking barefoot, and you step on a rock. Without thinking, you quickly lift your foot. You do not think about moving your foot. It just happens. *An automatic movement in response to a stimulus is called a* ***reflex.***

Reflexes are fast because, in most cases, the information goes only to the spinal cord. The information usually does not go to the brain. This fast response protects you because it takes less time to move away from harm.

A reflex often occurs before your brain knows your body is in danger. Once nerve signals travel from your spinal cord to your brain after the response, you feel pain.

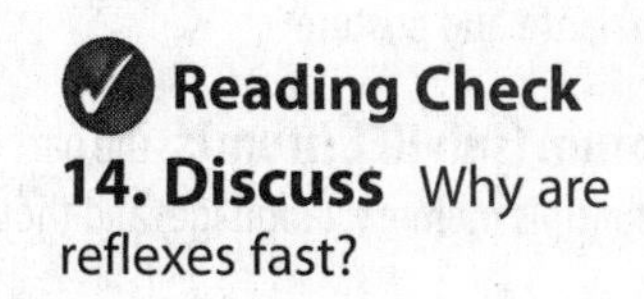

Reading Check

14. Discuss Why are reflexes fast?

____

____

Drugs

So far, you have learned that the nervous system can be affected by infections, diseases, and injuries. Drugs can also affect it. Drugs are chemicals that affect the body's functions. Many drugs affect the nervous system by speeding up or slowing down the communication between neurons.

Reading Check

15. Identify A chemical that affects how the body works is called which of the following? (Circle the correct answer.)

a. drug
b. neuron
c. cerebral fold

Some pain medicines slow the communication so much that they stop pain stimuli from being sent to the brain. A drug that slows down neuron communication is called a depressant.

Other drugs speed up communication between neurons. A drug that speeds up neuron communication is called a stimulant. Caffeine is a common stimulant.

The Nervous System and Homeostasis

Why do you shiver when you are cold? It is because your nervous system senses the cold temperature. It then signals your muscles to move quickly to warm your body. Doing so maintains homeostasis.

Your body maintains homeostasis by getting information from your environment and responding to it. Your nervous system senses changes in your environment. It then signals other systems, such as the digestive, the endocrine, and the circulatory systems, to make adjustments when needed.

Key Concept Check

16. Explain Give an example of how the nervous system works with another body system to maintain homeostasis.

____

____

____

____

After You Read

Mini Glossary

brain stem: the area of the brain that controls involuntary functions

central nervous system (CNS): the part of the nervous system that is made up of the brain and the spinal cord

cerebellum (ser uh BEH lum): the part of the brain that coordinates voluntary muscle movement and regulates balance and posture

cerebrum (suh REE brum): the part of the brain that controls memory, language, and thought

nervous system: the part of an organism that gathers, processes, and responds to information

neuron (NOO rahn): the basic functioning unit of the nervous system, also called a nerve cell

peripheral (puh RIH frul) nervous system (PNS): the part of the nervous system that has sensory neurons and motor neurons that transmit information between the CNS and the rest of the body

reflex: an automatic movement in response to a stimulus

spinal cord: a tubelike structure of neurons

stimulus (STIHM yuh lus): a change in an organism's environment that causes a response (plural, stimuli)

synapse (SIH naps): the gap between two neurons

1. Review the terms and their definitions in the Mini Glossary. Write two or three sentences that compare and contrast the brain stem and the cerebellum.

____

____

____

2. Fill in the missing parts of the table to describe the parts of the central nervous system.

Part of the CNS	What It Controls
Cerebrum	memory, language, thought, information from the senses
Cerebellum	
Brain stem	
Spinal cord	movement of information between the brain and other parts of the body

3. Look at the K-W-L chart that you made at the beginning of this lesson. Circle two facts you learned about the nervous system that you found interesting.

What do you think NOW?

Reread the statements at the beginning of the lesson. Fill in the After column with an A if you agree with the statement or a D if you disagree. Did you change your mind?

Log on to ConnectED.mcgraw-hill.com and access your textbook to find this lesson's resources.

Control and Coordination

The Senses

Before You Read

What do you think? Read the two statements below and decide whether you agree or disagree with them. Place an A in the Before column if you agree with the statement or a D if you disagree. After you've read this lesson, reread the statements to see if you have changed your mind.

Before	Statement	After
	3. A human has five senses that detect his or her environment.	
	4. The senses of smell and hearing work together.	

Key Concepts

- How do you learn about your environment?
- What is the role of the senses in maintaining homeostasis?

Read to Learn

You and Your Environment

Recall that your nervous system enables your body to receive information about your environment. It also processes that information and reacts to it.

Your nervous system is always responding to many types of stimuli. Your body has to receive a stimulus before it can respond to one. How does your body do this?

The **sensory system** *is the part of your nervous system that detects or senses the environment.* A human uses five senses to detect his or her environment. They are vision, hearing, smell, touch, and taste.

Imagine you are roasting marshmallows over a campfire. What senses are you using? You see the flames of the fire and hear the crackle of the wood burning. You smell the smoke and feel the warmth of the fire. You also taste the cooked marshmallows.

All parts of the sensory system have special structures called **receptors** *that detect stimuli.* Each of your five senses detects things about your environment by using different receptors.

Mark the Text

Build Vocabulary As you read, underline the words and phrases that you do not understand. When you finish reading, work with another student or your teacher to to make sure that you understand the things that you underlined.

Key Concept Check

1. Describe one way your senses help you learn about the environment.

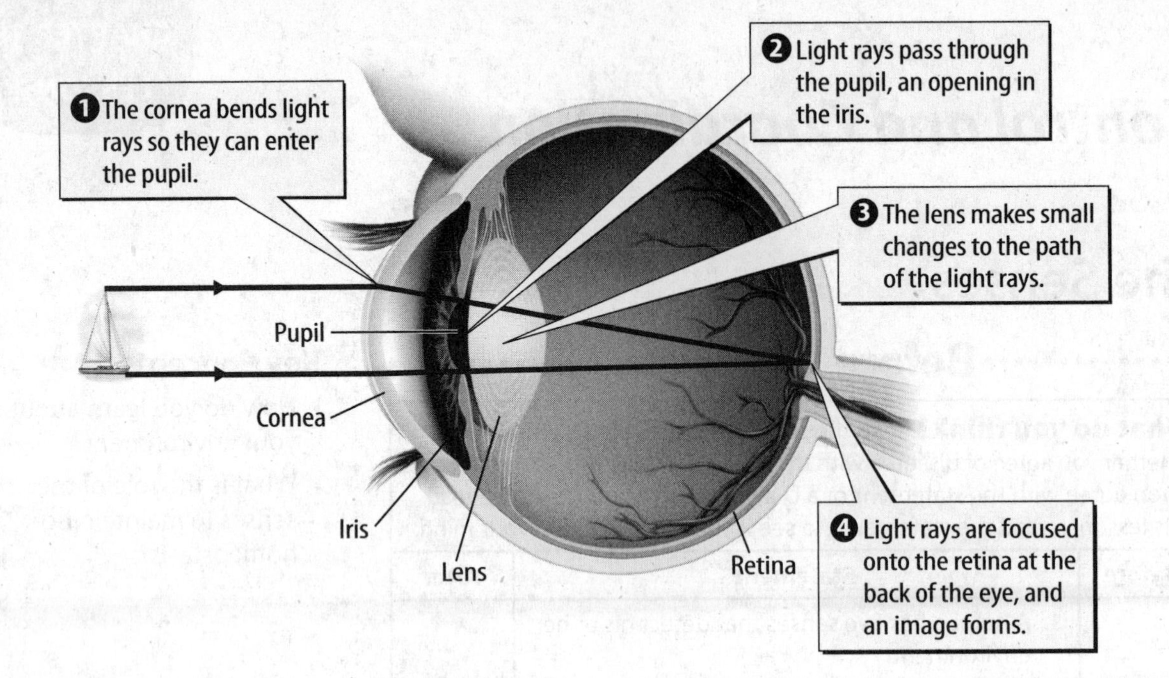

Visual Check

2. Identify Put an *X* on the figure where light is focused and forms an image.

FOLDABLES

Use three sheets of notebook paper as shown below to record information about the five senses.

Think it Over

3. Predict What do you think will happen to your pupils if you look out a window on a sunny day?

______

______

______

Vision

Your sense of vision lets you see things that are close, such as the words on this page. It also lets you see things that are far away, such as the stars in the night sky. Your eyes contain receptors for your sense of vision. Recall that receptors are used to detect the environment. The visual system uses photoreceptors in the eye to sense light and create vision.

The Parts of the Eye

The eye has many specialized parts. Look at the figure above as you read about how the eye creates vision.

1. Light enters the eye through the cornea (KOR nee uh). This thin membrane protects the eye and changes the direction of light rays.
2. Light goes through the pupil, an opening in the iris (I rus). The iris is the colored part of your eye. It controls how much light enters your eye by changing the size of the pupil. In bright light, your iris constricts and your pupil gets smaller. Less light is let into your eye. When there is little light, your iris relaxes and your pupil gets larger. More light is let into your eye.
3. Light then travels through the lens. The lens is a clear structure that works with the cornea to focus light.
4. Light rays are focused onto the retina (RET nuh), where an image forms. *The* ***retina*** *is an area at the back of the eye that has two types of cells—rod cells and cone cells—with photoreceptors.*

How You See

In order to see, light that enters your eyes has to be detected by the rods and cones in your retina. The rods and cones detect information about the colors and shapes of objects from the light that enters your eyes. The retina then sends that information as electric signals through the optic nerve to the brain. In the brain, the information is processed. The brain uses the information and creates a picture of what you are seeing.

SCIENCE USE V. COMMON USE

rod

Science Use one of the photoreceptors in the eye that picks up the shapes of objects

Common Use any long, cylinder-shaped object

Reading Check

4. Discuss Which parts of vision are rods and cones responsible for?

Focusing Light

Recall that the lens and the cornea focus light that enters the eye. These two parts work together to change the direction of the light that enters the eye. As you can see in the figure below, both the lens and the cornea are curved. These curves change the direction of the light and focus it onto the retina.

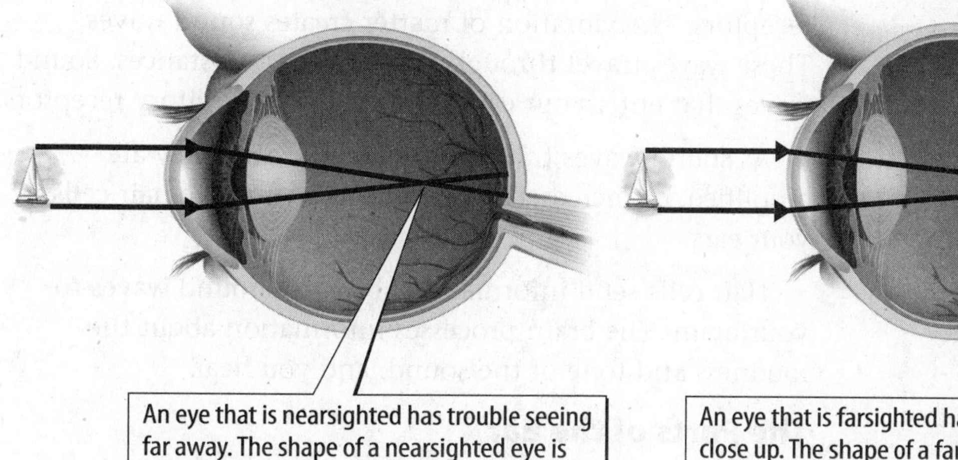

Why do some people need glasses to see well? Look again at the figure above. If the cornea or the lens is not curved exactly right, the person will have trouble focusing the images he or she sees.

A person who is nearsighted has trouble seeing things that are far away. A person is nearsighted if his or her eyes are longer than normal. A person who is farsighted has trouble seeing things that are nearby. A person is farsighted if his or her eyes are shorter than normal. Glasses or contact lenses are used to focus the light correctly so the person can see better.

Visual Check

5. Analyze How are the eyes of someone who is nearsighted or farsighted different from the eyes of someone with normal vision?

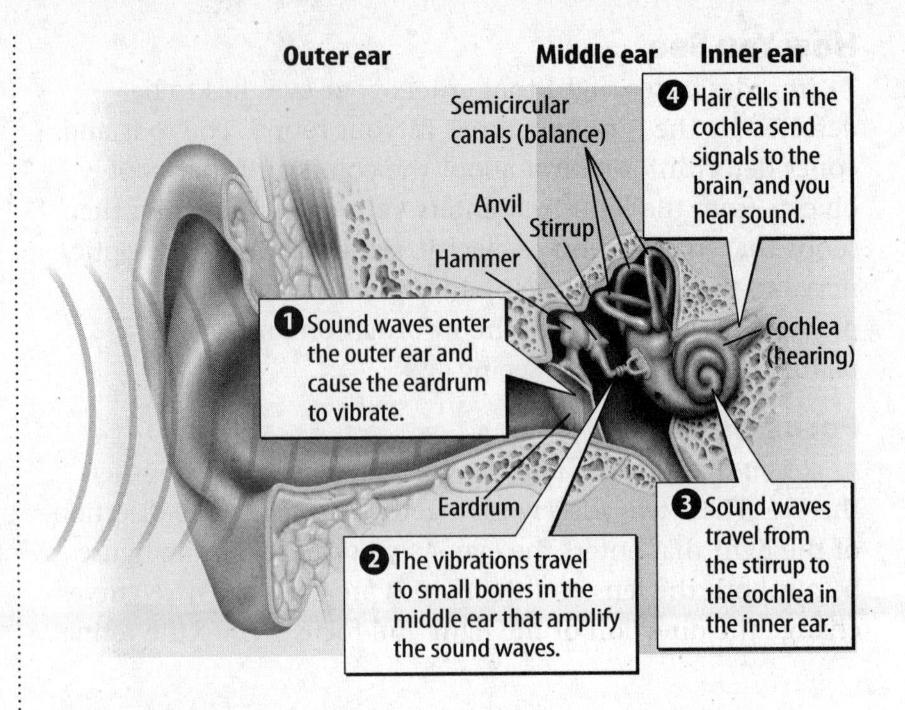

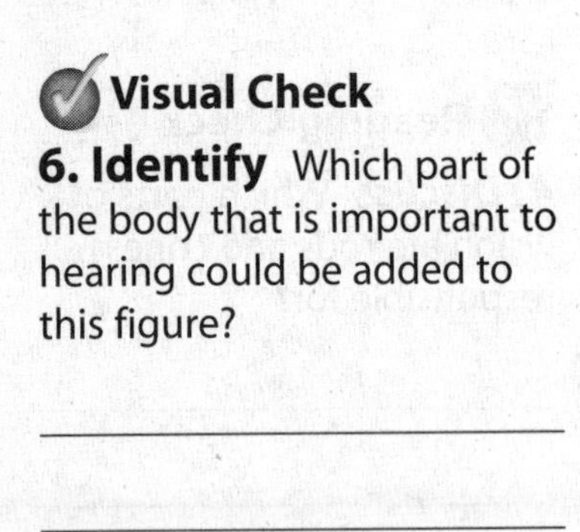

Visual Check

6. Identify Which part of the body that is important to hearing could be added to this figure?

______

______

______

Review Vocabulary

wave
a disturbance in a material that transfers energy without transferring matter

Hearing

To hear sound, you need auditory (AU duh tor ee) receptors. The vibration of matter creates sound waves. These waves travel through air and other substances. Sound waves that enter your ears are detected by auditory receptors.

As sound waves travel through your ears, they are amplified, or increased. The sound waves move hair cells in your ears.

Hair cells send information about the sound waves to your brain. The brain processes information about the loudness and tone of the sound, and you hear.

The Parts of the Ear

The ear has many parts that detect sound waves. Your ear has three areas: the outer ear, the middle ear, and the inner ear, as shown in the figure above. Each area has a special function.

The Outer Ear The outer ear includes all the parts of the ear that you can see. The outer ear collects sound waves that cause the eardrum to vibrate. *The* ***eardrum*** *is a thin membrane between the outer ear and the inner ear.*

The Middle Ear The middle ear is between the outer ear and the inner ear. The vibrations pass from the eardrum to the small bones in the middle ear—the hammer, the anvil, and the stirrup. The movement of these bones amplifies the sound waves.

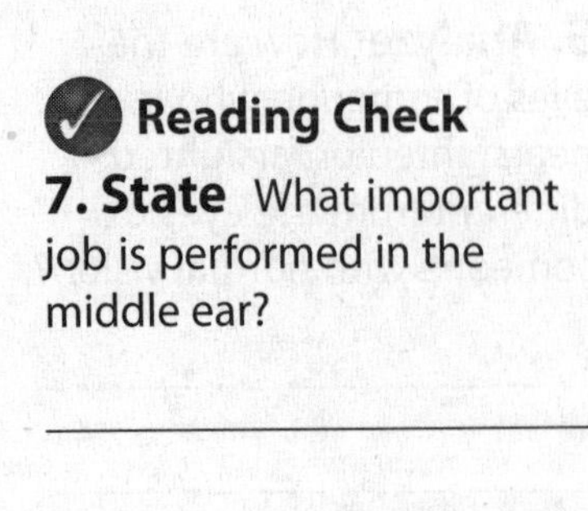

Reading Check

7. State What important job is performed in the middle ear?

______

______

______

The Inner Ear The inner ear is the part of the ear that detects sound waves. It changes sound waves into messages that are sent to the brain. Look again at the figure on the previous page. The inner ear contains the cochlea (KOHK lee ah). The cochlea looks like a snail shell and is filled with fluid. Sound waves that reach the cochlea cause the fluid to vibrate. The movement of the liquid in the cochlea is much like the movement of hot chocolate in a cup when you blow on it. When the fluid in the cochlea moves, it bends hair cells, which send messages to the brain for processing. You experience these signals as sound. ✓

The Ear and Balance

You learned that the ear detects sound waves. The inner ear sends messages about those waves to the brain. The inner ear has another function. Parts of the inner ear, called the semicircular canals, help you keep your balance.

The semicircular canals contain fluid and hair cells. This fluid moves the hair cells when you move your head. Information about that movement is sent to your brain. Your brain then signals muscles to move your head and body so you keep your balance.

Smell

With your eyes closed, you can tell the difference between an orange and a rotten egg. You can do this because of your sense of smell. Humans have hundreds of different receptors that detect odors, or smells. Some dogs have more than 1,000 different odor receptors.

Odors are molecules that chemical receptors in your nose can detect. These chemical receptors are called chemoreceptors (kee moh rih SEP turz). They send messages to your brain about odors. Your brain processes the information about odors. A smell might make you feel hungry. A smell might trigger a memory or feeling.

Taste

Your sense of taste also uses chemoreceptors. Chemoreceptiors on your tongue are called taste buds. When taste buds detect food or drink, they send messages to the brain to be processed. Taste buds can detect five different tastes: bitter, salty, sour, sweet, and umami (oo MAH mee). Umami is the taste of MSG (monosodium glutamate), a substance often used in processed food. ✓

✓ **Reading Check**

8. Name the three areas of the ear.

______

______

______

Think it Over

9. Apply If you had a problem with your inner ear, how might you feel?

______

______

______

✓ **Reading Check**

10. Name the five different tastes you can detect.

______

______

______

Visual Check

11. Identify (Circle the correct answer.) Taste buds are found

a. on cilia.

b. in sensory cells.

c. on the tongue.

12. Generalize Why doesn't food taste as good when you have a stuffy nose?

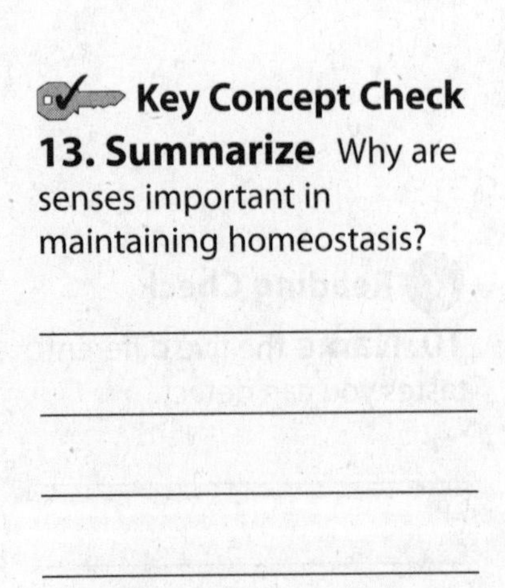

13. Summarize Why are senses important in maintaining homeostasis?

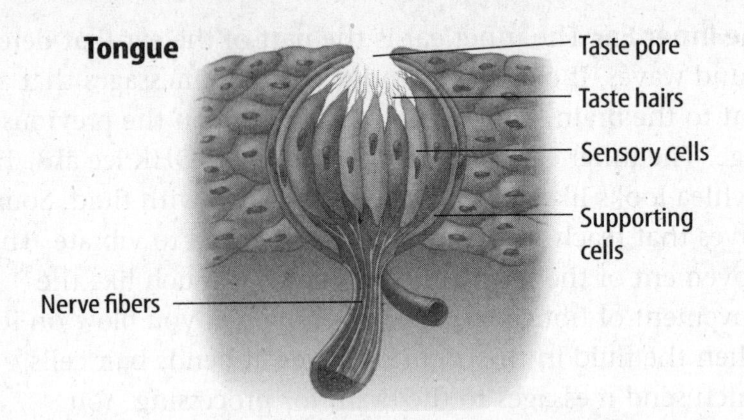

The figure above shows the tongue and the parts of a taste bud. Your senses of taste and smell work together to help you taste your food. Chemoreceptors detect the stimuli for taste and smell.

Touch

Like all the other senses, the sense of touch uses special receptors to detect the environment. Touch receptors are in your skin. They can detect temperature, pain, and pressure. Touch receptors send signals to the brain for processing.

These receptors are all over your body. Some areas, such as the palms of your hands and the soles of your feet, have lots of receptors. Other areas, such as the middle of your back, have fewer receptors. Just like the other four senses, touch receptors send messages to the brain for processing.

The Senses and Homeostasis

To survive, you need information about the environment. Your five senses get this information about the environment and send it to your nervous system. Your brain is then able to respond. It directs your body to maintain homeostasis.

Your senses are the first step in keeping your internal environment stable. They help you detect temperature changes and find food and water. They also help you avoid harmful environments.

After You Read

Mini Glossary

eardrum: a thin membrane between the outer ear and the inner ear

receptor: a special structure in the sensory system that detects stimuli from the environment

retina (RET nuh): an area at the back of the eye that includes two types of cells that contain photoreceptors—rod cells and cone cells

sensory system: the part of your nervous system that detects or senses the environment

1. Review the terms and their definitions in the Mini Glossary. Write two or three sentences that explain the role of the eardrum in hearing.

____

____

____

____

____

2. Fill in the following diagram to show the path of light in the eye.

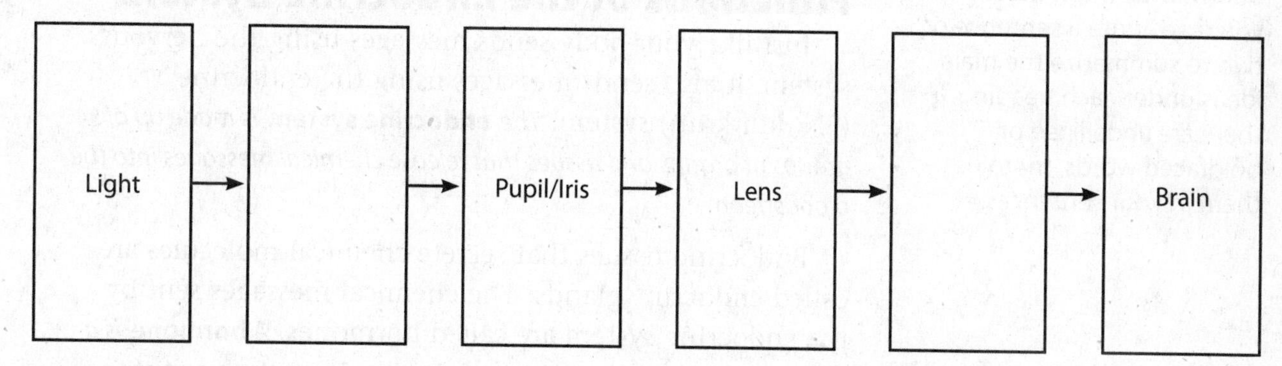

3. Name three of the five senses and give an example of each one being used.

____

____

____

What do you think NOW?

Reread the statements at the beginning of the lesson. Fill in the After column with an A if you agree with the statement or a D if you disagree. Did you change your mind?

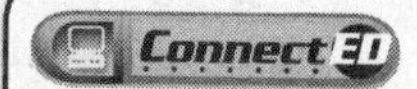

Log on to ConnectED.mcgraw-hill.com and access your textbook to find this lesson's resources.

CHAPTER 18
LESSON 3

Control and Coordination

The Endocrine System

Key Concepts
- What does the endocrine system do?
- How does the endocrine system interact with other body systems?

Before You Read

What do you think? Read the two statements below and decide whether you agree or disagree with them. Place an A in the Before column if you agree with the statement or a D if you disagree. After you've read this lesson, reread the statements to see if you have changed your mind.

Before	Statement	After
	5. Positive feedback systems in humans help maintain homeostasis.	
	6. Endocrine glands secrete hormones.	

Study Coach

Summarize Main Ideas As you read, write a sentence or two to summarize the main ideas under each heading. If there are underlined or boldfaced words, try to use them in your sentences.

Read to Learn

Functions of the Endocrine System

Just like your body sends messages using the nervous system, it also sends messages using the endocrine (EN duh krun) system. *The* **endocrine system** *is made up of groups of organs and tissues that release chemical messages into the bloodstream.*

Endocrine tissues that secrete chemical molecules are called endocrine glands. The chemical messages sent by the endocrine system are called hormones. *A* **hormone** *is a chemical that is produced by an endocrine gland in one part of an organism and is carried in the bloodstream to another part of the organism.*

Recall that the nervous system sends information quickly. Signals are sent to the cells. Information carried by the endocrine system is sent more slowly. Its messages have to travel through the bloodstream to other parts of the body.

The chemical messages sent by the endocrine system are usually sent to many more cells than the nervous system's messages. The messages also last longer. A message from the nervous system might cause a sudden body movement. A message from the endocrine system might signal your body to grow taller over a period of time.

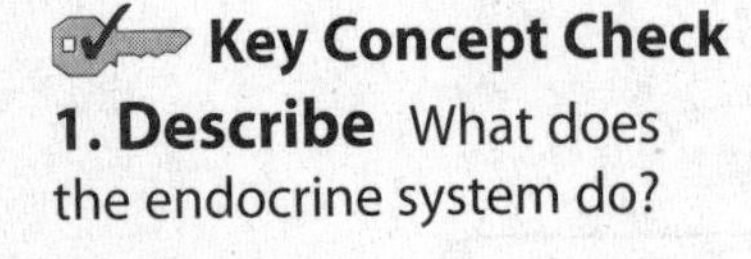

Key Concept Check
1. Describe What does the endocrine system do?

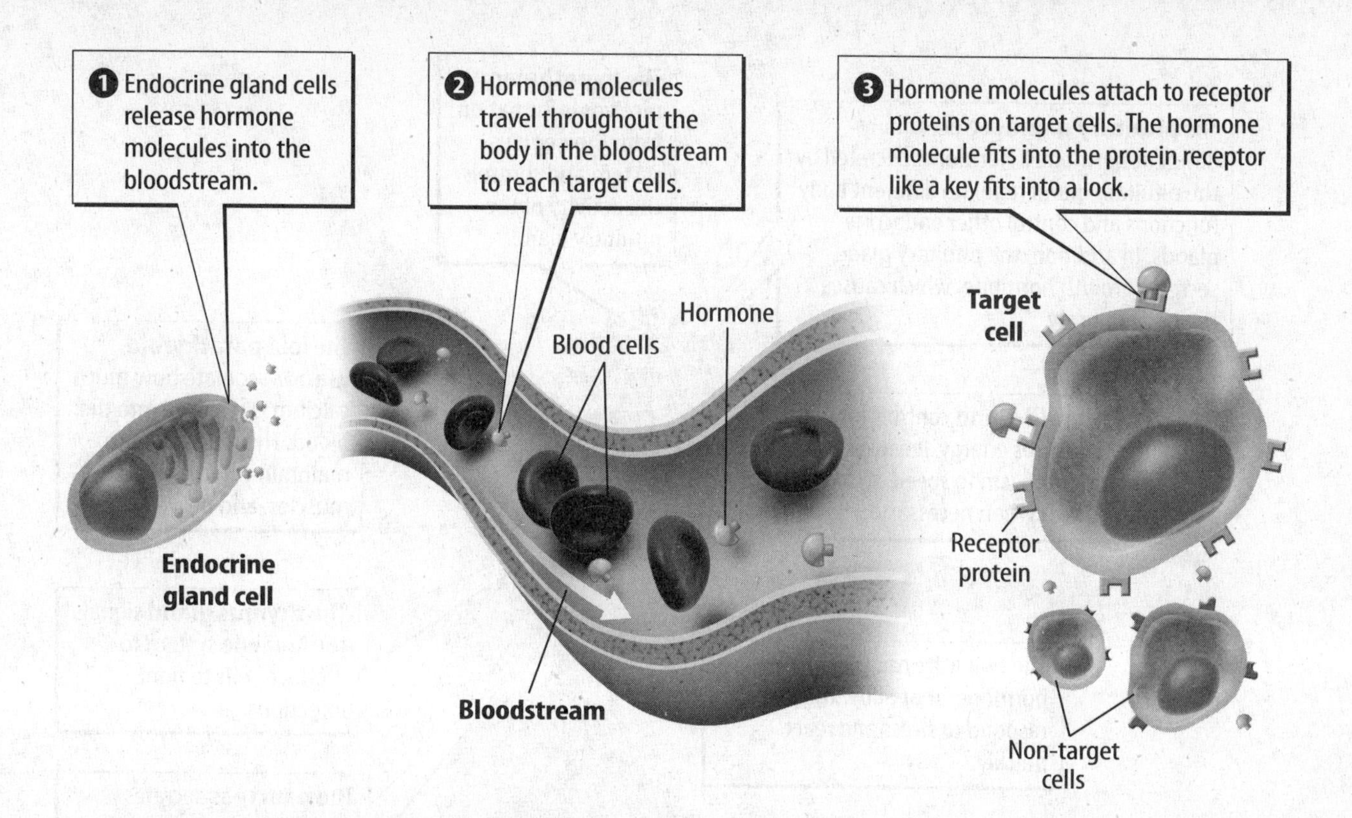

Endocrine Glands and Their Hormones

Hormones move from the endocrine glands to other parts of the body in the bloodstream. Hormones carry messages to specific cells called target cells. The hormone message changes the functions of organs and tissues. The target cells have certain receptor proteins on or inside them. The hormones recognize the receptor proteins of their target cells and attach to them. When a hormone finds its target cell, it binds to a receptor protein and delivers its chemical message. The target cell responds by taking the action called for in the message.

The figure above shows the path a hormone takes. The endocrine gland produces the hormone and it travels to the target cell where it delivers its message.

You may be familiar with some of the glands of the endocrine system. Ovaries in females and testes in males are endocrine glands. The hypothalamus is a gland that controls the activity of the pituitary gland. The pituitary gland is a gland that secretes hormones that control other endocrine glands. It also secretes growth hormone.

Visual Check

2. Analyze Why doesn't the hormone molecule attach to non-target cells?

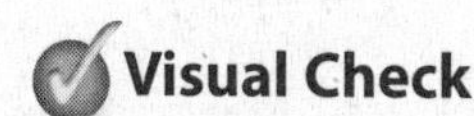

Use a two-column chart to take notes about each endocrine gland and what its hormones do.

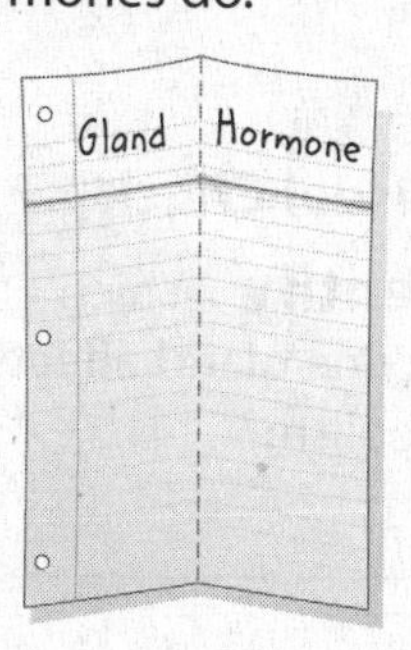

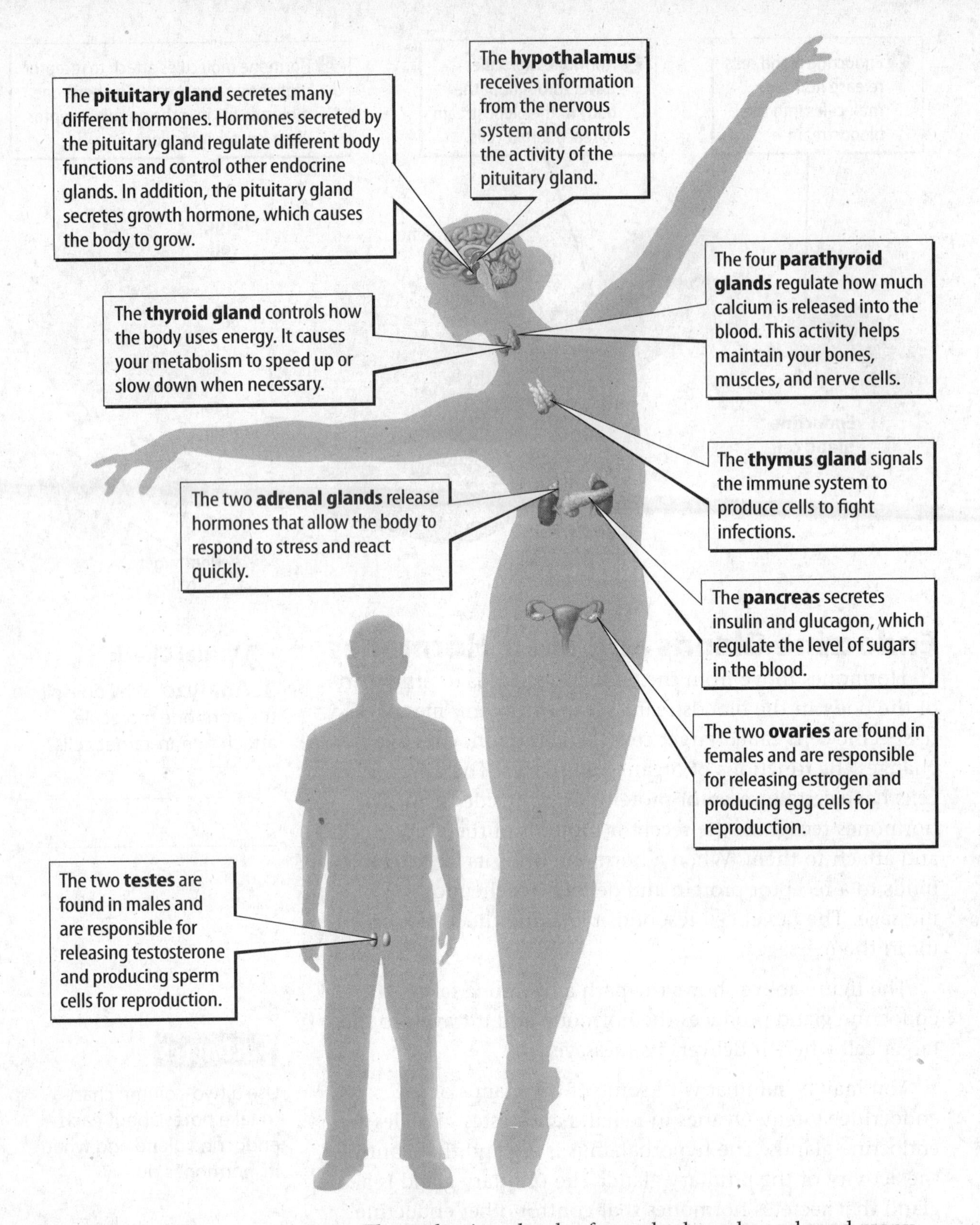

Visual Check

3. Identify Which endocrine gland affects metabolism?

____

The endocrine glands of your body make and send many different hormones to your body to maintain homeostasis. The hormones have many functions. The figure above shows the endocrine glands in the body and the hormones they make.

The Endocrine System and Homeostasis

Recall that the nervous system responds to changes in your environment and maintains homeostasis. The endocrine system also helps maintain homeostasis. Its glands release hormones in response to stimuli. These hormones change the function of other tissues and organs in the body to keep internal conditions stable.

Negative Feedback Systems

The way hormones maintain homeostasis is called negative feedback. **Negative feedback** *is a control system in which the effect of a hormone prevents further release of that hormone.* The endocrine system uses negative feedback and controls how much hormone a gland releases. The figure below shows one way in which negative feedback works.

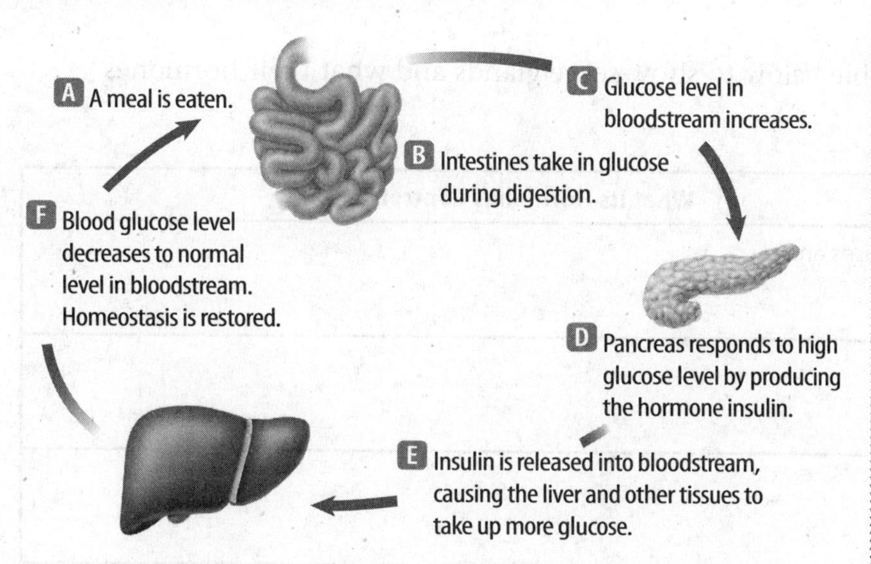

You can understand negative feedback by thinking about a thermostat in a house. A thermostat controls the temperature inside a house. When the temperature in the house drops below the preset temperature, the thermostat signals the furnace to turn on. When the house warms to the preset temperature, the thermostat signals the furnace to shut off.

Positive Feedback Systems

Negative feedback helps hormones maintain homeostasis in the body. Positive feedback does not. **Positive feedback** *is a control system in which the effect of a hormone causes the release of more of the hormone.* Because positive feedback does not help maintain homeostasis, your body uses fewer positive feedback systems.

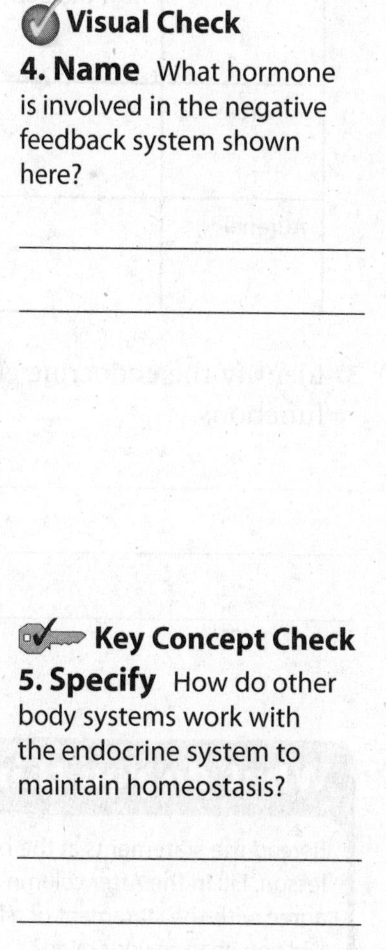

Visual Check

4. Name What hormone is involved in the negative feedback system shown here?

____

____

Key Concept Check

5. Specify How do other body systems work with the endocrine system to maintain homeostasis?

____

____

____

After You Read

Mini Glossary

endocrine (EN duh krun) system: groups of organs and tissues that release chemical messages into the bloodstream

hormone: a chemical that is produced by an endocrine gland in one part of an organism and is transported in the bloodstream to another part of the organism

negative feedback: a control system in which the effect of a hormone inhibits further release of the hormone

positive feedback: a control system in which the effect of a hormone causes the release of more of the hormone

1. Review the terms and their definitions in the Mini Glossary. In a sentence or two, explain the difference between positive feedback and negative feedback.

2. Fill in the blanks in the table below to show some glands and what their hormones control.

Gland	What Its Hormones Control
	how the body uses energy
Pancreas	
Adrenal	

3. Identify the endocrine glands that differ in males and females and discuss their functions.

What do you think NOW?

Reread the statements at the beginning of the lesson. Fill in the After column with an A if you agree with the statement or a D if you disagree. Did you change your mind?

Log on to ConnectED.mcgraw-hill.com and access your textbook to find this lesson's resources.

Reproduction and Development

The Reproductive System

Before You Read

What do you think? Read the three statements below and decide whether you agree or disagree with them. Place an A in the Before column if you agree with the statement or a D if you disagree. After you've read this lesson, reread the statements to see if you have changed your mind.

Before	Statement	After
	1. Reproduction ensures that a species survives.	
	2. The male reproductive system has internal and external parts.	
	3. The menstrual cycle occurs in males and females.	

Key Concepts
- What does the reproductive system do?
- How do the parts of the male reproductive system work together?
- How do the parts of the female reproductive system work together?
- How does the reproductive system interact with other body systems?

Read to Learn

Functions of the Reproductive System

The human reproductive system is a group of tissues and organs. Male and female reproductive cells join and form new offspring. Human males produce **sperm,** *the male reproductive cells.* Females produce **eggs,** *the female reproductive cells.*

During reproduction, a sperm joins with an egg, as shown in the figure below. Development of a human begins. This usually happens in the female's reproductive system. The female reproductive system then nourishes the developing human.

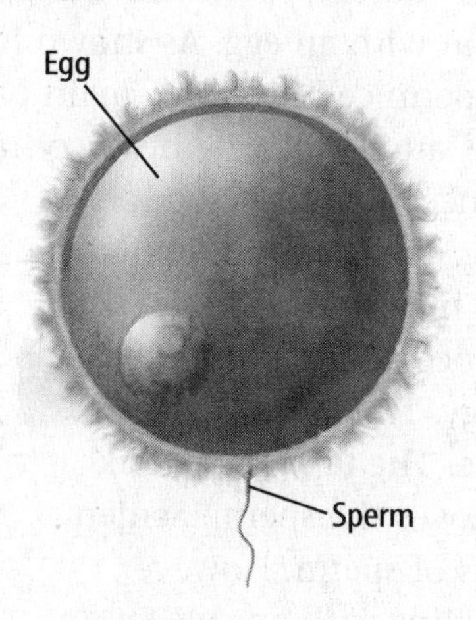

Mark the Text

Build Vocabulary Skim the lesson. On a sheet of paper, write any words that you do not know. After you read the lesson, look up the definitions for any words that you cannot define.

Key Concept Check
1. Explain What does the reproductive system do?

Visual Check
2. Label the reproductive cells in the figure as either *male* or *female.*

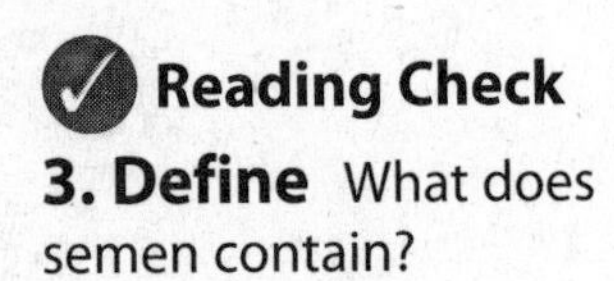

Reading Check
3. Define What does semen contain?

The Male Reproductive System

The main function of the male reproductive system is to produce and transport sperm to the female reproductive system. The parts of the male reproductive system are shown in the figure below. *The* **testis** (TES tihs; plural, testes) *is the male reproductive organ that produces sperm.* A male's two testes are inside a scrotum (SKROH tum). A scrotum is a saclike structure outside the body.

Sperm develop in the testes. Then they move to a tube called the sperm duct, where they are stored. During storage, sperm mature. They also develop the ability to swim. As mature sperm move through the male reproductive system, they mix with fluids produced by glands. *This mixture of sperm and fluids is called* **semen** (SEE mun). Semen contains nutrients for the sperm. Semen leaves the body through the penis. *The* **penis** *is a tubelike structure that delivers sperm to the female reproductive system.*

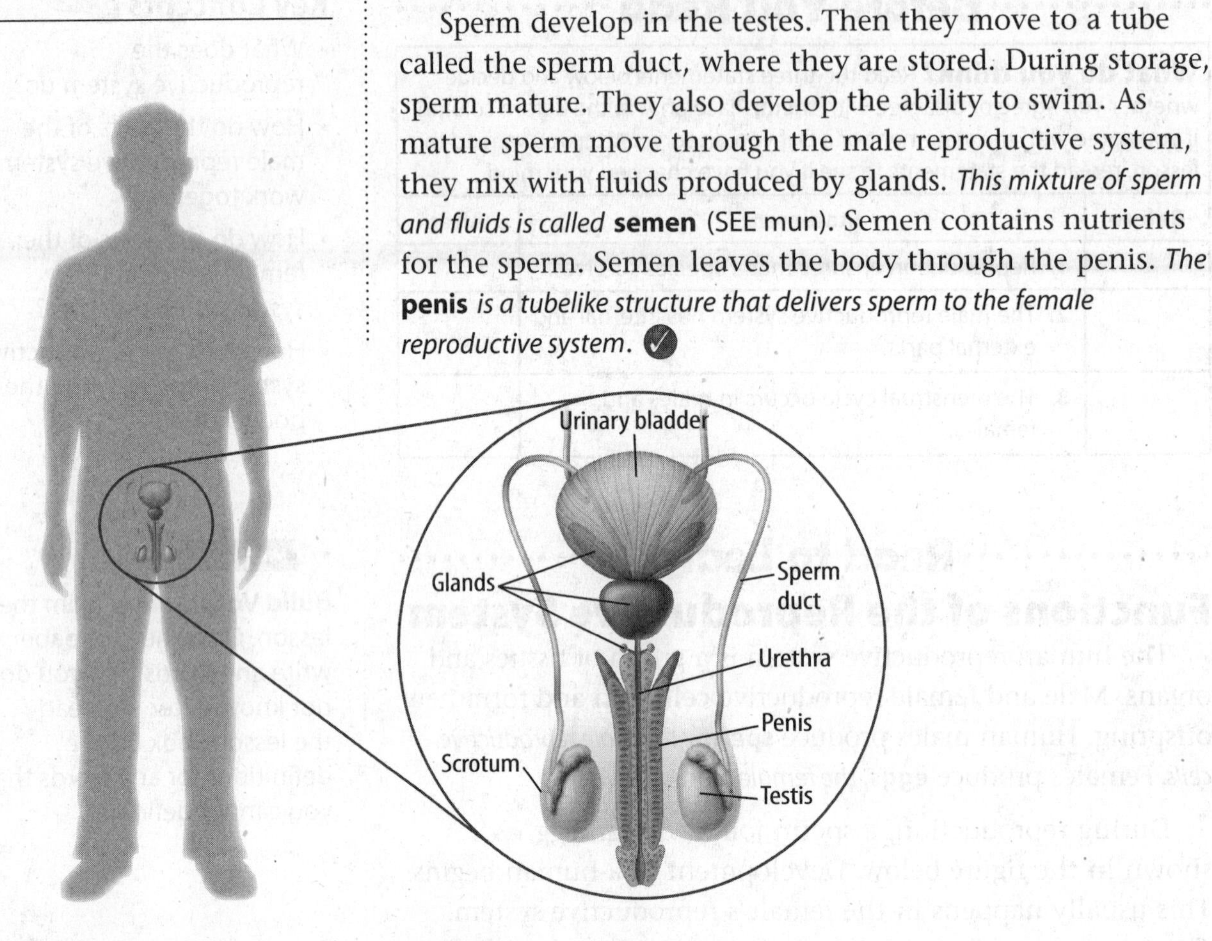

Visual Check
4. Specify How many different glands are shown in the male reproductive system?

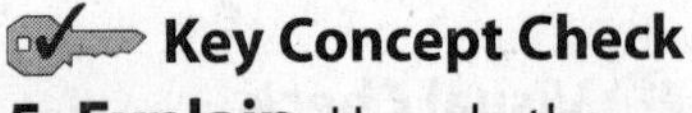

Key Concept Check
5. Explain How do the parts of the male reproductive system work together?

Sperm

When they are mature, sperm leave the body through the penis and can join with an egg. As shown in the figure to the right, a mature sperm cell has three main parts—a head, a midpiece, and a tail. The head contains DNA and substances that help the sperm join with the egg. The midpiece contains organelles called mitochondria. The mitochondria process food molecules and release energy. This energy helps a sperm's tail move. The tail whips back and forth and moves the sperm. Semen contains millions of sperm. However, only one sperm joins with an egg.

Head
Midpiece
Tail

The Female Reproductive System

Unlike in males, all parts of the female reproductive system are inside the body. The parts of the female reproductive system are shown in the figure below.

Eggs are female reproductive cells. Females produce immature eggs called oocytes (OH uh sites). *An* **ovary** (OH vah ree) *is an organ where oocytes are stored and mature.* A mature oocyte is called an egg, or ovum (plural, *ova*). Just as males have two testes, females have two ovaries.

About once a month, an ovary releases an egg. The egg enters a fallopian (fuh LOH pee un) tube. Short, hairlike structures called cilia move the egg through the fallopian tube toward the uterus. *The part of the female reproductive system that connects the uterus to the outside of the body is the* **vagina.** Sperm enter the female reproductive system through the vagina. The vagina is also called the birth canal because a baby moves through this structure during its birth.

If the egg is fertilized by a sperm, the uterus provides a nourishing environment for a fertilized egg's development. You will read more about how humans develop in the next lesson.

6. Contrast How is an oocyte different from an egg?

Visual Check

7. Locate Between which structures are the fallopian tubes located?

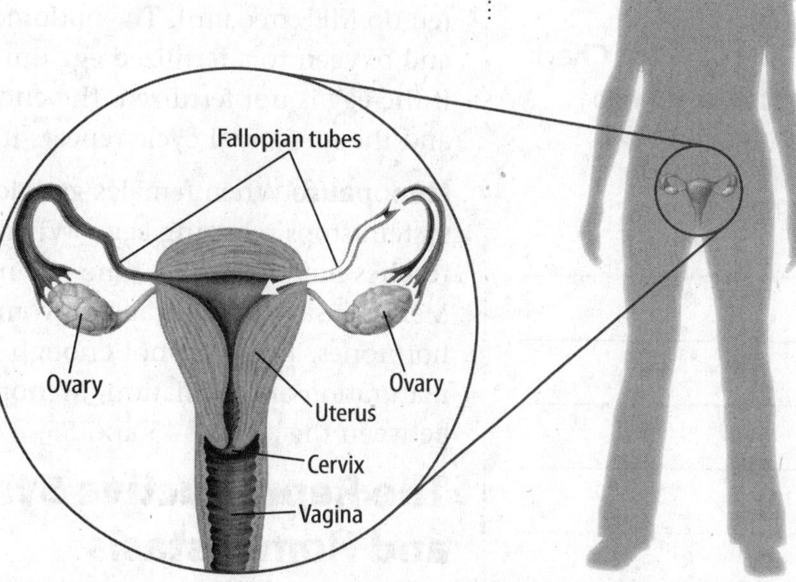

The Egg

Each oocyte in an ovary is surrounded by follicle cells. The follicle cells release hormones that help the oocytes develop into eggs. As they prepare to release eggs, the follicle cells increase in size. An egg contains DNA and substances that nourish it. A mature egg is round and about 2,000 times larger than a sperm. A female usually releases only one egg from an ovary about one time each month. ✓

Reading Check

8. Specify How many eggs are usually released at one time?

FOLDABLES

Use a three-tab book to organize information about the phases of the menstrual cycle.

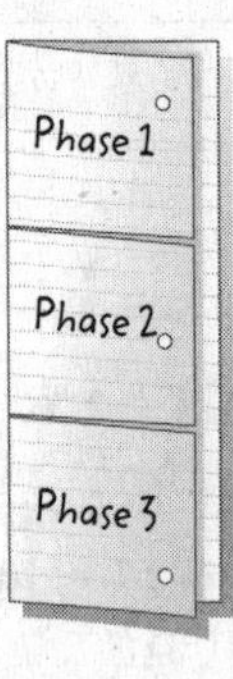

The Menstrual Cycle

An egg is released only when the uterus is prepared to nourish it. *The ovaries and the uterus go through reproductive-related changes called the* ***menstrual*** *(MEN stroo ul)* ***cycle.*** The menstrual cycle is caused by chemical signals called hormones. One menstrual cycle is about 28 days long. It has three phases.

Phase 1 The menstrual cycle begins with a process called menstruation (men stroo WAY shun). During menstruation, tissue, fluid, and blood cells pass from the uterus through the vagina and are removed from the body. Menstruation usually lasts about five days.

Phase 2 The tissue lining the uterus thickens. In the ovary, several oocytes begin maturing. Usually, only one egg survives. *Near the end of this phase, hormones cause an egg to be released from the ovary in a process called* ***ovulation.***

Phase 3 The tissue lining the uterus continues to thicken. *If sperm are present, the egg might join with a sperm in a process called* ***fertilization.***

The tissue lining of the uterus is called the endometrium (en do MEE tree um). The endometrium provides nutrients and oxygen to a fertilized egg during its early development. If the egg is not fertilized, the endometrium breaks down and the menstrual cycle repeats itself.

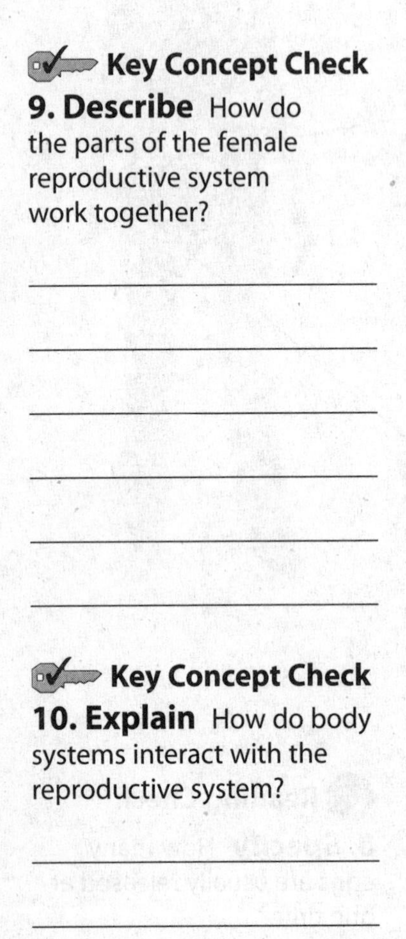

Key Concept Check
9. Describe How do the parts of the female reproductive system work together?

Key Concept Check
10. Explain How do body systems interact with the reproductive system?

Menopause When females get older, the reproductive system stops releasing eggs. When this happens, a woman reaches menopause—a time when the menstrual cycle stops. Menopause occurs because a woman's ovaries produce fewer hormones. There are not enough hormones to cause oocyte maturation and ovulation. Menopause usually happens between the ages of 45 and 55.

The Reproductive System and Homeostasis

The reproductive system does not help the body maintain homeostasis. However, reproduction ensures the survival of the human species. Hormones that are produced by the endocrine system control when eggs and sperm mature. The reproductive systems of males and females work together with the endocrine system.

After You Read

Mini Glossary

egg: the female reproductive cell

fertilization: the process by which a sperm joins with an egg

menstrual (MEN stroo ul) cycle: the reproductive-related changes that the ovaries and the uterus go through

ovary (OH vah ree): the organ where oocytes are stored and mature

ovulation: the process in which hormones cause an egg to be released from the ovary

penis: a tubelike structure that delivers sperm to the female reproductive system

semen (SEE mun): a mixture of sperm and fluids produced by several glands

sperm: the male reproductive cell

testis (TES tihs): the male reproductive organ that produces sperm (plural, testes)

vagina: the part of the female reproductive system that connects the uterus to the outside of the body

1. Review the terms and their definitions in the Mini Glossary. Choose three terms from the glossary and use them in a sentence that shows you understand what they mean.

____

____

2. Decide whether each structure listed in the chart is part of the male or female reproductive system. Make a check mark in the correct box.

Structure	Male	Female
a. oocyte	☐	☐
b. testis	☐	☐
c. sperm	☐	☐
d. ovary	☐	☐
e. uterus	☐	☐
f. scrotum	☐	☐

3. Review the words that you circled in the lesson. Choose one of these words and write its definition below.

____

____

What do you think NOW?

Reread the statements at the beginning of the lesson. Fill in the After column with an A if you agree with the statement or a D if you disagree. Did you change your mind?

Log on to ConnectED.mcgraw-hill.com and access your textbook to find this lesson's resources.

Reproduction and Development

Human Growth and Development

Key Concepts

- What happens during fertilization of a human egg?
- What are the major stages in the development of an embryo and a fetus?
- How do the life stages differ after birth?

Before You Read

What do you think? Read the three statements below and decide whether you agree or disagree with them. Place an A in the Before column if you agree with the statement or a D if you disagree. After you've read this lesson, reread the statements to see if you have changed your mind.

Before	Statement	After
	4. Eggs are fertilized in the ovary.	
	5. Lead is a nutrient that helps a fetus develop.	
	6. Puberty occurs during adolescence.	

Study Coach

Two-Column Notes Organize your notes into two columns. On the left, list the main idea in the section under each head. On the right, list the details that support the main idea. Use your notes to review the lesson.

Read to Learn

Stages of Development

At a family gathering, you might see people of every age. There are probably babies, children, teens, and adults. As people grow, they go through stages of development. The stages are infancy, childhood, adolescence (a duh LES unts), and adulthood. These stages are based on major developments that take place during each stage. Read about each stage in the table below.

Life Stage	Development
Infancy	The nervous system and muscular system develop rapidly.
Childhood	The abilities to speak, read, write, and reason develop.
Adolescence	A person becomes physically able to reproduce.
Adulthood	Growth of the muscular system and skeletal system stops.

1. Specify At which life stage does a person learn to read?

The time before a person is born is also divided into stages. The first stage before birth is fertilization.

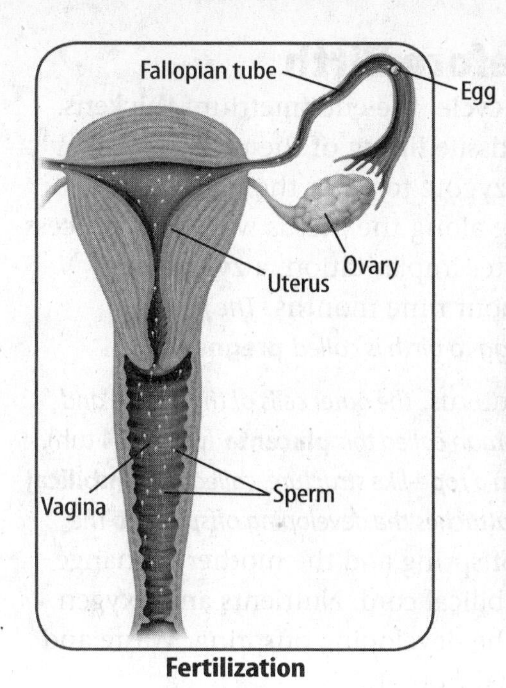

Fertilization

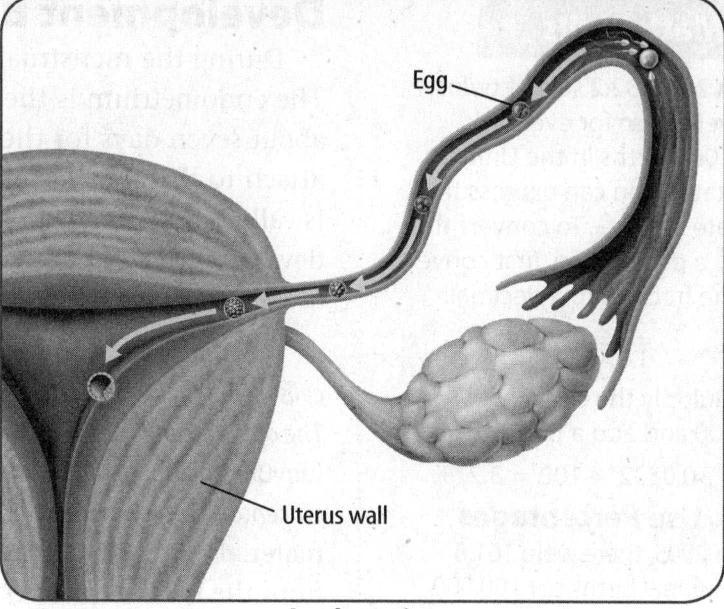

Implantation

Fertilization

The figure above on the left shows how fertilization occurs. If sperm enter the vagina, they can travel to the uterus and up into the fallopian tubes. The fallopian tubes are tubelike structures that connect the uterus to the ovaries.

Fertilization occurs in the fallopian tubes. A sperm contains substances that help it join with an egg. Once a sperm enters an egg, the egg's cell membrane stops other sperm from entering. When the nucleus of the sperm joins with the nucleus of the egg, fertilization is complete.

Zygote Formation

A fertilized egg is called a **zygote** (ZI goht). Human zygotes contain 46 chromosomes of DNA. Half of these chromosomes (23) come from the sperm cell, and half (23) come from the egg cell. This means that a zygote is a diploid cell. As shown above on the right, the zygote moves through the fallopian tube to the uterus. As it moves toward the uterus, the zygote undergoes mitosis and cell division many times. It develops into a ball of cells with a group of cells inside.

Multiple Births

The group of cells inside the zygote is called the inner cell mass. The inner cell mass can develop into a baby. Sometimes the inner cell mass divides in two. When one zygote contains two inner cell masses, identical twins can develop. Identical twins are the same gender and usually look alike. If two eggs are released and fertilized, fraternal twins can result. Fraternal twins can be different genders.

Visual Check

2. Locate Highlight the structure where fertilization occurs.

Key Concept Check

3. Explain What happens during fertilization?

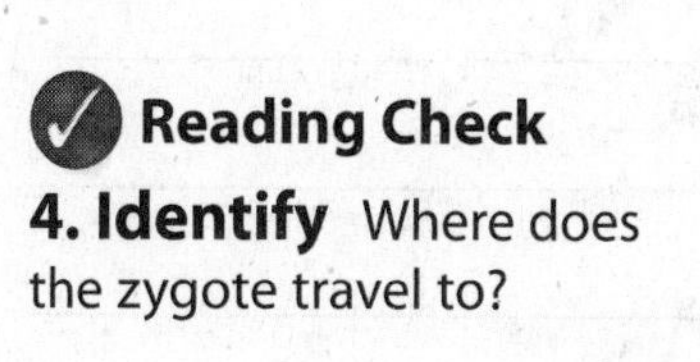

Reading Check

4. Identify Where does the zygote travel to?

Development Before Birth

During the menstrual cycle, the endometrium thickens. The endometrium is the tissue lining of the uterus. It takes about seven days for the zygote to enter the uterus and attach to this tissue lining along the uterus wall. This process is called implantation. After implantation, a zygote can develop into a baby in about nine months. *The period of development from fertilized egg to birth is called* **pregnancy.**

After attaching to the uterus, *the outer cells of the zygote and cells from the uterus form an organ called the* **placenta** (pluh SEN tuh). *The outer zygote cells also form a rope-like structure, called the* **umbilical** (um BIH lih kul) **cord,** *which attaches the developing offspring to the placenta.* The developing offspring and the mother exchange materials through the umbilical cord. Nutrients and oxygen from the mother pass to the developing offspring. Waste and carbon dioxide are removed from it.

From Zygote to Embryo

From the time the zygote attaches to the uterus until the end of the eighth week of pregnancy, it is called an **embryo.** During this time, cells divide, grow, and gain unique functions. As shown in the chart below, the heart, brain, limbs, fingers, and toes start to form. When the heart forms, the embryo develops a circulatory system.

From Embryo to Fetus

During the time between nine weeks and birth, the developing offspring is called a **fetus.** Organ systems begin to work. The fetus is now able to move its arms and legs. During the remaining time until birth, the fetus continues to grow in size. It can respond to sounds from outside the uterus, such as its mother's voice.

Prenatal Development

Time	Length	Development
5 weeks	7 mm	The heart, brain, and other organs start to develop. Arms and legs begin to bud.
8 weeks	2.5 cm	The heart is fully formed. The bones start to harden. Nearly all muscles have appeared.
14 weeks	6 cm	Growth and development continue.
16 weeks	15 cm	The fetus can make a fist. It can also make facial expressions.
22 weeks	27 cm	Footprints and fingerprints are forming.

Math Skills

In 2005, 32.2 sets of twins were born for every 1,000 births in the United States. You can express this rate as $\frac{32.2}{1,000}$. To convert this to a percentage, first convert the fraction to a decimal:

$$\frac{32.2}{1,000} = 0.0322$$

Multiply the decimal by 100 and add a percent sign.

$$0.0322 \times 100 = 3.22\%$$

5. Use Percentages In 2005, there were 161.8 multiple births per 100,000 live births. What percent of live births were multiple births?

Key Concept Check

6. Name What are the major stages in the development of a fetus?

Visual Check

7. Interpret A fetus can smile at about how many weeks?

Fetal Health

A fetus receives all nutrients from its mother. As a result, the fetus's growth and development depend on the food, water, and other nutrients that its mother eats and drinks. Other factors can also affect a fetus.

ACADEMIC VOCABULARY
affect
(verb) to influence or alter an outcome

Nutrition A pregnant woman must take in enough protein and vitamins for her fetus to grow and develop. Cells need nutrients, such as vitamins, and proteins for growth.

Environmental Factors A woman's uterus protects a fetus. However, the mother's contact with substances in the environment, such as smoke and chemicals, can harm her fetus. Heavy metals such as lead and mercury can affect the growth and development of the fetus.

8. Summarize What are some things that a pregnant woman can do to protect the health of her baby?

Drugs and Alcohol When a woman drinks alcohol during pregnancy, her fetus takes in the alcohol through the placenta. The baby that develops can be born with fetal alcohol syndrome.

A baby with fetal alcohol syndrome can have many problems. As illustrated in the figure below, the baby's brain might not develop properly. The baby might also have lifelong problems with growth, vision, and hearing. If a woman takes drugs such as cocaine or tobacco during pregnancy, these substances can also harm her fetus.

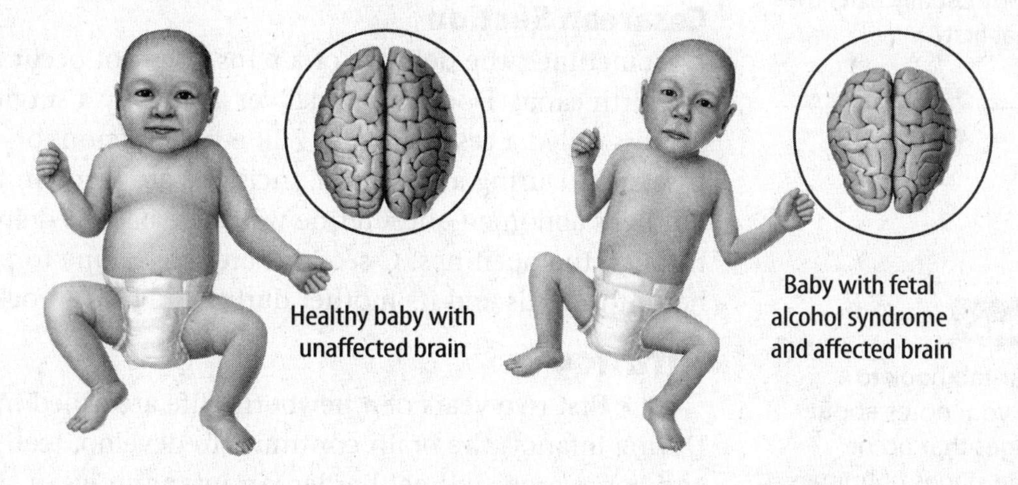

Birth

A fetus leaves its mother's body through a process called birth. Like the menstrual cycle, the birth of a fetus requires hormones. Hormones cause changes in the female reproductive system. These changes, called labor, help a fetus leave the uterus.

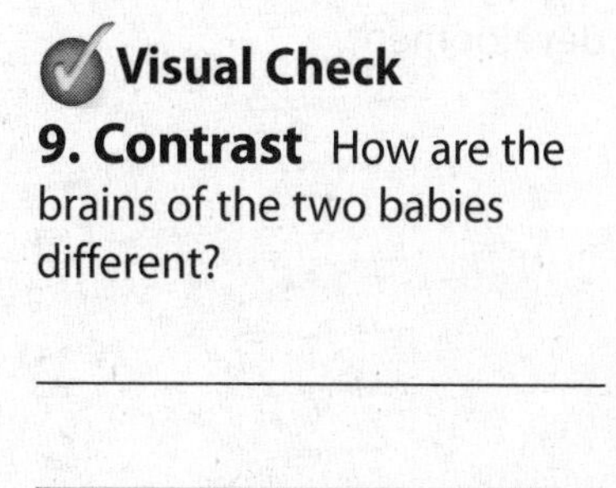

Visual Check

9. Contrast How are the brains of the two babies different?

Labor and Delivery

Labor begins when hormones that are released by the endocrine system cause muscles in the uterus to contract. Also, *a small structure between the uterus and the vagina, called the* **cervix** (SUR vihks), begins to open.

The figure below shows the birth process. As labor continues, the cervix opens wider. Muscles in the uterus contract faster and more strongly. The contractions push a fetus into the vagina, or birth canal, and out of the woman's body. After the fetus is delivered, the placenta breaks away from the uterus and also exits the woman's body through the vagina.

Reading Check

10. Specify Name three things that happen during labor and delivery.

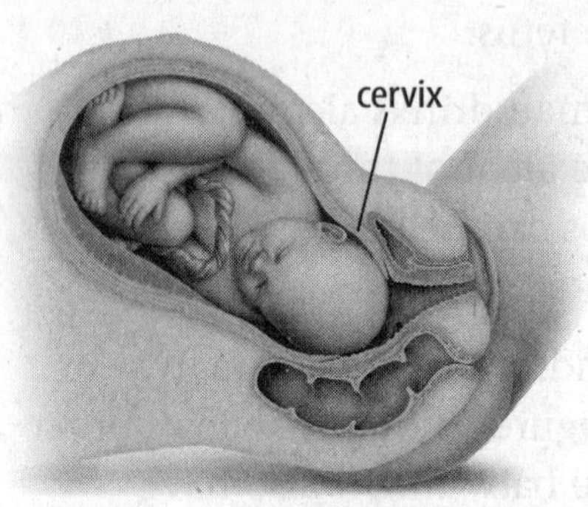

Stage 1 As the fetus moves into the birth canal, the opening to the uterus widens.

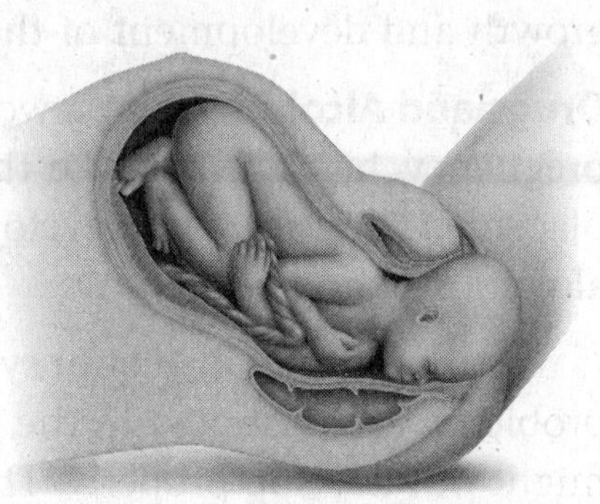

Stage 2 Muscle contractions help push the fetus out through the birth canal.

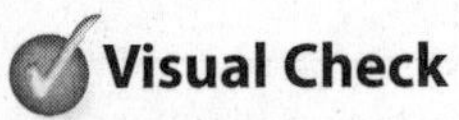

Visual Check

11. Interpret Based on this figure, which part of the baby's body usually exits the birth canal first?

Cesarean Section

Sometimes, the delivery of a fetus does not occur through the birth canal. Doctors can deliver a fetus by a surgical process called a cesarean (suh ZER ee un) section, or C-section. During a C-section, incisions are made in the mother's abdomen and uterine wall. The baby is delivered through the openings. C-sections are often done to prevent harm to a fetus and its mother during the birth process.

Infancy

The first two years of a newborn's life are called infancy. During infancy, the brain continues to develop, teeth form, and bones grow and get harder. An infant grows in size. Organ systems continue to mature. An infant learns to speak and begins to eat solid food.

Infants also develop the ability to move in different ways over time. The figure on the next page shows milestones that infants reach at different months of age. The milestones are events that mark progress as infants develop the ability to move around independently. Individual infants do not reach each milestone at exactly the same time.

FOLDABLES®

Use a four-tab book to organize your notes about the changes that occur during the stages of human development.

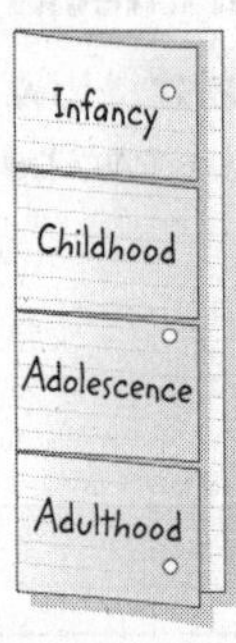

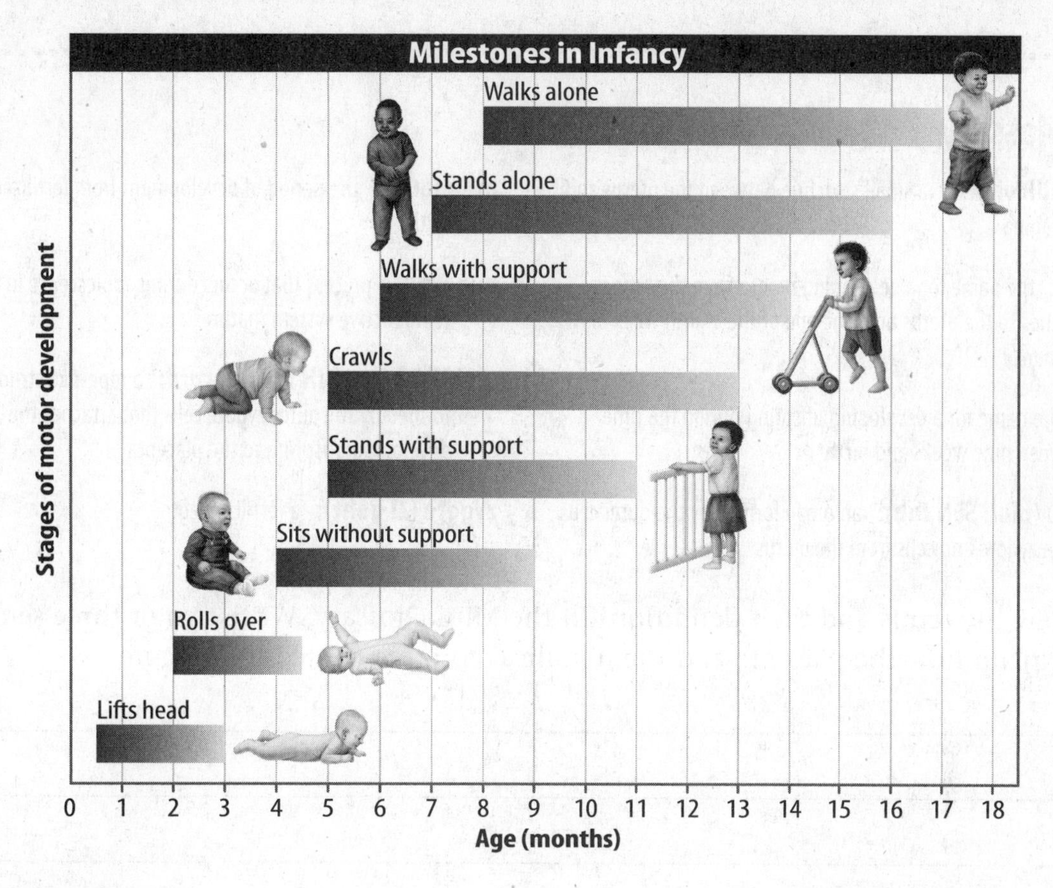

Childhood

The period following infancy is called childhood. During this time, the brain continues to grow and develop, and thinking improves. Muscles strengthen. Arms and legs grow longer.

Visual Check

12. Discover At about what age do infants sit without support?

Adolescence

Adolescence is a period of growth following childhood. *During adolescence, the reproductive system matures in a process called* **puberty** (PYEW bur tee). Hormones cause the changes that occur during puberty. In males, the voice deepens. Muscles increase in size. Hair grows on the face, in the pubic area, and under the arms. In females, breasts develop. Hair appears in the pubic area and under the arms. The buttocks and thighs gain fatty tissue.

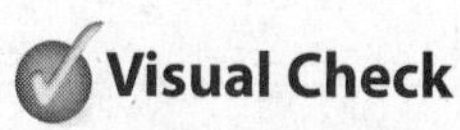

13. Express What substances in the body make puberty happen?

Adulthood and Aging

Adulthood is the period after adolescence that continues through old age. Adults will not grow taller. Aging is the process of changes in the body over time. Hair can turn white or gray and stop growing. The skin wrinkles. The skeletal system does not work as well. Vision and hearing decline. Bones become weaker, and the digestive system slows down.

Key Concept Check

14. Contrast How do the life stages after birth differ?

After You Read

Mini Glossary

cervix (SUR vihks): a small structure between the uterus and the vagina

embryo: the name for a developing zygote from the time it attaches to the uterus until the end of the eighth week of pregnancy

fetus: the name for a developing offspring during the time between nine weeks and birth

placenta (pluh SEN tuh): an organ formed by the outer cells of the zygote and cells from the uterus

pregnancy: the period of development from fertilized egg to birth

puberty: a process that occurs during adolescence in which the reproductive system matures

umbilical (um BIH lih kul) cord: a rope-like structure formed by the outer zygote cells that attaches the developing offspring to the placenta

zygote (ZI goht): a fertilized egg

1. Review the terms and their definitions in the Mini Glossary. Write two or three sentences to explain how the placenta and the umbilical cord are related to a zygote.

____

____

____

____

____

2. Use the flowchart below to show the order of events leading to birth. Complete the flowchart by writing each term below in the correct box.

embryo **fetus** **zygote** **fertilization** **labor**

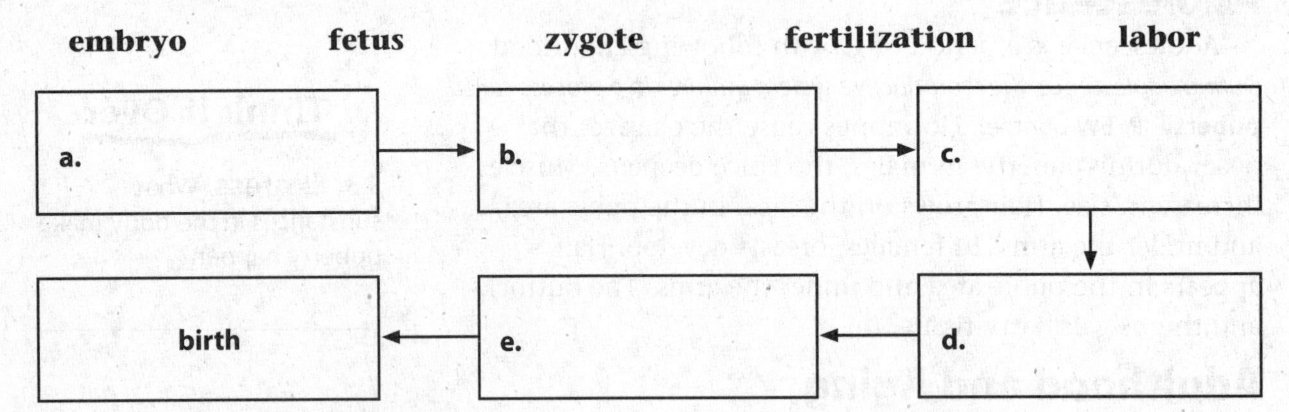

What do you think NOW?

Reread the statements at the beginning of the lesson. Fill in the After column with an A if you agree with the statement or a D if you disagree. Did you change your mind?

Log on to ConnectED.mcgraw-hill.com and access your textbook to find this lesson's resources.

Matter and Energy in the Environment

Abiotic Factors

Before You Read

What do you think? Read the two statements below and decide whether you agree or disagree with them. Place an A in the Before column if you agree with the statement or a D if you disagree. After you've read this lesson, reread the statements to see if you have changed your mind.

Before	Statement	After
	1. The air you breathe is mostly oxygen.	
	2. Living things are made mostly of water.	

Key Concept

- What are the nonliving parts of an environment?

Read to Learn

What is an ecosystem?

All organisms need both living and nonliving things to survive. Bees help the flowering plants reproduce. The flowers have nectar that the bees use to make honey. Flowers and bees, like all living organisms, also need nonliving things to survive, such as sunlight, air, and water.

An **ecosystem** *is all the living and nonliving things in a given area.* Ecosystems come in all sizes. An entire forest can be an ecosystem. A rotting log, a pond, a desert, an ocean, and your neighborhood are all ecosystems.

Biotic (bi AH tihk) **factors** *are the living things in an ecosystem.* Plants and animals are biotic factors. **Abiotic** (ay bi AH tihk) **factors** *are the nonliving things in an ecosystem.* Sunlight, water, and rocks are abiotic factors. 

Biotic factors and abiotic factors depend on each other. If one factor in an ecosystem is disturbed, other parts of the ecosystem are affected. During times of water shortages, or severe droughts, many fish in rivers and lakes will die. Animals that eat fish will have to find other food. This abiotic factor, lack of water, affects the biotic factors in the ecosystem, such as the fish and the other animals that feed on the fish.

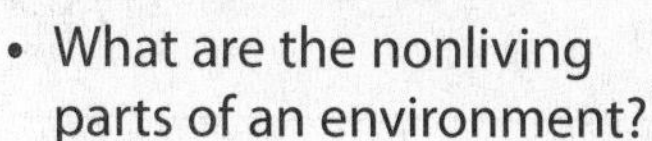

Make an Outline As you read, make an outline to summarize the information in this lesson. Use the main headings in the lesson as the main headings in the outline. Complete the outline using the information under each heading.

Reading Check

1. Contrast Underline the major difference between biotic and abiotic factors.

2. Explain Why would there be no food for people if there were no water or soil?

____

____

____

What are the nonliving parts of an ecosystem?

Think about the ecosystem around you. How do the abiotic factors in your own ecosystem affect you? You need sunlight for warmth. You need air to breathe. You need water to drink. Without sunlight, air, water, and soil, you would not have food. In the figure below, all of the nonliving parts of the environment affect all living things.

Visual Check

3. Identify Highlight the nonliving parts of the ecosystem that affect all living things.

The Sun

Almost all energy on Earth comes from the Sun. The Sun provides warmth and light. Many plants use sunlight and make food, as you will read about later. Sunlight is an abiotic factor. The Sun also affects two other abiotic factors—climate and temperature.

Climate

Polar bears live in the Arctic. The Arctic has a cold, dry climate. **Climate** *describes average weather conditions in an area over time.* Weather conditions include temperature, moisture, and wind. The graph below shows such conditions.

Climate is an abiotic factor that affects where organisms can survive. A desert climate is dry and often hot. A plant that needs a lot of water could not survive in a desert. A cactus can live in a desert because it can survive with little water.

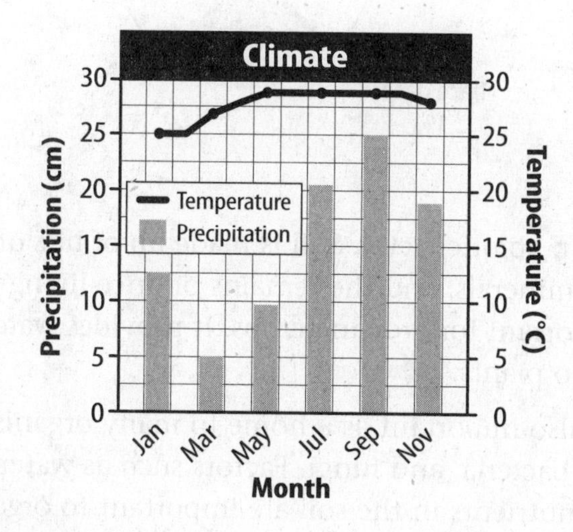

Temperature

Is it hot or cold where you live? Temperatures on Earth are different every day. Temperature is another abiotic factor that affects where organisms can survive. Some organisms, such as tropical birds, do well in hot temperatures. Other organisms, such as polar bears, do well in cold temperatures. Tropical birds do not live in cold ecosystems. Polar bears do not live in warm ecosystems.

Water

All life on Earth must have water. In fact, most organisms are made mostly of water. All organisms need water to grow and reproduce. Because of this, every ecosystem must contain some water to support life.

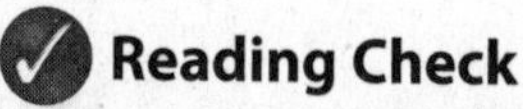

4. State For what do living things use energy from the Sun?

____

____

5. Apply In the graph at the left, which month shows both the highest temperature and the highest amount of precipitation?

____

FOLDABLES®

Use a six-door book to organize information about the abiotic parts of an ecosystem.

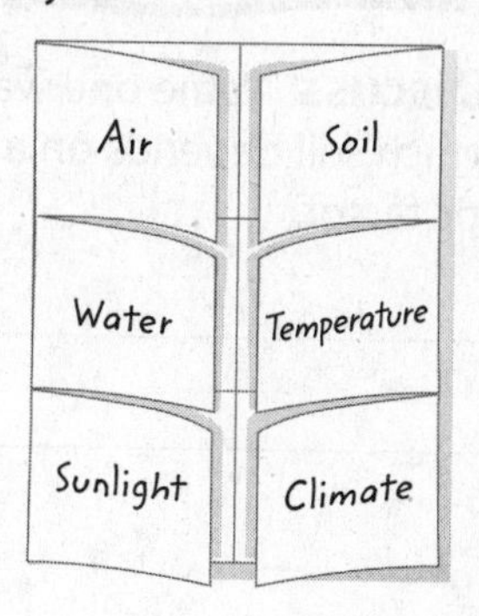

Atmosphere

The **atmosphere** (AT muh sfir) *is the layer of gases that surrounds Earth.* The atmosphere is an abiotic factor that is necessary for life. You interact with the atmosphere every time you take a breath.

The atmosphere is mostly nitrogen and oxygen, as shown in the circle graph below. The atmosphere provides oxygen that living things breathe. It also protects them against certain harmful rays from the Sun.

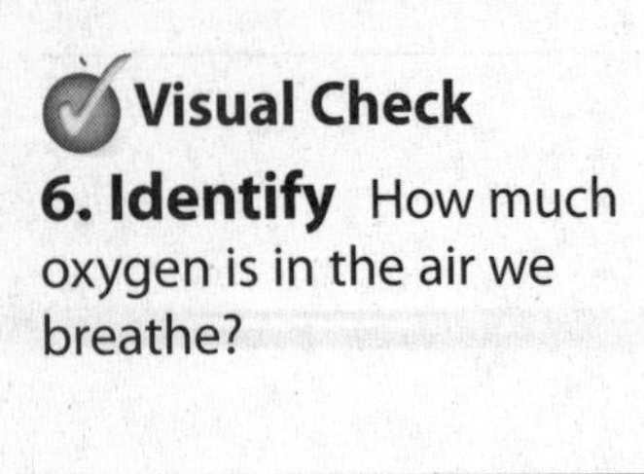

Visual Check

6. Identify How much oxygen is in the air we breathe?

____

Gases in Atmosphere

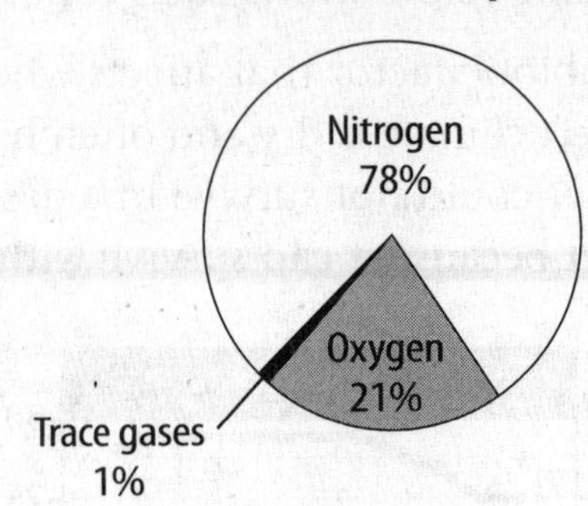

Soil

Soil is an abiotic factor. Soil is made up of bits of rocks, water, air, minerals, and the remains of once-living things. Soil is important for growing crops. It provides water and nutrients to plants.

Key Concept Check

7. Name abiotic factors that make up ecosystems.

____

____

____

Soil is also important as a home to many organisms, such as insects, bacteria, and fungi. Factors such as water, soil type, and nutrients in the soil are important to organisms that live in the soil. Organisms live in the soil that suits them best. Bacteria break down dead plants and animals. This returns nutrients to the soil. Earthworms and insects make small tunnels in the soil that water and air move through. Even dry soil in the deserts is home to living things.

Think it Over

8. Discuss Name one way in which soil depends on a biotic factor.

____

____

After You Read

Mini Glossary

abiotic (ay bi AH tihk) factors: the nonliving things in an ecosystem, such as sunlight and water

atmosphere (AT muh sfir): the layer of gases that surrounds Earth

biotic (bi AH tihk) factors: the living things in an ecosystem

climate: describes average weather conditions in an area over time

ecosystem: all the living things and nonliving things in a given area

1. Review the terms and their definitions in the Mini Glossary. Write a sentence that uses any three of the terms correctly.

____

____

2. Think of one of the ecosystems mentioned in the lesson or any other ecosystem you know about. Label the table below with the name of your ecosystem. Then fill in the table to show the abiotic and biotic factors in that ecosystem.

Ecosystem: ______________	
Abiotic Factors	**Biotic Factors**
1.	1.
2.	2.
3.	3.
4.	4.
5.	5.

3. How did the outline you made of the main ideas in this lesson help you review the material?

____

____

What do you think NOW?

Reread the statements at the beginning of the lesson. Fill in the After column with an A if you agree with the statement or a D if you disagree. Did you change your mind?

Log on to ConnectED.mcgraw-hill.com and access your textbook to find this lesson's resources.

Matter and Energy in the Environment

Cycles of Matter

Key Concept

- How does matter move in ecosystems?

Before You Read

What do you think? Read the two statements below and decide whether you agree or disagree with them. Place an A in the Before column if you agree with the statement or a D if you disagree. After you've read this lesson, reread the statements to see if you have changed your mind.

Before	Statement	After
	3. Carbon, nitrogen, and other types of matter are used by living things over and over again.	
	4. Clouds are made of water vapor.	

Read to Learn

Mark the Text

Identify the Main Point As you read, highlight the main point of each paragraph. Then use a second color to highlight a detail or example to help explain the main point.

How does matter move in ecosystems?

The water that you use to wash your hands might once have traveled through the roots of a tree in Africa. It might also have been part of a glacier in Alaska. How can this be? Water is used over and over again in ecosystems. It never stops moving. The same is true of carbon, oxygen, nitrogen, and other types of matter. Elements that move through one matter cycle might play a role in another cycle.

The Water Cycle

Water covers about 70 percent of Earth's surface. Water surrounds all of Earth's landmasses.

Most of Earth's water is in the oceans. Water is in rivers, streams, lakes, and underground. Water is also in the atmosphere, in icy glaciers, and in living things.

Water is always moving from Earth to the atmosphere and back again. This movement is called the water cycle. The water moves by three processes: evaporation, condensation, and precipitation.

Reading Check

1. Define In your own words, write a definition of the water cycle.

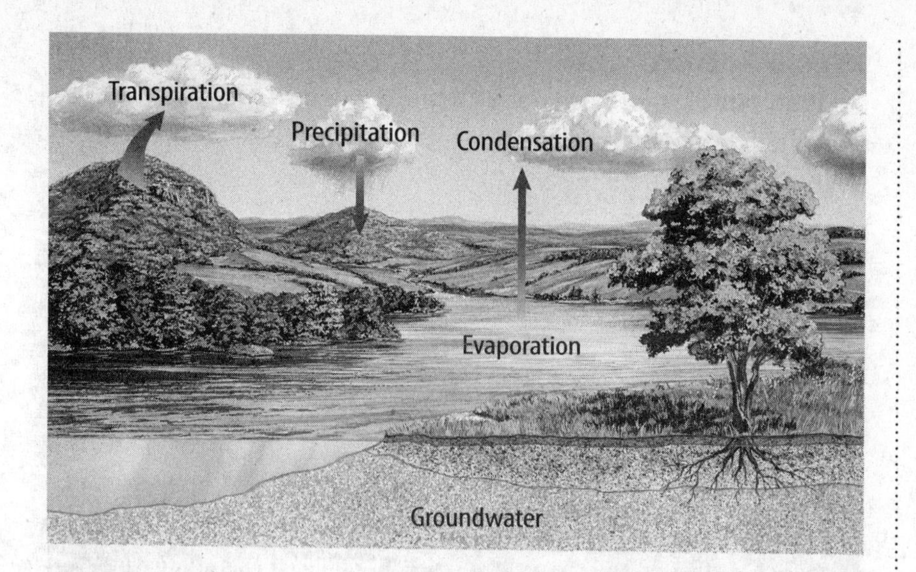

Evaporation

In the water cycle, as shown above, the Sun supplies the energy. As the Sun heats water on the surface of Earth, evaporation occurs. **Evaporation** (ih va puh RAY shun) *is the process during which liquid water changes into a gas called water vapor.* This water vapor rises into the atmosphere. Temperature, humidity, and wind affect how quickly water evaporates.

Water is also released from living things. Cellular respiration occurs in many cells. Water is a by-product of cellular respiration. This water leaves cells and enters the environment and atmosphere as water vapor. Transpiration is the release of water vapor from the leaves and stems of plants.

Condensation

The higher in the atmosphere you are, the cooler the temperature is. As water vapor rises in the atmosphere, it cools. This cooling causes condensation. **Condensation** (kahn den SAY shun) *is the process during which water vapor changes into liquid water.*

Clouds form as a result of condensation. Clouds are made of millions of tiny water droplets or ice crystals. These form when water vapor condenses on particles of dust and other substances in the atmosphere.

Visual Check

2. Identify Highlight the arrow showing which way water goes during transpiration.

Think it Over

3. Evaluate Why is condensation the opposite of evaporation?

___

___

___

Precipitation

Precipitation (prih sih puh TAY shun) *is water that falls from clouds to Earth's surface.* This water enters bodies of water or soaks into soil.

Precipitation can be rain, snow, sleet, or hail. Precipitation forms as water droplets or ice crystals join together in clouds. The droplets or crystals get so large and heavy that they fall to Earth as precipitation. Over time, living things use this precipitation, and the water cycle continues.

Key Concept Check

4. Describe What form does water take as it falls to Earth?

____

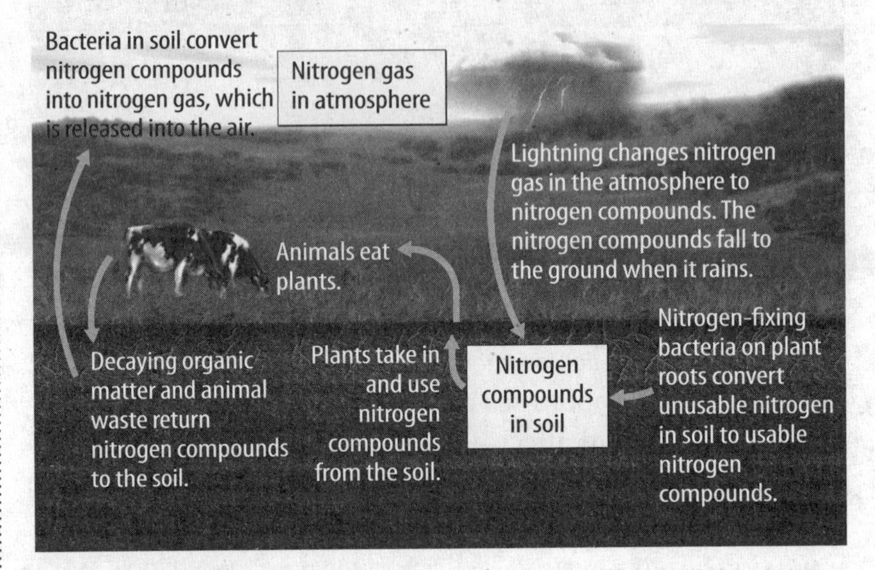

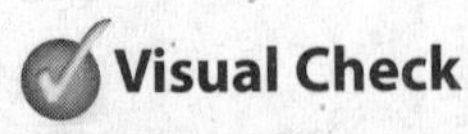

Visual Check

5. Identify What changes nitrogen gas in the air to nitrogen compounds that fall to the ground?

____

The Nitrogen Cycle

You know that water is necessary for life on Earth. The element nitrogen is also necessary for life. Nitrogen is part of proteins, which all organisms need to stay alive. It also is part of DNA, the molecule that contains genetic information. Like water, nitrogen cycles between Earth and the atmosphere and back again. The nitrogen cycle is shown above.

From the Environment to Organisms

You learned earlier that the atmosphere is mostly nitrogen. Plants and animals cannot use the form of nitrogen that is in the atmosphere. How do organisms get nitrogen into their bodies? *The process that changes nitrogen in the atmosphere into nitrogen compounds that living things can use is called* **nitrogen fixation** (NI truh jun • fihk SAY shun). Nitrogen from the atmosphere is changed into a different form with the help of bacteria that live in soil and water. These bacteria take in nitrogen from the atmosphere and change it into nitrogen compounds that other organisms can use.

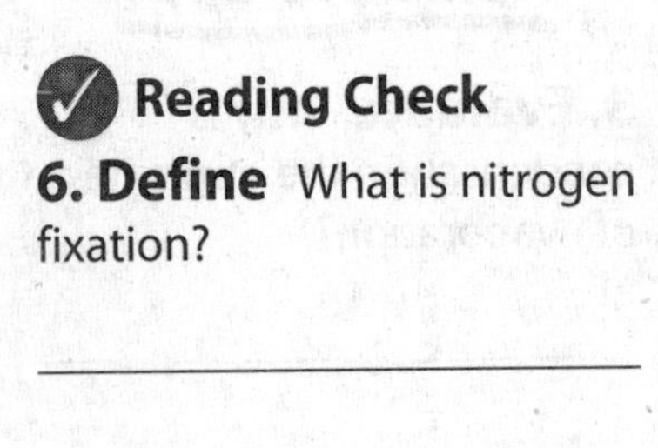

Reading Check

6. Define What is nitrogen fixation?

____

____

____

Plants and some other organisms take in this changed nitrogen from the soil and water. Then animals take in nitrogen when they eat these plants and other organisms.

From Organisms to the Environment

How does nitrogen from living things return to the environment? Some types of bacteria can break down the tissues of dead organisms. These bacteria help return the nitrogen in the tissues of those organisms to the environment. You've seen these bacteria at work in a decaying log or a rotten apple.

Nitrogen also returns to the environment in the waste products of organisms. Manure is a waste product of organisms. Farmers often spread manure on their fields as a way to provide nitrogen for crops.

The Oxygen Cycle

You learned that organisms need water and nitrogen. Almost all organisms also need oxygen for cellular processes that release energy. Oxygen is also a part of carbon dioxide and water, substances that are important to life. The picture below shows how oxygen cycles through an ecosystem.

ACADEMIC VOCABULARY
release
(verb) to set free or let go

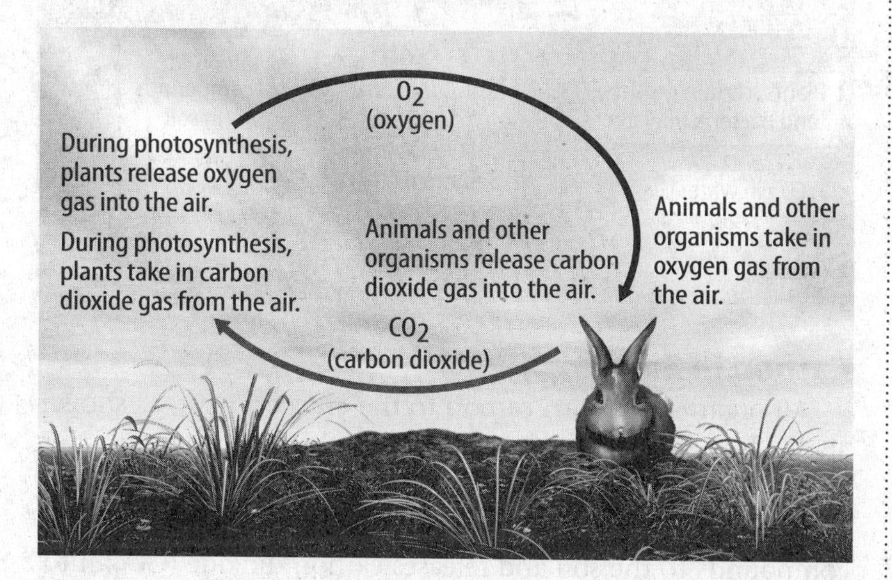

Visual Check

7. Explain What gas do humans take in and what gas do they release as part of the oxygen cycle?

Most of the oxygen in the atmosphere comes from photosynthesis. Earth's early atmosphere probably did not contain oxygen gas. Certain bacteria evolved that made their own food through photosynthesis. Oxygen gas is a by-product of photosynthesis. Over time, other photosynthetic organisms evolved. The amount of oxygen in Earth's atmosphere increased. Photosynthesis is the main source of oxygen in Earth's atmosphere today.

Humans and many other living organisms take in oxygen and release carbon dioxide during cellular processes. There are many other relationships between different types of matter in ecosystems.

Think it Over

8. Generalize Imagine that there are no plants on Earth. Would the planet have oxygen in its atmosphere? Why or why not?

The Carbon Cycle

All organisms contain carbon. Some organisms, including humans, get carbon from food. Other organisms, such as plants, get carbon from the atmosphere or bodies of water.

Carbon cycles through ecosystems like water, nitrogen, and oxygen do. It is used over and over again.

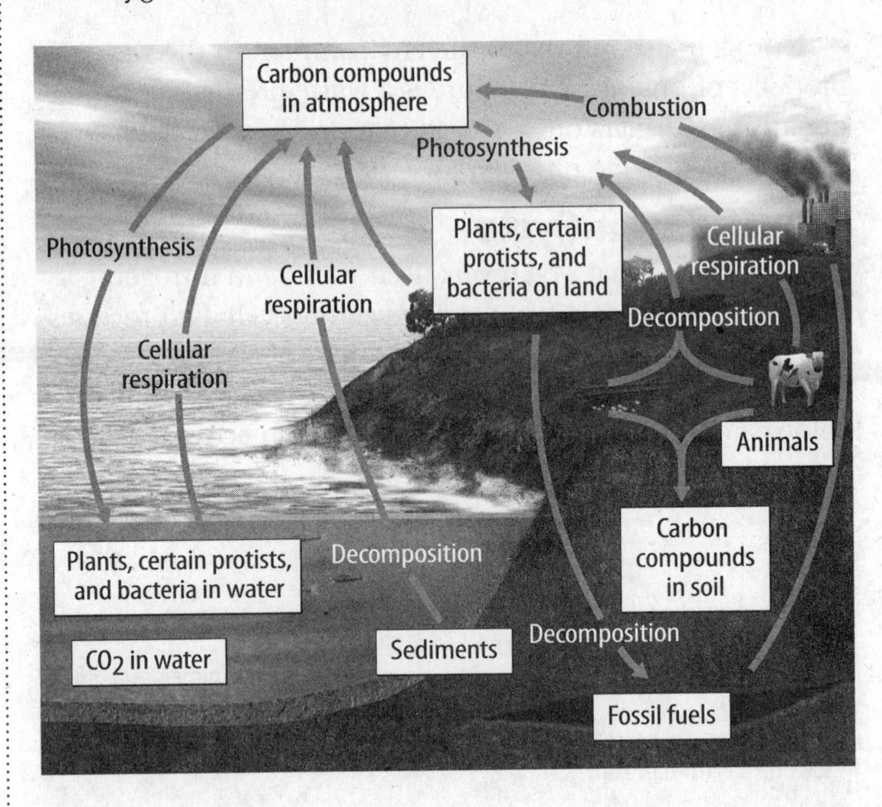

Visual Check

9. Explore Circle the processes that return carbon to the atmosphere.

Carbon in Soil

All organisms return carbon to the environment, as shown above. Like nitrogen, carbon can enter the environment when organisms die and decompose. Decomposition is the breaking down of dead plants and animals. This process returns carbon compounds to the soil and releases carbon dioxide (CO_2) into the atmosphere. Other organisms then use carbon dioxide. Carbon also is found in fossil fuels.

FOLDABLES®

Make a half-book to organize information about the biotic and abiotic parts of one cycle of matter.

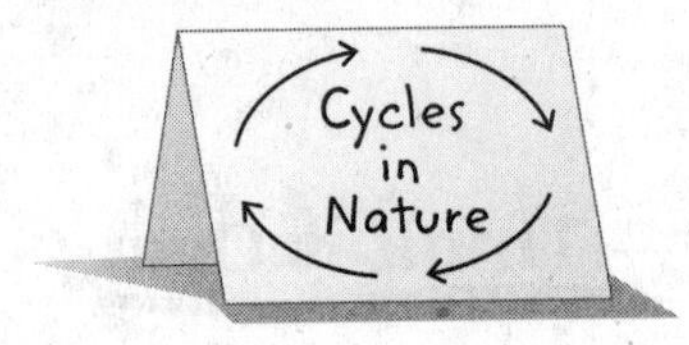

Carbon in Air

How do living things use the carbon dioxide that is in the air? Plants and other organisms make their own food through photosynthesis. They take in carbon dioxide and water to make sugars. When other organisms eat plants, they get carbon and energy. Carbon dioxide is a by-product of the cellular processes that break down sugars to release energy. Carbon dioxide enters the atmosphere to be used again.

The Greenhouse Effect

Carbon dioxide and other gases in the atmosphere absorb thermal energy from the Sun. This energy keeps Earth warm. This process is called the greenhouse effect.

How does the greenhouse effect help Earth? Life on Earth could not exist without the greenhouse effect. As you can see in the figure below, some of this energy is reflected back into space, and some passes through Earth's atmosphere. Greenhouse gases in the atmosphere absorb thermal energy that reflects off Earth's surface. The greenhouse effect helps keep Earth from becoming too hot or too cold.

How does the greenhouse effect harm Earth? While the greenhouse effect is necessary for life, a steady increase in greenhouse gases can harm ecosystems. For example, carbon is stored in fossil fuels, such as coal, oil, and natural gas. When people burn fossil fuels to heat homes, to power cars, or to provide electricity, carbon dioxide gas is released into the atmosphere. The amount of carbon dioxide in the air has increased because of both natural and human activities. The more greenhouse gases released, the greater the gas layer becomes and the more heat is absorbed.

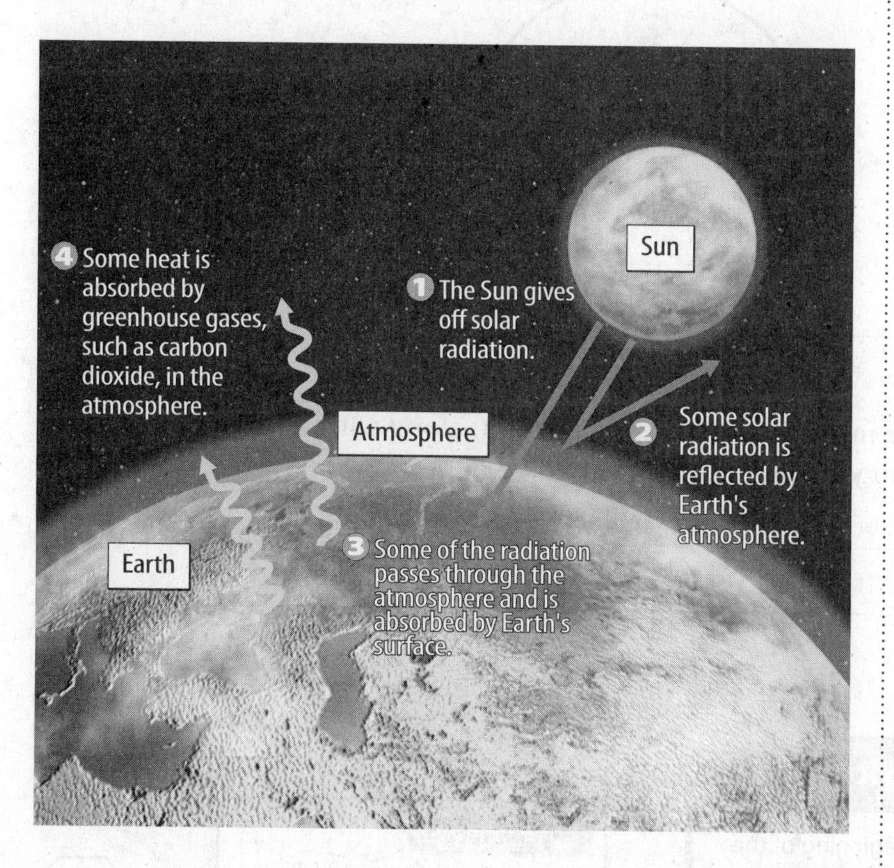

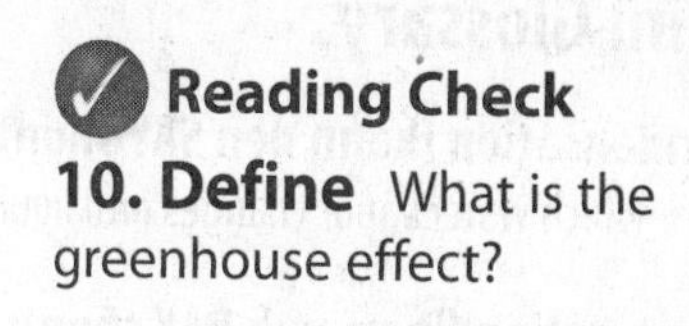

Reading Check

10. Define What is the greenhouse effect?

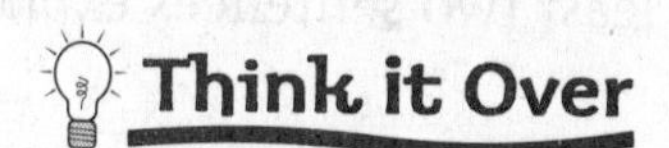

Think it Over

11. Describe Explain how the greenhouse effect benefits Earth.

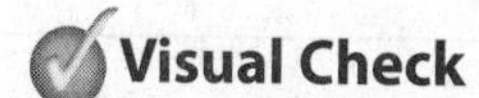

Visual Check

12. Explain What might happen if heat were not absorbed by greenhouse gases?

After You Read

Mini Glossary

condensation (kahn den SAY shun): the process during which water vapor changes into liquid water

evaporation (ih va puh RAY shun): the process during which liquid water changes into a gas called water vapor

nitrogen fixation (NI truh jun • fihk SAY shun): the process that changes atmospheric nitrogen into nitrogen compounds that are usable to living things

precipitation (prih sih puh TAY shun): water that falls from clouds to Earth's surface

1. Review the terms and their definitions in the Mini Glossary. Choose a term and write at least two sentences explaining how the term is linked to the nitrogen cycle.

2. On the lines in the graphic below, identify the processes that make up the water cycle.

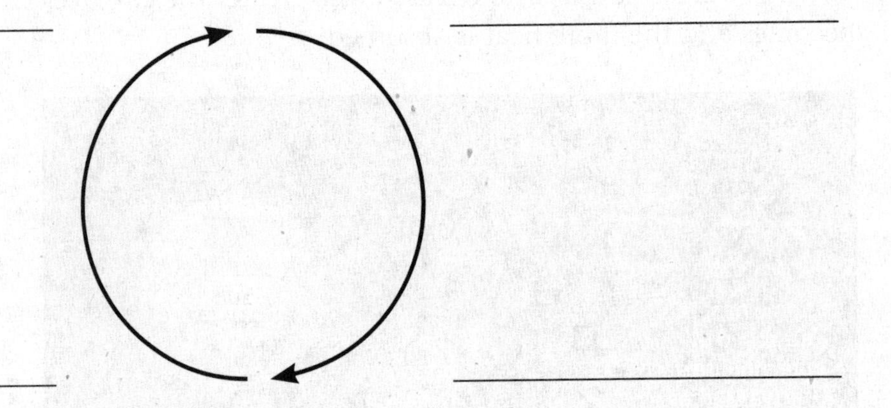

3. How did highlighting the main point of each paragraph, along with a detail or example to explain that point, help you understand what you read?

What do you think NOW?

Reread the statements at the beginning of the lesson. Fill in the After column with an A if you agree with the statement or a D if you disagree. Did you change your mind?

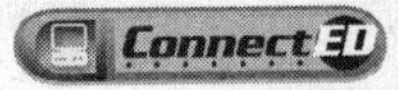

Log on to ConnectED.mcgraw-hill.com and access your textbook to find this lesson's resources.

Matter and Energy in the Environment

Energy in Ecosystems

Before You Read

What do you think? Read the two statements below and decide whether you agree or disagree with them. Place an A in the Before column if you agree with the statement or a D if you disagree. After you've read this lesson, reread the statements to see if you have changed your mind.

Before	Statement	After
	5. The Sun is the source for all energy used by living things on Earth.	
	6. All living things get their energy from eating other living things.	

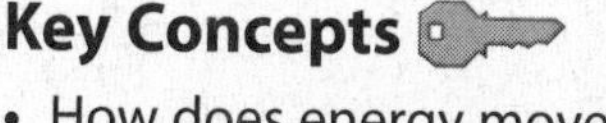

Key Concepts

- How does energy move in ecosystems?
- How is the movement of energy in an ecosystem modeled?

Read to Learn

How does energy move in ecosystems?

Think of an ecosystem. That ecosystem might seem peaceful. If you visit, though, you will notice that it is full of movement. Birds squawk and beat their wings. Plants sway in the breeze. Insects buzz. Animals run over leaves.

An organism requires energy to move. Growth, reproduction, and other life functions require energy. Most of the energy for life on Earth comes from the Sun.

Energy does not cycle through ecosystems like matter does. Energy flows in one direction. In most cases, energy flow starts with the Sun. It moves from one organism to another organism. Many organisms get energy by eating other organisms. Organisms can change energy into different forms of energy. Not all the energy an organism gets is used for life processes. Some is released into the environment as thermal energy. Energy cannot be created or destroyed. Its form can change, though. This idea is called the law of conservation of energy.

Mark the Text

Identify Important Words As you read this lesson, highlight all the words you do not understand. Then underline the part of the text that can help you learn what those words mean.

Key Concept Check

1. Explain How do the movements of matter and energy differ?

____

____

____

Producers

Producers make things. Living things that make their own food from materials in their environments are called producers. Most of these producers use the process of photosynthesis (foh toh SIHN thuh sus). Some use the process of chemosynthesis (kee moh SIHN thuh sus).

Photosynthesis Plants, algae, and some bacteria use photosynthesis to make their food, as shown below. **Photosynthesis** *is a series of chemical reactions that convert light energy, water, and carbon dioxide into the food energy molecule glucose and give off oxygen*. This process is part of the carbon cycle that you read about earlier in this chapter.

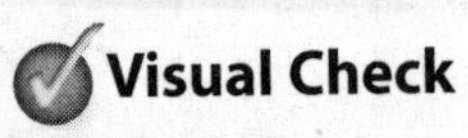

2. Identify What food do producers make in photosynthesis?

____

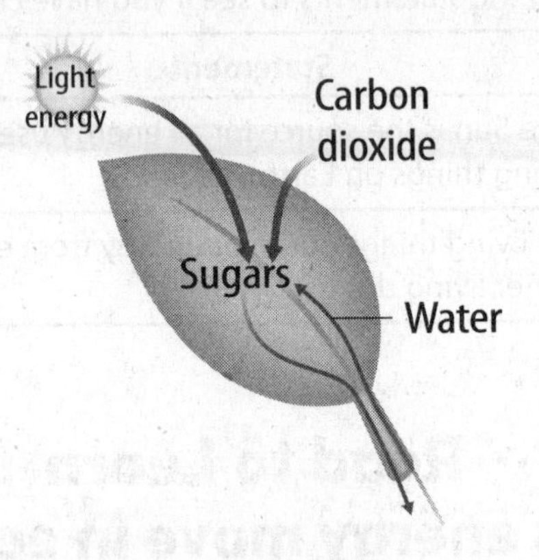

Photosynthesis

Chemosynthesis Some producers use chemosynthesis to make food. **Chemosynthesis** *is the process during which producers use chemical energy in matter rather than light energy and make food.* Chemosynthesis can occur on the deep ocean floor. Hydrothermal vents are outlets for compounds that contain hydrogen and sulfur, as well as thermal energy from inside Earth. Chemosynthetic bacteria that live there use the chemical energy in the compounds and produce food.

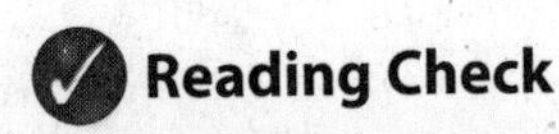

3. Explain During chemosynthesis, what do producers use to make food?

____

____

Consumers

Consumers do not make their own energy-rich food. They get their energy by consuming, or eating, other organisms.

Consumers can be classified by the type of food that they eat. The groups of consumers include herbivores, carnivores, omnivores, and detritivores.

Herbivores, Carnivores, and Omnivores Herbivores eat only producers. A deer eats only plants, so it is an herbivore. Carnivores eat other animals. Lions and wolves are carnivores. Omnivores eat both producers and other consumers. A bird that eats berries and insects is an omnivore.

4. Classify To which group of consumers do most humans belong?

____

Detritivores Detritivores (dih TRI tuh vorz) get their energy by eating the remains of other organisms. Some detritivores, such as insects, eat dead organisms. Detritivores, such as bacteria and mushrooms, feed on and help decompose dead organisms. They are often called decomposers. They produce carbon dioxide that enters the air. Some decayed matter enters the soil. In this way, detritivores help recycle nutrients through ecosystems. They also keep dead organisms from piling up in ecosystems.

Modeling Energy in Ecosystems

You learned that energy does not cycle through ecosystems. Instead, energy flows through ecosystems. As it flows, organisms might use some energy for life processes. Energy might also be stored in the bodies of organisms as chemical energy. When consumers eat these organisms, the chemical energy moves into the bodies of the consumers. Such a transfer of energy can be from a producer to a consumer. It can also be from one consumer to another consumer. With each transfer of energy, some energy changes to thermal energy. Any extra thermal energy enters the environment from the bodies of consumers.

Scientists use models to study this flow of energy through an ecosystem. The model they use depends on how many organisms they are studying.

Food Chains

A **food chain** *is a model that shows how energy flows in an ecosystem through feeding relationships.* In a food chain, arrows show the transfer of energy. A food chain is shown below.

FOLDABLES

Use each side of your pyramid book to organize information about one of the ways energy flows in an ecosystem.

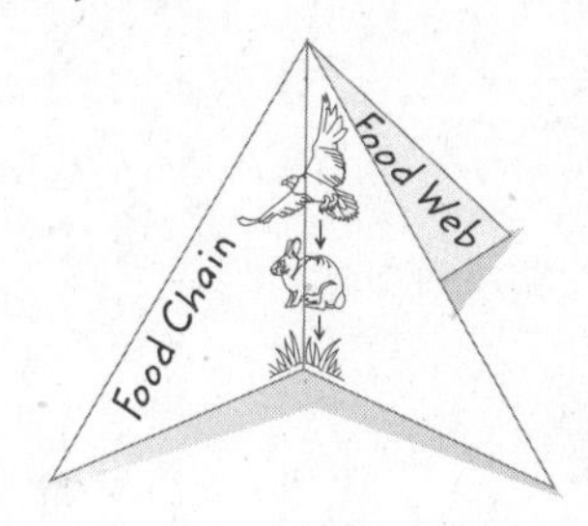

Key Concept Check

5. Explain How does a food chain model energy flow?

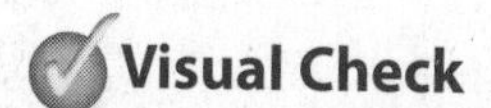

Visual Check

6. Identify Circle each producer and draw a square around each consumer in the food chain.

Food Webs

Imagine that you are working on a jigsaw puzzle of a prairie. The food chain is just one piece of that puzzle. It shows just one small part of the prairie. A food chain is like one piece of an ecosystem jigsaw puzzle. It can help you study parts of an ecosystem. It does not show the whole picture, though.

Look again at the food chain. The mouse might also eat the seeds of other producers, such as corn or trees. The snake might eat other animals, such as frogs, crickets, or earthworms. The hawk might also eat other animals, such as mice, squirrels, rabbits, or fish. The feeding relationships in this ecosystem are complex.

Scientists use a food web to study these feeding relationships. *A* ***food web*** *is a model of energy transfer that shows how food chains in a community are interconnected.* A food web is many overlapping food chains. The food web below shows the complex feeding patterns in an ecosystem. Arrows show how energy flows.

Key Concept Check

7. Name two models used to show the transfer of energy in an ecosystem.

____

____

Visual Check

8. Interpret Diagrams Draw a circle around the organism in the food web that receives energy from the greatest amount of producers and consumers.

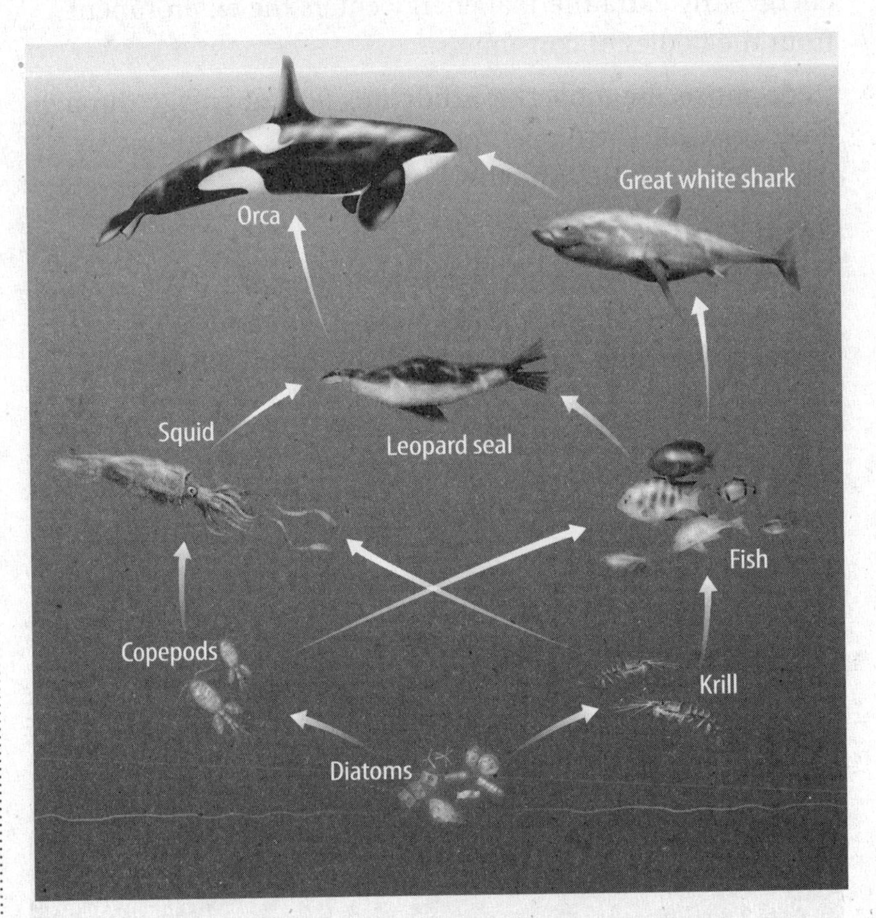

Energy Pyramids

Food chains and food webs show how energy moves in an ecosystem. They do not show how the amount of energy in an ecosystem changes, though. *An* **energy pyramid** *is a model that scientists use to show the amount of energy available in each step of a food chain.* The steps of an energy pyramid are called trophic (TROH fihk) levels.

Look at the energy pyramid below. Producers, such as plants, make up the bottom trophic level. Consumers, such as squirrels, that eat producers make up the next level. Consumers, such as hawks, that eat other consumers, make up the highest level.

Notice that as you move to a higher level, there is less energy available for consumers. Why? As you read earlier, organisms use some of the energy they get from food for life processes. During life processes, some energy is changed to thermal energy. The thermal energy is then transferred to the environment.

Trophic level 3
(1 percent of energy available)

Trophic level 2
(10 percent of energy available)

Trophic level 1
(100 percent of energy available)

Available energy decreases.

Math Skills

The first trophic level—producers—obtains energy from the Sun. They use 90 percent of the energy for their own life processes. Only 10 percent of the energy remains for the second trophic level—herbivores. Assume that each trophic level uses 90 percent of the energy it receives. Use the following steps to calculate how much energy remains for the next trophic level.

First trophic level gets 100 units of energy.

First trophic level uses 90 percent = 90 units

Energy remaining for second trophic level = 10 units

Second trophic level uses 90 percent = 9 units

Energy remaining for third trophic level = 1 unit

9. Use Percentages If the first trophic level receives 10,000 units of energy from the Sun, how much energy is available for the second trophic level?

Visual Check

10. Interpret Diagrams How does the amount of available energy change at each trophic level?

After You Read

Mini Glossary

chemosynthesis (kee moh SIHN thuh sus): the process during which producers use chemical energy in matter rather than light energy to make food

energy pyramid: a model that scientists use to show the amount of energy available in each step of a food chain

food chain: a model that shows how energy flows in an ecosystem through feeding relationships

food web: a model of energy transfer that scientists use to show how food chains in a community are interconnected

photosynthesis (foh toh SIHN thuh sus): a series of chemical reactions that convert light energy, water, and carbon dioxide into the food energy molecule glucose and give off oxygen

1. Review the terms and their definitions in the Mini Glossary. Write a sentence that describes one way in which photosynthesis and chemosynthesis differ.

2. Fill in the graphic organizer below. Name and describe the three models that scientists use to show the flow of energy through an ecosystem. Clues have been provided.

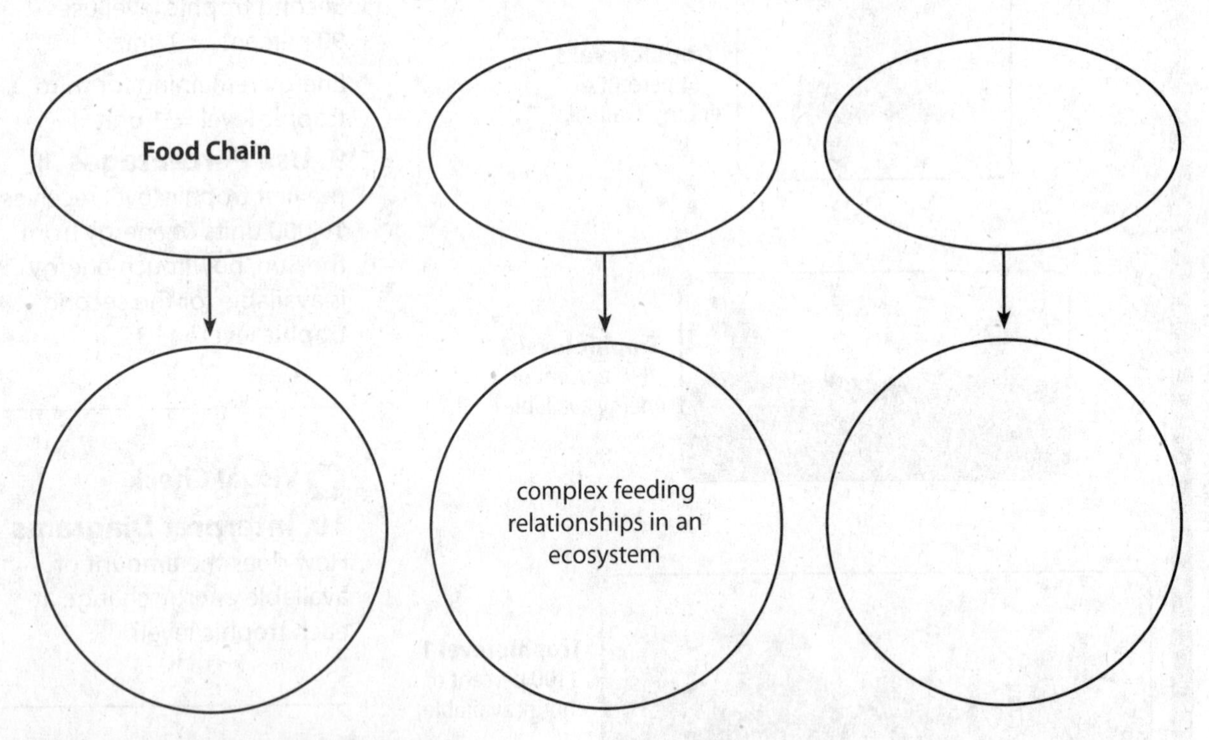

What do you think NOW?

Reread the statements at the beginning of the lesson. Fill in the After column with an A if you agree with the statement or a D if you disagree. Did you change your mind?

Log on to ConnectED.mcgraw-hill.com and access your textbook to find this lesson's resources.

Populations and Communities

Populations

Before You Read

What do you think? Read the two statements below and decide whether you agree or disagree with them. Place an A in the Before column if you agree with the statement or a D if you disagree. After you've read this lesson, reread the statements to see if you have changed your mind.

Before	Statement	After
	1. Some life exists in the ice caps of the North Pole and the South Pole.	
	2. A community includes all organisms of one species that live in the same area.	

Key Concepts 

- What defines a population?
- What factors affect the size of a population?

Read to Learn

The Biosphere and Ecological Systems

Earth's **biosphere** (BI uh sfihr) *is the parts of Earth and the surrounding atmosphere where there is life*. The biosphere includes all the land of the continents and islands. It also includes all of Earth's oceans, lakes, and streams. It includes the ice caps at the North Pole and the South Pole. Parts of the biosphere with large numbers of plants or algae often contain many other organisms.

Study Coach

Make Flash Cards Think of a quiz question for each paragraph. Write the question on one side of a flash card. Write the answer on the other side. Work with a partner to quiz each other using the flash cards.

What is a population?

The Kalahari Desert in Africa is a part of Earth's biosphere. Several groups of meerkats live there in a wildlife refuge. Meerkats are small mammals that live in family groups and help each other care for their young. Meerkats interact with each other for survival. They sleep underground in burrows. They hunt for food during the day. They stand upright to watch for danger and call out warnings to others.

Meerkats are part of an ecosystem. An ecosystem is a group of organisms that live together in an area at one time. It also includes the climate, soil, water, and other nonliving parts of the environment. The Kalahari Desert is one of many ecosystems that makes up Earth's biosphere. The study of all ecosystems on Earth is ecology.

Think it Over

1. Explain Why are meerkats considered to be part of an ecosystem?

Community The figure below shows a family group of meerkats on the bottom right. Many species besides meerkats also live in a wildlife refuge in the Kalahari Desert. They include scorpions, spiders, insects, snakes, birds, zebras, giraffes, lions, shrubs, grasses, small trees, and melon vines. All these plants and animals form a community. *A* **community** *is all the populations of different species that live together in the same area at the same time.*

Key Concept Check
2. Describe What defines a population?

____

____

____

Population All the family groups of meerkats that live in this refuge form a population. *A* **population** *is all the organisms of the same species that live in the same area at the same time.* A species is a group of organisms that have similar traits and are able to produce fertile offspring.

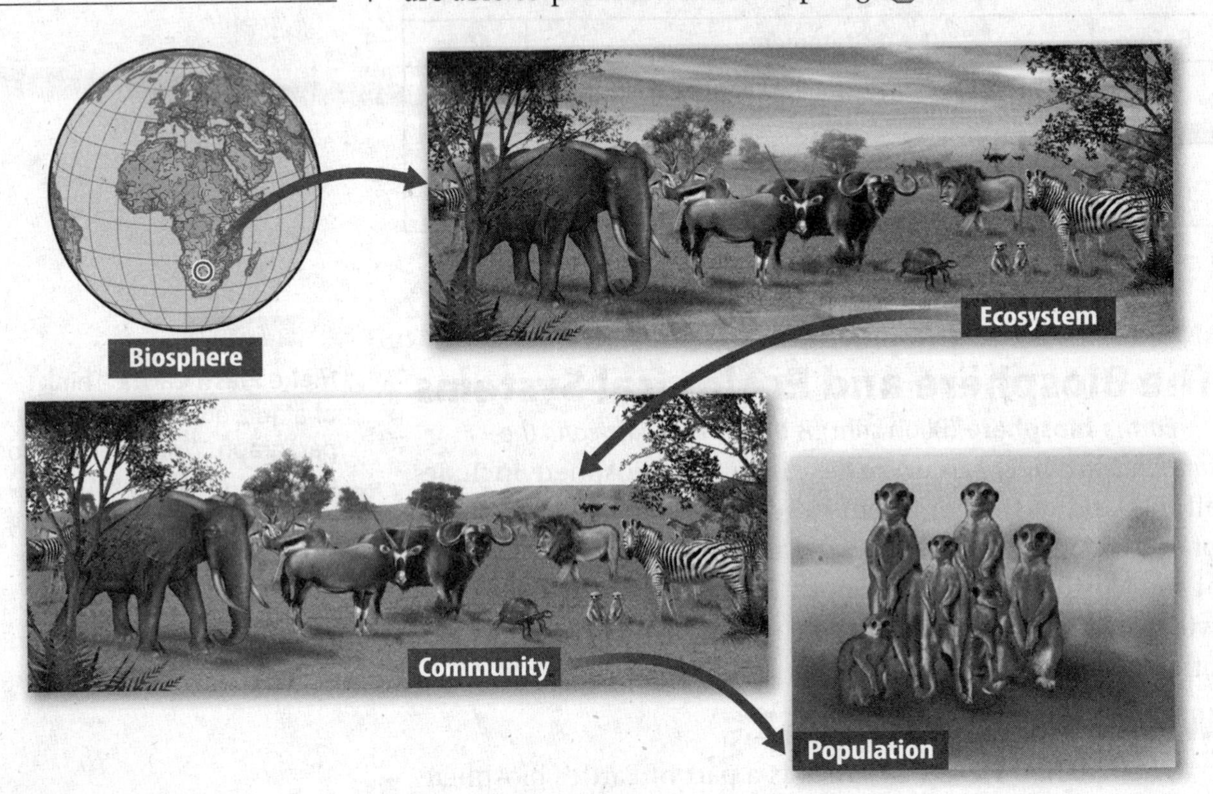

Visual Check
3. Identify Name three populations shown in the figure.

____

____

____

Competition

At times, there is not enough food for every organism in a community. Members of a population must compete with other populations and each other for enough food to survive. **Competition** *is the demand for resources, such as food, water, and shelter, in short supply in a community.* When there are not enough resources available to survive, there is more competition in a community. In the Kalahari Desert, where water is scarce, the meerkats compete with other animals for resources such as food and water.

Population Sizes

When there is less food available, a population of meerkats gets smaller. Female meerkats cannot raise as many young. Some meerkats might leave the area to find food elsewhere.

If there is plenty of food, the size of a population of meerkats grows larger. More meerkats survive to adulthood and live longer. Changes in environmental factors can result in changes to the size of a population.

Limiting Factors

Environmental factors, such as available food, water, shelter, sunlight, and temperature, are possible limiting factors for a population. *A* **limiting factor** *is anything that restricts the size of a population.* If there is not enough sunlight, green plants cannot make food by photosynthesis. A lack of green plants affects organisms that eat green plants.

Temperature is a limiting factor for some organisms. When the temperature drops below freezing, many organisms die because it is too cold for them to survive. Disease and predators—animals that eat other animals—can be limiting factors for organisms. Natural disasters such as fires and floods also limit the size of populations.

Key Concept Check
4. Specify What factors affect the size of a population?

Measuring Population Size

Measuring the size of a population can be difficult. Biologists often use the capture-mark-and-release method to count and observe animal populations. A population of lynx in Poland is counted and monitored using this method. To use the capture-mark-and-release method, biologists capture several animals of a species. They sedate the animals and put a radio collar on each one. Then they release the animals back into the wild. The radio collars help biologists estimate the size of the population and track the animals' movements.

Population density *is the size of a population compared to the amount of space available.* Biologists estimate population density by a sample count. Suppose you want to know how closely together Cumberland azaleas (uh ZAYL yuhz), a type of flower, grow in the Great Smoky Mountains National Park. Rather than counting every azalea shrub, you would count only the azalea shrubs in an area, such as 1 km^2. By multiplying the number of square kilometers in the park by the number of azaleas in 1 km^2, you would find the estimated population density of azalea shrubs in the entire park.

Reading Check
5. Describe two ways you can estimate population size.

Biotic Potential

Imagine that a population of raccoons has plenty of food, water, and den space. The population has no disease and is not in danger from other animals. The only limit to the size of this population is the number of offspring the raccoons can produce. **Biotic potential** *is the potential growth of a population if it could grow in perfect conditions with no limiting factors.* No population on Earth ever reaches its biotic potential because no ecosystem has an unlimited supply of natural resources.

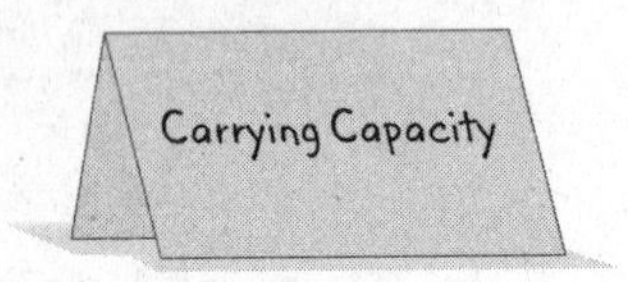

Make a half-book. Use it to take notes on the relationship between population size and carrying capacity.

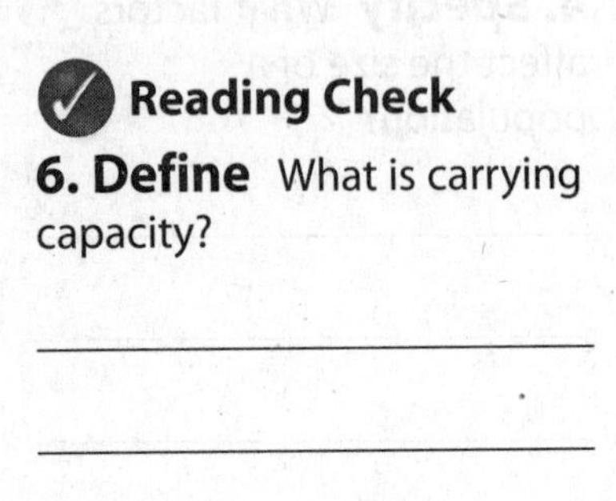

Carrying Capacity

What happens when a population reaches its biotic potential? It stops growing when the available resources in the ecosystem are used up. *The largest number of individuals of one species that an environment can support is the* **carrying capacity.** A population grows until it reaches the carrying capacity of an environment. Disease, space, food, and predators are some of the factors that limit the carrying capacity of an ecosystem.

The carrying capacity of an environment does not stay the same. It increases and decreases as the amount of available resources increases and decreases. At times, a population can briefly grow beyond the carrying capacity of an environment.

Reading Check

6. Define What is carrying capacity?

______

______

______

Overpopulation

Populations can grow so large that they cause problems for other organisms in the community. Overpopulation occurs when a population becomes larger than the carrying capacity of its ecosystem. For example, meerkats eat spiders. An overpopulation of meerkats causes a decrease in the size of the spider population in that community. Populations of birds and other animals that eat spiders also decrease when the number of spiders decreases.

Elephants in Africa's wild game parks present another example of overpopulation. Elephant herds searching for food can cause tree damage. They push over trees to feed on treetops. Other animals that use those trees for food and shelter must compete with the elephants. Also, the loss of trees can damage the soil. This might prevent other trees and plants from growing in that area.

Key Concept Check

7. Summarize
How can overpopulation affect a community?

______

______

______

After You Read

Mini Glossary

biosphere (BI uh sfihr): the parts of Earth and the surrounding atmosphere where there is life

biotic potential: the potential growth of a population if it could grow in perfect conditions with no limiting factors

carrying capacity: the largest number of individuals of one species that an environment can support

community: all the populations of different species that live together in the same area at the same time

competition: the demand for resources, such as food, water, and shelter, in short supply in a community

limiting factor: anything that restricts the size of a population

population: all the organisms of the same species that live in the same area at the same time

population density: the size of a population compared to the amount of space available

1. Review the terms and their definitions in the Mini Glossary. Write a sentence that explains how limiting factors and biotic potential are related.

____

____

2. Identify each example in the flowchart as a *population*, an *ecosystem*, or a *community*. Write the correct term in the box with its example.

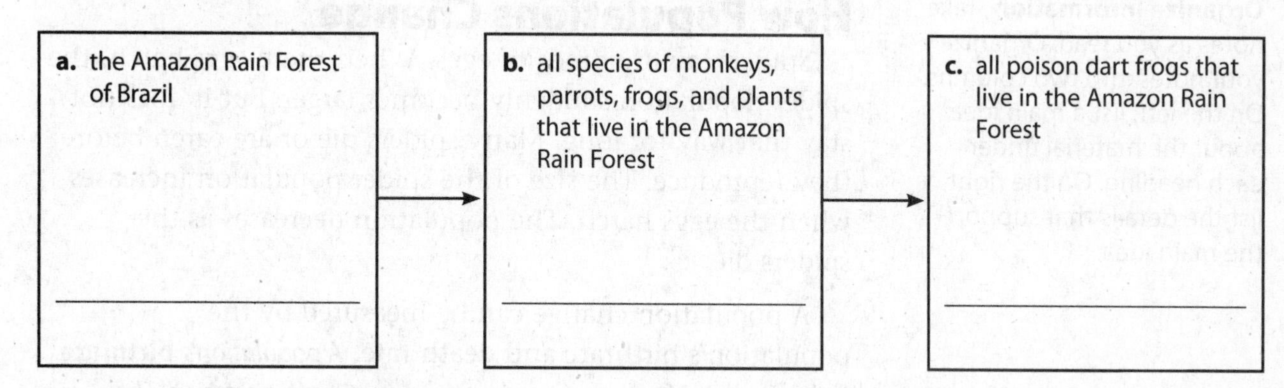

3. List two facts about population that you learned from your partner's flash cards.

____

____

____

What do you think NOW?

Reread the statements at the beginning of the lesson. Fill in the After column with an A if you agree with the statement or a D if you disagree. Did you change your mind?

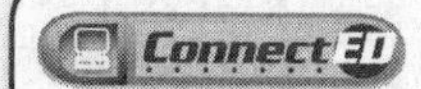

Log on to ConnectED.mcgraw-hill.com and access your textbook to find this lesson's resources.

Populations and Communities

Changing Populations

Key Concepts
- How do populations change?
- Why do human populations change?

Before You Read

What do you think? Read the two statements below and decide whether you agree or disagree with them. Place an A in the Before column if you agree with the statement or a D if you disagree. After you've read this lesson, reread the statements to see if you have changed your mind.

Before	Statement	After
	3. Some populations decrease in numbers because of low birthrates.	
	4. An extinct species has only a few surviving individuals.	

Study Coach

Organize Information Take notes as you read. Organize your notes into two columns. On the left, list a main idea about the material under each heading. On the right, list the details that support the main idea.

Read to Learn

How Populations Change

Spiders lay hundreds of eggs. When these eggs hatch, the spider population suddenly becomes larger. But it does not stay that way for long. Many spiders die or are eaten before they reproduce. The size of the spider population increases when the eggs hatch. The population decreases as the spiders die.

A population change can be measured by the population's birthrate and death rate. *A population's* **birthrate** *is the number of offspring produced over a given time period. The* **death rate** *is the number of individuals that die over the same time period.* If the birthrate is higher than the death rate, the population increases. If the death rate is higher than the birthrate, the population decreases.

Exponential Growth

A population in ideal conditions with unlimited resources grows in a pattern called exponential growth. During exponential growth, the larger a population gets, the faster it grows. For example, it takes *E. coli*, a type of microscopic bacterial organism, only 10 hours to grow from one organism to more than 1 million organisms. Exponential growth cannot continue for long. Limiting factors stop the population growth.

SCIENCE USE V. COMMON USE
exponential
Science Use a mathematical expression that contains a constant raised to a power, such as 2^3 or x^2

Common Use in great amounts

Population Size Decrease

Population size can increase, but it also can decrease. Several factors cause a population size to decrease.

- **Lack of resources** A mouse population might decrease in size during the winter because less food is available. Some mice will starve. More mice will die than will be born. When food is plentiful, the population usually increases.
- **Natural disasters** Floods, fires, hurricanes, and volcanic eruptions affect population size. They destroy habitats and food sources for organisms. The populations decrease.
- **Disease** The spread of disease causes populations to decrease. In the mid-1900s, a disease destroyed thousands of elm trees in the United States. The size of the population of elm trees decreased.
- **Predation** The hunting of animals for food is predation. Cats that live in barns feed on mice and reduce the mouse population. Predation reduces population size. ✓

Extinction Populations can decrease in numbers until they disappear. *An* **extinct species** *is a species that has died out and no individuals are left.* Extinctions can be caused by predation, natural disasters, or damage to the environment.

Scientists hypothesize that the extinction of the dinosaurs about 65 million years ago was caused by a meteorite crashing into Earth. The impact would have sent tons of dust into the atmosphere, blocking sunlight. Without sunlight, plants could not grow. Dinosaurs that ate plants probably starved.

Most extinctions involve fewer species. For example, about 700 years ago humans settled in New Zealand where a flightless bird called the giant moa lived. Humans hunted the moa for food. As the human population increased, the size of the moa population decreased. The species became extinct within 200 years.

Endangered Species Wild mountain gorillas are an endangered species. Just over 400 gorillas remain in the wild in Africa. *An* **endangered species** *is a species whose population is at risk of extinction.*

Threatened Species *A* **threatened species** *is a species at risk, but not yet endangered.* California sea otters are a threatened species. Worldwide, more than 4,000 species are classified as endangered or threatened. ✓

✓ **Reading Check**

1. Summarize What are four reasons that a population might decrease in size?

____

____

____

____

✓ **Reading Check**

2. Contrast What is the difference between an endangered species and a threatened species?

____

____

____

Movement

Populations change when organisms move from place to place. When an animal population becomes overcrowded, some individuals might move to find food or living space.

Plant populations also move from place to place. Seeds might be carried by the wind. Animals also help spread seeds.

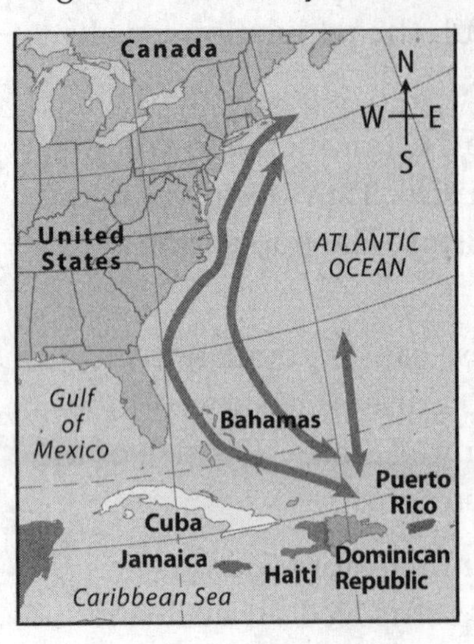

Migration Sometimes an entire population moves from one place to another. It later returns to its original location. **Migration** *is the instinctive seasonal movement of a population of organisms from one place to another.* Ducks, geese, and monarch butterflies are examples of organisms that migrate annually. As shown by the arrows in the map, humpback whales mate and give birth in warm ocean waters near the Bahamas during the winter. In the summer, they migrate north to find food.

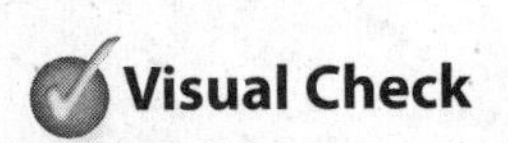

Visual Check

3. Locate Circle the northern area on the map where the whales migrate in summer.

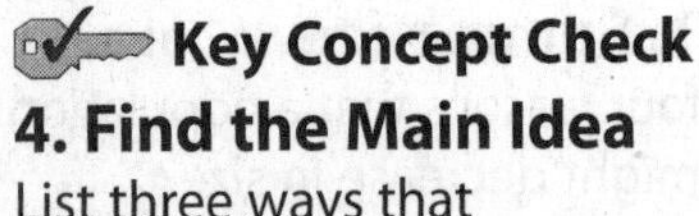

Key Concept Check

4. Find the Main Idea List three ways that populations change.

____

____

____

Human Population Changes

Birthrates, death rates, and movement also affect human population size. But unlike other species, humans have developed ways to increase the carrying capacity of their environment. Improved crop yields, domesticated farm animals, and timely methods of transporting foods and other resources enable people to survive in all types of environments.

The human population is growing quickly. Scientists estimate that there were about 300 million humans on Earth a thousand years ago. Today there are more than 6 billion humans on Earth. By 2050, there could be more than 9 billion. No one knows when the human population will reach Earth's carrying capacity. Some scientists estimate that Earth's carrying capacity is about 11 billion.

As the human population grows, people need more houses and roads. They clear more land for crops. This means less living space, food, and other resources are available for other species. Also, people use more energy to heat and cool homes; to fuel cars, airplanes, and other forms of transportation; and to produce electricity. This energy use adds pollution that affects other populations.

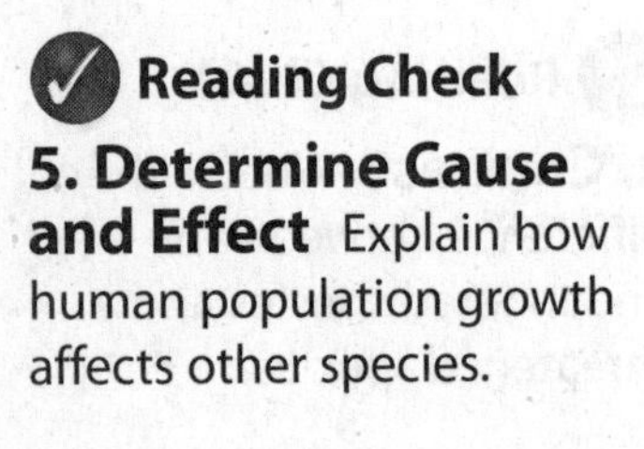

Reading Check

5. Determine Cause and Effect Explain how human population growth affects other species.

____

____

____

Population Size Increase

People are living longer today than people in past generations. More children reach adulthood. Recall that when the birthrate is higher than the death rate, the population grows. Several factors keep the human birthrate higher than the death rate.

Food For some people, finding food is easy. They simply go to the grocery store. Improved farming methods have helped farmers produce food for billions of people.

Resources Planes, trains, trucks, or boats transport fossil fuels, cloth, materials, food, and many other resources around the world. Today, people can get more resources because of better transportation methods.

Sanitation Only 100 years ago, diseases that spread through unclean water supplies and untreated waste caused many deaths. Better water treatment has reduced the spread of disease. Less-expensive and more-effective cleaning products are now available to help prevent the spread of organisms that cause disease.

Medical Care Modern medical care is keeping people alive and healthy longer than ever before. Scientists have developed vaccines, antibiotics, and other medicines to prevent and treat disease. As a result, human death rates have decreased.

Decreases in Human Population Size

Human populations in some parts of the world are decreasing in size. Diseases such as AIDS and malaria cause high death rates in some countries. Severe drought has caused crop failures and a lack of food. Floods, earthquakes, and other natural disasters can cause the deaths of thousands of people.

Population Movement

The size of a human population changes as people move from place to place. Humans might move when more resources are available in a different place. Immigration takes place when organisms move into an area. Typical population movements in the U.S. are shown in the graph.

Types of Moves, 2004–2005

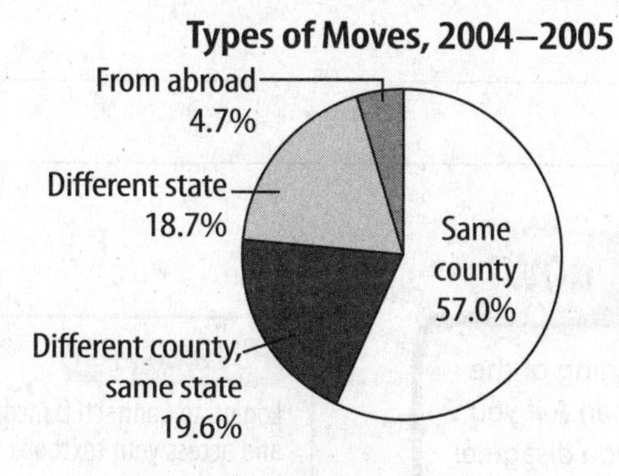

Source: U.S. Census Bureau, Current Population Survey, 2005 Annual Social and Economic Supplement.

FOLDABLES

Make a two-tab book to summarize why human populations change in size.

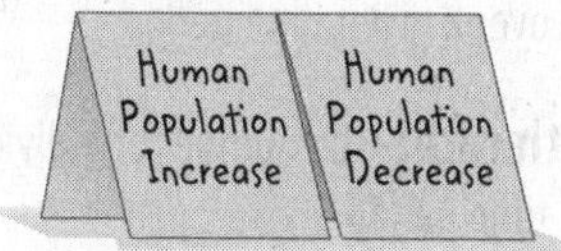

Math Skills

Graphs are used to make large amounts of information easy to interpret. The circle graph below represents 100 percent, and each segment represents one part making up the whole. The graph shows all the moves made by people in the United States during 2004–2005. The percentage of moves within the same county was 57 percent.

6. Interpreting a Graph

a. What percentage of the moves were from one state to another?

b. What percentage of the moves were within the same state?

Key Concept Check

7. Explain What makes human populations increase or decrease?

After You Read

Mini Glossary

birthrate: the number of offspring produced by a population over a given time period

death rate: the number of individuals that die over the same time period

endangered species: a species whose population is at risk of extinction

extinct species: a species that has died out and no individuals are left

migration: the instinctive seasonal movement of a population of organisms from one place to another

threatened species: a species at risk, but not yet endangered

1. Review the terms and their definitions in the Mini Glossary. Write a sentence that compares threatened species, endangered species, and extinct species.

____

____

2. Decide how each example would affect the human population size. Put a check mark in the correct box.

Example		The human population size would . . . increase	decrease	stay the same
a. A hurricane kills thousands of people in Thailand.	→	☐	☐	☐
b. Scientists discover a cure for cancer.	→	☐	☐	☐
c. The birthrate and death rate are equal in Germany this year.	→	☐	☐	☐
d. Many people flee their homeland to escape war.	→	☐	☐	☐

3. Explain how the relationship between birthrate and death rate affects population size.

____

____

What do you think NOW?

Reread the statements at the beginning of the lesson. Fill in the After column with an A if you agree with the statement or a D if you disagree. Did you change your mind?

Log on to ConnectED.mcgraw-hill.com and access your textbook to find this lesson's resources.

Populations and Communities

Communities

Before You Read

What do you think? Read the two statements below and decide whether you agree or disagree with them. Place an A in the Before column if you agree with the statement or a D if you disagree. After you've read this lesson, reread the statements to see if you have changed your mind.

Before	Statement	After
	5. No more than two species can live in the same habitat.	
	6. A cow is a producer because it produces food for other organisms.	

Key Concepts 
- What defines a community?
- How do the populations in a community interact?

Read to Learn

Communities, Habitats, and Niches

You learned that a community is made up of all the species that live in the same ecosystem at the same time. *The place within an ecosystem where an organism lives is its* **habitat.** A habitat provides all the resources an organism needs, including food and shelter. A habitat also has the right temperature, water, and other conditions the organism needs to survive.

Many species can live in the same habitat at the same time. This is possible because each species uses the habitat in a different way. *A* **niche** (NICH) *is what a species does in its habitat to survive.* For example, butterflies and ants can live in the same forest. The butterflies feed on flower nectar. The ants eat insects or plants. These species have different niches in the same environment.

Mark the Text

Build Vocabulary Read all the headings in this lesson and circle any word you cannot define. As you read the lesson, underline the part of the text that helps you define each circled word.

Key Concept Check
1. Define What is a community?

Energy in Communities

All organisms need energy to live. Consider a slow-moving sloth that sleeps 15 to 20 hours a day. Then consider a fast-moving squirrel monkey swinging through treetops. Sloths might seem not to use energy. However, sloths, squirrel monkeys, and all other organisms need energy to live. All living things use energy and carry out life processes such as growth and reproduction.

Energy Roles

How an organism gets energy is an important part of its niche. Almost all the energy available to living things comes from the Sun. There are exceptions, such as organisms that live near deep-sea vents. They obtain energy from chemicals such as hydrogen sulfide.

Producers *are organisms that get energy from the environment, such as sunlight, and make their own food.* For example, most plants are producers. They get their energy from sunlight. They use the process of photosynthesis (foh toh SIHN thuh sus) and make sugar molecules to use as food. Some producers live near deep-sea vents. They use hydrogen sulfide and carbon dioxide and make sugar molecules.

Reading Check

2. Classify Which of the following is an example of a producer? (Circle the correct answer.)

a. a plant

b. a bird

c. a deep-sea vent

Consumers *are organisms that get energy by eating other organisms.* Consumers are classified by the type of organisms they eat. Types of consumers are shown in the table.

Type of Consumer	What They Eat	Examples
Herbivores	producers such as plants	sloths, cows, and sheep
Carnivores	other consumers	harpy eagles, ants, lions, and wolves
Omnivores	producers and consumers	humans
Detritivores	dead organisms	some bacteria and some fungi

Visual Check

3. Apply Identify a producer, an herbivore, a carnivore, and an omnivore.

Energy Flow

A food chain is a way of showing how energy moves through a community. Think about a rain forest community. Energy flows from the Sun to a rain forest tree. The tree is a producer. It uses the light energy and grows, producing leaves and other plant parts. When a consumer eats leaves and other plant parts, energy moves to consumers. For example, a sloth eats the leaves of the tree and energy flows to the sloth. When the eagle eats the sloth, energy flows to the eagle. When the eagle dies, detritivores (dee TRI tuh vorz), such as bacteria, feed on its body. That food chain can be written like this:

Sun → leaves → sloth → eagle → bacteria

A food chain shows only part of the energy flow in a community. A food web, like the one on the next page, shows many food chains within a community. Notice that some of the food chains overlap.

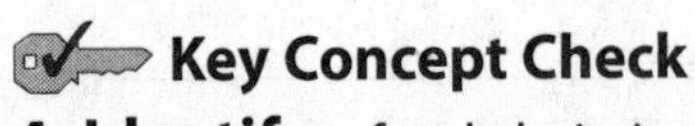 **Key Concept Check**

4. Identify a food chain in a community near your home. List the producers and consumers in your food chain.

Food Web

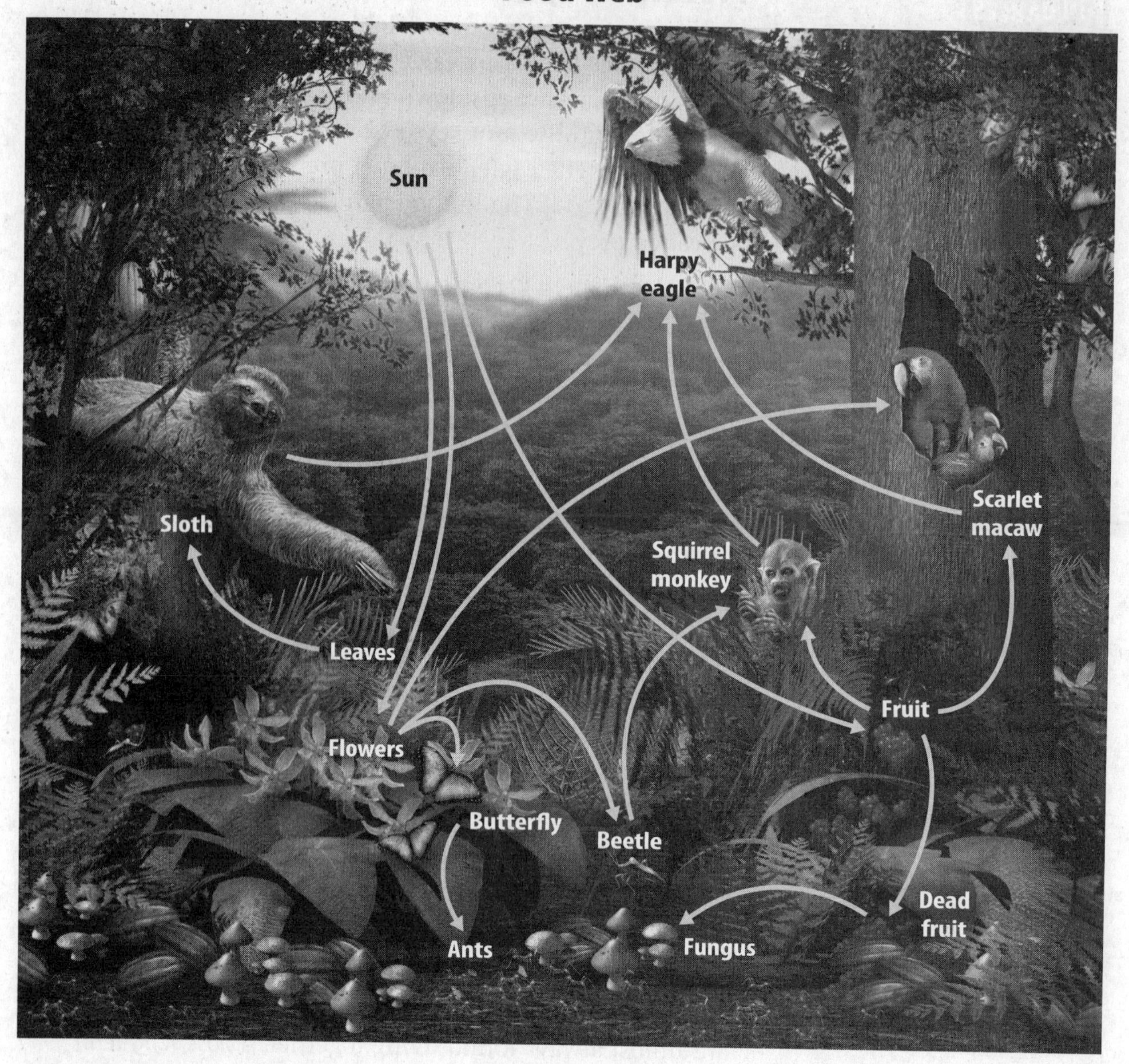

In the rain forest community above, the flowers provide food for a butterfly, a beetle, and a scarlet macaw. The flowers, then, are part of more than one food chain. Here are some food chains in which the flowers provide food:

Sun → flowers → butterfly → ants

Sun → flowers → beetle → squirrel monkey → harpy eagle

Sun → flowers → macaw → eagle

Relationships in Communities

The populations that make up a community interact with each other in many ways. Some species have feeding relationships. They either eat or are eaten by another species. Some species interact with another species to get the food or shelter they need.

Visual Check

5. Sequence List the members of two different food chains shown in the figure.

______

______

______

FOLDABLES

Make a three-tab book to organize information about the types of relationships that can exist among organisms within a community.

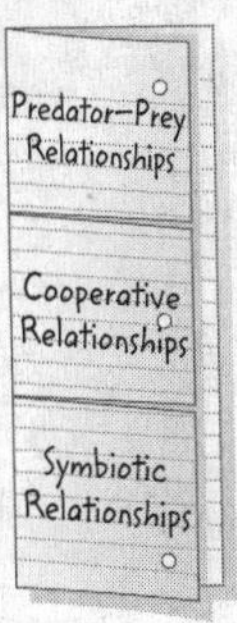

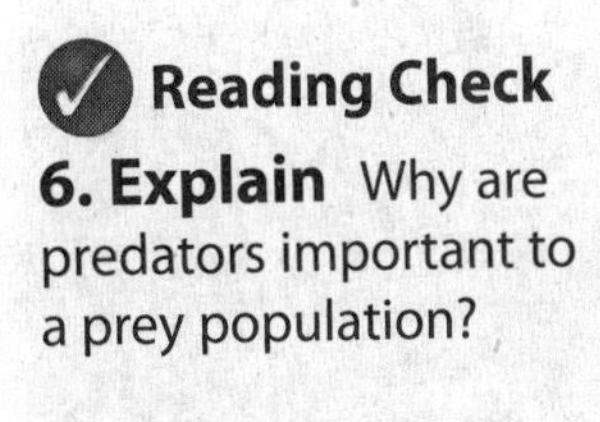

Reading Check

6. Explain Why are predators important to a prey population?

____

____

____

Think it Over

7. Define Cooperative relationships occur between ______. (Circle the correct answer.)

a. members of different species

b. members of the same species

c. members of any species that live together

Predator-Prey Relationships

Hungry squirrel monkeys fight over a piece of fruit. They do not notice the harpy eagle flying above them. Suddenly, the harpy eagle swoops down and grabs one of the monkeys. The eagles and the monkeys have a predator-prey relationship. The eagle, like other predators, hunts other animals for food. The hunted animals, such as the monkey, are called prey.

Predators help keep prey populations from growing too large. As you already learned, predators are one way that a prey population is kept from reaching the carrying capacity of the ecosystem. Predators often catch weak or injured members of a prey population. When the weak members are removed, more resources become available for the remaining members. This keeps the prey population healthy. ✓

Cooperative Relationships

The members of some populations work together in cooperative relationships for their survival. For example, leaf-cutter ants cooperate with each other and grow food. They work together to cut apart leaves and carry them to their underground nest. The ants do not eat the leaves. Instead, they eat the fungus that grows on the leaves.

Meerkats cooperate with each other as they raise young and watch for danger. Squirrel monkeys live in groups. They cooperate with each other as they hunt for food and watch for danger.

Symbiotic Relationships

Some species have such close relationships that they are almost always found living together. *A close, long-term relationship between two species that usually involves an exchange of food or energy is called* **symbiosis** (sihm bee OH sus). There are three types of symbiosis. They are mutualism, commensalism, and parasitism.

Mutualism *A symbiotic relationship in which both partners benefit is called* **mutualism.** Boxer crabs and sea anemones have a mutualistic partnership. Boxer crabs and sea anemones live in tropical coral reef communities. The crabs carry sea anemones in their claws. The sea anemones have stinging cells that help the crabs fight off predators. The sea anemones eat leftovers from the crabs' meals. Both partners benefit from the relationship.

Commensalism *A symbiotic relationship that benefits one species but does not harm or benefit the other is* **commensalism.** Plants called epiphytes (EH puh fites) grow on the trunks of trees and other objects. The roots of an epiphyte anchor it to the object. The plant absorbs nutrients from the air. The epiphytes benefit by getting living space and sunlight. The plants do not help or harm the tree. The trees are neither helped nor harmed by the plants. The epiphytes and trees have a commensal relationship.

Parasitism *A symbiotic relationship that benefits one species and harms the other is* **parasitism.** The species that benefits is the parasite. The species that is harmed is the host. Types of parasites and hosts are shown in the table below.

Parasitism

Parasite	Host	Result
heartworm	dog	causes a dog's heart to work harder and eventually fail
ringworm, toenail fungus	human	fungi feed on a protein in skin and nails
strangler fig	tree	sends roots through a tree and absorbs all of its nutrients, eventually killing the tree

Heartworms, tapeworms, fleas, and lice are parasites. They feed on a host organism, such as a human or a dog. The parasite benefits by getting food. The host is harmed by losing blood. The host is usually not killed, but it can be weakened. For example, heartworms in a dog can cause the dog's heart to work harder. The heart can fail after time, killing the host.

The fungi that cause ringworm and toenail fungus are other common parasites. The fungi feed on a protein in skin and nails.

Plants can be parasites too. The seeds of the strangler fig sprout on the branches of a host tree. The young fig sends roots into the tree and down into the ground below. The host tree provides nutrients to the fig and a trunk for support. The strangler fig grows quickly and can kill the host tree.

Visual Check

8. Identify Name two parasites that can affect humans.

Key Concept Check

9. Describe List five ways that species in a community interact.

After You Read

Mini Glossary

commensalism: a symbiotic relationship that benefits one species but does not harm or benefit the other

consumer: an organism that gets energy by eating other organisms

habitat: the place within an ecosystem where an organism lives

mutualism: a symbiotic relationship in which both partners benefit

niche (NICH): what a species does in its habitat to survive

parasitism: a symbiotic relationship that benefits one species and harms the other

producer: an organism that gets energy from the environment, such as sunlight, and makes its own food

symbiosis (sihm bee OH sus): a close, long-term relationship between two species that usually involves an exchange of food or energy

1. Review the terms and their definitions in the Mini Glossary. Use at least one term from the Mini Glossary in a sentence to explain why many species can live in the same habitat.

____

____

2. Identify each type of relationship described in the table. Write the name of the relationship next to its description.

Symbiosis		
Partner A	**Partner B**	**Type of Relationship**
Benefits	is harmed	**a.**
Benefits	benefits	**b.**
Benefits	does not benefit and is not harmed	**c.**

3. Describe some differences between producers and consumers.

____

____

What do you think NOW?

Reread the statements at the beginning of the lesson. Fill in the After column with an A if you agree with the statement or a D if you disagree. Did you change your mind?

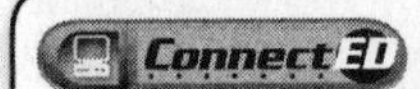

Log on to ConnectED.mcgraw-hill.com and access your textbook to find this lesson's resources.

Biomes and Ecosystems

Land Biomes

Before You Read

What do you think? Read the two statements below and decide whether you agree or disagree with them. Place an A in the Before column if you agree with the statement or a D if you disagree. After you've read this lesson, reread the statements and see if you have changed your mind.

Before	Statement	After
	1. Deserts can be cold.	
	2. There are no rain forests outside the tropics.	

Key Concepts 
- How do Earth's land biomes differ?
- How do humans impact land biomes?

Read to Learn

Land Ecosystems and Biomes

The living or once-living parts of an environment are the biotic parts. The biotic parts include people, trees, grass, birds, flowers, and insects. The nonliving parts of an environment are the abiotic parts. They include the air, sunlight, and water. The biotic parts of the environment need the abiotic parts to survive. The biotic and abiotic parts of an environment together make up an ecosystem.

Earth's continents have many different ecosystems. They range from deserts to rain forests. Scientists classify similar ecosystems in large geographic areas as biomes. *A* **biome** *is a geographic area on Earth that contains ecosystems with similar biotic and abiotic features.* You are already familiar with at least one of Earth's biomes—you live in one. ✓

Biomes Earth has seven major land biomes. Areas classified as the same biome have similar climates and organisms. In this lesson you will learn about each of these land biomes.

Mark the Text

Identify Main Ideas As you read this lesson, highlight the main ideas. Use the highlighted material to review the lesson.

✓ **Reading Check**

1. Define biome.

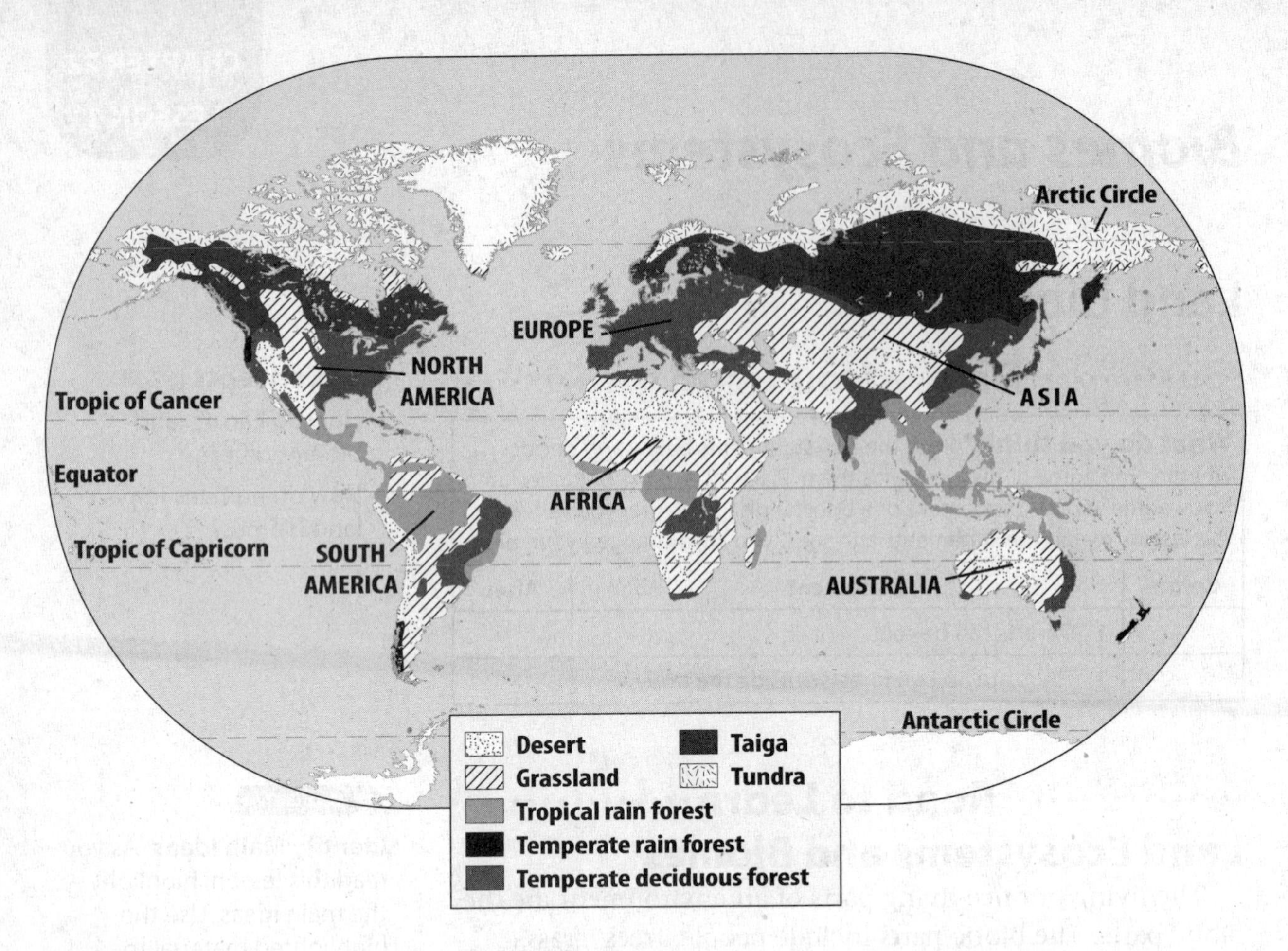

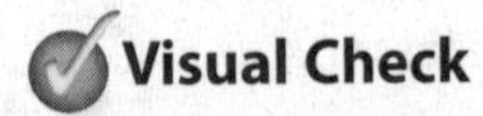

Visual Check

2. Locate Circle the areas where tropical rain forests are found.

Reading Check

3. State List four of the seven biomes found on Earth.

____

____

____

____

Earth's Biomes The seven types of land biomes—deserts, grasslands, tropical rain forests, temperate rain forests, temperate deciduous forests, taiga, and tundra—are shown above. Refer to this map as you read about each biome, its biodiversity, and the impact humans have on each of the biomes.

Desert Biome

Deserts *are biomes that receive very little rain*. Deserts are on almost every continent. They are Earth's driest ecosystems.

- Most deserts are hot during the day and cold at night. Others, like those in Antarctica, are always cold.
- After a rain, rainwater drains away quickly because the soil is thin and porous. Large patches of ground in deserts are bare.
- Average monthly temperatures in deserts in the U.S. range from 6°C to 34°C. Average monthly precipitation ranges from 1 cm to 2.5 cm.

Biodiversity

- Animals of the desert biome include lizards, bats, owls, woodpeckers, and snakes. Most animals are not active during the hottest parts of the day.
- Plants include spiny cacti and thorny shrubs. Shallow roots enable a plant to absorb water quickly. Some plants have accordion-like stems that expand and store water. Small leaves or spines reduce the loss of water.

Human Impact

- Cities, farms, and recreational areas located in deserts use valuable water.
- Desert plants grow slowly. When they are damaged by people or livestock, it can take many years for them to recover.

Grassland Biome

Grassland *biomes are areas where grasses are the dominant plants.* Grasslands are also called prairies, savannas, and meadows. Grasslands are referred to as the world's "breadbaskets." This is because wheat, corn, oats, rye, barley, and other important cereal crops are grasses. They grow well in these areas.

- Grasslands have a wet and a dry season.
- Deep, fertile soil supports plant growth.
- Grass roots form a thick mass, called sod, which helps soil absorb water and hold it during periods of drought.
- Average monthly temperatures in U.S. grasslands range from −2°C to 38°C. Average monthly precipitation ranges from 2 cm to 6 cm. ✓

Biodiversity

- Trees grow along moist banks of streams and rivers in grassland biomes. Wildflowers bloom during the wet season.
- In North America, large herbivores, such as bison and elk, graze on grasses. Insects, birds, rabbits, prairie dogs, and snakes find shelter in the grasses.
- Predators in the grasslands of North America include hawks, ferrets, coyotes, and wolves.
- Giraffes, zebras, and lions are found in the African savannas. Australian grasslands are home to kangaroos, wallabies, and wild dogs. ✓

Think it Over

4. Describe the benefit of shallow roots for a desert plant.

✓ Reading Check

5. Explain why grasslands are called "breadbaskets."

✓ Reading Check

6. Name What are the grasslands in Africa called?

Human Impact

- People plow large areas of grassland to raise cereal crops. This reduces habitats for wild species.
- Because of hunting and the loss of habitats, large herbivores, such as bison, are now uncommon in many grasslands.

Tropical Rain Forest Biome

The forests that grow near the equator are called tropical rain forests. These forests receive large amounts of rain and have dense growths of tall, leafy trees.

- Weather in tropical rain forest biomes is warm and wet all year.
- The soil is shallow and is easily washed away by rain.
- Less than 1 percent of the sunlight that reaches the top of forest trees reaches the forest floor.
- Half of Earth's species live in tropical rain forests. Most live in the canopy—the uppermost part of the forest.
- Average monthly temperatures range from 21°C to 32°C. Average monthly precipitation ranges from 14 cm to 26 cm.

Reading Check

7. Describe the weather in a tropical rain forest.

______

______

______

______

Biodiversity

- Few plants live on the dark forest floor.
- Vines climb the trunks of tall trees.
- Mosses, ferns, and orchids live on branches in the canopy.
- Insects make up the largest group of tropical animals. They include beetles, termites, ants, bees, and butterflies.
- Larger animals include parrots, toucans, snakes, frogs, flying squirrels, fruit bats, monkeys, jaguars, and ocelots.

Reading Check

8. Identify Where do most organisms live in a tropical rain forest? (Circle the correct answer.)

a. on the forest floors
b. in the canopy
c. deep in the soil

Human Impact

- People have cleared more than half of Earth's tropical rain forests for lumber, farming, and ranching. Poor soil does not support rapid growth of new trees in cleared areas.
- Some organizations are working to encourage people to use less wood harvested from rain forests.

Reading Check

9. Name four kinds of animals that live in tropical rain forests.

______

______

______

______

Temperate Rain Forest Biome

Regions of Earth between the tropics and the polar circles are **temperate** *regions.* Temperate regions have relatively mild climates with distinct seasons.

Several biomes are in temperate regions, including rain forests. Temperate rain forests are moist ecosystems located mostly in coastal areas. Temperate rain forests are not as warm as tropical rain forests.

- Winters are mild and rainy.
- Summers are cool and foggy.
- Soil is rich and moist.
- Average monthly temperatures of U.S. temperate rain forests range from 0°C to 23°C. Average monthly precipitation ranges from 2 cm to 36 cm. ✓

Biodiversity

- Forests are made up mostly of spruce, hemlock, cedar, fir, and redwood trees. These trees can grow very large and tall.
- Fungi, ferns, mosses, vines, and small flowering plants grow on the moist forest floor.
- Animals include mosquitoes, butterflies, frogs, salamanders, woodpeckers, owls, eagles, chipmunks, raccoons, deer, elk, bears, foxes, and cougars.

Human Impact

- Temperate rain forest trees are a source of lumber. Logging can destroy the habitat of forest species.
- Rich soil enables cut forests to grow back. Tree farms help provide lumber without destroying the habitats.

FOLDABLES®

Make a horizontal two-tab book to record information about desert and temperate rain forest biomes.

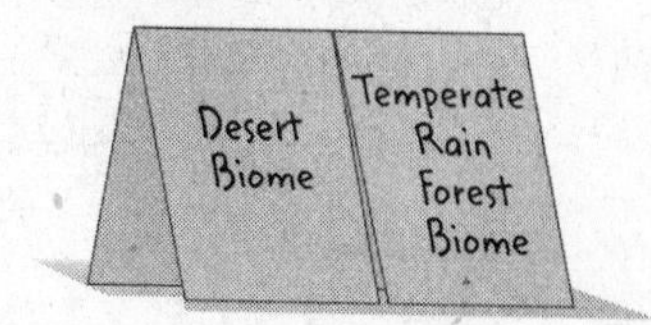

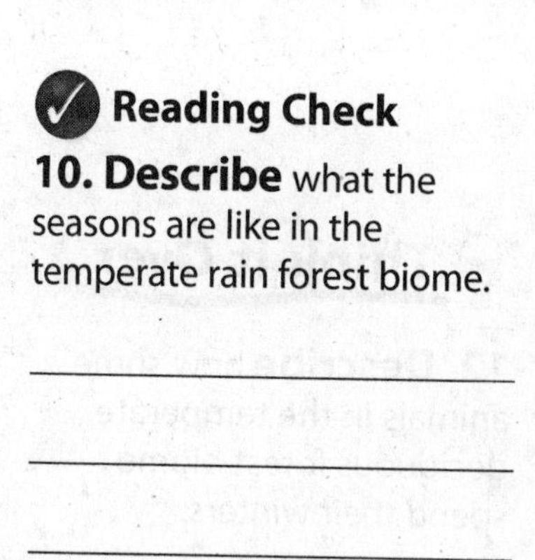

Reading Check

10. Describe what the seasons are like in the temperate rain forest biome.

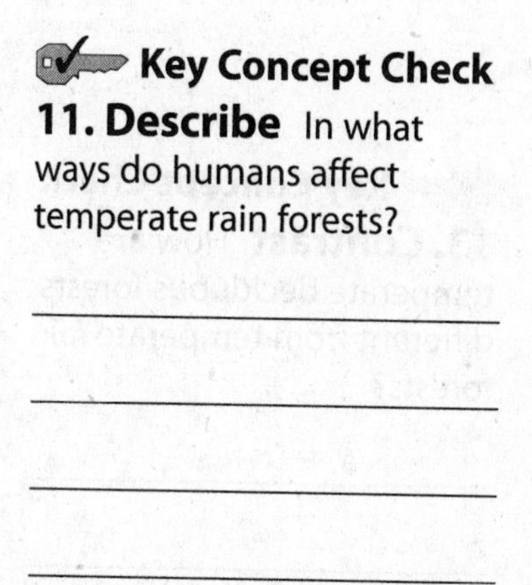

Key Concept Check

11. Describe In what ways do humans affect temperate rain forests?

Temperate Deciduous Forest Biome

Temperate deciduous forests grow in temperate regions where winter and summer climates have more variation than regions where temperate rain forests grow. Temperate deciduous forests are the most common forest ecosystems in the United States. They contain mostly deciduous trees. Deciduous trees lose their leaves in fall.

- Winter temperatures are often below freezing. Snow is common.
- Summers are hot and humid.
- Soil is rich in nutrients. A large variety of plants grow in great numbers in the soil.
- Average monthly temperatures in U.S. biomes range from –7°C to 29°C. Average monthly precipitation ranges from 4 cm to 10 cm.

Biodiversity

- Most plants, such as maples, oaks, birches, and other deciduous trees, stop growing during the winter and begin growing again in the spring.
- Animals include snakes, ants, butterflies, birds, raccoons, opossums, and foxes.
- Some animals, including chipmunks and bats, spend the winter in hibernation.
- Many birds and some butterflies, such as the monarch, migrate to warmer climates for the winter.

Human Impact

- Over the past several hundred years, humans have cleared thousands of acres of Earth's deciduous forests for farms and cities.
- Today, much of the clearing has stopped and some forests have regrown.

Think it Over

12. Describe how some animals in the temperate deciduous forest biome spend their winters.

____

____

Key Concept Check

13. Contrast How are temperate deciduous forests different from temperate rain forests?

____

____

____

____

Taiga Biome

A **taiga** (TI guh) *is a forest biome consisting mostly of cone-bearing evergreen trees.* The taiga biome exists only in the northern hemisphere. It occupies more space on Earth's continents than any other biome.

- Winters are long, cold, and snowy. Summers are short, warm, and moist.
- Soil is thin and acidic.
- Temperatures range from –16°C to 23°C. Precipitation ranges from 3 cm to 7 cm each month.

Biodiversity

- Evergreen trees, such as spruce, pine, and fir, are thin and shed snow easily.
- Animals include owls, mice, moose, bears, and other cold-adapted species.
- Abundant insects in summer attract many birds, which migrate south in winter.

Human Impact

- Tree harvesting reduces taiga habitat.

Tundra Biome

A **tundra** (TUN druh) *biome is cold, dry, and treeless.* Most tundra is located south of the North Pole. It also exists in mountainous areas at high altitudes.

- Winters are dark and freezing, and summers are short and cool. The growing season is 50–60 days long.
- Permafrost—a layer of permanently frozen soil—prevents deep root growth.
- Temperatures in the Canadian tundra range from –28°C to 4°C. Precipitation ranges from 3 cm a month to 23 cm a month.

Biodiversity

- Plants include shallow-rooted mosses, lichens, and grasses.
- Many animals hibernate or migrate south during the winter. A few animals, including lemmings, live in tundras year-round.

Human Impact

- Drilling for oil and gas can interrupt migration patterns of animals.

Reading Check

14. Name three kinds of trees found in the taiga biome.

Reading Check

15. Describe ways that humans can disturb the tundra biome.

After You Read

Mini Glossary

biome: a geographic area on Earth that contains ecosystems with similar biotic and abiotic features

desert: a biome that receives very little rain

grassland: a biome where grasses are the dominant plants

taiga (TI guh): a forest biome consisting mostly of cone-bearing evergreen trees

temperate: a region between the tropics and the polar circles

tundra (TUN druh): a biome that is cold, dry, and treeless

1. Review the terms and their definitions in the Mini Glossary. Write a sentence that describes what a biome is and how biomes can differ.

____

____

2. Fill in the table below to identify the plant and animal life in the different biomes.

	Desert	Tropical Rain Forest	Temperate Rain Forest	Temperate Deciduous Forest	Taiga	Tundra	Grassland
Animals		Insects, parrots, toucans, snakes, monkeys, frogs, jaguars		Snakes, birds, raccoons, opossums, chipmunks, foxes		Lemmings, few others year-round	
Plants	Shrubs, cacti		Spruce, hemlock, cedar, fir, redwood, fungi, ferns, mosses, vines	Maples, oaks, birches	Cone-bearing evergreen trees		Trees along rivers, wildflowers, cereal crops, grasses

What do you think NOW?

Reread the statements at the beginning of the lesson. Fill in the After column with an A if you agree with the statement or a D if you disagree. Did you change your mind?

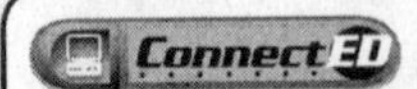

Log on to ConnectED.mcgraw-hill.com and access your textbook to find this lesson's resources.

Biomes and Ecosystems

Aquatic Ecosystems

Before You Read

What do you think? Read the two statements below and decide whether you agree or disagree with them. Place an A in the Before column if you agree with the statement or a D if you disagree. After you've read this lesson, reread the statements and see if you have changed your mind.

Before	Statement	After
	3. Estuaries do not protect coastal areas from erosion.	
	4. Animals form coral reefs.	

Key Concepts
- How do Earth's aquatic ecosystems differ?
- How do humans impact aquatic ecosystems?

Read to Learn

Aquatic Ecosystems

Water is full of life. There are four main types of water, or aquatic, ecosystems. They are freshwater, wetland, estuary, and ocean. Each type of ecosystem contains a unique variety of organisms. Whales, dolphins, and corals live only in ocean ecosystems. Trout live only in freshwater ecosystems. Many other organisms, such as birds and seals, depend on aquatic ecosystems for food and shelter.

Temperature, sunlight, and oxygen gas that is dissolved in the water are important abiotic factors in aquatic ecosystems. Fish and other aquatic species have adaptations that enable them to use the oxygen from the water. For example, the gills of a fish separate oxygen from water and move the oxygen into the bloodstream of the fish. Mangrove plants take oxygen in through small pores in their leaves and roots.

Salinity (say LIH nuh tee) is another important abiotic factor in aquatic ecosystems. **Salinity** *is the amount of salt dissolved in water*. Water in saltwater ecosystems has high salinity compared to water in freshwater ecosystems. Freshwater contains little salt.

Study Coach

Create a Quiz Create a quiz about the types of aquatic ecosystems. Exchange quizzes. After taking the quizzes, discuss the answers.

Math Skills

Salinity is measured in parts per thousand (PPT). One PPT water contains 1 g of salt and 1,000 g of water. Use proportions to calculate salinity. What is the salinity of 100 g of water with 3.5 g of salt?

$$\frac{3.5 \text{ g salt}}{100 \text{ g seawater}} = \frac{x \text{ g salt}}{1{,}000 \text{ g seawater}}$$

$$100x = 3500$$

$$x = \frac{3500}{100} = 35 \text{ PPT}$$

1. Use Proportions A sample contains 0.1895 g of salt per 50 g of seawater. What is its salinity?

Freshwater: Streams and Rivers

Freshwater ecosystems include streams, rivers, ponds, and lakes. Streams and rivers contain flowing freshwater. Streams are usually narrow, shallow, and fast-flowing. Rivers are larger, deeper, and flow more slowly.

- Streams form from underground springs or from runoff from rain and melting snow.
- Stream water is often clear because soil particles are quickly washed downstream.
- Oxygen levels in streams are high because air mixes into the water as it splashes over rocks.
- Rivers form when streams flow together.
- Soil that washes into a river from streams or nearby land can make the river water muddy. Soil also introduces nutrients, such as nitrogen, into rivers.
- Slow-moving river water has higher levels of nutrients and lower levels of dissolved oxygen than fast-moving water.

Biodiversity

- Willows, cottonwoods, and other water-loving plants grow along streams and on riverbanks.
- Species adapted to fast-moving water include trout, salmon, crayfish, and many insects.
- Species adapted to slow-moving water include snails and catfish.

Human Impact

- People take water from streams and rivers for drinking, doing laundry, bathing, irrigating crops, and industrial purposes.
- Hydroelectric plants use the energy in flowing water to generate electricity. Dams stop the natural flow of water.
- Runoff from cities, industries, and farms is a source of pollution.

Think it Over

2. Compare rivers and streams.

Reading Check

3. Name three organisms found in a river ecosystem.

Reading Check

4. Identify possible sources of pollution to a river or a stream.

Freshwater: Ponds and Lakes

Ponds and lakes contain freshwater that is not flowing downhill. These bodies of water form in low areas on land.

- Ponds are warm and shallow.
- Sunlight reaches the bottoms of most ponds.
- Pond water is often high in nutrients.
- Lakes are larger and deeper than ponds.
- Sunlight penetrates into the top few feet of lake water. Deeper water is dark and cold. 

Biodiversity

- Plants surround ponds and lake shores.
- Surface water in ponds and lakes contains plants, algae, and microscopic organisms that use sunlight for photosynthesis.
- Organisms living in shallow water along shorelines include cattails, reeds, insects, crayfish, frogs, fish, and turtles.
- Fewer organisms live in the deeper, colder water of lakes where there is little sunlight.
- Lake fish include perch, trout, bass, and walleye. 

Human Impact

- Humans fill in ponds and lakes with sediment to create land for houses and other structures.
- Runoff from farms, gardens, and roads washes pollutants into ponds and lakes. This disrupts the food webs that exist in these biomes.

Reading Check

5. Identify the statement that is true of ponds. (Circle the correct answer.)

a. Ponds have moving water.
b. Ponds have few nutrients.
c. Ponds are warm and shallow.

Reading Check

6. Explain why few organisms live in the deep water of lakes.

____

____

Key Concept Check

7. Contrast How do ponds and lakes differ?

____

____

____

Wetlands

Some types of aquatic ecosystems have mostly shallow water. **Wetlands** *are aquatic ecosystems that have a thin layer of water covering soil that is wet most of the time.* Wetlands contain freshwater, salt water, or both. They are among Earth's most fertile ecosystems.

- Freshwater wetlands form at the edges of lakes and ponds and in low areas on land. Saltwater wetlands form along ocean coasts.
- Nutrient levels and biodiversity are high.
- Wetlands trap sediments and purify water. Plants and microscopic organisms filter out pollution and waste materials.

Reading Check
8. Describe Where do wetlands form?

____

____

Biodiversity

- Plants that can live in wetlands include grasses and cattails. Few trees live in saltwater wetlands. Trees that can live in freshwater wetlands include cottonwoods, willows, and swamp oaks.
- Insects are numerous and include flies, mosquitoes, dragonflies, and butterflies.
- More than one-third of North American bird species use wetlands for nesting and feeding. Some of them are ducks, geese, herons, loons, warblers, and egrets.
- Other animals that depend on wetlands for food and breeding grounds include alligators, turtles, frogs, snakes, salamanders, muskrats, and beavers.

Think it Over
9. Discuss Why is it important to protect wetlands?

____

____

Human Impact

- In the past, people often thought wetlands were unimportant environments. Water was drained away to build homes and roads and to raise crops.
- Today, many wetlands are being preserved, and drained wetlands are being restored.

Key Concept Check
10. Explain How do humans impact wetlands?

____

____

____

Estuaries

Estuaries (ES chuh wer eez) *are regions along coastlines where streams or rivers flow into a body of salt water.* Most estuaries form along coastlines where freshwater in rivers mixes with salt water in oceans. The degree of salinity in estuary ecosystems varies. ✓

- Salinity depends on rainfall, the amount of freshwater flowing from land, and the amount of salt water pushed in by tides.
- Estuaries help protect coastal land from flooding and erosion. Like wetlands, estuaries purify water and filter out pollution.
- Nutrient levels and biodiversity are high.

Biodiversity

- Plants that grow in salt water include mangroves, pickleweeds, and seagrasses.
- Animals include worms, snails, and many species that people use for food, including oysters, shrimp, crabs, and clams.
- Striped bass, salmon, flounder, and many other ocean fish lay their eggs in estuaries.
- Many species of birds depend on estuaries for breeding, nesting, and feeding.

Human Impact

- Large portions of estuaries have been filled with soil to make land for roads and buildings.
- Destruction of estuaries reduces habitats for estuary species. It also exposes the coastline to flooding and storm damage. ✓

✓ **Reading Check**

11. Identify Where do estuaries form? (Circle the correct answer.)

a. in freshwater

b. in salt water

c. where freshwater and salt water mix

Make a two-tab book to use to compare how biodiversity and human impact differ in wetlands and estuaries.

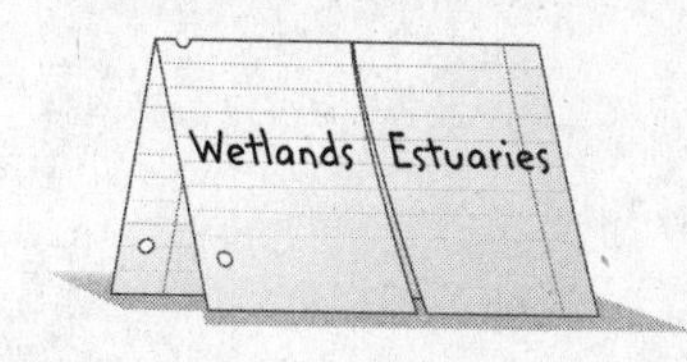

✓ **Reading Check**

12. Name problems that can come from destroying estuaries.

____

____

____

____

Ocean: Open Oceans

Most of Earth's surface is covered by ocean water with high salinity. The oceans contain different types of ecosystems. If you took a boat trip several kilometers out to sea, you would be in the open ocean. This is one type of ecosystem. The open ocean extends from the steep edges of continental shelves to the deepest parts of the ocean. A cross section of the open ocean is shown in the figure. The amount of light in the water depends on depth.

Visual Check

13. Draw arrows indicating the depth that sunlight reaches in the open ocean.

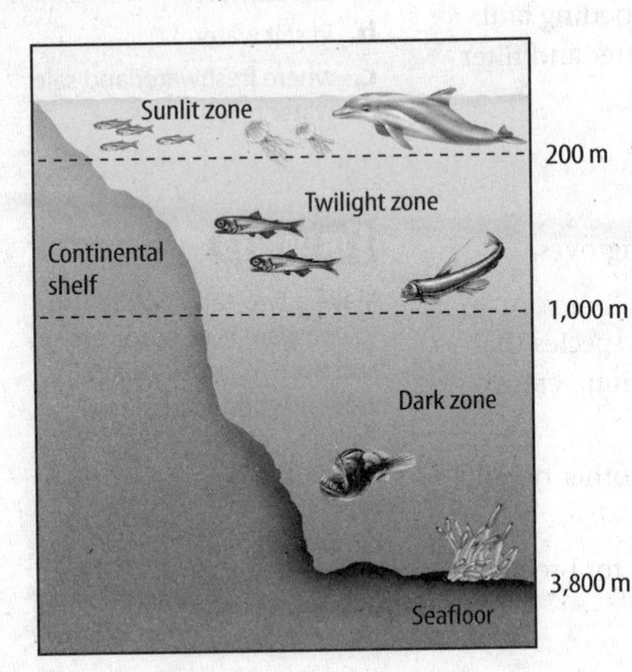

- Photosynthesis takes place only in the uppermost, or sunlit, zone. Very little sunlight reaches the deeper twilight zone. None reaches the deepest water, known as the dark zone.
- Decaying matter and nutrients float down from the sunlit zone, through the twilight and dark zones, to the seafloor.

Biodiversity

- Microscopic algae and other producers (organisms that make their own food) in the sunlit zone form the base of most ocean food chains. Other organisms that live in the sunlit zone include jellyfish, tuna, mackerel, and dolphins.
- Many species of fish stay in the twilight zone during the day and go to the sunlit zone at night to feed.
- Sea cucumbers, brittle stars, and other bottom-dwelling organisms feed on decaying matter that drifts down from above.
- Many organisms in the dark zone live near cracks in the seafloor where lava erupts and new seafloor forms.

Reading Check

14. Identify What is the base of most ocean food chains? (Circle the correct answer.)

a. microscopic algae

b. fish, such as tuna and mackerel

c. dolphins

Human Impact

- Overfishing threatens many ocean fish.
- Trash discarded from ocean vessels or washed into oceans from land is a source of pollution. Animals, such as seals, become tangled in plastic or mistake plastic for food.

Ocean: Coastal Oceans

Coastal oceans include several types of ecosystems, including continental shelves and the intertidal zones. *The* ***intertidal zone*** *is the ocean shore between the lowest low tide and the highest high tide.*

- Sunlight reaches the bottom of shallow coastal ecosystems.
- Nutrients washed in from rivers and streams add to the high biodiversity.

Biodiversity

- The coastal ocean is home to mussels, fish, crabs, dolphins, and whales.
- Intertidal species have adaptations for surviving exposure to air during low tides and to heavy waves during high tides.

Human Impact

- Oil spills and other pollution harm coastal organisms.

Ocean: Coral Reefs

Another ocean ecosystem with high biodiversity is coral reefs. *A* ***coral reef*** *is an underwater structure made from outside skeletons of tiny, soft-bodied animals called coral.*

- Most coral reefs form in shallow tropical oceans.
- Coral reefs protect coastlines from storm damage and erosion.

Biodiversity

- Coral reefs provide food and shelter for many animals, including parrotfish, groupers, angelfish, eels, shrimp, crabs, scallops, clams, worms, and snails.

Human Impact

- Pollution, overfishing, and harvesting of coral threaten coral reefs.

Reading Check

15. Describe Where is the intertidal zone located?

____

____

____

Think it Over

16. Discuss why organisms in the intertidal zone must have special adaptations.

____

____

____

Reading Check

17. Define coral reef.

____

____

____

After You Read

Mini Glossary

coral reef: an underwater structure made from skeletons of tiny, soft-bodied animals called coral

estuary (ES chuh wer ee): a region along a coastline where a stream or river flows into a body of salt water

intertidal zone: the ocean shore between the lowest low tide and the highest high tide

salinity: the amount of salt dissolved in water

wetland: an aquatic ecosystem with a thin layer of water covering soil that is wet most of the time

1. Review the terms and their definitions in the Mini Glossary. How is an estuary different from a wetland?

____

____

2. Fill in the chart below to review ocean ecosystems.

Ocean Ecosystem	Characteristics	Species That Live in the Ecosystem	One Human Impact
Open	Sunlit zone, twilight zone, dark zone	Algae, jellyfish, tuna, mackerel, dolphins, sea cucumbers, brittle stars, seals	
Coastal			Pollution
Coral Reef		Parrotfish, groupers, angelfish, eels, shrimp, crabs, scallops, clams, worms, snails	Harvesting of coral

3. You made your own quiz about aquatic ecosystems. How did writing quiz questions help you better understand aquatic ecosystems?

____

____

What do you think NOW?

Reread the statements at the beginning of the lesson. Fill in the After column with an A if you agree with the statement or a D if you disagree. Did you change your mind?

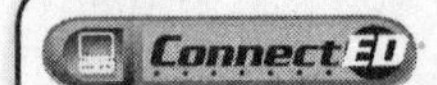

Log on to ConnectED.mcgraw-hill.com and access your textbook to find this lesson's resources.

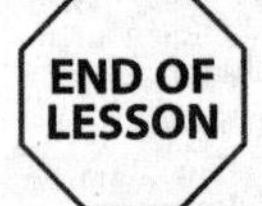

Biomes and Ecosystems

How Ecosystems Change

Before You Read

What do you think? Read the two statements below and decide whether you agree or disagree with them. Place an A in the Before column if you agree with the statement or a D if you disagree. After you've read this lesson, reread the statements and see if you have changed your mind.

Before	Statement	After
	5. An ecosystem never changes.	
	6. Nothing grows in the area where a volcano has erupted.	

Key Concepts 
- How do land ecosystems change over time?
- How do aquatic ecosystems change over time?

Read to Learn

How Land Ecosystems Change

Have you ever seen weeds growing up through cracks in a sidewalk? If they are not removed, the weeds will continue to grow. The crack will widen, making room for more weeds. Over time, the sidewalk will break apart. Shrubs and vines will sprout and grow. Their leaves and branches will grow large enough to cover the sidewalk. Eventually, trees could start to grow there.

Mark the Text

Identify Main Ideas
Highlight the main idea of each paragraph in the lesson. Reread the main ideas to review the lesson.

This process is an example of **ecological succession**—*the process of one ecological community gradually changing into another.* Ecological succession occurs in a series of steps. These steps can usually be predicted. For example, small plants usually grow in an ecosystem before trees do.

The final stage of ecological succession in a land ecosystem is a **climax community**—*a stable community that no longer goes through major ecological changes.* Climax communities differ depending on the type of biome they are in. In a grassland biome, mature grassland is the climax community. Climax communities are usually stable over hundreds of years. As plants die, new plants of the same species grow as long as the climate stays the same.

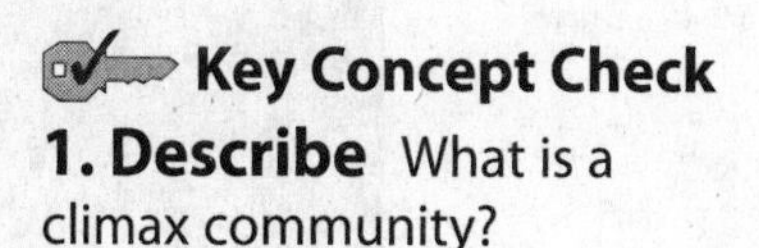

1. Describe What is a climax community?

Primary Succession

What do you think happens to a lava-filled landscape when a volcanic eruption is over? Volcanic lava eventually becomes new soil. This new soil supports plant growth. Ecological succession in new areas of land with little or no soil, such as on a lava flow, a sand dune, or exposed rock, is called primary succession. *The first species that colonize new or undisturbed land are* **pioneer species.** Lichens and mosses are pioneer species. The figure below shows what happens to an area during and after a volcanic eruption.

SCIENCE USE V. COMMON USE
pioneer
Science Use the first species that colonize new or undisturbed land
Common Use the first human settlers in an area

A. During a volcanic eruption, molten lava flows over the ground and into the water. After the eruption is over, the lava cools and hardens into bare rock.

B. Lichen spores carried on the wind settle on the rock. Lichens release acid that helps break down the rock and create soil. Lichens add nutrients to the soil as they die and decay.

C. Airborne spores from mosses and ferns settle on the thin soil and add to the soil when they die. The soil becomes thick enough to hold water. Insects and other small organisms begin living in the area.

D. After many years, the soil is deep and has enough nutrients for grasses, wildflowers, shrubs, and trees to grow. The new ecosystem provides habitats for animals. Eventually, a climax community develops.

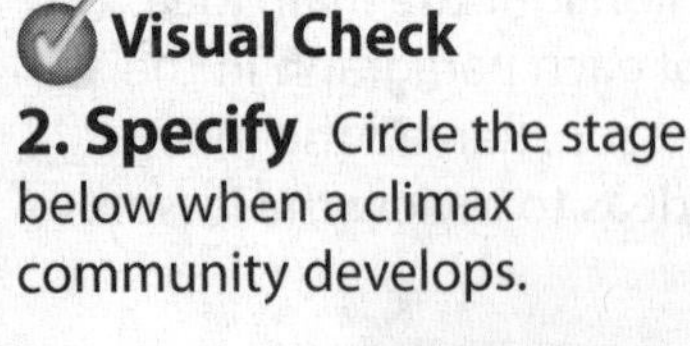
Visual Check
2. Specify Circle the stage below when a climax community develops.

A.

B.

C.

D.

Secondary Succession

Secondary succession occurs in areas where existing ecosystems have been disturbed or destroyed. One example is forestland in New England. Early colonists cleared land hundreds of years ago. Some of the cleared land was not planted with crops. This land gradually grew back to a climax forest community of beech and maple trees. The figure below shows what happens to an area of land as it is cleared and after it is cleared.

A. Settlers in North America cleared many acres of forests to plant their crops. On land where people stopped planting crops, the forests began to grow back.

B. Seeds of grasses, wildflowers, and other plants began to sprout and grow. Young shrubs and trees also started growing. These plants were habitats for insects and other small animals, such as mice.

C. White pines and poplars were the first trees in the area to grow tall. They provided shade and protection to trees that grow slower, such as beeches and maples.

D. Eventually, a climax community of beech and maple trees developed. As older trees die, new beech and maple trees grow and replace them.

Use a folded sheet of paper to describe and illustrate the before and after of secondary succession.

3. Identify Where does secondary succession occur?

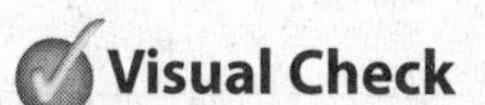

4. Specify Circle the stage below when trees begin growing again.

A.

B.

C.

D.

How Freshwater Ecosystems Change

Like land ecosystems, freshwater ecosystems change over time in a natural, predictable process. This process is called aquatic succession.

Aquatic Succession

As shown in the figures below, aquatic succession begins with a body of water, such as a pond. Over time, sediments and decaying organisms carried by rainwater and streams build up and create soil on the bottoms of ponds, lakes, and wetlands.

Soil begins to fill the body of water. Eventually the body of water fills completely with soil. The water disappears and the area becomes land.

Key Concept Check
5. Describe What happens to a pond, a lake, or a wetland over time?

____

____

6. Identify Which image shows soil being created? (Circle the correct answer.)

a. A
b. B
c. C

Eutrophication

Decaying organisms that fall to the bottom of a pond, a lake, or a wetland add nutrients to the water. **Eutrophication** (yoo troh fuh KAY shun) *is the process of a body of water becoming nutrient-rich*.

Eutrophication is a natural part of aquatic succession. However, humans also contribute to eutrophication. The fertilizers that farmers use on crops and waste from farm animals can be very high in nutrients. Other forms of pollution can also be very high in nutrients. When fertilizers and pollution run off into a pond or lake, concentrations of nutrients increase. Large populations of algae and microscopic organisms can grow using the nutrients. These organisms use up most of the dissolved oxygen in the water. Less oxygen is available for the fish and other organisms that live in the water. As a result, many of these organisms die. Their bodies decay and add to the buildup of the soil, speeding up succession. ✓

WORD ORIGIN
eutrophication
from Greek *eutrophos,* means "nourishing"

7. Name two causes of eutrophication.

____

____

After You Read

Mini Glossary

climax community: a stable community that no longer goes through major ecological changes

ecological succession: the process of one ecological community gradually changing into another

eutrophication (yoo troh fuh KAY shun): the process of a body of water becoming nutrient-rich

pioneer species: the first species that colonize new or undisturbed land

1. Review the terms and their definitions in the Mini Glossary. Write a sentence describing ecological succession. Use the term *climax community* in your sentence.

____

____

2. Fill in the table below to compare succession in aquatic ecosystems and land ecosystems.

	Aquatic Ecosystem	Land Ecosystem
How it changes		Grasses and other plants begin to grow. Young shrubs and trees then start growing.
What changes	Depth of water	
End result		Climax community

3. Summarize the characteristics of a climax community.

____

____

What do you think NOW?

Reread the statements at the beginning of the lesson. Fill in the After column with an A if you agree with the statement or a D if you disagree. Did you change your mind?

Log on to ConnectED.mcgraw-hill.com and access your textbook to find this lesson's resources.

Using Natural Resources

Earth's Resources

Before You Read

What do you think? Read the two statements below and decide whether you agree or disagree with them. Place an A in the Before column if you agree with the statement or a D if you disagree. After you've read this lesson, reread the statements to see if you have changed your mind.

Before	Statement	After
	1. The world's supply of oil will never run out.	
	2. You should include minerals in your diet.	

Key Concepts 

- What are natural resources?
- How do the three types of natural resources differ?

Read to Learn

Natural Resources

Where does the electricity for electric lights come from? It might come from a power plant that burns coal or natural gas. Or it might come from rooftop solar panels made with silicon, a mineral found in sand.

The smallest microbe and the largest whale rely on materials and energy from the environment. The same is true for humans. People depend on the environment for food, clothing, and fuels to heat and light their homes. *Parts of the environment that supply materials useful or necessary for the survival of living things are called* **natural resources.** Natural resources include land, air, minerals, and fuels. For example, trees and water are natural resources.

Study Coach

Building Vocabulary Work with another student to write a question about each vocabulary term in this lesson. Answer the questions and compare your answers. Reread the text to clarify the meaning of the terms.

Nonrenewable Resources

Do you travel in a vehicle that runs on gasoline? Do you drink soda from aluminum cans or water from plastic bottles? Gasoline, aluminum, and plastic are made from nonrenewable resources.

Nonrenewable resources *are natural resources that are being used up faster than they can be replaced by natural processes.* Nonrenewable resources form slowly, usually over thousands or millions of years. If they are used faster than they form, they will run out. Nonrenewable resources include fossil fuels and minerals.

Key Concept Check

1. Define What are natural resources?

FOLDABLES®

Make a horizontal tri-fold book and use it to identify similarities and differences among the types of resources.

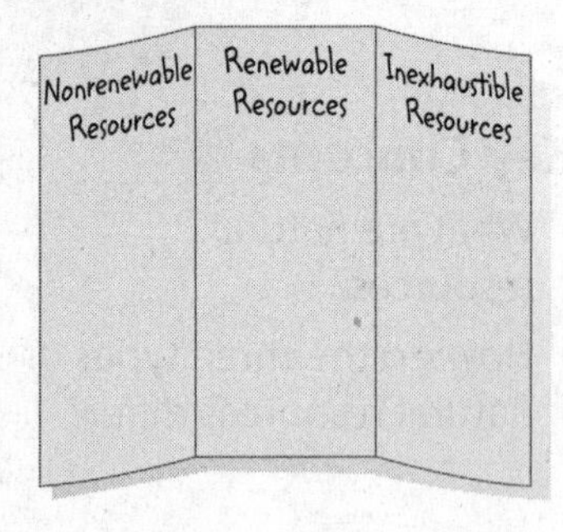

Fossil Fuels

Fossil fuels include coal, oil, and natural gas. The fossil fuels we use today formed from the decayed remains of organisms that died millions of years ago. Fossil fuels are forming all the time, but we use them much more quickly than nature replaces them.

Fossil fuels form underground. Coal is mined from the ground. Some coal mines are on the surface, and others are underground. The coal that is mined today formed from the decayed remains of trees, ferns, and other swamp plants that died 300–400 million years ago. The fossil fuels oil and natural gas are drilled from the ground.

Fossil fuels are mainly used as sources of energy. Many electric power plants burn coal or natural gas to heat water and make steam that powers generators. Natural gas also is used to heat homes and businesses. Gasoline, jet fuel, diesel fuel, kerosene, and other fuels are made from oil. Most plastics also are made from oil. ✓

✓ Reading Check

2. Identify three products that are made from oil.

____

____

____

Minerals

Have you ever added fertilizer to the soil around a plant? Fertilizers contain the minerals phosphorus and potassium. These minerals promote plant growth. Humans also need minerals for good health. Calcium and magnesium are two minerals the human body needs.

Minerals are nonliving substances found in Earth's crust. People use minerals for many purposes. The mineral gypsum is used in wallboard and cement. Silicon has many uses in industry. It is important for the manufacture of computers and other electronic devices. Copper is widely used in electrical wiring.

Uranium is a mineral that can be used as a source of energy. In a nuclear power plant, the nuclei of uranium atoms are split apart. This reaction is known as nuclear fission. Some of the energy that held the nuclei together is released as thermal energy. This thermal energy is then used to boil water and produce steam, which generates electricity.

Like fossil fuels, minerals are formed underground by geologic processes over millions of years. For that reason, most minerals are considered nonrenewable. Some minerals, such as calcium, are plentiful. Others, such as large rubies, are rare. ✓

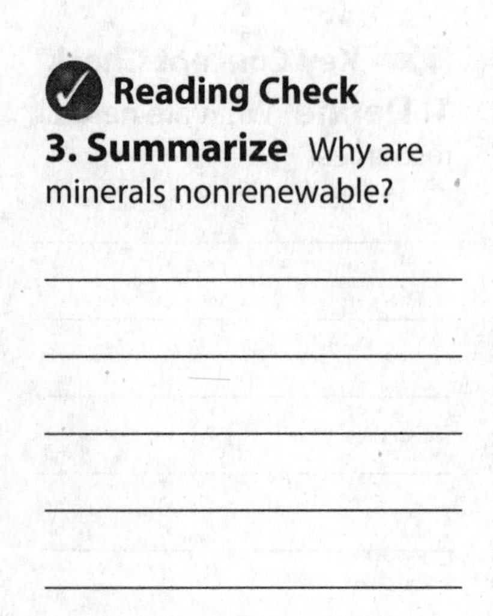

✓ Reading Check

3. Summarize Why are minerals nonrenewable?

____

____

____

____

____

____

Renewable Resources

Supplies of many natural resources are constantly renewed by natural cycles. The water cycle is an example. When liquid water evaporates, it rises into the atmosphere as water vapor. Water vapor condenses and falls back to the ground as rain or snow. Water is a renewable resource.

Renewable resources *are natural resources that can be replenished by natural processes at least as quickly as they are used.* These resources do not run out because they are replaced in a fairly short period of time. Renewable resources include water, air, land, and living things.

Renewable resources are replenished by natural processes. Still, they must be used wisely. If people use any resource faster than it is replaced, it becomes nonrenewable. A forest can be a nonrenewable resource if the trees are cut down faster than they can be replaced. ✓

Reading Check

4. Compare In what way are renewable and nonrenewable resources similar?

Air

Green plants produce almost all of the oxygen in the air we breathe. Oxygen is a product of photosynthesis. Remember that photosynthesis is a series of chemical reactions in plants that use energy from light and produce sugars. Without plants, Earth's atmosphere would not contain enough oxygen to support most forms of life.

Think it Over

5. Generalize Do you think humans could survive if most of the plants on Earth died? Why or why not?

Air also contains carbon dioxide (CO_2), which plants need for photosynthesis. CO_2 is released into the air when dead plants and animals decay, when fossil fuels or wood are burned, and as a product of cellular respiration in plants and animals. Cellular respiration is a series of chemical reactions that convert energy from food into a form usable by cells. Without CO_2, photosynthesis would not be possible.

Land

Fertile soil is an important resource. Topsoil is the upper layer of soil that contains most of the nutrients plants need. Gardeners know that topsoil can be replenished by the decay of plant material. The carbon, nitrogen, and other elements in the decomposing plants become available for the growth of new plants.

Topsoil can be classified as a renewable resource. However, if it is carried away by water or wind, it can take hundreds of years to rebuild. Land resources also include wildlife and ecosystems such as forests, grasslands, deserts, and coral reefs. ✓

Reading Check

6. Consider How is topsoil replenished by natural processes?

Math Skills

Converting a ratio to a percentage often makes it easier to visualize a set of numbers. For example, in 2007, 101.5 quadrillion units (quads) of energy were used in the United States. Of that, 6.813 quads were produced from renewable energy sources. What percentage of U.S. energy was produced from renewable energy sources?

Set up a ratio of the part over the whole.

$$\frac{6.813 \text{ quads}}{101.5 \text{ quads}}$$

Rewrite the fraction as a decimal.

$$\frac{6.813 \text{ quads}}{101.5 \text{ quads}} = 0.0671$$

Multiply by 100 and add %.

$$0.0671 \times 100 = 6.71\%$$

7. Use Percentages Of the 101.5 quads of energy used in 2007, 0.341 quads were from wind energy. What percentage of U.S. energy came from wind?

____

Key Concept Check

8. Differentiate How do inexhaustible resources differ from renewable and nonrenewable resources?

____

____

____

____

Water

Can you imagine a world without water? All organisms require water to live. People need a reliable supply of freshwater for drinking, washing, and irrigating crops. People also use water to run power plants and factories. Oceans, lakes, and rivers serve as major transportation routes and recreational areas. They are important habitats for many species, including some that people depend on for food.

Most of Earth's surface is covered by water. But only a small amount of that water is freshwater that people use. Freshwater is renewed through the water cycle. The total amount of water on Earth always remains the same.

Has your community ever been asked to conserve water because of a drought? A drought can cause a shortage in the supply of freshwater. In many large cities, water is transported from hundreds of miles away to meet the needs of residents. In some parts of the world, people must travel long distances every day to get water.

Inexhaustible Resources

An ***inexhaustible resource*** *is a natural resource that will not run out, no matter how much of it people use.* Energy from the Sun, solar energy, is inexhaustible. So is wind, which is generated by the Sun's uneven heating of Earth's lower atmosphere. Another inexhaustible resource is thermal energy from within Earth.

Solar Energy

Without the Sun's energy, life as it is on Earth would not be possible. If you've studied food chains, you know that energy from the Sun is used by plants and other producers during photosynthesis to make food. Consumers are organisms that get energy by eating producers or other consumers. The energy in food chains is always traced back to the Sun.

Solar energy can be used in many ways. Greenhouses trap thermal energy. They make it possible to grow warm-weather plants in cool climates. Solar cookers concentrate the Sun's thermal energy to cook food. Large solar-power plants provide electricity to many homes.

Solar energy also can be used to heat water in individual homes, as shown in the figure at the top of the next page. The hot water can be stored in a tank until it is needed.

Solar Energy from an Inexhaustible Resource

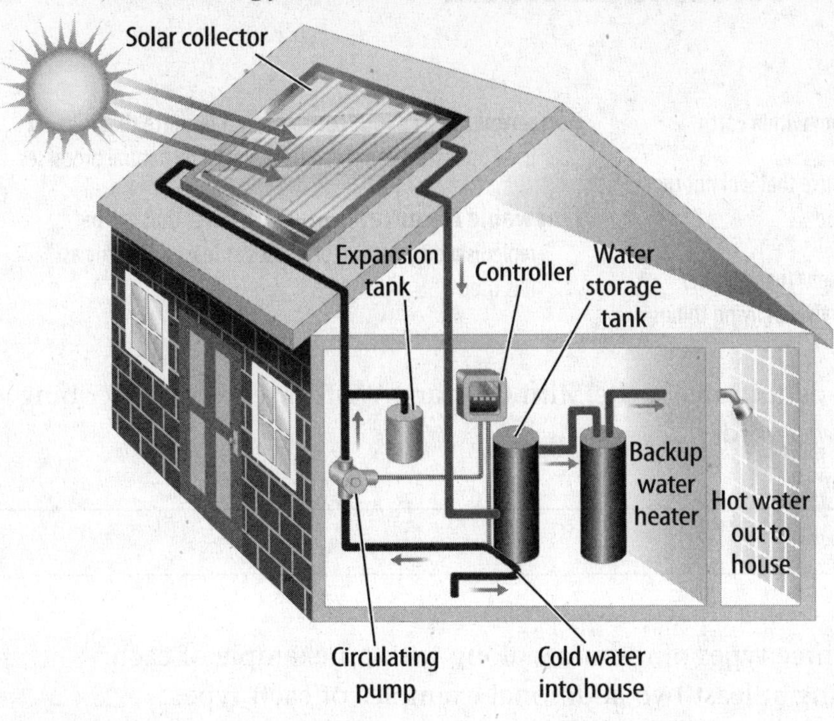

Wind Power

Sailboats, kites, and windmills are powered by wind. Wind is the movement of air over Earth's surface. Wind is an inexhaustible resource produced by the uneven heating of the atmosphere by the Sun.

If you live in an area that has strong winds, you might have seen giant wind turbines at work. In areas with frequent, strong winds, turbines can be used to produce electricity. ✔

Geothermal Energy

Another type of inexhaustible resource is geothermal energy. **Geothermal energy** *is thermal energy from within Earth.* Molten rock that rises close to the surface of Earth's crust is called magma. Pockets of magma in some parts of Earth's crust heat underground water and rocks.

The heated water produces steam, which is used in geothermal power plants to generate electricity. People who live in California—as well as in other parts of the world—rely on geothermal energy to produce a large amount of their electricity. ✔

✔ Visual Check

9. Identify In which part of the system is water heated by the Sun?

____

____

✔ Reading Check

10. Describe What causes wind?

____

____

____

✔ Reading Check

11. Recognize What are three types of inexhaustible resources?

____

____

____

After You Read

Mini Glossary

geothermal energy: thermal energy from within Earth

inexhaustible resource: a natural resource that will not run out, no matter how much of it people use

natural resource: a part of the environment that supplies materials useful or necessary for the survival of living things

nonrenewable resource: a natural resource that is being used up faster than it can be replaced by natural processes

renewable resource: a natural resource that can be replenished by natural processes at least as quickly as it is used

1. Review the terms and their definitions in the Mini Glossary. Write a sentence describing natural resources in your own words.

2. The table below identifies three types of resources along with an example of each. Complete the table by writing at least two additional examples of each type.

Inexhaustible Resource	Nonrenewable Resource	Renewable Resource
geothermal energy	oil	water

3. Review the questions you wrote as you read the lesson. Select one, and write the answer below without referring to the text.

What do you think NOW?

Reread the statements at the beginning of the lesson. Fill in the After column with an A if you agree with the statement or a D if you disagree. Did you change your mind?

Log on to ConnectED.mcgraw-hill.com and access your textbook to find this lesson's resources.

Using Natural Resources

Pollution

Before You Read

What do you think? Read the two statements below and decide whether you agree or disagree with them. Place an A in the Before column if you agree with the statement or a D if you disagree. After you've read this lesson, reread the statements to see if you have changed your mind.

Before	Statement	After
	3. Global warming causes acid rain.	
	4. Smog can affect human health.	

Key Concepts 
- How does pollution affect air resources?
- How does pollution affect water resources?
- How does pollution affect land resources?

Read to Learn

What is pollution?

What happens when smoke gets in the air or toxic chemicals leak into soil? Smoke is a mixture of gases and tiny particles that make breathing difficult, especially for people who have health problems. Toxic chemicals that leak into soil can kill plants and soil organisms. These substances cause pollution. **Pollution** *is the contamination of the environment with substances that are harmful to life.*

Most pollution occurs because of human actions, such as burning fossil fuels or spilling toxic materials. However, pollution also can come from natural disasters. Wildfires create smoke. Volcanic eruptions send ash and toxic gases into the atmosphere. Regardless of its source, pollution affects air, water, and land resources.

Air Pollution

Many large cities issue alerts about air quality when air pollution levels are high. On such days, people are asked to avoid activities that contribute to air pollution, such as driving, using gasoline-powered lawn mowers, or cooking on charcoal grills. To avoid breathing problems, people also are advised to exercise in the early morning when the air is cleaner. Air pollution that can affect human health and recreational activities can be caused by ozone loss, photochemical smog, global warming, and acid precipitation.

Mark the Text

Main Ideas and Details
Highlight the main idea of each paragraph. Highlight two details that support each main idea with a different color. Use your highlighted copy to review what you studied in this lesson.

FOLDABLES®

Make a horizontal three-tab book and use it to explain the effects of pollution.

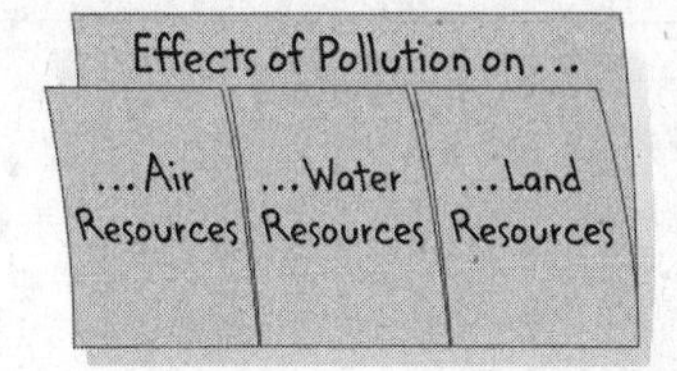

Ozone Loss

Ozone is a molecule comprised of three oxygen atoms. In the upper atmosphere, it forms a protective layer around Earth. *The* **ozone layer** *prevents most harmful ultraviolet (UV) radiation from reaching Earth.* UV radiation from the Sun can cause cancer and cataracts. It can also damage crops.

In the 1980s, scientists warned that Earth's protective ozone layer was getting thinner. The problem was caused mainly by chlorofluorocarbons (CFCs). CFCs are compounds used in refrigerators, air conditioners, and aerosol sprays.

Governments around the world have phased out the use of CFCs and other ozone-depleting gases. Because compounds such as CFCs are no longer widely used, the ozone layer is expected to recover within several decades.

Reading Check

1. Define What is ozone?

Reading Check

2. Explain Why have many countries banned the use of CFCs?

Photochemical Smog

Sunlight reacts with waste gases from the burning of fossil fuels and forms a type of air pollution called **photochemical smog.** As shown in the figure below, smog makes the air dark. It also can smell bad.

Photochemical smog is formed of particles and gases that irritate the respiratory system, making it hard for people to breathe. Smog can worsen throughout the day as chemicals react with sunlight.

One of the gases in smog is ozone. In the upper atmosphere, ozone is helpful. But in the lower atmosphere, it is a pollutant that can harm plants and animals and cause lung damage.

Visual Check

3. Identify What activities contribute to the formation of smog?

Smog

Sunlight

Photochemical smog

Waste gases produced by vehicles burning fossil fuels

Global Warming

You might have heard news reports about the melting of glaciers and sea ice. Earth is getting warmer. **Global warming** *is the scientific observation that Earth's average surface temperature is increasing.* Global warming can change Earth's climate in many ways. It can

- change weather conditions;
- change ecosystems and food webs;
- increase the number and severity of floods and droughts; and
- increase coastal flooding as sea ice melts and sea levels rise.

Data indicate that Earth's average surface temperature and increases in atmospheric carbon dioxide (CO_2) follow the same general trend. CO_2 is a greenhouse gas. This means it traps thermal energy, helping to keep Earth warm. Greenhouse gases occur naturally. Without them, Earth would be too cold to support life. But human activities add greenhouse gases to the atmosphere, especially CO_2 from the burning of fossil fuels. Most scientists, including those on the United Nations Intergovernmental Panel on Climate Change, agree that increases in atmospheric CO_2 are contributing to global warming.

Acid Precipitation

Gases produced by the burning of fossil fuels also create other forms of air pollution, including acid precipitation. **Acid precipitation** *is acidic rain or snow that forms when waste gases from automobiles and power plants combine with moisture in the air.* Coal-burning power plants produce sulfur dioxide gas that combines with moisture to form sulfuric acid. Cars and trucks produce nitrous oxide gases that form nitric acid. Acid precipitation pollutes soil and can kill plants, including trees. It also contributes to water pollution and can damage buildings.

Water Pollution

Have you ever seen a stream covered with thick green algae? The stream might have been polluted with fertilizers from nearby lawns or farms. It might contain chemicals from nearby factories. Water pollution can come from chemical runoff and other agricultural, residential, and industrial sources.

Reading Check

4. Describe four possible effects of global warming.

ACADEMIC VOCABULARY

occur
(verb) to appear or happen

Key Concept Check

5. Summarize How does pollution affect air resources?

Wastewater

You probably already know that you should not pour paint or used motor oil into storm drains. In most cities, rainwater that flows into storm drains goes directly into nearby waterways. Materials that go in the storm drain, including grease and oil washed from the street, can contribute to water pollution.

The wastewater that drains from showers, sinks, and toilets contains harmful viruses and bacteria. To safeguard health, this wastewater usually is purified in a sewage-treatment plant before it is released into streams or used to irrigate crops. In some parts of the world, little or no sewage treatment occurs. People who live in these places might have to use polluted water. ✓

Wastewater that comes from industries and mining operations also contains pollutants. It requires treatment before it can be returned to the environment. Even after treatment, some harmful substances might remain and impact water quality. ✓

Runoff and Sediments

When it rains, water can flow over the land. This water, called runoff, flows across lawns and farmland. Along the way, it picks up pesticides, herbicides, and fertilizers.

Runoff carries these pollutants into streams, where they can harm insects, fish, and other organisms. Runoff also carries sediment particles into streams. Too much sediment can damage stream habitats, clog waterways, and cause flooding.

Land Pollution

Have you ever helped clean up litter? Foam containers, plastic bags, bottles, cans, and even furniture and appliances get dumped along roadsides. Litter is more than an eyesore. It can pollute soil and water and disturb wildlife. Sources of land pollution include homes, farms, industry, and mines.

Agriculture

Farmers use pesticides and other agricultural chemicals to help plants grow. But these chemicals become pollutants if they are used in excess or disposed of improperly.

Herbicides kill weeds. But if they flow into streams, they can kill algae and plants and harm fish and amphibians. Some farming practices contaminate soil. Irrigation water contains salts that can build up in soil that is irrigated on a regular basis.

✓ **Reading Check**

6. Consider Why is household wastewater treated in sewage-treatment plants?

✓ **Reading Check**

7. Name three sources of wastewater.

Key Concept Check

8. Summarize How does pollution affect water resources?

Industry and Mining

Many industrial facilities, including oil refineries and ore processors, produce toxic wastes. For example, power plants that burn coal produce coal ash sludge as waste. The sludge contains mercury, lead, arsenic, and other potentially harmful metals. If toxic wastes like these are incorrectly stored or disposed of, they can contaminate soil and water. The health of people, plants, and wildlife can be affected.

The mining of fossil fuels and minerals can disturb or destroy entire ecosystems. Some coal-mining techniques can release toxic substances that were buried in rock. After the coal has been removed, the area can be restored. But it is difficult or impossible to replace the original ecosystem.

Key Concept Check

9. Summarize How does pollution affect land resources?

After You Read

Mini Glossary

acid precipitation: acidic rain or snow that forms when waste gases from automobiles and power plants combine with moisture in the air

global warming: the scientific observation that Earth's average surface temperature is increasing

ozone layer: a protective layer around Earth that prevents most harmful ultraviolet (UV) radiation from reaching Earth

photochemical smog: a type of air pollution formed when sunlight reacts with waste gases from the burning of fossil fuels

pollution: the contamination of the environment with substances that are harmful to life

1. Review the terms and their definitions in the Mini Glossary. Write a sentence comparing and contrasting acid precipitation and photochemical smog.

2. Use the graphic organizer below to identify at least two sources of water pollution and two sources of land pollution.

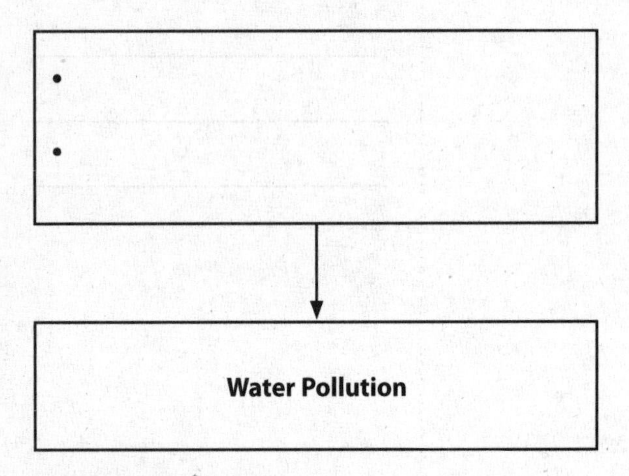

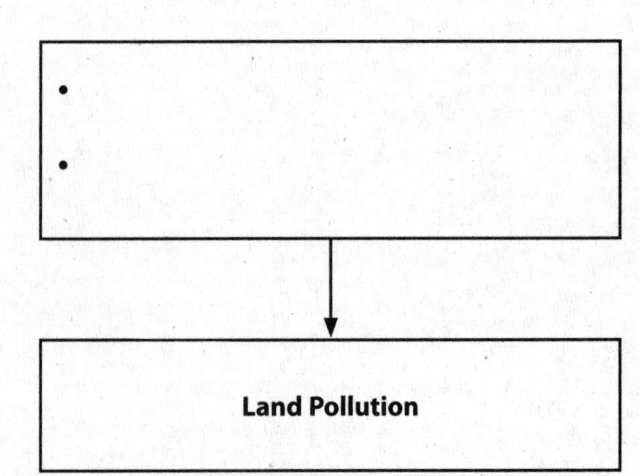

3. What do you think would happen to Earth's surface temperature if atmospheric CO_2 levels decreased? Why?

What do you think NOW?

Reread the statements at the beginning of the lesson. Fill in the After column with an A if you agree with the statement or a D if you disagree. Did you change your mind?

Log on to ConnectED.mcgraw-hill.com and access your textbook to find this lesson's resources.

Using Natural Resources

Protecting Earth

Before You Read

What do you think? Read the two statements below and decide whether you agree or disagree with them. Place an A in the Before column if you agree with the statement or a D if you disagree. After you've read this lesson, reread the statements to see if you have changed your mind.

Before	Statement	After
	5. Oil left over from frying potatoes can be used as automobile fuel.	
	6. Hybrid electric vehicles cannot travel far or go fast.	

Key Concepts
- How can people monitor resource use?
- How can people conserve resources?

Read to Learn

Monitoring Human Impact on Earth

As the human population increases, so does its impact on the planet. Scientists, governments, and concerned citizens around the world are working to identify environmental problems, educate people about them, and help find solutions.

Scientists collect data on environmental conditions by placing detectors on satellites, aircraft, high-altitude balloons, and ground-based monitoring stations. For example, the United States and the European Union have launched satellites into orbit around Earth. These satellites gather data on greenhouse gases, the ozone, ecosystem changes, melting glaciers and sea ice, climate patterns, and ocean health. The U.S. Environmental Protection Agency (EPA) is a government organization that watches the health of the environment and looks for ways to reduce the impact of humans. The EPA enforces environmental laws and supports research. It also identifies superfund sites—abandoned areas that have been contaminated by toxic wastes—and develops plans to clean them.

Mark the Text

Building Vocabulary As you read, underline the words and phrases that you do not understand. When you finish reading, discuss these words and phrases with another student or your teacher.

Key Concept Check

1. State How can people monitor resource use?

______

Developing Technologies

Many technologies have been developed to protect Earth's resources. These advances often focus on saving energy and reducing pollution.

FOLDABLES®

Make a small shutterfold book and use it to identify technology and methods that protect natural resources.

Monitoring Natural Resources

Conserving Natural Resources

Reading Check

2. Explain How does using CFLs help the environment?

Visual Check

3. Summarize How do CFCs affect ozone molecules?

Water-Saving Technologies

It takes energy to clean water and to transport it to homes and businesses. So technologies that conserve water also save energy. Low-flow showerheads and toilets as well as drip irrigation systems help reduce water use.

Energy-Saving Technologies

Saving energy can make Earth's supply of fossil fuels last longer. Using renewable energy sources reduces fossil fuel use. Some of these sources are expensive, but designs are improving and costs are falling. Solar electricity might soon cost the same as electricity produced by burning fossil fuels. Burning fewer fossil fuels also creates less pollution.

Other energy-saving advances include compact fluorescent lightbulbs (CFLs). They use about one-fourth the energy of incandescent bulbs and can last ten times longer. In 2007, Americans who switched to CFLs reduced greenhouse gas emissions by an amount equal to removing 2 million cars from the road.

CFC Replacements

CFCs thin the ozone layer because the chlorine atoms in CFC molecules react with sunlight to destroy ozone, as shown in the figure below. All CFCs soon will be phased out and replaced with chemicals that do not contain chlorine. Replacements include hydrofluorocarbons (HFCs) and perfluorocarbons (PFCs). Even after CFCs are no longer in use, it will take decades for the ozone layer to recover.

CFCs

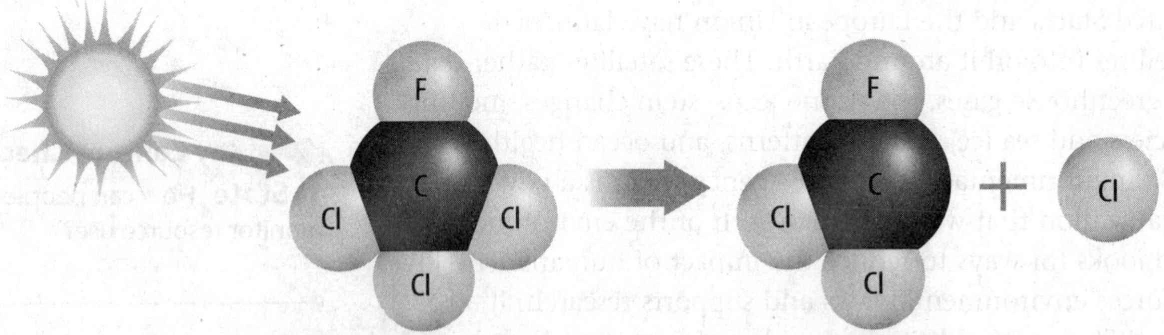

Sunlight reacts with a CFC molecule, causing a chlorine atom to break away.

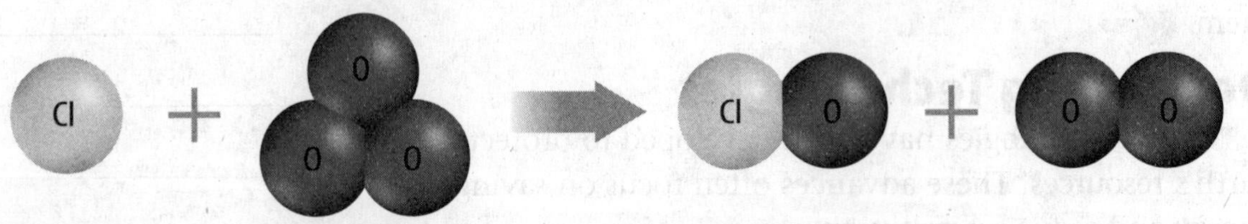

The chlorine atom reacts with and breaks apart an ozone molecule.

Alternative Fuels

Gasohol and biodiesel are alternative fuels that help reduce humans' use of fossil fuels. These alternative fuels also help reduce air pollution. 

Gasohol Gasohol is a mixture of 90 percent gasoline and 10 percent ethanol. Ethanol is alcohol made from corn, sugarcane, or other plants. Using gasohol in gasoline engines helps reduce emissions of carbon monoxide. Carbon monoxide is an air pollutant that contributes to smog.

The carbon in ethanol comes from plants instead of fossil fuels. So using gasohol can help reduce emissions that contribute to global warming.

Biodiesel Biodiesel is an alternative fuel that is made from renewable resources, primarily vegetable oils and animal fats. Biodiesel can even be made from oil that is left over from frying foods in restaurants!

Biodiesel can be burned in diesel engines in farm and industrial machinery, trucks, and cars. It produces fewer pollutants than regular diesel fuel, and it reduces CO_2 emissions by 78 percent.

Automobile Technologies

If you were buying a car, you would want to know how many miles it can travel per gallon of fuel. This measurement is called miles per gallon (mpg).

The higher a car's mpg rating, the less pollution it will add to the environment. A car with a high mpg rating also will use up fewer fossil-fuel resources.

HEVs Many people have decided to purchase a hybrid electric vehicle (HEV). HEVs combine a small gasoline engine with an electric motor powered by batteries.

HEVs run on battery power as much as possible. They get a boost from the gasoline engine for longer trips, higher speeds, and steep hills. The gasoline engine also charges the vehicle's batteries.

HEVs usually get excellent gas mileage. They can get up to twice the mileage of a regular car. Some recent models of HEVs get close to 50 mpg.

Reading Check

4. Name two alternative fuels that help reduce humans' use of fossil fuels.

______

______

SCIENCE USE V. COMMON USE

hybrid

Science Use an offspring of two animals or plants of different breeds or species

Common Use something that has two different components performing essentially the same task

Think it Over

5. Analyze Which do you think would probably be cheaper to own: a car that gets 30 mpg or one that gets 20 mpg? Why?

______

______

______

______

Visual Check
6. Draw a circle around the power sources in the hybrid vehicle.

Reading Check
7. Compare HEVs and FCVs.

Reading Check
8. Define What is sustainability?

Hybrid Vehicle

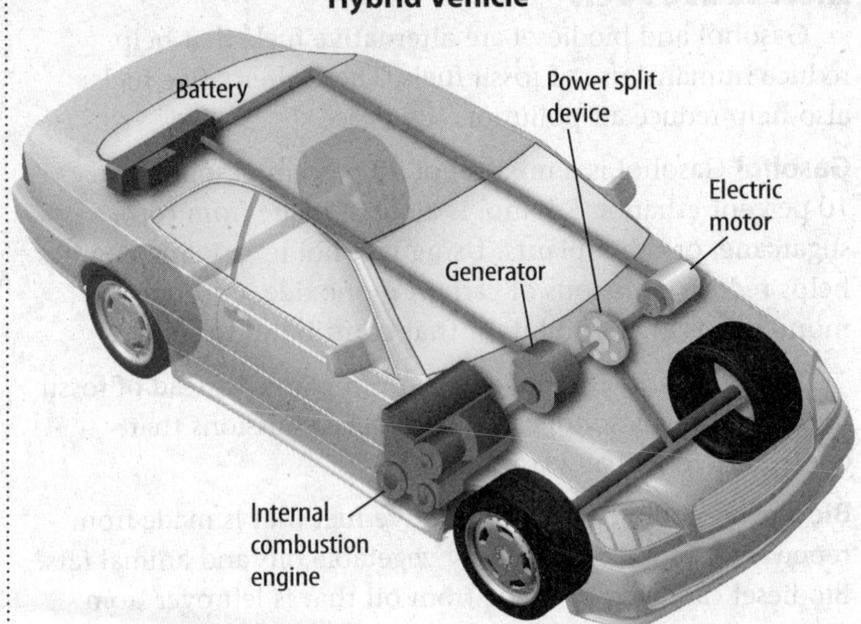

The figure above is a diagram of a typical hybrid electric vehicle. The battery powers the electric motor. The small gasoline engine can provide additional power.

FCVs In the future, another automobile alternative might be a fuel-cell vehicle (FCV). Inside a fuel cell, oxygen from the air chemically combines with hydrogen to produce electricity. The primary waste product is water. Tailpipe emissions from FCVs are nearly pollution-free. However, obtaining hydrogen fuel requires using methane or other fossil fuels. Researchers are looking for alternatives.

Making a Difference

Do you turn off the lights when you leave a room and recycle bottles and cans? If so, you are helping reduce your impact on the environment. You can help protect Earth's resources in other ways as well. Possibilities include cleaning up a stream, educating others about environmental issues, and analyzing the choices you make as a consumer.

Sustainability

When people talk about environmental issues, they often use the word *sustainability*. **Sustainability** *means meeting human needs in ways that ensure future generations also will be able to meet their needs.* When you turn off the lights as you leave a room, you are saving energy—and you are also helping to ensure a sustainable future. Other actions that promote a sustainable future include planting trees, composting, and picking up litter.

Restore and Rethink

Restoring damaged habitats and ecosystems to their original state is one way to make a difference. For example, picking up trash can restore water habitats.

You also can rethink the way you perform everyday activities. Instead of riding in a vehicle to nearby places, you could ride your bike or walk.

Reduce and Reuse

You can reduce the amount of waste you create by reducing the amount of material you use. For example, you might avoid buying products with too much packaging.

Another way to reduce the amount of waste you create is to bring your own bags when you go shopping. Carrying your purchases in reusable bags, rather than using the plastic or paper bags the store offers, can help save energy and reduce waste.

Reusing items also helps reduce waste. Instead of buying new, reuse something that will work just as well. You also can donate used items to charities or sell them.

Recycle

If an item cannot be reused, you might be able to recycle it instead of throwing it away. **Recycling** *is manufacturing new products out of used products*. The recycling process reduces waste and extends our supply of natural resources.

Computers and Electronics Computers and other electronics are examples of items that can be recycled. For example, these products contain valuable metals that can be used again. They also contain toxic materials that can contribute to pollution. So recycling also helps ensure that toxins are properly disposed of.

Compost Leaves, grass clippings, and vegetable scraps can be recycled by composting. In a compost pile, these materials decay into nutrient-rich soil that can be put back into the garden.

Buy Recycled Separating recyclables from the rest of the trash is just one step. To keep the cycle going, people need to buy and use recycled products. You can find shoes, clothing, paper, and carpets made from recycled materials.

Reading Check

9. Consider How does using your bike for transportation benefit the environment?

Reading Check

10. State How does recycling help the environment?

Key Concept Check

11. Explain How can you conserve resources?

After You Read

Mini Glossary

recycling: manufacturing new products out of used products

sustainability: meeting human needs in ways that ensure future generations also will be able to meet their needs

1. Review the terms and their definitions in the Mini Glossary. Write a sentence explaining how recycling contributes to sustainability.

____

____

____

2. Write the correct letter in each box within the graphic organizer to compare and contrast gasohol and biodiesel.

a. made from vegetable oils
b. burned in gasoline engines
c. made from gasoline and ethanol
d. burned in diesel engines
e. reduces harmful pollutants
f. alternative fuel

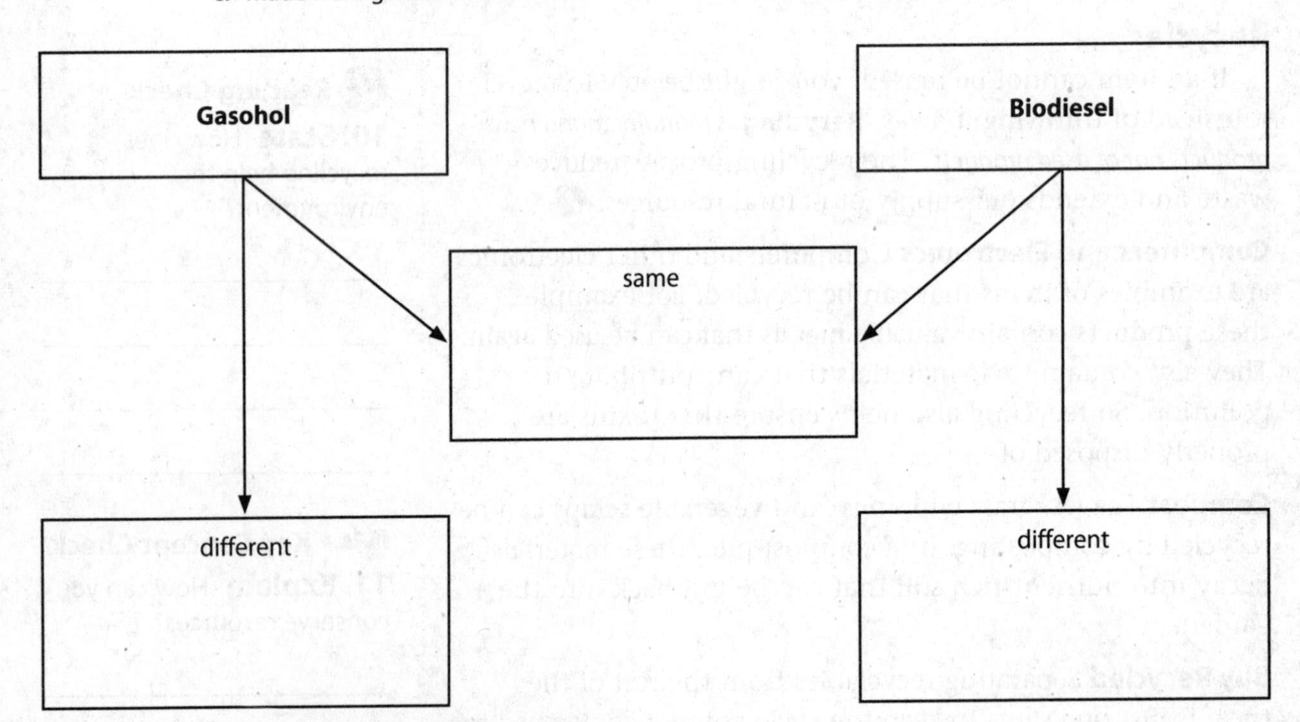

What do you think NOW?

Reread the statements at the beginning of the lesson. Fill in the After column with an A if you agree with the statement or a D if you disagree. Did you change your mind?

Log on to ConnectED.mcgraw-hill.com and access your textbook to find this lesson's resources.

PERIODIC TABLE OF THE ELEMENTS

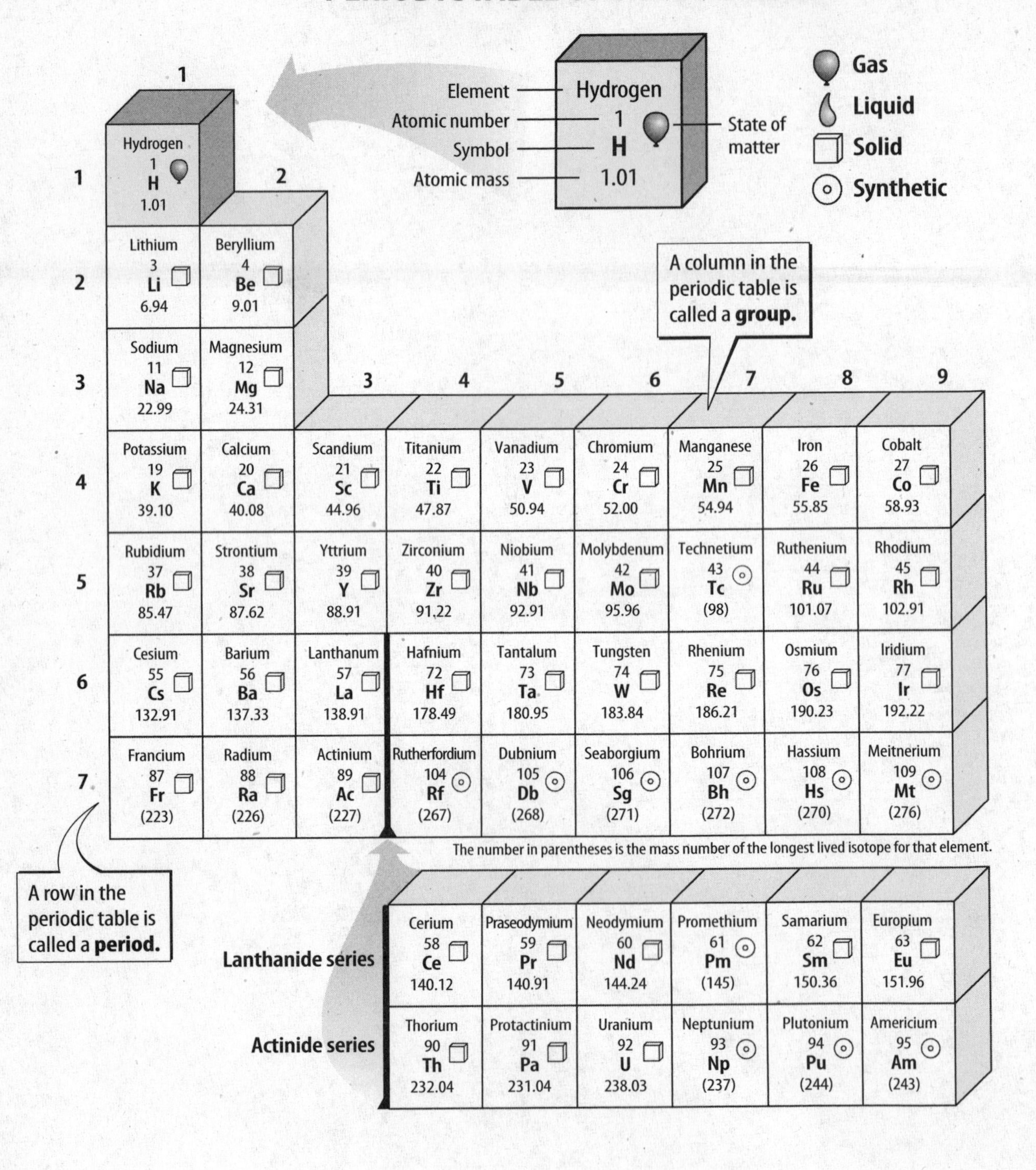

Period	1	2	3	4	5	6	7	8	9
1	Hydrogen 1 H 1.01								
2	Lithium 3 Li 6.94	Beryllium 4 Be 9.01							
3	Sodium 11 Na 22.99	Magnesium 12 Mg 24.31							
4	Potassium 19 K 39.10	Calcium 20 Ca 40.08	Scandium 21 Sc 44.96	Titanium 22 Ti 47.87	Vanadium 23 V 50.94	Chromium 24 Cr 52.00	Manganese 25 Mn 54.94	Iron 26 Fe 55.85	Cobalt 27 Co 58.93
5	Rubidium 37 Rb 85.47	Strontium 38 Sr 87.62	Yttrium 39 Y 88.91	Zirconium 40 Zr 91.22	Niobium 41 Nb 92.91	Molybdenum 42 Mo 95.96	Technetium 43 Tc (98)	Ruthenium 44 Ru 101.07	Rhodium 45 Rh 102.91
6	Cesium 55 Cs 132.91	Barium 56 Ba 137.33	Lanthanum 57 La 138.91	Hafnium 72 Hf 178.49	Tantalum 73 Ta 180.95	Tungsten 74 W 183.84	Rhenium 75 Re 186.21	Osmium 76 Os 190.23	Iridium 77 Ir 192.22
7	Francium 87 Fr (223)	Radium 88 Ra (226)	Actinium 89 Ac (227)	Rutherfordium 104 Rf (267)	Dubnium 105 Db (268)	Seaborgium 106 Sg (271)	Bohrium 107 Bh (272)	Hassium 108 Hs (270)	Meitnerium 109 Mt (276)

The number in parentheses is the mass number of the longest lived isotope for that element.

Lanthanide series	Cerium 58 Ce 140.12	Praseodymium 59 Pr 140.91	Neodymium 60 Nd 144.24	Promethium 61 Pm (145)	Samarium 62 Sm 150.36	Europium 63 Eu 151.96
Actinide series	Thorium 90 Th 232.04	Protactinium 91 Pa 231.04	Uranium 92 U 238.03	Neptunium 93 Np (237)	Plutonium 94 Pu (244)	Americium 95 Am (243)